NANOSCOPY
AND
MULTIDIMENSIONAL OPTICAL
FLUORESCENCE MICROSCOPY

NANOSCOPY

AND
MULTIDIMENSIONAL OPTICAL
FLUORESCENCE MICROSCOPY

EDITED BY

ALBERTO DIASPRO

CRC Press
Taylor & Francis Group
Boca Raton London New York

CRC Press is an imprint of the
Taylor & Francis Group, an informa business

A CHAPMAN & HALL BOOK

MATLAB® and Simulink® are trademarks of the MathWorks, Inc. and are used with permission. The MathWorks does not warrant the accuracy of the text or exercises in this book. This book's use or discussion of MATLAB® and Simulink® software or related products does not constitute endorsement or sponsorship by the MathWorks of a particular pedagogical approach or particular use of the MATLAB® and Simulink® software.

CRC Press
Taylor & Francis Group
6000 Broken Sound Parkway NW, Suite 300
Boca Raton, FL 33487-2742

First issued in paperback 2019

© 2010 by Taylor & Francis Group, LLC
CRC Press is an imprint of Taylor & Francis Group, an Informa business

No claim to original U.S. Government works

ISBN-13: 978-1-4200-7886-2 (hbk)
ISBN-13: 978-0-367-38421-0 (pbk)

Library of Congress Cataloging-in-Publication Data

Nanoscopy and multidimensional optical fluorescence microscopy / edited by Alberto Diaspro.
     p. ; cm.
    Includes bibliographical references and index.
    ISBN 978-1-4200-7886-2 (hardcover : alk. paper)
    1. Fluorescence microscopy. 2. Nanotechnology. I. Diaspro, Alberto, 1959- II. Title.
    [DNLM: 1. Microscopy, Fluorescence--methods. 2. Cells--ultrastructure. 3. Imaging,
Three-Dimensional--methods. 4. Nanotechnology--methods. QH 212.F55 N186 2010]

    QH212.F55N36 2010
    502.8'2--dc22
                                        2009027797

Visit the Taylor & Francis Web site at
http://www.taylorandfrancis.com

and the CRC Press Web site at
http://www.crcpress.com

Dedication

This book is dedicated with immense love to Teresa, Claudia, and our sweet puppy, Sissi. A special thank you to my grandfather, Mario, who bought me my first oscilloscope (which I still use) in 1978, and who gave me a subscription to Nature *when I was still a high-school student.*

Contents

Foreword ix

Preface.. xiii

Editor xv

Contributors xvii

1 STED Microscopy with Compact Light Sources 1-1
   Lars Kastrup, Dominik Wildanger, Brian Rankin, and Stefan W. Hell

2 Nonlinear Fluorescence Imaging by Saturated Excitation 2-1
   Nicholas I. Smith, Shogo Kawano, Masahito Yamanaka, and Katsumasa Fujita

3 Far-Field Fluorescence Microscopy of Cellular Structures at Molecular
   Optical Resolution.. 3-1
   *Christoph Cremer, Alexa von Ketteler, Paul Lemmer, Rainer Kaufmann, Yanina
   Weiland, Patrick Mueller, M. Hausmann, Manuel Gunkel, Thomas Ruckelshausen,
   David Baddeley, and Roman Amberger*

4 Fluorescence Microscopy with Extended Depth of Field............................. 4-1
   Kai Wicker and Rainer Heintzmann

5 Single Particle Tracking.. 5-1
   Kevin Braeckmans, Dries Vercauteren, Jo Demeester, and Stefaan C. De Smedt

6 Fluorescence Correlation Spectroscopy 6-1
   Xianke Shi and Thorsten Wohland

7 Two-Photon Excitation Microscopy: A Superb Wizard for
   Fluorescence Imaging 7-1
   Francesca Cella and Alberto Diaspro

8 Photobleaching Minimization in Single- and Multi-Photon
   Fluorescence Imaging 8-1
   Partha Pratim Mondal, Paolo Bianchini, Zeno Lavagnino, and Alberto Diaspro

9 Applications of Second Harmonic Generation Imaging Microscopy........... 9-1
   Paolo Bianchini and Alberto Diaspro

10 Green Fluorescent Proteins as Intracellular pH Indicators..........................10-1
Fabio Beltram, Ranieri Bizzarri, Stefano Luin, and Michela Serresi

11 Fluorescence Photoactivation Localization Microscopy...........................11-1
Manasa V. Gudheti, Travis J. Gould, and Samuel T. Hess

12 Molecular Resolution of Cellular Biochemistry and Physiology by
FRET/FLIM 12-1
Fred S. Wouters and Gertrude Bunt

13 FRET-Based Determination of Protein Complex Structure at
Nanometer Length Scale in Living Cells..13-1
Valericǎ Raicu

14 Automation in Multidimensional Fluorescence Microscopy: Novel
Instrumentation and Applications in Biomedical Research.........................14-1
Mario Faretta

15 Optical Manipulation, Photonic Devices, and
Their Use in Microscopy..15-1
*G. Cojoc, C. Liberale, R. Tallerico, A. Puija, M. Moretti, F. Mecarini, G. Das,
P. Candeloro, F. De Angelis, and E. Di Fabrizio*

16 Optical Tweezers Microscopy: Piconewton Forces in Cell and
Molecular Biology 16-1
Francesco Difato, Enrico Ferrari, Rajesh Shahapure, Vincent Torre, and Dan Cojoc

17 *In Vivo* Spectroscopic Imaging of Biological Membranes and Surface
Imaging for High-Throughput Screening 17-1
Jo L. Richens, Peter Weightman, Bill L. Barnes, and Paul O'Shea

18 Near-Field Optical Microscopy: Insight on the Nanometer-Scale
Organization of the Cell Membrane...18-1
Davide Normanno, Thomas van Zanten, and María F. García-Parajo

Index..Index-1

Foreword

As a fundamental process that mediates our interaction with the world around us, all scientists rely in some way on light to peer into the workings of the universe. Originally, this investigation involved nothing but the naked eye. Later, some individuals figured out how to use curved glass to manipulate light and enlarge the image of distant or microscopic objects, and make the invisible visible. By bending light to their will, scientists have made and continue to make marvelous discoveries. But light is like a complex woman and although her nature makes the science of optics possible, she can also confound those who try to bend her to their will.

Unlike astronomers who must resign themselves to collecting and analyzing the light that happens to arrive from their objects of study, biologists are blessed with a far more intimate relationship with light and the specimens they are investigating. This dynamic allows them to dictate the properties of light as she falls on their sample. They can even introduce probes into the specimen that dictate how light will react when she encounters them. Fluorescent probes, for example, will give her a fleeting embrace and then release her after stealing a small portion of her energy. She departs displaying a warmer color than she arrived with, while her partner may return to his original state, or he may be altered in a way that he reacts very differently to the next photon he meets.

Physicists, chemists, and biologists have learned how to serve as matchmakers and choreographers of this dance between photons and their partners, fluorophores. By carefully orchestrating the properties of these partners, they can use these probes to obtain detailed insights into the environments where light encounters them. The techniques researchers have developed to arrange this dance are now indispensable tools in the ongoing quest to unravel the processes of life.

Alberto Diaspro has been choreographing light's dance for over 20 years, and in *Nanoscopy and Multidimensional Optical Fluorescence Microscopy*, he has assembled a diverse group of experts to explain the methods they use to coax light to reveal biology's secrets. The highlight of the book is the coverage of nanoscopy, or super-resolution microscopy. The rapid advances in this area over the past few years and the promise the method holds for obtaining new insights into biology prompted *Nature Methods* to choose this recent revolution in fluorescence microscopy as the Method of the Year in 2008.

This is the first book to extensively cover the rapidly developing field of nanoscopy. Four complementary chapters describe how scientists have exploited the properties of light and her fluorophore partners to overcome the resolution limit of conventional light microscopy that has hampered its use for examining nanoscale details in living cells. In Chapter 3, Cremer and colleagues review some of the early work in this area and their more recent advances, while in Chapter 1, Kastrup and colleagues discuss their recent advances in using new compact light sources to implement nanoscopy based on stimulated emission depletion. In Chapter 2, Smith and colleagues present a tactic for coaxing more spatial information from a standard fluorescence microscope that requires little more than modulating the intensity of the incident light. In Chapter 11, Gudheti and colleagues provide practical guidance for implementing a class of nanoscopy that sprang into prominence in 2006, triggering a groundswell of interest and launching nanoscopy on a path reminiscent to that of two-photon microscopy 16 years earlier. Two-photon microscopy is now a pillar of biological microscopy techniques and Cella and Diaspro briefly discuss its principles and implementation in Chapter 7.

When light interacts with her partner she can have a long-lasting effect on him that makes him unresponsive to the advances of later photons. In the most extreme case, this unresponsiveness is permanent and the fluorophore remains dark and alone. While this process of photobleaching plays an indispensable role in some imaging methods, most microscopists go to great extremes to minimize these damaging trysts. Mondal and colleagues delve into some of the most recent methods to minimize photobleaching in Chapter 8, but emphasize that more work is needed in this important area.

Photobleaching can be particularly problematic for single-particle-tracking techniques like those discussed in Chapter 5 by Braeckmans and colleagues. In this method, the researcher must balance the amount of light necessary to obtain an adequate response from individual fluorophores with the need to keep them active long enough to follow their movements for an extended period of time in a living cell. If the researcher is successful, they can obtain detailed information on the movements of single molecules that would be impossible to obtain using the ensemble techniques that dominate microscopy.

While single-particle tracking is powerful, the challenges for the user are considerable. An alternative for some biological questions is fluorescence correlation spectroscopy (FCS), which limits light to one or more stationary focal points in the sample, then waits for fluorophores to pass through light's embrace. This approach limits the problem of photobleaching and simplifies data analysis. Shi and Wohland discuss the principles and implementation of FCS and related methods in Chapter 6.

In most of these methods, investigators often limit their observations to light coming from a single plane in a sample. But biological samples are inherently three dimensional. In Chapter 4, Wicker and Heintzmann discuss methods for overcoming this limitation, from the most basic method to emerging methods under active development.

Regardless of whether one is looking at a single plane in a sample, a three-dimensional volume, or just a single point, fluorescence microscopy is heavily dependent on the quality of the available fluorophores and how they react to light. The desirable properties of green fluorescent protein (GFP) have made it the most ubiquitous fluorescent probe in use today. Normally, the pH sensitivity of GFP compromises its usefulness and developers have worked hard to engineer the protein and limit this effect. pH, however, has important biological effects. As Bizzarri and colleagues discuss in Chapter 10, the pH sensitivity of GFP makes the protein a natural light-mediated sensor for probing the role of pH in cellular function.

So far we have considered the choreography of light and its partners in samples to be a private affair where interactions between fluorophores are undesirable. But promiscuous interactions can be valuable when properly choreographed. If two or more fluorophores are carefully matched, the first fluorophore embraced by a photon can transfer his acquired energy to his partner who will release it in the form of a photon that is distinguishable from a photon released by the first fluorophore. Because the energy transfer is distance dependent, it can be used to obtain molecular-scale distance information about the two partners. Wouters and Bunt review the basics of this fluorescence resonance energy transfer technique and its use to obtain insights into cellular systems in Chapter 12, and Raicu details its use for investigating the structure of protein complexes in Chapter 13.

Light is not limited to the common molecular-scale interactions exemplified by her absorption and emission from fluorophores. When light similar to that used for two-photon imaging interacts with appropriate nonfluorescent molecules in a sample, it can scatter and combine in a process called second harmonic generation. As described in Chapter 9 by Bianchini and Diaspro, this process can be used for imaging and has a number of unique characteristics that can be very useful to biologists. Light can also flirt with nonfluorescent molecules through a variety of other mechanisms, and with much larger objects she can even exert mechanical forces. These and other emerging nonfluorescence microscopy methods are introduced in Chapters 15 and 16 by Cojoc and colleagues, and in Chapter 17 by Richens and colleagues. The phenomenon of light-based force generation can be exploited to allow the physical trapping and movement of a variety of objects ranging from beads to living cells and cellular organelles.

Combining this technique with fluorescence microscopy provides an unparalleled ability to manipulate and visualize biological samples.

Historically, the scientists choreographing this dance between light and her partners would spend long periods gazing at their specimens through a microscope objective or in carefully arranging the encounters. The arrival of cameras decreased the amount of time spent peering through an objective and computers recently made it possible to fully automate image acquisition. As explained by Faretta in Chapter 14, these developments allow a new class of investigation using high-throughput microscopy-based screening. These methods are becoming essential for dealing with the complexity of living systems under investigation by modern biologists.

Nanoscopy and Multidimensional Optical Fluorescence Microscopy highlights the work of physicists, chemists, and biologists at the forefront of efforts to choreograph the dance between light and matter by developing methods capable of illuminating biological function. Although based largely on well-established optics theory, many of the methods herein were unimagined by the founders of modern microscopy. Reading this book shows not only how far microscopy has advanced but also how many developments are surely yet to come.

Daniel Evanko

Preface

"One more cup of coffee for the road, One more cup of coffee 'fore I go. To the valley below."
Bob Dylan, 1975, "One More Cup of Coffee"

It was in 1968 that I discovered my passion for microscopy. Later, thanks to a paper by David Agard and John Sedat, given to me by Massimo Grattarola, I discovered the world of optical sectioning microscopy under the supervision of Bruno Bianco and Francesco Beltrame. In 1990, at Drexel University with my family, my wife Teresa, and my six-year-old daughter Claudia, I was invited by Stanley Zietz to give a lecture on three-dimensional optical sectioning microscopy. During the lecture, I noticed my daughter writing, or rather doodling, something. I later discovered she was taking notes, as shown in the figure below, reporting the "double-cone" and the "optical transfer function." I collected these notes thinking that one day I could publish them in a book!

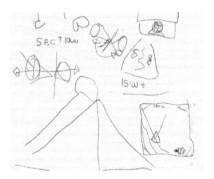

During the course of my career, I have been very fortunate to have been allowed to continue experimenting with microscopes: from wide-field optical microscopy and scanning probe microscopy to acoustic microscopy and multi-photon microscopy. In recent years, I have taken a keen interest in optical nanoscopy. This term can be used for a variety of approaches; however, the following statement in a paper published by Stefan Hell et al. (Far-field optical nanoscopy, *Science*, 2007, 316, 1153) strikes me as poetic: "In 1873, Ernst Abbe discovered what was to become a well-known paradigm: the inability of a lens-based optical microscope to discern details that are closer together than half of the wavelength of light. However, for its most popular imaging mode, fluorescence microscopy, the diffraction barrier is crumbling." His use of the word, "crumbling," in this context was particularly appropriate and appealing to me. I am, as well, convinced that microscopy is growing in diverse, complementary directions bringing with it a strong multidisciplinary background. I hope that this book can answer questions about optical nanoscopy and multidimensional fluorescence microscopy and stimulate scientists from every field. Methods and techniques reported here should provide sufficient details for readers to design and perform experiments on their own, thereby prompting new advances in the field. You will also find a brilliant summary of the field in the foreword by Daniel Evanko.

I will complete this brief introduction by quoting from Richard Feynman's seminal lecture delivered on December 29, 1959, at the annual meeting of the American Physical Society: "We have friends in other fields—in biology, for instance. We physicists often look at them and say, 'You know the reason you fellows are making so little progress?' '[Actually, I don't know any field where they are making more rapid progress than they are in biology today.]' 'You should use more mathematics, like we do.' They could answer us—but they're polite, so I'll answer for them: 'What you should do in order for us to make more rapid progress is to make the electron microscope 100 times better.'" Although this goal has been realized by now with the optical microscope, our story has only just begun. Happy and fruitful reading!

Alberto Diaspro
University of Genoa and Italian Institute of Technology

MATLAB® and Simulink® are registered trademarks of The MathWorks, Inc. For product information, please contact:

The MathWorks, Inc.
3 Apple Hill Drive
Natick, MA, 01760-2098 USA
Tel: 508-647-7000
Fax: 508-647-7001
E-mail: info@mathworks.com
Web: www.mathworks.com

Editor

Alberto Diaspro received his doctoral degree in electronic engineering from the University of Genoa, Italy. He is a university professor in applied physics and the head of the nanophysics unit of the Italian Institute of Technology, Genoa, Italy; he has also founded the Laboratory for Advanced Microscopy, Bioimaging and Spectroscopy (LAMBS). He is a member of the advisory committees for PhD courses in biotechnology (University of Genoa) and medical nanotechnology (University of Milan, Italy). Currently, he is serving as the president-elect of the European Biophysical Societies' Association (EBSA) and on the International Relations Committee of the Biophysical Society. He is also an active member of the Optical Society of America, the International Society for Optical Engineering, the Italian Society of Microscopical Sciences, and the Italian Physical Society. His professional activities have included co-organizing a world conference on microscopy and an annual meeting of the EBSA. He is the author of more than 170 publications and has been an invited speaker at more than 200 international conferences. He has been a visiting scientist at Drexel University, Philadelphia, Pennsylvania; the Autonomous University of Madrid, Spain; the Academy of Sciences of Czech Republic, Prague, Czech Republic; the Polytechnic University of Bucharest, Romania; and the University of Illinois at Urbana–Champaign, Urbana, Illinois.

Contributors

Roman Amberger
Kirchhoff-Institute for Physics
University of Heidelberg
Heidelberg, Germany

David Baddeley
Kirchhoff-Institute for Physics
University of Heidelberg
Heidelberg, Germany
and
Department of Physiology
University of Auckland
Auckland, New Zealand

Bill L. Barnes
Department of Physics
University of Exeter
Exeter, United Kingdom

Fabio Beltram
NEST Laboratory
Scuola Normale Superiore
and
Italian Institute of Technology
    Research Unit
and
National Council of Research
Pisa, Italy

Paolo Bianchini
University of Genoa
Genoa, Italy

Ranieri Bizzarri
NEST Laboratory
Scuola Normale Superiore
and
Italian Institute of Technology
    Research Unit
and
National Council of Research
Pisa, Italy

Kevin Braeckmans
Laboratory of General
    Biochemistry and Physical
    Pharmacy
Ghent University
Ghent, Belgium

Gertrude Bunt
Department of Neuro- and
    Sensory Physiology
University of Göttingen
Göttingen, Germany

P. Candeloro
BioNEM Laboratory
Magna Graecia University
Catanzaro, Italy
and
Center for Molecular
    Physiology of the Brain
Göttingen, Germany

Francesca Cella
Department of Physics
University of Genoa
Genoa, Italy

Dan Cojoc
Advanced Technology and
    Nanoscience National
    Laboratory
National Institute for the
    Physics of Matter
National Research Council
Trieste, Italy

G. Cojoc
BioNEM Laboratory
Magna Graecia University
Catanzaro, Italy

Christoph Cremer
Kirchhoff-Institute for Physics
and
Institute for Pharmacy and
    Molecular Biotechnology
and
Bioquant Center
University of Heidelberg
Heidelberg, Germany
and
Institute for Molecular
    Biophysics
University of Maine
Orono, Maine

G. Das
BioNEM Laboratory
Magna Graecia University
Catanzaro, Italy

F. De Angelis
BioNEM Laboratory
Magna Graecia University
Catanzaro, Italy

Stefaan C. De Smedt
Laboratory of General
    Biochemistry and Physical
    Pharmacy
Ghent University
Ghent, Belgium

Jo Demeester
Laboratory of General
    Biochemistry and Physical
    Pharmacy
Ghent University
Ghent, Belgium

E. Di Fabrizio
BioNEM Laboratory
Magna Graecia University
Catanzaro, Italy

Alberto Diaspro
Department of Physics
University of Genoa
and
Italian Institute of Technology
Genoa, Italy

Francesco Difato
Department of Neuroscience
    and Brain Technologies
Italian Institute of Technology
Genoa, Italy

Mario Faretta
Department of Experimental
    Oncology
European Institute of Oncology
Milan, Italy

Enrico Ferrari
Advanced Technology and
    Nanoscience National
    Laboratory
National Institute for the
    Physics of Matter
National Research Council
Trieste, Italy

Katsumasa Fujita
Department of Applied Physics
Osaka University
Osaka, Japan

María F. García-Parajo
Institute for Bioengineering of
    Catalonia
and
Catalan Institute for Research
    and Advanced Studies
Barcelona, Spain
and
Center for Biomedical Research
    Network in Bioengineering,
    Biomaterials and
    Nanomedicine
Zaragoza, Spain

Travis J. Gould
Department of Physics and
    Astronomy
and
Institute for Molecular
    Biophysics
University of Maine
Orono, Maine

Manasa V. Gudheti
Department of Physics and
    Astronomy
and
Institute for Molecular
    Biophysics
University of Maine
Orono, Maine

Manuel Gunkel
Kirchhoff-Institute for Physics
University of Heidelberg
Heidelberg, Germany

M. Hausmann
Kirchhoff-Institute for Physics
University of Heidelberg
Heidelberg, Germany

Rainer Heintzmann
Randall Division of Cell and
    Molecular Biophysics
King's College London
London, United Kingdom

Stefan W. Hell
Department of
    NanoBiophotonics
Max Planck Institute for
    Biophysical Chemistry
Göttingen, Germany

Samuel T. Hess
Department of Physics and
    Astronomy
and
Institute for Molecular
    Biophysics
University of Maine
Orono, Maine

Lars Kastrup
Department of
    NanoBiophotonics
Max Planck Institute for
    Biophysical Chemistry
Göttingen, Germany

Rainer Kaufmann
Kirchhoff-Institute for Physics
University of Heidelberg
Heidelberg, Germany

Shogo Kawano
Department of Frontier
    Biosciences
Osaka University
Osaka, Japan

Alexa von Ketteler
Kirchhoff-Institute for Physics
University of Heidelberg
Heidelberg, Germany

Zeno Lavagnino
Department of Physics
University of Genoa
Genoa, Italy

Paul Lemmer
Kirchhoff-Institute for Physics
University of Heidelberg
Heidelberg, Germany

C. Liberale
BioNEM Laboratory
Magna Graecia University
Catanzaro, Italy

Stefano Luin
NEST Laboratory
Scuola Normale Superiore
Pisa, Italy

F. Mecarini
BioNEM Laboratory
Magna Graecia University
Catanzaro, Italy

Partha Pratim Mondal
Department of Biological
    Engineering
Massachusetts Institute of
    Technology
Cambridge, Massachusetts

M. Moretti
BioNEM Laboratory
Magna Graecia University
Catanzaro, Italy

Patrick Mueller
Kirchhoff-Institute for Physics
University of Heidelberg
Heidelberg, Germany

Davide Normanno
Institute of Bioengineering of
    Catalonia
Barcelona, Spain

Paul O'Shea
Cell Biophysics Group
and
Institute of Biophysics, Imaging
    & Optical Science
School of Biology
University of Nottingham
Nottingham, United Kingdom

A. Puija
BioNEM Laboratory
Magna Graecia University
Catanzaro, Italy

Valericǎ Raicu
Department of Physics
and
Department of Biological
    Sciences
University of Wisconsin–
    Milwaukee
Milwaukee, Wisconsin

Brian Rankin
Department of
    NanoBiophotonics
Max Planck Institute for
    Biophysical Chemistry
Göttingen, Germany

Jo L. Richens
Cell Biophysics Group
and
Institute of Biophysics, Imaging
    & Optical Science
School of Biology
University of Nottingham
Nottingham, United Kingdom

Thomas Ruckelshausen
Kirchhoff-Institute for Physics
University of Heidelberg
Heidelberg, Germany

Michela Serresi
NEST Laboratory
Scuola Normale Superiore
and
Italian Institute of Technology
    Research Unit
Pisa, Italy

Rajesh Shahapure
International School for
    Advanced Studies
Trieste, Italy

Xianke Shi
Department of Chemistry
National University of
    Singapore
Singapore, Singapore

Nicholas I. Smith
WPI Immunology Frontier
    Research Center
Osaka University
Osaka, Japan

R. Tallerico
BioNEM Laboratory
Magna Graecia University
Catanzaro, Italy

Vincent Torre
International School for
    Advanced Studies
Trieste, Italy
and
Italian Institute of Technology
Genoa, Italy

Dries Vercauteren
Laboratory of General
    Biochemistry and Physical
    Pharmacy
Ghent University
Ghent, Belgium

Peter Weightman
Department of Physics
University of Liverpool
Liverpool, United Kingdom

Yanina Weiland
Kirchhoff-Institute for Physics
University of Heidelberg
Heidelberg, Germany

Kai Wicker
Randall Division of Cell and
   Molecular Biophysics
King's College London
London, United Kingdom

Dominik Wildanger
Department of
   NanoBiophotonics
Max Planck Institute for
   Biophysical Chemistry
Göttingen, Germany

Thorsten Wohland
Department of Chemistry
National University of
   Singapore
Singapore, Singapore

Fred S. Wouters
Department of Neuro- and
   Sensory Physiology
Institute for Physiology and
   Pathophysiology
University of Göttingen
Göttingen, Germany
and
Center for Molecular
   Physiology of the Brain
Göttingen, Germany

Masahito Yamanaka
Department of Frontier
   Biosciences
Osaka University
Osaka, Japan

Thomas van Zanten
Institute of Bioengineering of
   Catalonia
Barcelona, Spain

1

STED Microscopy with Compact Light Sources

Lars Kastrup
Dominik Wildanger
Brian Rankin
Stefan W. Hell

*Max Planck Institute for
Biophysical Chemistry*

1.1   Introduction 1-1
1.2   Far-Field Fluorescence Nanoscopy 1-2
1.3   New Light Sources 1-5
References 1-11

1.1 Introduction

Fluorescence microscopy is an invaluable and extremely popular tool in biological sciences due to its ability to noninvasively image the interior of cells in all spatial dimensions. Attaching fluorophores to the molecules of interest with antibody labeling or by genetic modification is highly specific, enabling sensitive measurements with low background. Until recent years, however, the resolution of fluorescence microscopy has been conceptually and practically limited by the wave nature of light to about half the wavelength of the light used to form the image. Ernst Abbe [1], Émile Verdet, and Lord Rayleigh described this diffraction limit quantitatively, giving the lateral extent of the light distribution in the focus, Δr, as $\Delta r \approx \lambda/(2n\sin\alpha)$, where λ is the wavelength of light used, n the index of refraction of the surrounding medium, and α the half-angle subtended by the outermost rays focused by the lens. Thus, the details of structures smaller than this dimension could not be seen by the far-field optical microscope, preventing the observation of interactions between cellular constituents that take place below the diffraction limit.

Due to the fundamental nature of diffraction, optical imaging was considered incapable of delivering subdiffraction resolution and, accordingly, progress in terms of resolution was made only within the bounds of diffraction. To this end, the numerical aperture was increased by using optimized lens designs or by combining two lenses as pursued in 4Pi microscopy [2–4] and I^5M [5]. Also, illumination was conducted with shorter wavelengths (thus eventually leaving the optical domain). While confocal [6–9] microscopy and multiphoton microscopy [10] introduced optical sectioning, they did not overcome the diffraction barrier. It was not until the invention of near-field scanning microscopy (NSOM), a concept designed in 1928 [11] and realized in the mid-1980s [12,13], that optical imaging with subdiffraction resolution became possible. However, both the original implementations and more modern concepts sensing the near field [14,15] with optical metamaterials [16] are surface bound and are thus limited in terms of applications. The diffraction barrier as a resolution limitation for optical microscopy operating with regular lenses, i.e., in the far-field, remained.

1.2 Far-Field Fluorescence Nanoscopy

Stimulated emission depletion (STED) [17] and the related concept of ground-state depletion micros-copy [18], introduced in 1994, fundamentally overcome the far-field diffraction limit in fluorescence microscopy. The underlying idea is to switch the marker molecules in a controlled fashion between a bright (fluorescent) state and a dark state using a beam of light featuring an intensity minimum, e.g., at the focal center, such as to confine the bright, fluorescence-emitting state to the intensity minimum and keep the fluorophore in the ground state in the focal periphery. In the case of STED microscopy, the switching is performed by utilizing the process of stimulated emission. In order to switch off the abil-ity of the markers to fluoresce, a second beam of light inducing stimulated emission (called the STED beam) is superimposed to the excitation beam. It operates at a wavelength that does not excite the mol-ecules but, rather, efficiently induces stimulated emission, thereby switching off the ability of the dye to fluoresce (Figure 1.1).

The STED beam instantly forces (any) potentially excited molecules back to their ground state thus precluding fluorescence emission. Markers lose their ability to fluoresce via the emission of a stimulated copy photon, but because the STED wavelength is spectrally excluded from detection, this photon and the many photons of the stimulating beam are not detected. To increase the resolution, the switching has to be spatially selective, meaning that the area from which fluorescence can occur is confined by the STED beam. Therefore, the intensity distribution of the STED beam features a central minimum (e.g., like a torus). At the very center of the focus, the switching-off STED beam does not affect the molecules and they remain fluorescent, but at a distance $d/2$ from the minimum, the local intensity of the STED beam is high enough to switch off all the molecules. Molecules remain bright within a radius $d/2$, which defines the resolution and can be inferred from [19]

$$d \approx \lambda / \left(2NA\sqrt{1 + I/I_s} \right) \tag{1.1}$$

with

   I representing the maximum intensity of the STED beam
   I_s representing the intensity for which the fraction 1/e of the molecules are switched by the STED
      beam

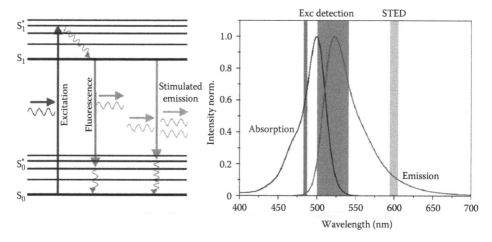

FIGURE 1.1 (See color insert following page 15-30.)   The Jablonski diagram (left) shows the processes utilized in STED microscopy. The spectrum (right) shows the corresponding wavelength inducing excitation and stimulated emission. As shown, the STED wavelength, and hence photons created via stimulated emission, are excluded from the detection band.

The image is recorded by moving both beams simultaneously through the sample and detecting the fluorescence. This renders a high-resolution image of the structures directly by the physical phenomena used, without further processing.

Figure 1.2 illustrates the resolution gain as the STED intensity I is increased. The amount of fluorescence detected or, likewise, the probability for an excited molecule to reside in the excited state and to emit a fluorescence photon rather than undergoing stimulated emission is, in first approximation, an exponentially decaying function of I. Once the STED intensity is beyond a certain value, the remaining fluorescence emission is negligible. The insets in Figure 1.2 show an overlay of the excitation focus (green) and a toroidal STED focus weighted by the probability to inhibit fluorescence emission at three different values of I/I_s (0, 5, and 15). The red areas—which converge toward the focal center with increasing I/I_s —indicate where fluorescence cannot occur due to the action of the STED beam. Consequently, the size of the remaining central region defines the resolution of the microscope, which is only limited by the available laser power and the properties of the dye. In practice, a resolution down to ~5 nm is achievable [20] today. By employing fast beam scanning systems, it is also possible to record high-resolution movies of living cells at video rate [21].

In order to give the STED focus its particular shape, an additional optical element must be placed in the STED beam. Such a specifically engineered phase filter (PF) adds some spatially varying phase information onto the incoming (flat) wave front such that, upon focusing, the fields interfere destructively in the focal center. For a given PF, the electric field near the focus can be calculated using vectorial diffraction theory [22,23]:

$$\mathbf{E}(r,z,\varphi) = iC \int_0^\alpha \int_0^{2\pi} \sqrt{\cos\theta}\, \mathbf{P}(\theta,\phi) \exp\left[ikn(z\cos\theta + r\sin\theta)\cos(\phi - \varphi) \right] \sin\theta \, d\theta \, d\phi \qquad (1.2)$$

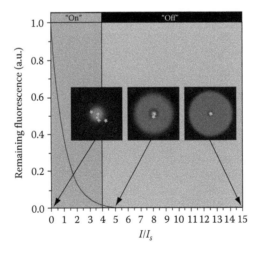

FIGURE 1.2 (See color insert following page 15-30.) Fluorescence as a function of STED intensity I (normalized to the saturation intensity I_s which is characteristic of the dye used) and resolution scaling. Beyond a certain value of I/I_s, the probability for fluorescence to occur is negligible for all practical purposes. With increasing power of the STED laser beam the region where this value is reached converges toward the focal center. As a result, nearby features that are illuminated simultaneously by the diffraction-limited excitation focus can be separated because only those features in the center of the focus remain fluorescent.

where

$P(\theta, \phi)$ describes the complex polarization state of the incident electromagnetic field including the phase information due to the PF

$E(r, z, \varphi)$ is the electric field at the (cylindrical) coordinate (r, z, φ)

n is the refractive index

$k = 2\pi n/\lambda$ is the wavenumber

α is the semiaperture angle

C is a constant

The focal intensity is then given by $I(r, z, \varphi) = |E(r, z, \varphi)|^2$.

Carrying out the integration in Equation 1.2 with a constant phase yields the normal diffraction-limited point-spread function (PSF) shown in Figure 1.3A. The two binary $(0/\pi)$ PFs shown in Figure 1.3B and C lead to two shifted intensity maxima in the focal plane and along the optic axis, respectively. Accordingly, these PFs can be used to enhance the resolution in one lateral direction or (mostly) in the axial direction. The advantage of the binary PFs is that they are relatively easy to fabricate by vapor deposition through a mask. However, the vortex PF shown in Figure 1.3D has the advantage that it enables a two-dimensional resolution enhancement with just one STED beam. The linear increase in phase along the azimuth generates a toroidal STED PSF, which suppresses fluorescence in two directions. Since vortex phase plates have become commercially available, they have been adopted for many STED applications. Unlike the binary PF, which needs to be used with linear or even arbitrary polarization, the vortex phase plate requires clean circular polarization.

Several STED beams may also be combined in a single setup to combine lateral and axial resolution enhancements or simply to further increase the resolution in one direction. The STED beams may be superimposed both in a coherent or an incoherent way, depending on the setup. An obvious choice

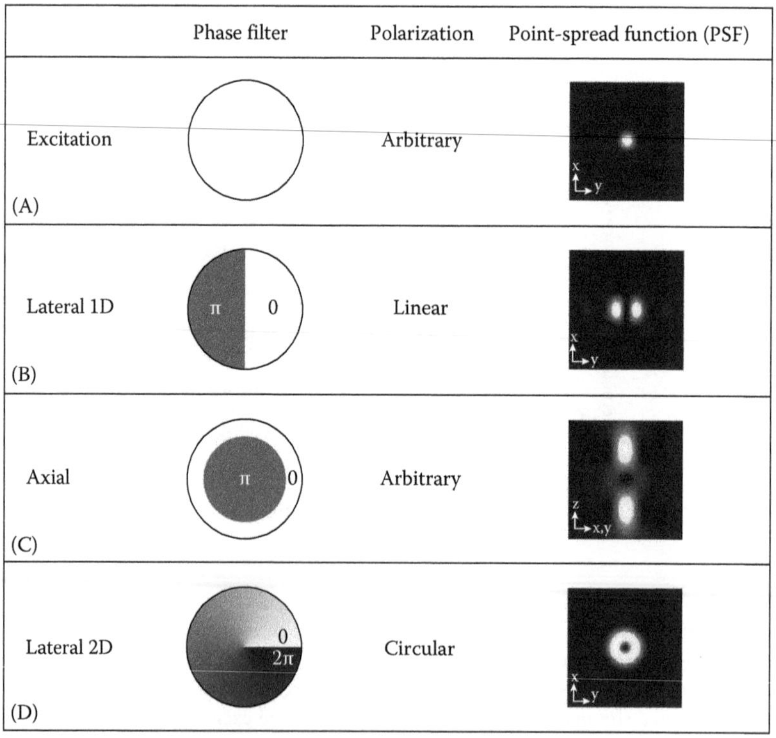

	Phase filter	Polarization	Point-spread function (PSF)
Excitation (A)		Arbitrary	
Lateral 1D (B)		Linear	
Axial (C)		Arbitrary	
Lateral 2D (D)		Circular	

FIGURE 1.3 PFs and resulting point-spread functions (PSFs).

is to use the vortex PF in combination with the semicircular phase plate (C) to achieve a high resolution in 3D. In this case, the two STED foci are overlapped incoherently. A more sophisticated scheme employing both coherent and incoherent superpositions of multiple STED beams has been implemented in a 4Pi microscope to achieve an isotropic resolution of ~40 nm (iso-STED) [24].

As the key to high resolution is the switching of the markers, STED is only one member of the family of targeted switching methods [25–28]. In general, resolution beyond the diffraction limit can be achieved by utilizing a marker that can be toggled between two interconvertible states A and B, one of which is bright (meaning that it provides a detectable signal) and the other dark. One of the transitions, say A → B has to be light driven but then the reverse process A ← B may be spontaneous. This generalization is known as the reversible saturable fluorescent transitions (RESOLFT) concept [29]. It shows that the key ingredient is the marker. Alternative RESOLFT implementations have been demonstrated so far with standard fluorescent dyes (GSD) [30] and photoswitchable proteins [31].

All techniques mentioned so far were based on targeted or ensemble switching. A different nanoscopy approach uses the random photoswitching of individual fluorescent molecules. By switching only a small, sparsely distributed fraction of the markers into the bright state one can assure that only one molecule is fluorescent within the diffraction limited area. Although the fluorescent signature of these molecules is still diffraction limited, their exact position can be determined with high fidelity. After being imaged, the molecules are switched off and the next subset is stochastically switched to the bright state. The whole image is composed by adding the successively imaged and processed frames leading to a map of localized points representing the positions of the markers. In contrast to the ensemble concepts, the number of emitted photons determines the achievable resolution $d \approx \lambda/(2NA\sqrt{N})$ [32,33], where N denotes the number of detected photons per fluorophore molecule. The localization accuracies of 10 nm are easily achieved with common dyes. As with the ensemble switching, many implementations (PALM, STORM, PALMIRA, GSDIM) of stochastic switching have been developed. In conclusion, it can be stated that imaging beyond the diffraction limit is based on switching markers between their molecular states.

1.3 New Light Sources

In this chapter, we review recent developments in laser technology that have substantially extended the range of laser sources available for STED microscopy. The basic problem is that because typical organic fluorophores exhibit excited state lifetimes in the 1–10 ns range one needs high rates of stimulated emission (approx. 10^{10}–10^{12} s^{-1}) in order to achieve quantitative fluorescence suppression. Given typical cross sections of stimulated emission (~10^{-17} cm^2), this translates into irradiances of a several hundred MW/cm^2, which are most conveniently realized by focusing (ultra)short laser pulses to a diffraction-limited spot. The use of pulsed laser sources keeps the time-averaged optical power low and, in many cases, compares favorably regarding photobleaching—in particular when low repetition rates are used. It has been shown that long time intervals between successive pulses allow for a relaxation of the potentially populated triplet state thus eliminating triplet state photobleaching pathways [34]. Accordingly, STED microscopes have been most frequently implemented using mode-locked laser systems, which have rendered their setup both costly and labor intensive. Therefore, in order to make STED microscopy amenable to a wider community, there is a strong demand for alternative laser sources suitable to implement STED instruments at lower cost and with less pulse preparation effort.

Driven by this demand, it has been shown that the use of pulsed illumination is not strictly mandatory: using a CW laser power of 825 mW at a wavelength of 730 nm, spatial resolutions down to 30 nm were realized [35]. While CW lasers are available in a wide range of wavelengths and at a substantially lower price compared to pulsed laser systems, the downside of their use is the relatively high time-averaged power required to achieve high resolutions. The reason for the high time-averaged power is the simple fact that the fluorophores are constantly in use; there is no time left for a "break," as is the case in pulsed excitation. On the other hand, it should be noted that, due to the fact that the photons used for STED are not bunched in pulses, the focal intensities used in CW STED microscopy (including the I_s)

are lower than in pulsed mode implementations. This is totally unlike multiphoton microscopy, because stimulated emission is a "linear" phenomenon in terms of the number of photons used; it is a single-photon-induced event. The effects of the high powers acting on the specimen are not fully explored, yet detrimental side effects such as optical trapping and, potentially, an increased cell toxicity must be taken into consideration.

In the past few years, exciting new light sources have emerged that make use of phototonic crystal fibers (PCF) in order to generate pulses with a broad spectral width, potentially covering the whole visible region and NIR region. The generation of these supercontinua in bulk media, conventional and tapered optical fibers has been known since the 1970s [36] when spectral broadening was observed as high-energy (μJ–mJ) laser pulses propagated through these media. Since then, the mechanisms responsible for the spectral broadening have been investigated in great detail. It has become clear that Raman scattering, four-wave mixing [37], and soliton formation [38] are the predominant processes involved but that their individual contributions depend on the particular experimental conditions. However, it was not until the introduction of PCFs and the availability of compact mode-locked fiber lasers that broad spectra could be produced efficiently with compact and reliable sources. Supercontinuum lasers are now commercially available. Initially, their pulse energies were not high enough to drive stimulated emission at rates sufficient to achieve significant resolution enhancement and, consequently, they were first used only to excite fluorescence [39,40]. Meanwhile, they provide spectral power densities of up to 1.5 nJ/nm in the visible region and NIR region (Figure 1.4A). For a 20 nm spectral band (as is typically used for the STED beam), this translates to pulse energies of 30 nJ that is even ahead of amplified Ti:sapphire laser systems.

The basic design of a STED microscope built with a single supercontinuum laser as the light source is shown in Figure 1.5. In order to derive the excitation and the STED beams, the supercontinuum laser beam is first split with a dichroic mirror (DC). Further spectral filtering is achieved by employing monochromators (MC) in both beams. For the STED beam, a spectral band of ~20–25 nm with very steep edges is preferable in order to provide sufficient pulse energy and to avoid excitation caused by the STED beam. As described in [41], a prism-based MC fulfills these needs satisfactorily. The described 4*f*-arrangement provides an almost rectangular spectrum with an efficient suppression of other wavelengths. The resulting STED pulse was found to be ~90 ps long, which is suitable to be used without further pulse compression or expansion. For the excitation beam, the spectral requirements are much lower, and standard interference band-pass filters or an acousto-optic tunable filter (AOTF) can be

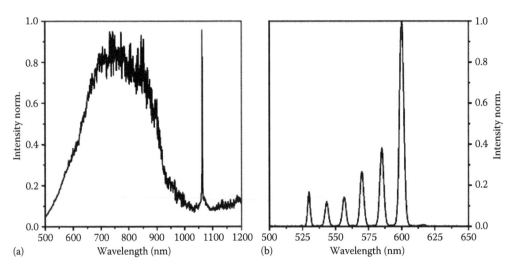

FIGURE 1.4 (a) Spectrum of a supercontinuum laser source and (b) comb spectrum generated by stimulated Raman scattering (SRS).

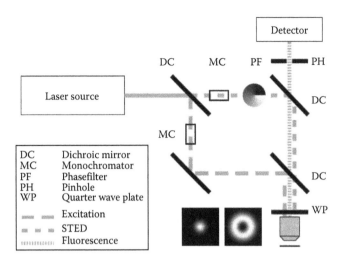

FIGURE 1.5 Simplified setup of a STED setup operating with a broadband laser source. The insets show the excitation and STED PSFs, respectively.

employed. Both beams are spatially filtered with polarization maintaining single mode fibers. At the fiber output, the STED beam is collimated and passes through the PF. Most commonly, a vortex PF like the one shown in Figure 1.3D is used, which leads to a toroidal focus shape. An additional quarter wave plate (WP) inserted into the laser beam affords, upon careful alignment, circular polarization as is required for the vortex PF. The excitation light leaving the fiber is separately collimated. Both beams are coupled into the objective with DCs. The fluorescence that emerges from the sample is collected with the same lens, passes through the DCs, is focused through a confocal pinhole (PH) and detected with an avalanche photodiode (APD).

In order to gain resolution enhancement, the excitation and STED PSFs must be carefully optimized to minimize aberrations and must be overlapped in space and time. This is most conveniently done by scanning, in reflection mode, gold nanoparticles (80 nm) dispersed on a glass surface. A scan of a single gold bead directly renders the PSF, which can be analyzed and optimized in terms of aberrations. Subsequently, alternating scans with the excitation and the STED beam allow one to monitor both PSFs while overlapping them in space. The temporal alignment is then performed by maximizing the fluorescence suppression in a bulk dye solution by changing the optical path lengths of either of the two beams.

Single dye molecules or fluorescent nanoparticles dispersed on a glass surface provide well-defined test samples that allow the performance evaluation of the microscope. As was shown in [41], the supercontinuum STED microscope can be adapted to virtually every fluorescent dye and, using various commercial fluorescent nanoparticles, resolutions between 30 and 40 nm could be readily achieved at the wavelength combinations 532 nm/650 nm, 570 nm/700 nm and 630 nm/745 nm (exc/STED). These results promised a very versatile instrument also for biological applications. Figure 1.7 shows samples of the immunolabeled microtubular network of PtK2 cells. Obviously, the STED image reveals single filaments where the confocal counterpart shows only blurred bright areas. A further improvement of the setup led to a resolution in the range of 20 nm. Another example (Figure 1.6) shows immunostained neurofilaments.

The images shown in Figures 1.6 and 1.7 were acquired using a single STED beam, which enhances the resolution to about 20–25 nm in the focal plane. However, the axial resolution remains at the confocal level of about 600 nm. As a result, the effective PSF is shaped much like a cigar, which is unfavorable particularly if thicker specimens are to be imaged. In order to make the detection volume more isotropic, a second STED beam fitted with a PF as shown in Figure 1.3C was added to the setup. The two STED beams are orthogonally polarized such that they can be easily combined with a polarizing beam splitter. They are superimposed incoherently, i.e., they do not interfere but rather their

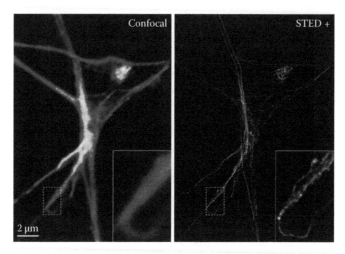

FIGURE 1.6 (See color insert following page 15-30.)   Fluorescence images of immunostained neurofilaments (light subunit, ATTO 590). Clearly visible, the STED (right) image shows more structural details than the confocal (left) counterpart. The STED image is deconvolved to further increase the contrast. (From Wildanger, D. et al., *Opt. Express.*, 16, 9614, 2008. With permission.)

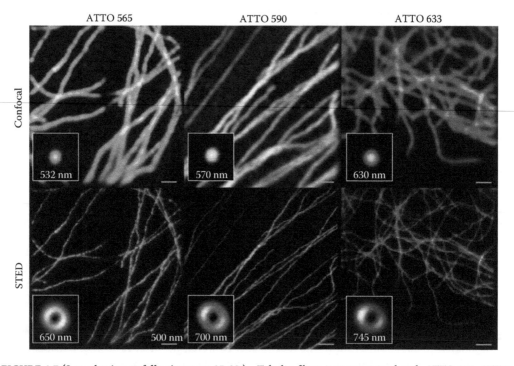

FIGURE 1.7 (See color insert following page 15-30.)   Tubulin fibers immunostained with ATTO 565, ATTO 590, and ATTO 633. The upper panels show confocal images obtained by employing excitation wavelengths of 532, 570, or 630 nm. The lower panels show the corresponding STED images. Depending on the dye, the central STED wavelength was adjusted to 650, 700, and 750 nm. Obviously, the STED images reveal single fibers where the confocal counterpart is blurred. (From Chi, K.R., *Nat. Meth.*, 6, 15, 2009. With permission.)

intensities add up. Because of the axial confinement due to the second beam, the effective focal spot becomes more isotropic. With this configuration it was possible to acquire a whole stack of images of microtubules. A series of consecutive frames from this stack is shown in Figure 1.8 demonstrating a slicing capability with ~110 nm resolution.

Beside the possibility to adapt the system to arbitrary dyes, the microscope can be readily upgraded to a two-color high-resolution microscope allowing colocalization studies on the nanometer scale.

Up to now laser systems with sufficient pulse energies only operate between 1 and 5 MHz. These low repetition rates have the advantage that the microscope operates under T-REX modalities leading to a significantly reduced photobleaching of the employed fluorescent markers. Unfortunately, these low repetition rates limit the image acquisition speed and make these laser sources unsuitable for live cell imaging. Scaling supercontinuum lasers to higher repetition rates remains challenging because

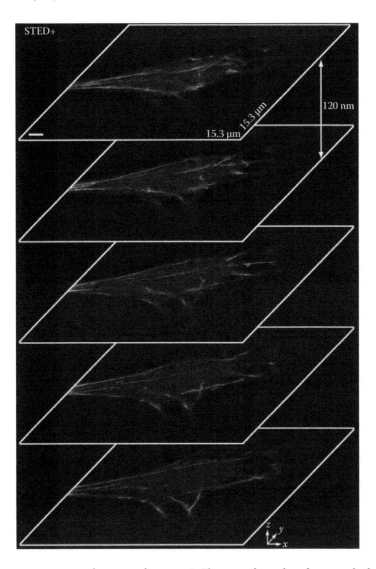

FIGURE 1.8 STED microsopy with superresolution in 3D. The image shows slices from a stack of immunostained tubulin in PTK2 cells. As dye ATTO 590 (Exc. 570 nm; STED 700 nm) was used. Shown is every second slice corresponding a distance of 120 nm. Obviously, structures appear and disappear from one slice to the next, proving an axial resolution of less than 120 nm. (From Wildanger, D. et al., *J. Microsc.*, 236, 35, 2009. With permission.)

only a small portion (~2%) of the continuous spectrum is actually used for the operation of the microscope. Increasing the repetition rate while conserving the pulse energy leads to average powers beyond practicability. Therefore, the ideal laser source for STED does not provide a broad continuum but distinct narrow lines in which the laser power is concentrated.

A technique which provides line spectra is stimulated Raman scattering (SRS). As has been shown in several publications [42–45], SRS in optical fibers can be used to generate a comb spectrum with multiple lines separated by a constant energy (Figure 1.4B). When the pump laser pulse reaches a threshold optical intensity in a fiber, the number of photons with energy $\hbar\omega_p$ in the pump beam which scatter inelastically in the core becomes significant, yielding lower energy, Stokes-shifted photons of energy $\hbar\omega_p$.

Four-wave mixing and dopants can modify the frequency shift of the generated Stokes line. Once a Stokes line itself reaches the threshold intensity for SRS, it becomes the pump wavelength for an additional Stokes line and a cascaded process generates the characteristic spectrum. Longer fiber lengths increase the gain at the Stokes wavelengths quasi-exponentially until the pump wavelength is depleted, the gain is balanced by fiber losses at the pump and Stokes wavelengths, or both. The optimal choice of fiber length balances Stokes gain and fiber losses to yield the widest comb spectrum with the highest power for a given pump intensity. In contrast to a supercontinuum spectrum, the optical power is concentrated in a finite number of lines such that, in order to achieve a certain spectral energy density, the total optical power is comparatively low.

STED microscopy with an SRS source was tested using a similar setup as shown in Figure 1.5 [46]. The source was built from a passively Q-switched, 60 kHz microchip laser providing 1 ns pulses at 532 nm and with a pulse energy of 0.5 µJ. The collimated laser beam was coupled into a 4 µm core diameter single mode optical fiber with a 410 nm cutoff wavelength. With a coupling efficiency of 50%, peak optical intensities on the order of 500 MW/cm^2 were reached in the fiber. Six Stokes lines were generated using a 50 m fiber patch, extending from the pump wavelength at 532 nm up to 620 nm (Figure 1.4B). Instead of an MC, the light from the fiber was selected with a band-pass filter. Due to energy fluctuations caused by passive Q-switching of the pump laser, not all pulses had the requisite energy to produce the most red-shifted lines of the comb spectrum, and the repetition rate fell for each subsequent Stokes line, down to 18 kHz for 620 nm. The excitation light was provided by electronically synchronized picosecond-pulsed diode lasers operating at wavelengths 440 or 470 nm.

To demonstrate the usability of light from the entire comb spectrum for STED resolution enhancement, from the fundamental wavelength to the last Stokes peaks, STED images were obtained using 532 nm and the last three Stokes lines, at 588, 604, and 620 nm. Measurements are shown in Figure 1.9. Self-made silica beads resolved at 532 nm were marked with the fluorophore ATTO 425 (ATTO-TEC GmbH, Siegen, Germany) and treated with DABCO (1,4-Diazabicyclo[2.2.2]octan) (Roth, Karlsruhe, Germany) to counteract photobleaching. Using the remaining three STED wavelengths, different samples of yellow–green fluorescent beads (Invitrogen, Carlsbad, California, USA) of sizes 24 and 43 nm were imaged, which contain a fluorophore unspecified by the manufacturer. To demonstrate biological imaging with 604 nm STED light neurofilaments labeled with ATTO 532 were imaged. Average pulse energies at the back aperture of the objective lens varied from 0.4 to 2.7 nJ, and pixel dwell times for image acquisition were typically between 15 and 30 ms, with pixel sizes of 10–20 nm. To obtain a reliable estimate of resolution, the images of 30–60 isolated beads from a single measurement were superimposed by aligning the positions of maximum signal from each bead image using in-house software, averaging over the photon noise associated with each bead image. Using a Gaussian functional dependence for the effective PSF, full width half-maximum (FWHM) values of 70–80 nm were obtained, averaged over the x, y, and both diagonal orientations in the image. Using the bead size together with the measured FWHM values from the bead images yielded effective PSF values varying from 58 nm (588 and 603 nm STED) to 78 nm (620 nm STED). The lower resolution at 620 nm can be explained by the smaller cross section for stimulated emission of the fluorophore at this wavelength, rendering STED quenching less efficient. Thus, even using this simple proof-of-principle setup, it is clear that light generated in this manner is suitable for achieving resolutions well below the diffraction barrier with STED.

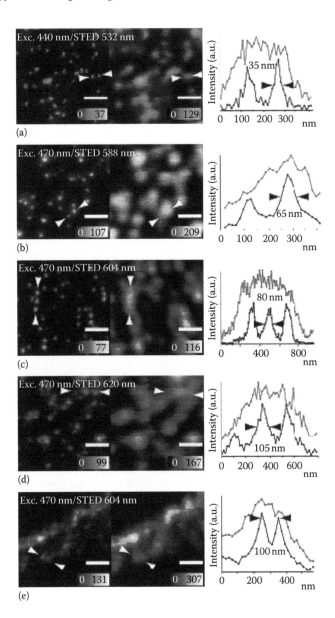

FIGURE 1.9 (See color insert following page 15-30.)   STED measurements are shown at left, with corresponding confocal measurements of the same site in the sample center, and line-profile measurements, at sites indicated by arrows, at right. Scale bars 500 nm. Excitation and STED wavelengths are indicated. (a) 20–30 nm silica beads labeled with ATTO 425. (b and c) 40 nm beads. (d) 20 nm beads. (e) Neurofilaments labeled with ATTO 532. (From Rankin, B.R. et al., *Opt. Lett.*, 33, 2491, 2008. With permission.)

References

1. Abbe, E. (1873). Beiträge zur Theorie des Mikroskops und der mikroskopischen Wahrnehmung. *Arch. Mikr. Anat. 9*, 413–468.
2. Hell, S.W. (1990/1992). Double-scanning confocal microscope. European Patent 0491289.
3. Hell, S. and Stelzer, E.H.K. (1992). Properties of a 4Pi-confocal fluorescence microscope. *J. Opt. Soc. Am. A 9*, 2159–2166.

4. Hell, S. and Stelzer, E.H.K. (1992). Fundamental improvement of resolution with a 4Pi-confocal fluorescence microscope using two-photon excitation. *Opt. Commun. 93,* 277–282.

5. Gustafsson, M.G.L., Agard, D.A., and Sedat, J.W. (1999). I^{5}M: 3D widefield light microscopy with better than 100 nm axial resolution. *J. Microsc. 195,* 10–16.

6. Minsky, M. (1961). Microscopy apparatus. US Patent 3,013,467.

7. Brakenhoff, G.J., Blom, P., and Barends, P. (1979). Confocal scanning light microscopy with high aperture immersion lenses. *J. Microsc. 117,* 219–232.

8. Wilson, T. and Sheppard, C.J.R. (1984). *Theory and Practice of Scanning Optical Microscopy* (New York: Academic Press).

9. Sheppard, C.J.R. and Kompfner, R. (1978). Resonant scanning optical microscope. *Appl. Optics 17,* 2879–2882.

10. Denk, W., Strickler, J.H., and Webb, W.W. (1990). Two-photon laser scanning fluorescence microscopy. *Science 248,* 73–76.

11. Synge, E.H. (1928). A suggested method for extending microscopic resolution into the ultra-microscopic region. *Philos. Mag. 6,* 356.

12. Pohl, D.W., Denk, W., and Lanz, M. (1984). Optical stethoscopy: Image recording with resolution λ/20. *Appl. Phys. Lett. 44,* 651–653.

13. Lewis, A., Isaacson, M., Harootunian, A., and Murray, A. (1984). Development of a 500 A resolution light microscope. *Ultramicroscopy 13,* 227–231.

14. Shelby, R.A., Smith, D.R., and Schultz, S. (2001). Experimental verification of a negative index of refraction. *Science 292,* 77–79.

15. Smolyaninov, I.I., Hung, Y.-J., and Davis, C.C. (2007). Magnifying superlens in the visible frequency range. *Science 315,* 1699–1701.

16. Pendry, J.B. (2000). Negative refraction makes a perfect lens. *Phys. Rev. Lett. 85,* 3966–3969.

17. Hell, S.W. and Wichmann, J. (1994). Breaking the diffraction resolution limit by stimulated emission: Stimulated emission depletion fluorescence microscopy. *Opt. Lett. 19,* 780–782.

18. Hell, S.W. and Kroug, M. (1995). Ground-state depletion fluorescence microscopy, a concept for breaking the diffraction resolution limit. *Appl. Phys. B 60,* 495–497.

19. Westphal, V. and Hell, S.W. (2005). Nanoscale resolution in the focal plane of an optical microscope. *Phys. Rev. Lett. 94,* 143903.

20. Rittweger, E., Han, K.Y., Irvine, S.E., Eggeling, C., and Hell, S.W. (2009). STED microscopy reveals color centers with nanometric resolution. *Nat. Photon.,* doi:10.1038/NPHOTON.2009.1032.

21. Westphal, V., Rizzoli, S.O., Lauterbach, M.A., Kamin, D., Jahn, R., and Hell, S.W. (2008). Video-rate far-field optical nanoscopy dissects synaptic vesicle movement. *Science 320,* 246–249.

22. Born, M. and Wolf, E. (2002). *Principles of Optics,* 7th Edition (Cambridge: Cambridge University Press).

23. Richards, B. and Wolf, E. (1959). Electromagnetic diffraction in optical systems II. Structure of the image field in an aplanatic system. *Proc. R. Soc. Lond. A 253,* 358–379.

24. Schmidt, R., Wurm, C.A., Jakobs, S., Engelhardt, J., Egner, A., and Hell, S.W. (2008). Spherical nano-sized focal spot unravels the interior of cells. *Nat. Methods 5,* 539–544.

25. Hell, S.W. (2003). Toward fluorescence nanoscopy. *Nat. Biotechnol. 21,* 1347–1355.

26. Hell, S.W., Jakobs, S., and Kastrup, L. (2003). Imaging and writing at the nanoscale with focused visible light through saturable optical transitions. *Appl. Phys. A 77,* 859–860.

27. Hell, S.W. (2007). Far-field optical nanoscopy. *Science 316,* 1153–1158.

28. Hell, S.W. (2008). Microscopy and its focal switch. *Nat. Methods 6,* 24–32.

29. Hell, S.W., Dyba, M., and Jakobs, S. (2004). Concepts for nanoscale resolution in fluorescence microscopy. *Curr. Opin. Neurobiol. 14,* 599–609.

30. Bretschneider, S., Eggeling, C., and Hell, S.W. (2007). Breaking the diffraction barrier in fluorescence microscopy by optical shelving. *Phys. Rev. Lett. 98,* 218103.

31. Hofmann, M., Eggeling, C., Jakobs, S., and Hell, S.W. (2005). Breaking the diffraction barrier in fluorescence microscopy at low light intensities by using reversibly photoswitchable proteins. *Proc. Natl. Acad. Sci. USA 102*, 17565–17569.

32. Heisenberg, W. (1930). *The Physical Principles of the Quantum Theory* (Chicago, IL: Chicago University Press).

33. Thompson, R.E., Larson, D.R., and Webb, W.W. (2002). Precise nanometer localization analysis for individual fluorescent probes. *Biophys. J. 82*, 2775–2783.

34. Donnert, G., Keller, J., Medda, R., Andrei, M.A., Rizzoli, S.O., Lührmann, R., Jahn, R., Eggeling, C., and Hell, S.W. (2006). Macromolecular-scale resolution in biological fluorescence microscopy. *Proc. Natl. Acad. Sci. USA 103*, 11440–11445.

35. Willig, K.I., Harke, B., Medda, R., and Hell, S.W. (2007). STED microscopy with continuous wave beams. *Nat. Methods 4*, 915–918.

36. Dudley, J.M., Genty, G., and Coen, S. (2006). Supercontinuum generation in photonic crystal fiber. *Rev. Mod. Phys. 78*, 1135.

37. Coen, S., Chau, A.H.L., Leonhardt, R., Harvey, J.D., Knight, J.C., Wadsworth, W.J., and Russell, P.S.J. (2002). Supercontinuum generation by stimulated Raman scattering and parametric four-wave mixing in photonic crystal fibers. *J. Opt. Soc. Am. B 19*, 753–764.

38. Hilligsøe, K.M., Paulsen, H.N., Thøgersen, J., Keiding, S.R., and Larsen, J.J. (2003). Initial steps of supercontinuum generation in photonic crystal fibers. *J. Opt. Soc. Am. B 20*, 1887–1893.

39. McConnell, G. (2004). Confocal laser scanning fluorescence microscopy with a visible continuum source. *Opt. Express 12*, 2844–2850.

40. Auksorius, E., Boruah, B.R., Dunsby, C., Lanigan, P.M.P., Kennedy, G., Neil, M.A.A., and French, P.M.W. (2008). Stimulated emission depletion microscopy with a supercontinuum source and fluorescence lifetime imaging. *Opt. Lett. 33*, 113–115.

41. Wildanger, D., Rittweger, E., Kastrup, L., and Hell, S.W. (2008). STED microscopy with a supercontinuum laser source. *Opt. Express 16*, 9614–9621.

42. Stolen, R.H., Ippen, E.P., and Tynes, A.R. (1972). Raman oscillation in glass optical waveguide. *Appl. Phys. Lett. 20*, 62–64.

43. Gao, P.-j., Nuie, C.-j., Yang, T.-l., and Su, H.-z. (1981). Stimulated Raman scattering up to 10 orders in an optical fiber. *Appl. Phys. A 24*, 303–306.

44. Rosman, G. (1982). High-order comb spectrum from stimulated Raman scattering in a silica-core fibre. *Opt. Quant. Electron. 14*, 92–93.

45. Agrawal, G. (2006). *Nonlinear Fiber Optics*, 4th edn. (Burlington, MA: Academic Press).

46. Rankin, B.R., Kellner, R.R., and Hell, S.W. (2008). Stimulated-emission-depletion microscopy with a multicolor stimulated-Raman-scattering light source. *Opt. Lett. 33*, 2491–2493.

47. Wildanger, D., Medda, R., Kastrup, L., and Hell, S.W. (2009). A compact STED microscope providing 3D nanoscale resolution. *J. Microsc., 236*, 35–43.

48. Chi, K.R. (2009). Super-resolution microscopy: breaking the limits. *Nat. Meth. 6*, 15-18.

2

Nonlinear Fluorescence Imaging by Saturated Excitation

2.1 Introduction: Methods to Improve Fluorescence
 Microscopy Resolution 2-1
2.2 Saturated Excitation (SAX) Microscopy for High-
 Resolution Fluorescence Imaging 2-3
 Overview and Principles of SAX Microscopy • Imaging
 Properties and Effective Point Spread Function of SAX
 Microscopy • Confirmation of the Onset and Nature of the
 Saturation Nonlinearity Effect • High-Resolution Saturated
 Fluorescence Microscopy of Biological Samples
2.3 Discussion and Perspectives 2-13
References 2-15

Nicholas I. Smith
Shogo Kawano
Masahito Yamanaka
Katsumasa Fujita
Osaka University

2.1 Introduction: Methods to Improve Fluorescence Microscopy Resolution

The development of the microscope has been constantly driven by the need to image smaller and smaller details in samples of interest. This is as true in fluorescence microscopy as it is in other types of imaging. The imaging capabilities of the fluorescence microscope have improved over the last few decades, with particularly productive new techniques emerging over the last 15 years or so. There are numerous approaches to improve optical resolution, and in this chapter, we will briefly review methods that involve time-resolution and/or saturation processes to improve on conventional optical fluorescence microscopy. We will then turn to the main topic of this chapter: how to improve optical resolution in fluorescence microscopy by taking advantage of the nonlinearity that is brought on during the phenomenon of saturated excitation. The optical fluorescence microscope is a good base from which to work toward high-resolution microscopy since it works in combination with chemical- or protein-specific fluorescent labels, can observe specimens in living conditions, and can image three-dimensional regions inside a sample. In combination with attempts to improve resolution, the fluorescent labels that are used in such microscopy have also undergone steady improvement and have added abilities allowing fundamentally new imaging possibilities such as dynamic ion concentration change and cell membrane potential imaging (Giepmans et al. 2006, Johnson 1998).

From the early 1990s onward, several methods of significantly improving resolution have been developed and continue to improve. One method of note is stimulated-emission depletion (STED) (Hell and Wichmann, 1994). This method can increase the lateral resolution to beyond the diffraction limit in far-field fluorescence microscopy, and works by exploiting STED in the sample (Klar and Hell, 1999).

It requires two laser beams, one for fluorescence excitation and one for depleting the excitation of the surrounding area. This second so-called STED beam typically has a zero in the beam intensity at the center of the focus, and is used to deplete the excited states of the molecules in the outer region of the focal spot. The depletion is done by stimulated emission, which has the advantage that the emitted fluorescence and the STED beam have the same wavelength, and both may be eliminated by the use of a suitable optical filter. This leaves a subdiffraction limited spot at the center of the focal spot, where fluorescence can be generated by the excitation beam. The excitation beam must, therefore, be of a different wavelength. If the STED beam intensity is high enough, saturation of the STED process will occur, which can further reduce the effective dimensions of the fluorescence-emitting spot at the center of the beams. This technique has been shown to be able to produce an improvement in the resolution of close to 10 times, when compared with conventional optical microscopy of fluorescent samples (Hell 2003). The saturation effect comes into play in STED microscopy by enhancing the depletion of the excitation of areas around the center of the focal spot. Saturation typically has the effect of "squaring off" the edges of a smooth function (which is in this case the intensity profile of the STED beam) and effectively increases the abruptness of the transition between the depleted region in the sample and the fluorescence-emitting region at the center of the focal spot. For STED, it is the geometrical effect of the saturation that improves the resolution above what would be possible with standard beam-shaping techniques. If the STED beam intensity profile appears as a donut shape, and the fluorescence signal of interest is generated from the region inside the donut hole, then the resolution can be arbitrarily increased by decreasing the size of the hole. The optimization of the beam-forming optics alone cannot reduce the hold at the center of the STED beam but if the intensity of the STED beam increases and saturates, then the shoulders of the STED beam encroach further and further on the donut hole, reducing the effective size of the spot that generates the fluorescence signal.

Saturation has also been applied to enhance an imaging mode, which is already superior to conventional wide-field microscopy in its own right. A relatively simple method by which to improve the resolution is to use structured illumination. This technique uses the interference of multiple beams to enhance the resolution in the axial, lateral, or in a combination of these directions (Bailey et al. 1993, Gustafsson et al. 2008). The improvement in resolution is generally limited to twice the resolution of the conventional imaging mode, given the same wavelength and numerical aperture. This can be further improved, however, by allowing or in fact forcing the structured illumination pattern to saturate (Heintzmann et al. 2002). The saturation creates nonlinearity in the excitation and emission relationship. Such resulting nonlinearities can typically be exploited in microscopy. The saturation distorts the spatial distribution of the light intensity pattern. As in the case of one-dimensional time-based signals, distortion creates additional harmonics, and the distortion of a two-dimensional light distribution creates additional harmonics that effectively increase the spatial frequencies present in the illumination pattern (Gustafsson 2005). Typical distortion modes that occur when the illumination pattern is saturated will introduce more than one harmonic, and the higher harmonics can be used to achieve higher order improvements in resolution. The process is limited by the fact that many of the high harmonics are below the noise, which is inherent in the illumination or the detection system itself, and are therefore relatively difficult to measure. Nevertheless, those harmonics that do exceed the noise floor can be put to use to provide relevant image information and higher resolution. The technique has produced dramatic images with resolutions as high as 50 nm or better (Gustafsson 2005), and has been used to demonstrate very high-resolution imaging in cultured cells (Schermelleh et al. 2008). A merit of this technique is that the imaging can be done in wide-field mode (i.e., not requiring a scanned microscope stage or laser spot) so that the imaging rate can potentially be faster than other scan-based methods, although this is offset somewhat by the computational requirements of the imaging mode.

More recently, the time domain has started to play a role in improving resolution. Two notable methods with some similarities are photo-activated localization microscopy (PALM) (Betzig et al. 2006) and stochastic optical reconstruction microscopy (STORM) (Huang et al. 2008). In these techniques, time plays a role in improving the resolution by the fact that the separation of signals from individual fluorophore emissions takes place over time and this can be used to aid the reconstruction of the fluorophore distribution. In the sample, the emission of any given fluorescent molecule is not coherent or correlated with the emission from neighboring molecules. This means that, in principle, each emitter is an independent body, and since the physical dimensions of the fluorophore are on the order of a few nanometers, if the signal emitted from each fluorophore can be independently captured in time, the precise location of the fluorophore can be reconstructed. The excitation light conditions are set so that at any given moment, only a few fluorophores in the field are excited, and those fluorophores are separated in space. For samples that are not moving in time, the imaging process can take place over long time periods, and extremely high-resolution images can be reconstructed from the overall fluorescence signals captured over time. The exposure time periods for PALM fluorescence images were originally several hours, but have recently been optimized and the imaging rate is approaching timescales of several minutes or less per frame (Shroff et al. 2008).

While correlation-based spectroscopy (FCS) (Magde et al. 1972), fluorescence lifetime imaging (FLIM) (Lakowicz et al. 1992), and fluorescence recovery after photobleaching (FRAP) (Sprague et al. 2004) methods are well known and make explicit use of the time-dependent behavior of fluorescence molecules, there has been relatively little exploitation of the time domain in the search for methods to improve resolution. Although methods, such as STED, use synchronized pulse lasers, and PALM or similar techniques require a careful management of the amount of data per time frame, the time domain remains mostly unexplored in the sense that it exists as an additional degree of freedom with which to optimize the resolution. Recently, photoswitchable proteins have offered the potential to modify the fluorescent probe and increase the resolution in a manner similar to the STED technique described above (Hofmann et al. 2005). Also in 2005, Enderlein et al. reported a method of increasing the spatial resolution of fluorescence microscopy by utilizing the intensity-dependent ground state to triplet state transition of the fluorophore (Enderlein et al. 2005). This technique is called dynamic saturation optical microscopy (DSOM) and requires fast temporal measurements of the fluorescence decay. The intensity dependence of the process allows an ability to differentiate between the signals detected, which were generated from different regions of the focal spot. This knowledge allows the retention of signals generated from the center of the focal spot. This allows the improvement in spatial resolution. It was also reported that the technique works for both Gaussian and Bessel beams, and that the Bessel beam can produce a further improvement in resolution, albeit at the cost of higher side peaks that can degrade the overall imaging performance. However, with the nonlinearity present in the imaging mode, side peak contributions to the main signal tend to be decreased (in common with most other nonlinear imaging methods).

2.2 Saturated Excitation (SAX) Microscopy for High-Resolution Fluorescence Imaging

2.2.1 Overview and Principles of SAX Microscopy

Recent work has shown the possibilities of a new concept in fluorescence resolution enhancement, which makes explicit use of the time domain. In general, to improve resolution, some techniques are required to introduce nonlinearity in the relationship between the excitation light and the emitted and detected fluorescence. Methods such as PALM and STORM use a different approach, and attempt to track the individual fluorescence emission from isolated fluorophores, but the majority of methods discussed earlier use experimental procedures designed to induce nonlinearity. Once nonlinearity is present, it can be exploited, often without theoretical limitation, to improve the imaging-mode resolution. The key to exploiting the nonlinearity lies in the fact that the spatial distribution of excitation, emission, and/or the distribution of molecular states in the sample at the focus will change in shape as well as intensity when the incoming light intensity changes. Generating some type of nonlinearity in the excitation–emission relationship is therefore an important part of resolution improvement. In fluorescence microscopy with relatively low excitation light levels, the fluorescent molecules in the light path are excited by the incoming light at a rate that is close to linearly proportional to the excitation light intensity. However, since the number of actual fluorescent molecules in the light path is fixed, when the excitation light intensity is sufficiently high, all available molecules may be emitting fluorescence, and the fluorescence emission rate cannot rise even if the incoming light intensity is increased. At this point, the fluorescence process is completely saturated. Ignoring, for the moment, complications such as photobleaching and interstate transitions, the transition from linear to nonlinear will occur gradually, with the amount of saturation-based nonlinearity itself being proportional to the excitation light intensity. This last point is worthy of further consideration. Typically, saturation in fluorescence microscopy is considered to be a process that degrades the imaging performance, decreases resolution (by increasing the effective size of the point spread function), and increases photobleaching, and should therefore be avoided. However, since the saturation itself is dependent on the excitation intensity, if the saturation signal itself is detected, it can be used to improve the resolution.

In saturated excitation (SAX) microscopy, excitation is achieved by focused scanning laser, and detection is done with a single channel detector, in the same manner as a typical scanning confocal microscope (Cogswell et al. 1992). The incident light intensity is modulated by an acousto-optical modulator (AOM), at a well-defined frequency (ω), and the detected output can be correspondingly demodulated at either the same frequency ω or at multiples of the base frequency (i.e., 2ω, 3ω...). For an ideal and completely linear system, where the emission is proportional to the excitation, the output will look like a scaled version of the input. The demodulation of the detected output at any frequency other than the input frequency (ω) will then produce no signal for a linear system. However, if some nonlinearity exists, the harmonic components will appear as multiples of ω and therefore the 2ω, 3ω, and higher order harmonics will contain some nontrivial signal. The focused laser beam also has a spatially dependent intensity distribution, generally with a peak at the center of the focal spot. When the excitation intensity is high enough to generate saturation of the fluorescence emission, the demodulated harmonic frequencies can be detected. Since the saturation itself is a function of the intensity, the strongest saturation will occur when the excitation intensity is at its highest. For a typical Gaussian beam, this occurs at the center of the focal spot. By detecting the 2ω, 3ω, or higher order harmonics, at the onset of saturation, the harmonic signals will contain saturation signals that are generated from the center of the focal spot. In this way, the time-demodulated harmonic signals resulting from the induced nonlinearity in the fluorescence response can be used to form an image. This image will have higher spatial resolution than the fluorescent image that is formed by the detection of the signal from the entire focal spot, since it has a strong bias toward detecting signals from the very center of the focal spot. Practically, the incident light modulation can be done by an AOM controlled by a computer-generated control signal. The demodulation can be done by a lock-in amplifier, which has the additional benefit of being able to record signals that are below the noise floor, by averaging over a large number of cycles. The conversion of a microscope from conventional fluorescence confocal microscopy to SAX confocal microscopy requires very little additional hardware, namely the AOM and lock-in amplifier, which means that the technique may find wide use as a simple means of generating higher resolution images.

2.2.2 Imaging Properties and Effective Point Spread Function of SAX Microscopy

The improvement in imaging resolution by the use of saturated excitation and harmonic demodulation of the detected light can be calculated. Using either a simple two-level model system (Fujita et al. 2007a), or a more complex multi-state model with inter-system crossings and multiple singlet and triplet states (Fujita et al. 2007b, Eggeling et al. 2005), the probabilities of the molecules existing in any given state can be calculated using rate equations for the transitions between these states. Since many of the transitions (such as stimulated emission) are intensity dependent, the ensemble of molecular states at a given time and for a defined laser intensity can be well approximated by the intensity-dependent rate equations (Kawano et al. 2009). Note that the saturation effect is not particularly dependent on the concentration of the fluorescent molecules, so long as the overall light intensity is not significantly affected by the total amount of absorption. This is because the presence of a greater or lesser number of molecules simply results in a larger or smaller amount of sources that may emit saturated fluorescence. The saturation itself depends on the probability of any individual molecule being in a state that can emit fluorescence, and this in turn depends primarily on the intensity of the light, and not on the presence or absence of neighboring molecules. This is an important point for the use of this technique in imaging, because if the saturation were significantly concentration dependent, then the spatial resolution would depend on the sample molecular distribution, rendering the technique ineffectual as a quantitative measurement method.

Figure 2.1 shows calculation results using such a model for Rhodamine-6G molecules excited by 532 nm light (Kawano et al. 2009). The results show that the emission intensity is proportional to the excitation at intensities below 4.0×10^5 W/cm², and saturation becomes dominant at around 10^6 W/cm² excitation intensity.

The saturation effect is clear in Figure 2.1, and as the excitation–emission relationship becomes more nonlinear, the harmonic components rise in strength (Yamanaka et al. 2008). The effective response function of the demodulation can also be calculated by Fourier transforming the harmonics of the output signal (Figure 2.2). When the excitation intensity is expressed as $\Phi(t)$, the sinusoidal modulated excitation intensity can be expressed as

$$\Phi(t) = \Phi_0 \frac{1}{2}(1 + \cos(\omega t)) \tag{2.1}$$

where
   Φ_0 is the average intensity
   ω is the modulation frequency

In the calculation, the modulation frequency 10 kHz was used with Rhodamine-6G molecule parameters used in the model.

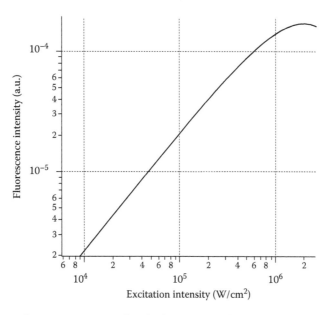

FIGURE 2.1 Theoretical saturation response for Rhodamine-6G molecules in solution excited by 532 nm excitation. The relationship between the fluorescence intensity and the excitation intensity becomes nonlinear when saturation occurs at high excitation intensities. (From Kawano, S. et al., *Proc. SPIE*, 7184, 718415, 2009. With permission.)

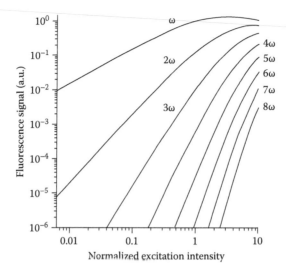

FIGURE 2.2 The calculated demodulated response of fluorescence intensity, where the incident light modulation frequency was 10 kHz, and the output was demodulated at the harmonics of the incident light frequency. The sample for the calculation was Rhodamine-6G molecules in solution.

The calculations show that the response becomes increasingly nonlinear, as the excitation intensity increases. Each harmonic of the modulation frequency has a corresponding increase in the order of the nonlinear response. The nth harmonic has an nth-order nonlinear response, which can be used to increase the spatial frequency in a similar manner to the use of harmonics in second harmonic generation (SHG) microscopy and similar to the higher photon-number modes in multiphoton microscopy. The advantage of the SAX technique is that it is not dependent on particular geometrical properties of the sample, as is the case for SHG. The use of SAX microscopy does require high excitation intensities compared to typical fluorescence microscopy, but the intensities are still below the laser intensities used for pulsed-mode multiphoton microscopy. The higher order modes of SAX microscopy have additional benefits over those in SHG or multiphoton microscopy: the higher order modes occur at the same light wavelength as the excitation beam. The higher spatial resolution that occurs with higher harmonics is then relative to the excitation wavelength, and the excitation wavelength can be chosen so that the single photon absorption is optimized. This is in contrast to the case of multiphoton or SHG microscopy, where the excitation wavelength is usually twice as long as the wavelength that will be detected, thereby reducing the total gain in resolution that occurs via the nonlinearity.

The effective point spread and optical transfer functions under saturated excitation condition were calculated (Kawano et al. 2009). The spatial resolution of laser scanning microscopy is limited by the size of the focus spot, and the focus spot dimensions are always nonzero due to the diffraction limit. This is what limits the spatial resolution of the majority of optical microscopes. This relationship between an ideal point and the image that can be obtained from such a point is expressed as the point spread function. The point spread function (PSF) defines the resolution limit, and an improvement in decreasing the size of the PSF is an ideal way to track the improvement in the microscope resolution. Confocal microscopes, where a pinhole is used before the detector to reject light from out-of-focus planes in the sample, have a specially defined point spread function. It is given as the multiplication of the illumination distribution and the detection distribution, as shown by

$$h_{\text{conf}}(x, y, z) = h_{\text{ill}}(x, y, z) \cdot h_{\text{det}}(x, y, z) \tag{2.2}$$

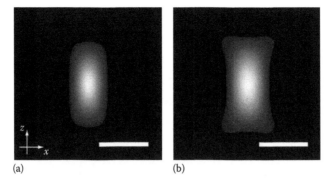

FIGURE 2.3 PSF of confocal fluorescence microscopy with Rhodamine-6G parameters. (a) The excitation intensity is 10^4 W/cm². (b) The excitation intensity is 10^6 W/cm². The scale bar is 500 nm in length. (From Fujita, K. et al., *Proc. SPIE,* 6443, 64430Y, 2007a. With permission.)

where the overall PSF of the confocal imaging mode is given by $h_{conf}(x, y, z)$, the illumination PSF is $h_{ill}(x, y, z)$, and $h_{det}(x, y, z)$ is the detection point spread function.

When the intensity begins to saturate, the effective size of the point PSF grows. This is only true, however, for the PSF calculated for the signal that is demodulated at the same frequency as the incoming light modulation frequency. This is shown in Figure 2.3, where panel (a) shows the PSF of the fundamental frequency ω, calculated for relatively low excitation intensities (10^4 W/cm²), and the correspondingly larger PSF, calculated for the much higher excitation intensity 10^6 W/cm².

Looking at the overall dimensions in Figure 2.3, the saturated PSF is larger than the PSF at low excitation intensities. This would seem counterproductive, since a smaller PSF is usually desirable for optimizing the resolution. However, the dimensions of the *saturated* portion of the PSF are smaller than the overall dimensions of the unsaturated PSF. This means that if we extract the saturated components, we obtain signals from a smaller region of the sample, and hence, improve the resolution. The modulated fluorescence signal at the center of the PSF saturates first, and exhibits strong harmonic components, while the modulated fluorescence signal at the edges of the PSF does not undergo saturation and therefore does not generate temporal harmonics. The improved resolution in the SAX imaging mode can be achieved by detecting the harmonic signal, generated only at the saturated regions. We can also calculate the demodulated PSF by modulating the excitation intensity of the PSF and demodulating the PSF at each point. The entire effective PSF of the SAX imaging mode, denoted by $h_{conf-sat}(n, x, y, z)$, can then be expressed as the following equation:

$$h_{conf-sat}(n, x, y, z) = h_{dem}(n, x, y, z) \cdot h_{det}(x, y, z) \tag{2.3}$$

where
$h_{dem}(n, x, y, z)$ is the demodulated PSF
n represents the nth harmonic component

The detection system PSF is independent of the harmonic number and is denoted by $h_{det}(x, y, z)$. To calculate $h_{dem}(n, x, y, z)$, we modulate the excitation intensity of the PSF, $h_{ill}(x, y, z)$ in time at a frequency ω. In this case, the modulated PSF gains an additional variable to cover the time dimension. The demodulated fluorescence PSF can then be obtained by Fourier transforming the modulated fluorescence PSF at the frequency nω. This is shown in Figure 2.4, where the improved spatial resolution is evident at increasing harmonic demodulation frequencies. Note that these are calculations, and therefore do not include the noise that would be present in the demodulation harmonic PSFs measured from experimental data. Figure 2.5 shows the intensity profiles of the harmonic effective PSFs for a more quantitative evaluation of the resolution improvement.

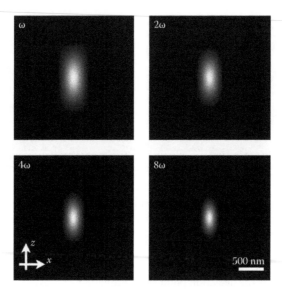

FIGURE 2.4 PSFs for increasing harmonic demodulation frequencies. The size decreases from the conventional PSF measured at the fundamental frequency (ω), second-order harmonic (2ω), fourth-order harmonic (4ω), eighth-order harmonic (8ω). (From Fujita, K. et al., *Proc. SPIE,* 6443, 64430Y, 2007a. With permission.)

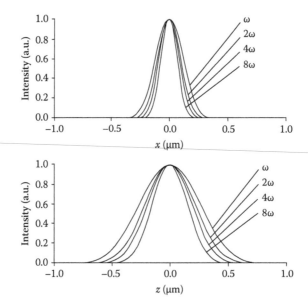

FIGURE 2.5 The intensity profiles of the effective PSFs for increasing harmonic demodulation frequencies. The size decreases in the lateral (x) and axial (z) directions, when compared to the conventional PSF, measured at the fundamental frequency (ω), second-order harmonic (2ω), fourth-order harmonic (4ω), eighth-order harmonic (8ω). (Reprinted with permission from Fujita, K. et al., *Phys. Rev. Lett.,* 99, 228105, 2007b. Copyright 2007 by the American Physical Society.)

Theoretically, there is no limitation on the order of harmonic that can be obtained by demodulation, and therefore, there is no theoretical limitation on the increase in spatial resolution that can be achieved using the SAX technique. Practically, the limitations are imposed by the finite signal-to-noise ratio, with the noise appearing in harmonics obstructing the measurement of the high spatial resolution signals that are required to reconstruct the final image. The lock-in amplifier provides an easy and efficient means by which to minimize this problem, and further developments in detection may again increase the sensitivity in detecting harmonic distortion. Realistically, at present, detection of the 2nd, 3rd, or higher order harmonic demodulation can be carried out without special care taken with signal-to-noise issues, as we will see in the experimental results shown below.

2.2.3 Confirmation of the Onset and Nature of the Saturation Nonlinearity Effect

While it is fairly easy to detect saturation by detecting the nonlinear response in the excitation–emission relationship of fluorescence, detecting the harmonic components that result from distortion of the time-modulated intensity is slightly more complex. To generate the modulation, an AOM is ideal, so long as the modulator itself has inherent distortion that is lower than the harmonic of choice to be detected in the experiment. The output fluorescence is detected by a photomultiplier tube with a confocal pinhole. The detected signal is then fed to a lock-in amplifier, which can operate at the fundamental frequency or multiples thereof, to detect the harmonic components. The AOM and lock-in amplifier are both synchronized by a master signal provided by a PC. An example of a system capable of performing SAX microscopy is shown in Figure 2.6.

To demonstrate the detection of the saturation harmonics, a sample solution of 1 μM Rhodamine B was irradiated with an incident light modulation frequency of 10 kHz. The detected signal was demodulated at both 10 and 20 kHz, and the peak intensity of the modulated incident beam was varied. The results are shown in Figure 2.7, and the increasing strength of the detected second harmonic with the increase in peak incident intensity is clear.

The results show that the overall response remains linear (shown by the adherence to the straight dashed line of ω) until the peak incident intensity is high enough for saturation to occur. When the intensity is increased above this threshold, the rate of increase in second harmonic is higher than the increase in the fundamental frequency. This means that the strength of the saturated components rises rapidly after the onset of saturation, which makes the task of detection feasible so long as the overall light intensity is sufficient. The fluorescence signal demodulated at the fundamental frequency (10 kHz) is linear in response to the excitation power until the onset of saturation. Conversely, the fluorescence

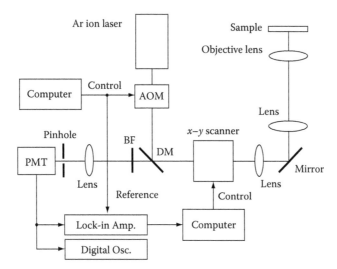

FIGURE 2.6 An example of a setup for SAX microscopy. The synchronization between the AOM and lock-in Amp, the *x*–*y* scan control, and the collection of image data could also be performed by a single computer. Abbreviations are as follows. BF, band-pass filter; DM, dichroic mirror; PMT, photomultiplier tube; AOM, acousto-optical modulator.

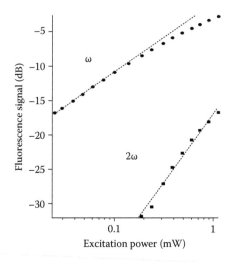

FIGURE 2.7   The onset of saturation and second-order dependence of the 2ω harmonic frequency on the excitation power. The incident light beam was modulated at a fundamental frequency ω = 10 kHz, and was demodulated at the same frequency or at twice the fundamental frequency. (From Fujita, K. et al., *Proc. SPIE*, 6443, 64430Y, 2007a. With permission.)

signal demodulated at the second-order harmonic frequency (20 kHz) is proportional to the square of the laser power. This clearly shows the second-order nature of the saturated components and the benefits for using the phenomenon as a technique for increasing the resolution of the microscope.

The generation of higher harmonics typically requires higher intensities of excitation light. An example of this can be seen using a 10 μM solution of Rhodamine-6G, and Fourier transforming the detected fluorescence emission. Figure 2.8 shows the growth of higher order harmonics with higher incidence laser intensity.

The imaging possibilities of the technique are shown by using test beads immersed in a solution of fluorescent dye. This is the inverse of the usual test method for fluorescence microscopy and was used to test the improvement in resolution without needing to compensate for the effects of photobleaching. The beads were polystyrene beads immersed in a solution of Rhodamine B.

The fluorescence image obtained using the same modulation and demodulation frequency of 10 kHz is shown in Figure 2.9a. The pixel dwell time in the image was about 100 μs, and the laser power at the sample was 240 μW. Panel (b) shows the equivalent image where the demodulation frequency was 2ω (20 kHz). The pixel dwell time for the 2ω image was increased to 1 ms. In the comparison, we can see several points of note. The signal-to-noise ratio is generally worse for harmonic demodulation, since the signal level is lower and residual distortion in the AOM, lock-in amp, and control signal will also appear in the harmonics of the fundamental frequency. The improvement in resolution is clear; the beads in the image demodulated at the fundamental frequency suffer from blurring. This is as a result of the combination of the aberration of the excitation PSF due to the refractive index contributions from out of focus planes, as well as the overall lower resolution of the PSF at the fundamental frequency. The image where the demodulation frequency was 2ω shows clear improvements in the resolution of the transition between the two beads (shown in the line plot), and also in the sharpness of the transitions between beads and surrounding fluorescent media.

One particular advantage of using the modulated saturated excitation with harmonic demodulation to improve the spatial resolution is that it not restricted to a particular dye or fluorescence technique. It can be applied in any imaging mode where fluorescence can be saturated. Figure 2.10 shows the demodulation of harmonics up to the fifth order in four different fluorescent stains. The results show that one of the main difficulties in extracting higher order components is not distortion in the modulation or

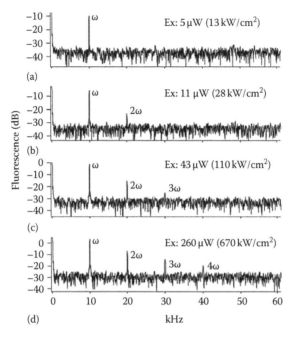

FIGURE 2.8 Fourier transforms of the harmonic distortion components, which rise with increasing excitation intensity. The laser power and intensity is shown for each data trace, and the background noise is shot noise from the photomultiplier tube. The incident light beam was modulated at a fundamental frequency $\omega = 10\,\mathrm{kHz}$, and the sample was a $10\,\mu\mathrm{M}$ solution of Rhodamine-6G. (Reprinted with permission from Fujita, K. et al., *Phys. Rev. Lett.*, 99, 228105, 2007b. Copyright 2007 by the American Physical Society.)

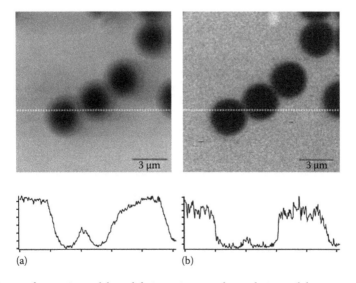

FIGURE 2.9 The use of saturation and demodulation to improve the resolution and the resistance to aberration in a confocal scanning microscope. The left image shows polystyrene beads immersed in a solution of Rhodamine B. The fluorescence was detected using demodulation at the fundamental frequency $\omega = 10\,\mathrm{kHz}$. The right-hand image shows the signal detected using second-order demodulation at 2ω. The line plots show sections through the objects in the image, and demonstrate the ability to achieve greater resolution, and stronger resistance to aberrations that degrade the PSF. (From Fujita, K. et al., *Proc. SPIE*, 6443, 64430Y, 2007a. With permission.)

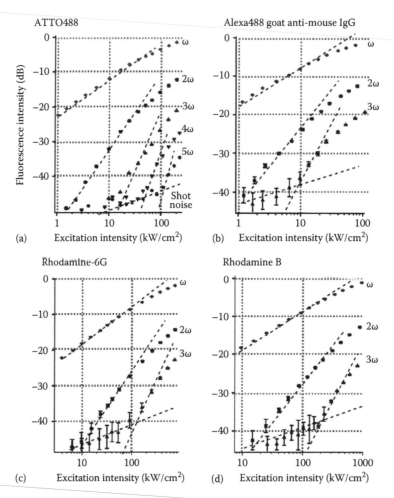

FIGURE 2.10 The onset of saturation, and the rise in second (2ω) and third (3ω) and demodulation harmonics in four different fluorescent stains. The incident light beam was modulated at a fundamental frequency $\omega = 10\,\mathrm{kHz}$, and was demodulated at the same frequency, at twice the fundamental frequency, and three or more times the fundamental frequency, up to the detection limit. All four dyes show that the detection of at least the third demodulation harmonic is feasible, and that up to the fifth demodulation harmonic can be detected. (From Yamanaka, M. et al., *J. Biomed. Optics*, 13, 050507, 2008. http://spie.org/x648.html?product_id=819809. With permission.)

detection system but is actually shot noise from the photomultiplier tube, shown as the dashed line on the graphs, and which has a dependence on the excitation intensity on the order of 0.5.

2.2.4 High-Resolution Saturated Fluorescence Microscopy of Biological Samples

The SAX technique can also be applied to biologically relevant samples, such as cultured cells stained with a fluorescent label. To demonstrate the potential of SAX microscopy, we used fixed HeLa cells, where the microtubules were stained by ATTO488. The experimental setup was the same as shown in Figure 2.6, with a 488 nm excitation wavelength and a modulation frequency of $\omega = 10\,\mathrm{kHz}$. The higher order components were demodulated at 20 kHz, and they are shown in panel (a) of Figure 2.11. The image data resulting from demodulation at the fundamental frequency are shown in panel (b). The line profiles of the fluorescence intensity distribution are shown for each demodulation frequency, below the images. In order to compare the best-case scenarios for demodulation, each

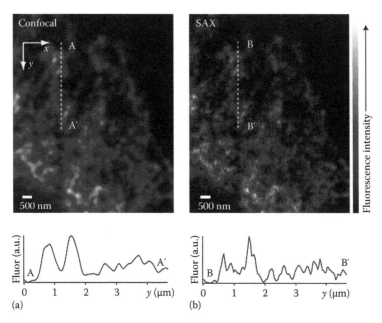

FIGURE 2.11 Stained microtubules in HeLa cells. The images show x–y sections, and were taken with a modulation frequency of $\omega=10\,\mathrm{kHz}$, and (a) demodulated at 2ω, which corresponds to typical confocal fluorescence microscopy, or (b) demodulated at 2ω, where the higher frequencies present in SAX microscopy can be used to increase the imaging resolution.

image was taken with the intensity optimized for the demodulation harmonic that is shown. For the image demodulated at the fundamental frequency ω, the excitation intensity was $2\,\mathrm{kW/cm^2}$. This intensity is low enough that saturation is negligible. The spatial resolution should therefore be the same as conventional confocal fluorescence microscopy. For the demodulation at 2ω, the excitation intensity was raised to $66\,\mathrm{kW/cm^2}$ providing significant saturation (see Figure 2.10, for example). The resulting images show that the demodulation at 2ω provides a higher spatial resolution. From the line profiles through the images, we can see that some peaks that are resolved in the 2ω image are not visible in the image demodulated at ω. There is a corresponding drop in the signal-to-noise level of the image, but the noise does not obscure any parts of the image, which are visible for demodulation at ω. There are no post-processing or noise removal techniques applied to the images; each frame consists of the raw data acquired with a pixel dwell time of $0.3\,\mathrm{ms}$. The image sizes are 256 pixels along each side, and were captured over a total time of $20\,\mathrm{s}$ per frame.

As shown above, in the calculation for the effective PSF, the SAX technique does not require any additional modification to achieve resolution improvement in the axial direction as well as in the lateral direction. Figure 2.12 shows x–z images from the same sample, with the same 2ω and ω demodulation conditions as Figure 2.11. The intensity profiles show that more peaks are resolved in the image demodulated at 2ω, and the widths of the peaks are also decreased. Peak widths in the 2ω image are approximately 1.25 times more narrow than the peaks in the image demodulated at ω. The image size was 256 (x) by 128 (z) pixels, with a dwell time of $0.3\,\mathrm{ms}$ per pixel and a total time for an image acquisition of $10\,\mathrm{s}$.

2.3 Discussion and Perspectives

In this chapter, we discussed different techniques for improving the imaging resolution in fluorescence microscopy. As a field, fluorescence microscopy has undergone rapid expansion in the number

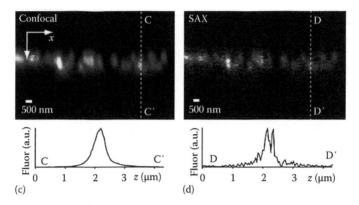

FIGURE 2.12 (See color insert following page 15-30.) Stained microtubules in HeLa cells. The images show x–z sections, and were taken with a modulation frequency of $\omega = 10\,\text{kHz}$, and (c) demodulated at 2ω, which corresponds to typical confocal fluorescence microscopy, or (d) demodulated at 2ω, where the higher frequencies present in SAX microscopy can be used to increase the imaging resolution.

of methods available to improve the resolution, sensitivity, and/or quality of the images. This chapter has focused on using saturated excitation in combination with time-modulated incident light and harmonic demodulation of the detected signal. This is novel, since saturated excitation has typically been a phenomenon to be avoided in laser scanning fluorescence microscopy since the deformation of the PSF usually decreases the spatial resolution when using typical fluorescence detection. When harmonic demodulation of the detected fluorescence is used, however, it becomes possible to exploit the nonlinear response of the saturation process. It is then a simple step to apply the nonlinear response to improve the spatial resolution of the imaging mode. The process is limited by the shot noise, which is unavoidable but can be minimized. Results shown in this chapter demonstrate that the demodulation of frequencies up to the fifth harmonic or higher are practically feasible, since the intensity of an nth-order demodulation harmonic increases in proportion to the excitation laser intensity raised to the power of n, while the shot noise increases only with the square root of the laser intensity. The generation of higher order harmonics is limited by the amount of total power that the sample can sustain.

Along with the issue of sample damage, the technique is also limited by the photobleaching of the fluorescent molecules in the sample. The typical laser intensity required for SAX microscopy is higher than the intensities used for conventional fluorescence microscopy, and the effect of photobleaching cannot be ignored. The images shown in this chapter do demonstrate that it is possible to use the SAX technique on real samples without photobleaching. The resistance of the sample to photobleaching varies significantly between samples, depending primarily on the fluorescent dye used to stain the sample. New dyes are being developed with an increasing resistance to photobleaching, which is one factor that should make this point less of a concern in the future. Additionally, although all results shown in this chapter use continuous-wave excitation laser sources, it may also be very useful to use pulsed laser excitation for SAX microscopy. The pulsed laser light produces extremely high light intensity for the same number of photons incident on the sample. This means that the probability of saturation will be much higher for the same average laser power. We expect that this additional improvement will increase the effective spatial resolution of the imaging mode and decrease the requirement to consider the photobleaching of the sample (Donnert et al. 2007). In addition to decreasing photobleaching, the pulsed laser excitation may also reduce other undesirable effects in the sample such as the generation of reactive oxygen species and the laser modification of biologically relevant targets in the sample (Jay 1988, Tanabe et al. 2005).

As a final note, it is worth pointing out the significance of the fact that SAX microscopy does not require any particular limitation on the fluorescent dye, microscope optics, laser source type, laser wavelength, or sample type. It can be implemented as a pure time-domain modulation and demodulation of the input and output signals. This means that a wide variety of existing microscopes can be converted to SAX usage without any modification of the microscope itself. While photobleaching considerations encourage us to consider dyes that are resistant to photobleaching, there is no special limitation on the spectral properties of the fluorescent dye. This is a significant advantage of SAX microscopy over many competing techniques. The imaging frame rate can also be high speed, operating at the same rate as conventional scanning fluorescence microscopy. The relative simplicity of the implementation and the wide range of applications to which it can be immediately applied to confirms the potential to have a large impact in biology and medicine.

References

Bailey, B., Farkas, D. L., Taylor, D. L., and Lanni, F. 1993. Enhancement of axial resolution in fluorescence microscopy by standing-wave excitation. *Nature* 366: 44–48.

Betzig, E., Patterson, G. H., Sougrat, R. et al. 2006. Imaging intracellular fluorescent proteins at nanometer resolution. *Science* 313: 1642–1645.

Cogswell, C. J., Hamilton, D. K., and Sheppard, C. J. R. 1992. Colour confocal reflection microscopy using red, green and blue lasers. *Journal of Microscopy* 165: 103–117.

Donnert, G., Eggeling, C., and Hell S. W. 2007. Major signal increase in fluorescence microscopy through dark-state relaxation. *Nature Methods* 4(1): 81–86.

Eggeling, C., Volkmer, A., and Seidel, C. A. 2005. Molecular photobleaching kinetics of Rhodamine 6G by one- and two-photon induced confocal fluorescence microscopy. *Chemphyschem* 6(5): 791–804.

Enderlein, J. 2005. Breaking the diffraction limit with dynamic saturation optical microscopy. *Applied Physics Letters* 87(9): 1–3.

Fujita, K., Kawano, S., Kobayashi, M., and Kawata, S. 2007a. High-resolution laser scanning microscopy with saturated excitation of fluorescence. *Proceedings of SPIE* 6443: 64430Y.

Fujita, K., Kobayashi, M., Kawano, S., Yamanaka, M., and Kawata, S. 2007b. High-resolution confocal microscopy by saturated excitation of fluorescence. *Physical Review Letters* 99(22): 228105.

Giepmans, B. N. G., Adams, S. R., Ellisman, M. H., and Tsien, R. Y. 2006. The fluorescent toolbox for assessing protein location and function. *Science* 312: 217–224.

Gustafsson, M. G. L. 2005. Nonlinear structured-illumination microscopy: Wide-field fluorescence imaging with theoretically unlimited resolution. *Proceedings of the National Academy of Sciences of the United States of America* 102(37): 13081–13086.

Gustafsson, M. G., Shao, L., Carlton, P. M. et al. 2008. Three-dimensional resolution doubling in wide-field fluorescence microscopy by structured illumination. *Biophysical Journal* 94(12): 4957–4970.

Heintzmann, R., Jovin, T. M., and Cremer, C. 2002. Saturated patterned excitation microscopy—A concept for optical resolution improvement. *Journal of the Optical Society of America A: Optics and Image Science, and Vision* 19(8): 1599–1609.

Hell, S. W. 2003. Toward fluorescence nanoscopy. *Nature Biotechnology* 21(11): 1347–1355.

Hell, S. W. and Wichmann, J. 1994. Breaking the diffraction resolution limit by stimulated emission: stimulated-emission-depletion fluorescence microscopy. *Optics Letters* 19(11): 780–782.

Hofmann, M., Eggeling, C., Jakobs, S., and Hell, S. W. 2005. Breaking the diffraction barrier in fluorescence microscopy at low light intensities by using reversibly photoswitchable proteins. *Proceedings of the National Academy of Sciences of the United States of America* 102(49): 17565–17569.

Huang, B., Wang, W., Bates, M., and Zhuang, X. 2008. Three-dimensional super-resolution imaging by stochastic optical reconstruction microscopy. *Science* 319: 810–813.

Jay, D. G. 1988. Selective destruction of protein function by chromophore assisted laser inactivation. *Proceedings of the National Academy of Sciences of the United States of America* 85: 5454–5458.

Johnson, I. 1998. Fluorescent probes for living cells. *Histochemical Journal* 30: 123–140.

Kawano, S., Yamanaka, M., Fujita, K., Smith, N. I., and Kawata, S. 2009. Imaging properties of saturated excitation (SAX) microscopy. *Proceedings of SPIE* 7184: 718415.

Klar T. A. and Hell S. W. 1999. Subdiffraction resolution in far-field fluorescence microscopy. *Optics Letters* 24(14): 954–956.

Lakowicz, J. R., Szmacinski, H., Nowaczyk, K., Berndt, K. W., and Johnson, M. 1992. Fluorescence lifetime imaging. *Analytical Biochemistry* 202(2): 316–330.

Magde, D., Elson, E., and Webb, W. W. 1972. Thermodynamic fluctuations in a reacting system measurement by fluorescence correlation spectroscopy. *Physical Review Letters* 29(11): 705–708.

Schermelleh, L., Carlton, P. M., Haase, S. et al. 2008. Subdiffraction multicolor imaging of the nuclear periphery with 3D structured illumination microscopy. *Science* 320(5881): 1332–1336.

Shroff, H., Galbraith, C. G., Galbraith, J. A., and Betzig, E. 2008. Live-cell photoactivated localization microscopy of nanoscale adhesion dynamics. *Nature Methods* 5(5): 417–423.

Sprague, B. L., Pego, R. L., Stavreva, D. A., and McNally, J. G. 2004. Analysis of binding reactions by fluorescence recovery after photobleaching. *Biophysical Journal* 86(6): 3473–3495.

Tanabe, T., Oyamada, M., Fujita, K., Dai, P., Tanaka, H., and Takamatsu, T. 2005. Multiphoton excitation-evoked chromophore-assisted laser inactivation using green fluorescent protein. *Nature Methods* 2: 503–505.

Yamanaka, M., Kawano, S., Fujita, K., Smith, N. I., and Kawata, S. 2008. Beyond the diffraction-limit biological imaging by saturated excitation microscopy. *Journal of Biomedical Optics* 13(05): 050507, http://spie.org/x648.html?product_id=819809.

3

Far-Field Fluorescence Microscopy of Cellular Structures at Molecular Optical Resolution

3.1   Introduction 3-2
The Abbe-Limit • Light Optical Analysis of Biostructures by
Enhanced Resolution • Topics Covered

3.2   Approaches to Superresolution of Fluorescence-Labeled
Cellular Nanostructures A: Focused and Structured
Illumination 3-4
Superresolution by Focused Illumination I: Confocal Laser Scanning
4Pi-Microscopy • Superresolution by Focused Illumination II:
Stimulated Emission Depletion Microscopy • Superresolution by
Structured Illumination Microscopy

3.3   Approaches to Superresolution of Fluorescence-Labeled
Cellular Nanostructures B: Basic Principles of Spectrally
Assigned Localization Microscopy 3-8
Basic Ideas • Basic Experiments • Virtual SPDM I: SPDM with
a Small Number of Spectral Signatures • Virtual SPDM II: SPDM
with a Large Number of Spectral Signatures • "Proof-of-Principle"
SPDM Experiments • Principles of Experimental SPDM/SALM
with a Large Number of Spectral Signatures

3.4   Approaches to Superresolution of Fluorescence-Labeled
Cellular Nanostructures C: Experimental SALM/SPDM
with a Large Number of Spectral Signatures 3-18
Microscope Setup • Software for Data Registration and
Evaluation • Specimen Preparation • Experimental SPDM$_{Phymod}$
Nanoimaging of the Distribution of emGFP-Tagged Tubulin
Molecules in Human Fibroblast Nuclei • Experimental SPDM$_{Phymod}$
Nanoimaging of GFP-Labeled Histone Distribution in Human Cell
Nuclei • 3D-Nanoimaging by Combination of SPDM and SMI
Microscopy

3.5   Discussion...3-28
Perspectives for Nanostructure Analysis in Fixed
Specimens • Perspectives for In Vivo Imaging at the Nanometer
Scale • Perspectives for Single Molecule Counting

Acknowledgments...3-31

References 3-31

Christoph Cremer
Alexa von Ketteler
Paul Lemmer
Rainer Kaufmann
Yanina Weiland
Patrick Mueller
M. Hausmann
Manuel Gunkel
Thomas
  Ruckelshausen
David Baddeley
Roman Amberger

*Kirchhoff-Institute
for Physics*

A basic goal of biophysics is to quantitatively understand cellular function as the consequence of the fundamental laws of physics governing a system of extreme complexity. For this, the knowledge of its molecules and biochemical reactions is of utmost importance. In addition, however, more detailed spatial information about nanostructures involved is necessary. A serious problem to analyze cellular nanostructures by far-field light microscopy is the conventional optical resolution restricted to about 200 nm laterally and 600 nm axially. Various recently introduced laseroptical "nanoscopy" methods such as SI-, 4Pi-, and STED microscopy have allowed a significant improvement of the spatial analysis far beyond these limits. Here, the focus will be on a complementary approach, "spectrally assigned localization microscopy" (SALM). SALM is based on labeling "point-like" objects (e.g., single molecules) with different spectral signatures (see the following text), spectrally selective registration and high-precision localization monitoring by far-field fluorescence microscopy. The basic condition is that in a given observation volume defined, e.g., by the full-width-at-half-maximum (FWHM) of the point-spread-function (PSF) of the optical system used, at a given time and for a given spectral registration mode, only one such object (e.g., a single molecule) is registered. According to the type of fluorophores used (e.g., photostable, or photoconvertable, or stochastically convertible), various SALM procedures have been described (e.g., FPALM, PALM, PALMIRA, RPM, SPDM, STORM, dSTORM) and a lateral spatial resolution in a few tens of nm range has been realized. In particular, in this chapter a SALM procedure will be described, which allows "nanoimaging" of large numbers of molecules at high intracellular densities by spectral precision distance/position determination microscopy (SPDM) and in combination with widely used fluorescent proteins and synthetic dyes in standard media, including physiological conditions. The technique called SPDM$_{Phymod}$ (SPDM with physically modified fluorophores) is based on excitation intensity–dependent reversible photobleaching and the induction of "fluorescent bursts" of the excited molecules at stochastic "onset" times. In this case, the "spectral signature" can be defined as the time duration between the start of the exciting illumination and the time of the single molecule fluorescence bursts induced ("onset time"). Since such onset times can be stochastically distributed over a periods of many seconds up to the range of minutes, in a given specimen, e.g., a cell, thousands of different spectral signatures can be created. Presently, SPDM$_{Phymod}$ techniques (also called reversible photobleaching microscopy, RPM) have been used to determine the intracellular spatial location of single molecules at a density up to ca. 1000 molecules/μm^2 of the same type with an estimated best localization precision of 2 nm (at an excitation wavelength of $\lambda_{exc} = 488$ nm). Distances in the range of 10–30 nm were nanoscopically resolved between such individual fluorescent molecules. In combination with structured illumination, a 3D effective optical resolution of 40–50 nm was realized (ca. one-tenth of the exciting wavelength). As original applications, nanoimaging of the distribution of tubulin proteins in human interphase fibroblasts and histone proteins in mitotic HeLa cells are reported.

3.1 Introduction

3.1.1 The Abbe-Limit

In the 1830s, light microscopy made possible the discovery of the cell as the fundamental unit of life, thus initiating one of the great revolutions of human science (for review, see Cremer 1985). In the development of modern cell biology and its biomedical applications, however, analysis methods using visible light microscopy approaches often played a secondary role, compared with biochemical techniques and DNA-based procedures. A major reason for this was the achievable optical resolution thought to be principally restricted to a limit of a few hundreds of nm laterally and about 600 nm axially. Consequently, the nanostructure of cellular machines (e.g., responsible for molecular transport, DNA replication, transcription and repair, RNA splicing, protein synthesis and degradation, cellular signaling, cell–cell contacts) was not accessible to light microscopy. While ultrastructural methods, in particular electron microscopy, allowed many essential discoveries, light microscopy has many complementary advantages (Rouquette et al. 2009). In addition to the biomolecular machines mentioned, other nanostructures play important roles. For example, the spatial organization of specific chromatin domains with a size down to the 100 nm range are regarded as essential for gene regulation (Cremer and Cremer 2001, Misteli 2007, Solovei et al. 2009).

Since the famous publication of Ernst Abbe (Abbe 1873), the optical resolution limit of several hundred nm was regarded to be due to the wave nature of light and thus insurmountable. Based on the diffraction theory of light, Abbe postulated for a specific color a "specific smallest distance… which never can be significantly smaller than half a wavelength of blue light," i.e., about 200 nm. A very similar conclusion was obtained in 1896 by Lord Rayleigh, "tracing the image representative of a mathematical point in the object, the point being regarded as self-luminous. The limit depends upon the fact that owing to diffraction the image thrown even by a perfect lens is not confined to a point, but distends itself over a patch or disk of light of finite diameter" (Rayleigh 1896). From this, Rayleigh concluded "that the smallest resolvable distance ε is given by $\varepsilon = \lambda/2 \sin \alpha$, λ being the wavelength in the medium where the object is situated, and α the divergence-angle of the extreme ray (the semi-angular aperture) in the same medium." Thus, the same light optical (lateral) resolution limit of about 200 nm (often simply known as "Abbe-limit") is obtained in both theories. It may be noted that in the original publication of Lord Rayleigh, a factor of 0.5 is indeed given instead of the usual 0.61.

3.1.2 Light Optical Analysis of Biostructures by Enhanced Resolution

While techniques for light optical analysis of biostructures by enhanced resolution (LOBSTER) based on laseroptical scanning techniques using near-field illumination have been developed since two decades (Lewis et al. 1984, Pohl et al. 1984), a variety of laseroptical far-field LOBSTER techniques based on fluorescence excitation have been developed to circumvent the "Abbe/Rayleigh-limit" of 200 nm (abbreviated in the following as "Abbe-limit") at least in one direction.

3.1.3 Topics Covered

In this chapter, we focus on two basic approaches to surpass the "Abbe-limit." The first approach is based on "point-by-point" illumination and "point-by-point" detection using an appropriately focused laser beam and refers to the optical resolution achievable by using one spectral signature only (for the definition of spectral signature, see following text). In this approach, the optical resolution is given by the full-width-at-half maximum (FWHM) of the point spread function (PSF) (FWHM_{PSF}), i.e., essentially by the diameter of the focused laser beam. In this context, for completeness we briefly also mention important related concepts of "point-spread-function engineering" like stimulated emission depletion (STED) and structured illumination microscopy (SIM), which will be discussed in detail elsewhere.

The second approach is based on the localization of "optically isolated" point sources of fluorescence (down to individual molecules) with appropriate (multiple) spectral signatures, assuming a given FWHM_{PSF}. In this case, the achievable optical resolution is intimately connected with the localization accuracy of such optically isolated point sources. Such a criterion is thought to be a general parameter to express optical resolution (Albrecht et al. 2001, Van Aert et al. 2006).

Since the localization accuracy is dependent on the FWHM_{PSF}, both approaches may be combined to result in further improved cellular "nanoimaging." Currently, the two approaches have become "state of the art," and the optical ideas behind them now appear to be self-evident. Nonetheless, their present success has been the result of a long development, having already started in the 1950s. Therefore, some references will also be made to the historical perspective.

3.2 Approaches to Superresolution of Fluorescence-Labeled Cellular Nanostructures A: Focused and Structured Illumination

3.2.1 Superresolution by Focused Illumination I: Confocal Laser Scanning 4Pi-Microscopy

3.2.1.1 Principles of "Point-by-Point" Scanning

In light microscopy, an object may be considered as an arrangement of a number of point sources at given spatial positions (x, y, z) absorbing, emitting, or scattering light. The ultimate goal of light microscopic structure analysis is to identify the point sources and to determine their positions as precisely as possible. In conventional microscopy, for each of these point sources the Abbe sine condition must be fulfilled in order to produce from each of these object points a sharp image, and hence a sharp image of the entire object. In terms of geometrical optics, it relates the angles α_1, α_2 of any two rays of light emitted by an object point source P to the angles β_1, β_2 of the same rays reaching the image plane (all angles relative to the optical axis of the microscope system) by the equation $\sin \alpha_1/\sin \alpha_2 = \sin \beta_1/\beta_2$. A further basic idea in the Abbe/Rayleigh theory of optical resolution is the assumption that all point sources are registered simultaneously. Consequently, two neighboring point sources of the same color can be separated from each other only if their diffraction patterns (with their maxima positions fulfilling the Abbe sine condition) are separated in such a way that the maxima of the two diffraction patterns can be identified independently from each other. In terms of the object, this leads to the above-mentioned relationships. These conditions, however, are not necessary to obtain an image of an object, i.e., an information about the positions (x, y, z) of its light-emitting/absorbing/scattering elements: As an alternative to the simultaneous registration of the light-emitting elements of the object, one may reconstruct the image in a sequential manner "point by point." For each of the sequentially imaged point sources, the Abbe condition might be fulfilled "one after the other"; it might even be possible that for a single point source, the Abbe condition is not necessary to obtain an information about its position. For example, one might use a focused beam of light for transmission, scattering, or fluorescence excitation and register the transmission/scattering/fluorescence signal obtained in any way convenient with or without the Abbe sine condition. In this case, the positional information is given by the knowledge about the position of the illuminating "spot," and this may also be obtained from a blurred image; the optical resolution achievable is given by the diameter of the illuminating "spot."

3.2.1.2 Confocal Scanning

In an attempt to improve the microspectrophotometry of nucleic acids, instead of illuminating a large area of an object, Naora (1951) used an iris diaphragm as an "excitation pinhole" in front of an incoherent light source to illuminate a small part (diameter 1–5 μm) of a cellular object. The transmitted light was then transferred through an objective lens to a photomultiplier tube, cutting out the remaining stray light by a second iris diaphragm ("detection pinhole") placed in the image plane (Naora 1951). In such a device, the position of the small cellular area studied would be given by the position of the first iris diaphragm, and the light distribution in the plane of the second diaphragm would not necessarily have to form a "sharp image" according to the sine condition; for position determination, it would in principle be enough to register a measurable signal.

Whereas the idea of sequential "point-by-point" imaging has been well known and applied for a century in the telegraphic transfer of images, in light microscopy this idea has been explicitly described in the patent application of Marvin Minsky on a stage scanning confocal microscope based on transmitted/reflected light using a conventional, (i.e., incoherent) light source (Minsky 1957). In 1968, Petran et al. constructed a tandem scanning reflected light microscope (Petran et al. 1968).

3.2.1.3 Confocal Laser Scanning

After the advent of appropriate laser sources, the confocal concept has been further developed to confocal laser scanning microscopy (CLSM) (Davidovits and Egger 1972, Sheppard and Wilson 1978, Brakenhoff et al. 1979) and to CLSM using fluorescence (Cremer and Cremer 1978, Brakenhoff et al. 1985, Stelzer et al. 1991, for review see Pawley 2006). In these "conventional" confocal laser scanning fluorescence microscopes, one objective lens of high numerical aperture was used for focusing and detection; consequently, the limiting optical resolution (in any direction relative to the optical axis) was at best around 0.5 λ_{exc} (excitation wavelength).

3.2.1.4 Confocal Laser Scanning Fluorescence 4Pi-Microscopy: Basic Ideas

First ideas to overcome the Abbe-limit of ~200 nm optical resolution using a confocal 4Pi-laser scanning fluorescence microscopy approach were discussed already in the 1970s (Cremer and Cremer 1978, see also Cremer and Cremer 2006a, b). In this early 4Pi-microscopy concept based on experimental results with diffraction-limited focusing and fluorescence excitation of coherent ultraviolet light (Cremer et al. 1974), the object is illuminated "point by point" by a focused laser beam to a diameter below the Abbe-limit using a "4π-geometry." The fluorescence emitted from each object point is registered via a detection pinhole in the image plane, which excludes contributions outside the central maximum of the diffraction pattern produced by a point source. The individual fluorescence signals are used to electronically construct an image with improved optical resolution. It was suggested that by 4π-focusing, the focal diameter could be reduced to a minimum at least in one direction. It was speculated that this goal might be realized by focusing the incident laser beam from all sides ("4π" geometry), thus increasing the aperture angle substantially beyond conventional possibilities. To realize a full 4π-geometry, it was proposed to generate a "4π-point-hologram," either experimentally from a source with a diameter much below the wavelength of the light emitted from such a source, or to produce it on the basis of calculations. This "4π hologram" should replace a conventional optical lens (numerical aperture, NA < 1.5), which allows focusing only down to the Abbe-limit. In addition, the use of plane point-holograms for focusing the exciting beam in the laser-scanning microscope was suggested to be advantageous at least with respect to the much larger working distance available.

In the original "4π"-concept, a "hologram" was generally defined as a device to produce the boundary conditions, which together with suitable illumination conditions yields the appropriately reconstructed waves, leaving explicitly open questions such as production, material problems, or direction, amplitudes and coherence of the incident and the reconstructed waves; furthermore, it was assumed that the amplitudes and the incident angles of the coherent waves falling on the 4π-point-hologram may be varied almost independently from each other.

Since in the electromagnetic theory of light, the consequences of complex boundary conditions can be solved only numerically, it still appears to be difficult to make precise conclusions on the ultimate physical limits of focusing possible by such an approach. However, in the case of fluorescence excitation for a hypothetical continuous wave, monochromatic spherical wavefront of constant intensity is focused in a full 4π-geometry, theoretically in the far field (distance r from focusing device $(2\pi/\lambda) \cdot r > 2$, Chen 2009) a limiting focus diameter of about one-third of the wavelength may be obtained (Hell 2007). Assuming a wavelength $\lambda_{exc} = 488$ nm and a refraction index of $n = 1.5$, this would result in a lateral focal diameter significantly smaller than the typical "Abbe-limit" (200 nm) and an axial diameter substantially smaller than 200 nm. In the confocal registration mode assumed, still somewhat smaller FWHMs of the 4Pi-PSF would be obtained. Furthermore, it was noted (Cremer and Cremer 1978) that for improved optical resolutions, it would already be advantageous if in this way only one focal diameter could be reduced to a minimum.

3.2.1.5 Confocal Laser Scanning Fluorescence 4Pi-Microscopy: Experimental Realization

The first experimental confocal laser scanning 4Pi-microscope was designed and realized in the early 1990s (Hell 1990, Hell and Stelzer 1992): Instead of holograms, two opposing high numerical aperture lenses were used to concentrate two opposing laser beams constructively in a joint single focus (Figure 3.1).

Presently, confocal laser scanning 4Pi fluorescence microscopy applying either continuous wave visible laser light or femtosecond-pulsed infrared laser wavelengths for excitation (Hell et al. 1994, Hänninen et al. 1995) has become an established light optical "nanoscopy" method (Egner et al. 2002, Baddeley et al. 2006, Bewersdorf et al. 2006). An axial optical resolution down to the 100 nm regime was experimentally realized, i.e., about six to seven times better than in conventional confocal microscopy, and thus about two times smaller than the lateral "Abbe-limit." By combination of such a 4Pi-microscope with microaxial tomography techniques (Bradl et al. 1994, 1996) even an isotropic optical resolution considerably below the Abbe-limit would be possible.

Recently, superresolution in the macromolecular range was experimentally realized using present 4Pi-microscopy equipment: In combination with spectrally assigned localization microscopy (SALM) (see following text), a commercially available 4Pi-microscope allowed the analysis of single cellular protein complexes: The structure of the nuclear pore complex (NPC) was resolved much better by SALM-4Pi-microscopy than by confocal fluorescence microscopy (Hüve et al. 2008). In single NPCs, distances as small as 76 ± 12 nm between two epitopes labeled with different spectral signatures were resolved.

3.2.1.6 Perspectives of Superresolution by Focusing

While an axial optical resolution in the 100 nm range provided a substantial progress compared to all previous fluorescence microscopy techniques, such figures are still far from the desired molecular resolution. One alternative would be to use shorter wavelengths for 4Pi-fluorescence excitation, e.g., by frequency doubling light from a 514 nm laser source (Cremer et al. 1974). Apart from this, the question of the absolute limits of 4Pi focusing has been put forward and presented as an open problem: "The ... speculations on focusing by '4π-point-holograms' are only intended to allude to a method which perhaps might be used to enhance the resolving power of a laser-scanning microscope..." (Cremer and Cremer 1978).

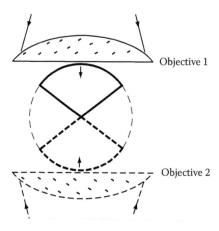

FIGURE 3.1 Scheme of illumination in confocal laser scanning 4Pi-microscopy. To further "sharpen" the scanning laser focus used for "point-by-point" excitation of fluorescence (as in conventional confocal laser scanning fluorescence microscopy), ideally the laser light is focused from all sides ("4Pi geometry"). As an approximation, in a present experimental realization two opposing high numerical aperture objective lenses are used for constructive focusing. (Reprinted with permission from Hell, S.W. et al., *Appl. Phys. Lett.*, 64, 1335, 1994. Copyright 1994 American Institute of Physics.)

Since focusing by a continuous wave spherical wave front of constant intensity and linear polarization as assumed above is just one of the many possible illumination conditions, the theoretical resolution limits of confocal 4Pi-laser scanning fluorescence microscopy still deserve further theoretical and experimental studies. For example, recently it has been demonstrated that using a radially polarized laser beam, the experimentally observed spot size for a numerical aperture NA=0.9 was about 35% below the theoretical limit for linearly polarized light; using specific photosensitive layers, even smaller spot sizes may be realized (Dorn et al. 2003). Since in the vicinity of the focus, effects such as the polarization properties of the electromagnetic field start to play a dominant role, the absolute limits of 4Pi-focusing considering an appropriate spatial, temporal, and phase distribution of the illuminating field and including novel possibilities of attosecond laser physics (Silberberg 2001) still appear to be not fully explored. For example, using extremely short pulses with corresponding ultrashort coherence lengths, conditions might be conceived where constructive interference by 4Pi-focusing may be limited to a region with a diameter even considerably smaller than the 100 nm range possible by spherical waves with continuous wave visible light laser sources. Another possibility concerns focusing of light beyond the diffraction limit using plasmonic "lens" devices (Chen 2009, Chen et al. 2009), i.e., including plasmonic "boundary conditions" in the design of focusing elements. Such an approach would be compatible with the idea to use plane "point-holograms" (Cremer and Cremer 1978, for the definition of "hologram" see previous text). Even if in such cases, the working distances would not be substantially increased compared to the near field, it might be advantageous compared with the tip devices used in present near-field scanning optical microscopy (NSOM). In may be noted that in the 1978 considerations, such an arrangement has not been excluded.

3.2.2 Superresolution by Focused Illumination II: Stimulated Emission Depletion Microscopy

In experimental 4Pi-microscopy, so far a 3D optical resolution around $200 \times 200 \times 100\,nm^3$ was realized. This means that cellular nanostructures with spatial features below about 100 nm would still remain unresolved. The STED microscopy conceived by Stefan Hell (Hell and Wichmann 1994) and realized in his laboratory in the following decade surpassed this magical limit of visible light microscopy substantially and presently has found numerous applications in high resolution cell biology (Hell 2007, Nägerl et al. 2008, Schmidt et al. 2008, Westphal et al. 2008). The basic idea of STED microscopy is to reduce the size of the region excited to fluorescence by the focusing laser beam further by using a very short excitation pulse, which is immediately followed by a "depletion" pulse ("STED pulse") acting in the vicinity of the center of the fluorescent region. This is done in such a way that the STED pulse forms a "doughnut" like shape around the center of the fluorescent region. As a consequence, at appropriate illumination intensities fluorescence is detected from a much smaller region. Since the position of this smaller fluorescent region can be known with an accuracy limited by the fluorescence photon statistics (i.e., down to the subnanometer level) due to the scanning mechanism, the fluorescence signal obtained can now be assigned to this smaller region; hence, the optical resolution may be improved further. STED microscopy was the first implementation of more general concepts like RESOLFT (REversible Saturable OpticaL Fluorescence Transitions) or "Ground State Depletion Microscopy" (Hell 2007, 2009). In the case that the FWHM of the STED-PSF is smaller than the distance between two STED excitable molecules in the region of interest (ROI) studied, single molecules can be discriminated and their positions may be individually determined with an accuracy much better then the STED-FWHM.

3.2.3 Superresolution by Structured Illumination Microscopy

Combining advanced optical and computational methods, also other modes for considerable resolution improvement become possible. One of these additional possibilities is to illuminate the object with an appropriate pattern of light; moving either the object or the pattern, at each relative position an image of the object is taken by a highly sensitive CCD camera. Using complex but well-established algorithms in the Fourier space, from such images it becomes possible to reconstruct an image of the object with enhanced effective optical resolution (Heintzmann and Cremer 1999, Frohn et al. 2000, Gustafsson 2005, Gustafsson et al. 2008). Recently, even a 3D optical resolution enhancement and its application to the analysis of structures at the nuclear membrane has become possible (Schermelleh et al. 2008).

Spatially modulated illumination (SMI) far-field light microscopy (Failla et al. 2002a,b, Baddeley et al. 2007) is another possibility to use an SI to improve spatial analysis. It is based on the creation of a standing wave field of laser light (Bailey et al. 1993). This can be realized in various ways, e.g., by focusing coherent light into the back focal planes of two opposing objective lenses of high numerical aperture. The fluorescence-labeled object is placed between the two lenses and moved axially in small steps (e.g., 20 or 40 nm) through the standing wave field. At each step, the excited fluorescence is registered by a highly sensitive CCD camera. This procedure allowed to measure the diameter of individual fluorescent subwavelength-sized small objects down to a few tens of nm, and to determine axial distances between "point-like" fluorescent objects down to the range of a few tens of nm and with a precision in the 1 nm range (Albrecht et al. 2002). Several biophysical application examples indicated the usefulness of SMI-"nanoscopy" for the study of the size of individual chromatin regions far below the Abbe-limit (Hildenbrand et al. 2005, Mathee et al. 2007, Reymann et al. 2008) and of individual transcription factor complexes (Martin et al. 2004). In the latter case, the SMI results obtained were compatible with the electron microscopic images obtained from the same type of specimens. Although SMI nanosizing as such did not allow an increase in optical resolution, it made it possible to deliver specific spatial information typically obtained by ultrastructural methods only.

In combination with SALM, SMI-microscopy allowed to analyze cellular nanostructures with an effective three-dimensional (3D) optical resolution of 40–50 nm or one-tenth of the exciting wavelength (Lemmer et al. 2008) (see following text).

3.3 Approaches to Superresolution of Fluorescence-Labeled Cellular Nanostructures B: Basic Principles of Spectrally Assigned Localization Microscopy

The basis of SALM as a far-field fluorescence microscopy "nanoimaging" approach using biocompatible temperature conditions is the independent localization of "point-like" objects excited to fluorescence emission either by a focused laser beam or by non-focused illumination, on the principle of "optical isolation," i.e., the localization is achieved by appropriate spectral features. This approach has been referred to as "spectral precision distance (or position determination) microscopy" (SPDM). SPDM/SALM was already conceived and realized in proof-of-principle experiments in the 1990s (Cremer et al. 1996, 1999, Bornfleth et al. 1998, Edelmann et al. 1999, Rauch et al. 2000, Esa et al. 2000, 2001) and in related studies (Weiss 1999, Lacoste et al. 2000, Schmidt et al. 2000, Heilemann et al. 2002) using the term "colocalization." In the following, the discussion will be focused on this method, which recently has allowed far-field fluorescence microscopy of cellular structures at molecular optical resolution.

3.3.1 Basic Ideas

The principle of SPDM/SALM (Cremer et al. 1996, 1999) is shown schematically in Figure 3.2: As an example, let us assume three closely neighboring "point-like" targets (diameter much smaller than the $FWHM_{PSF}$) t_1, t_2, and t_3 (e.g., three molecules) in a cell with distances much smaller than 1 $FWHM_{PSF}$. In case the targets have the same spectral signature, the diffraction pattern overlaps in such a way that the individual target positions cannot be resolved.

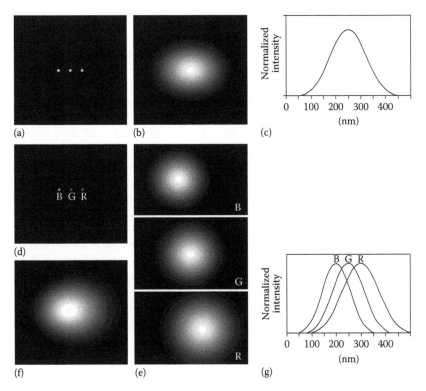

FIGURE 3.2 Principle of SPDM. As an example, three point-like objects are located in the xy-plane within 50 nm distance to the middle one. Furthermore, they are labeled with the same spectral signature in (a), and with three different, unique spectral signatures in (d). The computed system responses of the objects in (a) and (d) (parameters for computation are NA=1.4 and system magnification of 63×) are shown in (b), (e), and (f). In (f), the different spectral components are imaged together (all targets are in a "bright" state whereas in (e) the same signals are recorded independently from each other: In the individual images B, G, R, one object only is in the "bright" state while the other objects situated within the same observation volume of diameter FWHM$_{PSF}$ are in the "dark" state. Line scans through the objects in (b) and (e) are shown in (c) and (g). Figures according to Cremer et al. (1999). (From Kaufmann, R. et al., *Proc. SPIE*, 7185, 71850J-1, 2009. With permission.)

In the SPDM approach, the three targets are assumed to have been labeled with three different fluorescent spectral signatures spec$_1$, spec$_2$, and spec$_3$. For example, t_1 is labeled with spec$_1$, t_2 with spec$_2$, and t_3 with spec$_3$. The registration of the images (using focused or non-focused optical devices with a given FWHM$_{PSF}$) is performed in a spectrally discriminated way so that in a first image stack IM$_1$, a spec$_1$ intensity value I_1 is assigned to each voxel v_k ($k=1, 2, 3,...$) of the object space; in a second image stack IM$_2$, a spec$_2$ intensity value I_2 is assigned to each voxel v_k of the object space; and in a third image stack IM$_3$, a spec$_3$ intensity value I_3 is assigned to each voxel v_k of the object space. The positions of the objects/molecules are assigned to a position map. In other words, in the first image stack IM$_1$ only t_1 is in a "bright" state, and t_2, t_3 are in a "dark" state; in the second image stack IM$_2$, only t_2 is in a "bright" state, and t_1 and t_3 are in a "dark" state; in the third image stack IM$_3$, t_3 is in a "bright" state, and t_1 and t_2 are in a "dark" state. It is clear that this concept can be extended to any number of targets with any number of spectral signatures.

3.3.2 Basic Experiments

In early "proof-of-principle" experiments using CLSM at room temperature, the positions (xyz) and mutual Euclidean distances of small sites on the same DNA molecule labeled with three different spectral signatures yielded a lateral effective optical resolution of about 30 nm (approx. 1/16th of the wavelength used) and approx. 50 nm axial effective resolution (Esa et al. 2000): Sites with a lateral distance of 30 nm and a 3D distance of 50 nm were still resolved; their (xyz) positions were determined independently from each other with an error of a few tens of nm, including the correction for optical aberrations. It was noted that the SPDM strategy "can be applied to more than two or three closely neighbored targets, if the neighboring targets $t_1, t_2, t_3, ..., t_n$ have sufficiently different spectral signatures $spec_1, spec_2, ..., spec_n$. Three or more spectral signatures allow true structural conclusions. Furthermore, it is clear that essentially the same SPDM strategy can be applied also in all cases where the distance between targets of the same spectral signature is larger than FWHM" (Cremer et al. 1999). Compared with related concepts (Burns et al. 1985, Betzig 1995, van Oijen et al. 1998), in the SPDM approach the focus of application was (a) to perform superresolution analysis by the direct evaluation of the positions of the spectrally separated objects; (b) to adapt the method to far-field fluorescence microscopy at temperatures in the 300 K range ("room temperature"); (c) to specify spectral signatures to include all kinds of fluorescent emission parameters suitable, from absorption/emission spectra to fluorescence lifetimes (Heilemann et al. 2002) to luminescence in general, and to any other method (Cremer et al. 2001) allowing "optical isolation." Optical isolation means that at a given time and registration mode, the distance between two objects/molecules is larger than the $FWHM_{PSF}$. Early concepts (Cremer and Cremer 1972) even included the use of photoswitching molecules from a state A to a state B in an observation volume in a reversible or irreversible way to improve the gain of nanostructural information. In the case of fluorescence emission spectra, this is realized by switching point-like objects "On" and "Off" by a corresponding use of filters (Esa et al. 2000). Presently, a variety of such photoswitching modes has become available (Hell 2009).

The original concept of SPDM (Figure 3.2) was further developed by Bornfleth et al. (1998), both theoretically and experimentally. Problems like model-based segmentation of the point sources using the microscopic PSF, or the influence of the photon statistics and the calibration of chromatic aberrations in the case of spectral emission differences in the spectral signatures were discussed in detail; first experimental "proof-of-principle" high precision measurements using fluorescent calibration objects were performed. In addition, it was realized that the SPDM localization approach would allow providing nanoscale images. For a quantitative visualization of the effects of various experimental conditions on the achievable resolution and image quality, numerical modeling calculations of SPDM ("virtual SPDM") were performed.

3.3.3 Virtual SPDM I: SPDM with a Small Number of Spectral Signatures

Figure 3.3 (Bornfleth et al. 1998, Fig. 3f) gives an example. A simulation of 3D microscopic images was performed using a model system, which consisted of an ellipsoidal cell volume with half axis a, $b = 5.2\,\mu m$ in the object plane (x, y), and $c = 3.1\,\mu m$ in the direction of the optical axis (z) ; 500 FITC spots (corresponding to an assumed labeling with fluorescein as spectral signature) (1) and 500 Cy5 spots (corresponding to an assumed labeling with Cy5 as spectral signature) (2) were randomly distributed in the ellipsoidal volume. The images were convoluted with experimentally measured PSFs for FITC and Cy5, respectively, and noise (Gaussian, multiplicative, additive, photon) was added. Between 98% and 99% of all spots were identified correctly by the algorithm, and hence their positions were determined.

In the case of using two spectral signatures with different fluorescence emission spectra (and hence a chromatic shift error), a standard deviation of the error of the 3D distance around $\varepsilon = 65$–75 nm was obtained, depending on the photon count statistics. Using spectral signatures with the same fluorescence emission spectrum (e.g., fluorescence lifetimes (Cremer et al. 1996)), chromatic shift errors are excluded. In this case (perfect ability to correct for chromatic shifts), under the same conditions as mentioned earlier, 3D distance errors between nearest neighbors around $\varepsilon = 40$–55 nm were determined from the SPDM model calculations. Such distance errors corresponded to the localization errors of

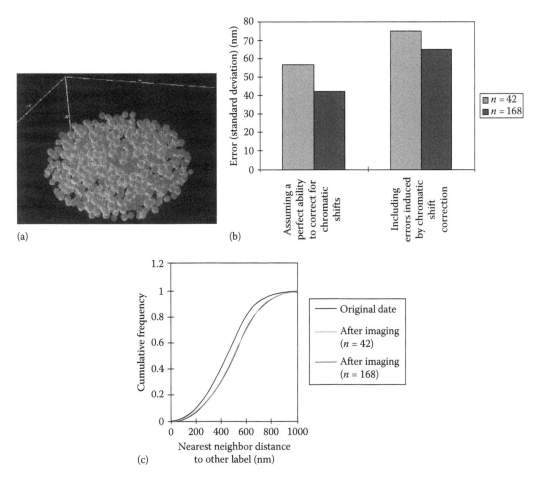

FIGURE 3.3 Potential of SPDM for nanoscale imaging of multiple intracellular objects. (a) Simulated model nucleus ($V = 350\,\mu m^3$) with 1000 randomly distributed objects with two spectral signatures (500 each). (b) Standard deviation (nm) of the error of nearest neighbor 3D distances (spectral signature 1 to spectral signature 2) obtained by a volume segmentation algorithm and comparison with the "true" positions from the model data. The right two columns of (b) show the realization of the two spectral signatures by fluorophores with different fluorescent emission spectra (FITC and Cy3 assumed), using experimentally determined values for chromatic shift determination. The left two columns of (b) show the realization of the two spectral signatures by fluorophores with the same fluorescent emission spectra. (c) Ordinate; cumulative frequency; abscissa; nearest neighbor 3D distance to object with other label (FITC and Cy5 assumed). "Original data"; distance determination from the original model data before convolution and segmentation. After imaging, $n = 42$, $n = 168$: Calculated results for the number of photons per voxel detected for an object. (From Bornfleth, H. et al., *J. Microsc.*, 189, 118, 1998. With permission.)

about $1/1.41\ \varepsilon \sim 46$–$53\,nm$ assuming a chromatic shift (FITC/Cy5) and to 28–39 nm (no chromatic shift), respectively. These predictions of localization errors based on very "conservative" assumptions were considerably higher than obtained by current methods (see following text); nonetheless, the simulations confirmed the potential of SPDM for the nanoscale imaging of multiple targets distributed in cells. As an example, Figure 3.3c gives the cumulative frequency of the nearest neighbor distance between the model objects of Figure 3.3a assumed to be labeled with two spectral signatures (in this case FITC and Cy5). Since these distances were determined in 3D, the graph indicates a considerable number of objects below the Abbe-limit that were correctly identified and localized.

Furthermore, using simulated data, Bornfleth et al. (1998) found that the distances of two objects of the same spectral signature of 1.3 times the FWHM resulted in a bias of 0.25 FWHM while from 1.9 FHWMs onward, localization errors were dominated by noise. Below 1.3 FWHMs, spots of the same spectral signature were not segmented correctly; consequently, the localization algorithm used resulted in the condition that for using SPDM as a tool of nanoscale imaging of a large number of object point in a cell, the requirement of optical isolation is sufficiently fulfilled if the distance between two objects of the same spectral signature is $d_{optisol} > 1.3$ FWHM.

The higher the number of distinguishable spectral signatures, the more pronounced is the potential of nanoscale imaging.

Figure 3.4 gives a numerical modeling example for the resolution potential if molecular complexes with each of seven spectral signatures are assumed. Macromolecular complexes (in this case chromatin nanostructures) were assumed to consist of a linear fiber composed of a number of labeled sections of

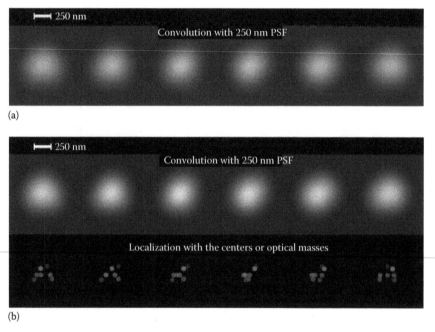

FIGURE 3.4 Numerical modeling of an SPDM image assuming seven spectral signatures. (a) Following convolution with the PSF$_{conv}$ of a conventional high NA epifluorescence microscope, each of the complexes can be identified even in the case that the entire complex is labeled with the same spectral signature; however, no internal structure can be detected. (b, top portion) In this case, each section was assumed to be labeled with a different spectral signature. Simultaneous registration of the fluorescence signals obtained from different spectral signatures, convolution with PSF$_{conv}$ and a precise calibration of optical/chromatic aberrations again results in each of the complexes discriminated from each other; from the different distributions of the "colors" representing the various spectral signatures, by visual inspection the existence of different internal structures (in this case projections of the same molecular folding) may be deduced. However, no detailed conclusions are possible. (b, lower portion) After labeling each section with a different spectral signature as above (b, top) but assuming independent registration of each spectral signature and evaluation according to the principles of SPDM (compare Figure 3.2), the bary centers ("fluorescence intensity gravity centers") clearly indicate different (projected) nanostructures. The further repetition of such SPDM resolved complexes (e.g., 50,000 complexes in an $70 \times 70\,\mu m^2$ field of view with minimum complex–complex distances of $0.3\,\mu m$) would result in the formation of an image with a substantially increased optical resolution. (From Cremer, C. et al., Principles of spectral precision distance confocal microscopy for the analysis of molecular nuclear structure, in *Handbook of Computer Vision and Applications*, Jähne, B. et al., Eds., Vol. 3, Academic Press, San Diego, CA, 1999, 839–857.)

equal linear length. The minimum distance between any two elements of neighboring complexes was assumed to be larger than 250 nm (corresponding to the FWHM of the PSF_{conv} a conventional epifluorescence microscope). Whereas conventional epifluorescence imaging (upper row) allowed to count the number of complexes and to determine their positions, the SPDM method indicated that even fine nanostructural details may be revealed (lower row).

3.3.4 Virtual SPDM II: SPDM with a Large Number of Spectral Signatures

To test the effects of localization precision and the density of fluorescence molecules on the appearance of the "nanoimages" obtained by SPDM assuming a large number of spectral signatures, "virtual SPDM" was performed using a large number of spectral signatures. For this, an excitation wavelength $\lambda_{exc} = 488$ nm and a numerical aperture NA = 1.4 was assumed (FWHM = 200 nm). The simulated original image was convoluted with

$$I(r) = I_0 \left(\frac{2J_1\big((2\pi/\lambda)\text{NA }r\big)}{(2\pi/\lambda)\text{NA }r} \right)^2$$

where
> $I(r)$ is the intensity in the diffraction image of a point-like light source as a function of the radius r (normalized to the object plane)
> J_1 is the first-order Bessel function
> λ is the average wavelength of the emitted light, $\lambda = 509$ nm
> NA is the numerical aperture of the objective lens producing the diffraction image
> I_0 is the intensity at $r = 0$, NA = 1.4

Example 1: Virtual SPDM of a concentric ring structure. For this, an "original" object was calculated containing a central circular area of ca. 280 nm diameter, surrounded by concentric rings with decreasing thickness/distance from 70 to 20 nm (Figure 3.5 left, above). For optimal structural resolution, a sufficiently large number of point sources (representing single molecules) were assumed to have the optical characteristics required for SPDM_{Phymod} and were assumed to fulfill the Nyquist sampling theorem.

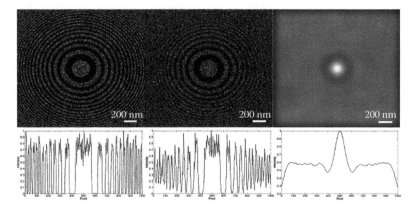

FIGURE 3.5 Virtual SPDM of concentric ring structures. *Left*; original structure; *middle*; SPDM image (5 nm localization accuracy); *right*; original structure after convolution with a conventional epifluorescence PSF. *Above*; images; *below*; result of horizontal intensity line plots across the centers of the images.

In the simulated original structure, between the rings no other objects were assumed; in consequence, a horizontal intensity line plot across the simulated structure allows an estimate of the object modulation $M_{object}=(I_{max}-I_{min})/(I_{max}+I_{min})$ as a function of the horizontal coordinate (in pixel). Since I_{min} between the rings is zero, the object obtained in this way gives a value $M_{object}=1$ down to a ring-to-ring distance of about 20 nm (peripheral rings). The convolution of the "original structure" of Figure 3.5 (upper left) with the PSF_{conv} of a high numerical aperture (NA = 1.4) epifluorescence microscope resulted in the image (Figure 3.5 upper right); the central area was well discriminated while the location of the concentric rings became increasingly obscure. The intensity line plot (Figure 3.5 below right) indicated M_{image} values around 0.67 for the central area and around 0.04 for the ring structures. In the SPDM modus (Figure 3.5 below middle; 5 nm localization accuracy assumed), however, all rings remained clearly distinguishable; M_{image} was 1.0 for the inner structures; even in the periphery (distance between the rings about 20 nm), M_{image} was still around 0.5. This indicates that lines with distances around 20 nm can still clearly be resolved by $SPDM_{Phymod}$. Resolved ring-to-ring distances of 20 nm correspond to a spatial frequency of $\nu_R=50/\mu m$.

In addition, extensive SPDM simulations of line structures with variable number of lines per μm and variable localization accuracies were performed (Lemmer, P. et al., manuscript in preparation); the resulting modulation transfer functions for SPDM (and corresponding modes of spectrally assigned localization microscopy) were fully equivalent to a highly increased optical resolution; for example, a localization accuracy of 5 nm corresponds to a cutoff spatial frequency of around 100/μm; a localization accuracy of 2.5 nm corresponds to a cutoff spatial frequency of around 200/μm or an optical resolution of 5 nm ($=1/100 \lambda_{exc}$). The corresponding minimum distances to be resolved may be estimated to be around 12 and 6 nm, respectively.

Example 2: Virtual SPDM of a structure with variable localization accuracy and molecule density. While the first example for virtual SPDM presented earlier showed that under conditions of high molecule density and localization accuracy, SPDM is equivalent to an optical resolution with highly increased spatial frequency, ν_R, this second example is intended to show the effects of a given molecule density and localization accuracy on the structural resolution of objects with various shapes.

For this, the four letters SPDM were assumed to have an overall length of 2 μm, a height of 0.37 μm, and a total area of 0.16 μm² (Figure 3.6a). The "letters" were simulated to consist of various numbers (168–5538) of single molecules with a "blinking" emission corresponding to $SPDM_{Phymod}$.

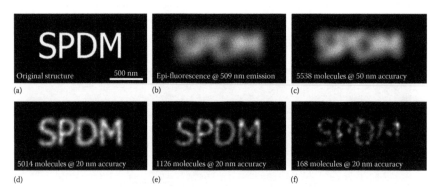

FIGURE 3.6 Example for the simulation of complex nanostructures by $SPDM_{Phymod}$. The total area of the SPDM letters was 0.16 μm². (a) Original structure; (b) conventional epifluorescence image; (c) SPDM image assuming a total of 5538 molecules randomly distributed in the SPDM structure and 50 nm localization accuracy; (d) SPDM image assuming a total of 5014 molecules randomly distributed in the SPDM structure and 20 nm localization accuracy; (e) SPDM image assuming a total of 1126 molecules randomly distributed in the SPDM structure and 20 nm localization accuracy; (f) SPDM image assuming a total of 168 molecules randomly distributed in the SPDM structure and 20 nm localization accuracy. All calculations were performed for NA = 1.4, $\lambda_{em}=509$ nm emission wavelength. (From Kaufmann, R. et al., *Proc. SPIE*, 7185, 71850J-1, 2009. With permission.)

The result is presented in Figure 3.6b. The diffraction image shows some alterations in the intensity distribution and gives an estimate of the overall size of the "structure" in Figure 3.6a; a reconstruction of the original "object," however, is clearly not possible by assuming a "conventional" resolution (FWHM$_{PSF}$ = 190 nm).

In Figure 3.6c, the image is shown for the application of the SPDM procedure assuming a high number (N_{mol} = 5538 corresponding to a molecule density of 34,612/μm^2) of single molecules/point sources causing the light emission of the object; compared with Figure 3.6b, in this case the resolution gain obtained is very small. This is due to the relatively low localization accuracy of only 50 nm assumed for each molecule; it demonstrates the importance of high precision measurements of molecule positions. If a localization accuracy of 20 nm were assumed and a similar number of molecules as in the low accuracy case (Figure 3.6d), the individual letters of the "SPDM" nano-object are clearly resolved, even the thickness of the individual letters of about 70 nm are also resolved. Figure 3.6e give examples for the importance of appropriate molecule numbers detected: Approximately 1/5 of the molecules were needed for good resolution (N_{mol} = 1126 corresponding to a molecule density of 7038/μm^2), and 20 nm localization accuracy was still sufficient to clearly resolve all letters in the original structure (Figure 3.6e). Lowering the number of molecules to 1/30 of the original value (N_{mol} = 168 corresponding to a molecule density of 1050/μm^2) resulted in a loss of spatial information. Nonetheless, a large amount of nanostructural details were still resolved, from the clear discrimination of the four "main complexes" (letters) to the identification of the "main complexes" ("S," "P," "D"). Even the "M" can be suggested.

To summarize, the simulations show (a) that the evaluation algorithms used indeed allow the extraction of structural information far below the conventional resolution limit of epifluorescence microscopy; (b) that the density of molecules contributing to SPDM$_{Phymod}$ is a critical parameter, depending on the structure to be studied.

3.3.5 "Proof-of-Principle" SPDM Experiments

Both first theoretical analyses and first "proof-of-principle" experiments with calibration objects indicated the usefulness of the SPDM concept for nanoscale imaging beyond localization alone.

Figure 3.7 shows an experimental "proof-of-principle" example (Esa et al. 2000, Cremer 2003) from cellular biophysics using CLSM. In this case, in cell nuclei three sites on a single large macromolecule were labeled: The DNA molecule of a translocation chromosome was labeled at three specific sites, each site with a unique spectral signature, based on different fluorescence excitation and emission spectra. The fluorescence signals were imaged confocally at the same time but registered separately from each other via appropriate filters, and evaluated according to the above-described principles of SPDM. The smallest distances resolved were about 30 nm laterally (ca. 1/16 of the exciting wavelength) and about 50 nm axially (ca. 1/10 of the exciting wavelength), experimentally demonstrating the possibility to measure distances and positions of "point-like" fluorescent down to the macromolecular range. It is obvious (compare Figure 3.3) that this method of SPDM can be extended to all objects where the minimum distance between objects of the same spectral signature (in this case excitation/emission spectrum) is larger than the FWHM of the PSF of the optical system used; furthermore, it is obvious that in principle the number of such spectral signatures can be increased; in appropriately labeled objects, the result would be analogous to an image expected at a substantially smaller FWHM (Figure 3.6).

3.3.6 Principles of Experimental SPDM/SALM with a Large Number of Spectral Signatures

In the last few years, experimental approaches based on principles of SPDM and related methods (SALM) have been considerably improved by several groups, especially by using photoswitchable fluorochromes (Schneider et al. 2005). By imaging fluorescent bursts of single molecules after light

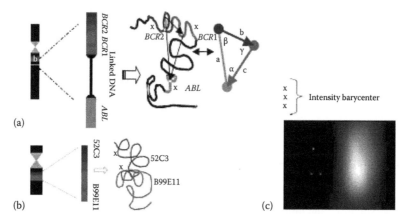

FIGURE 3.7 SPDM: "Proof-of-Principle" experiment in a cellular biophysics application. Analysis of genome nanostructure by three-color confocal microscopy and SPDM. (a) *From left to right (schematically)*: Linear mapping of three probes labeled with different spectral signatures and covering the BCR–ABL fusion region in bone marrow cells of patients with chronic myelogeneous leukemia (CLM); a complex folding of the 3D chromatin structure of this region in nuclei carrying the fusion region; the triangles resulting from the intensity barycenters of the three sequence sites labeled with three spectral signatures, after application of the SPDM procedure. (From Esa, A. et al., *J. Microsc.*, 199, 96, 2000. With permission.) (b) *From left to right (schematically)*: Linear mapping of two probes covering the BCR region on the intact chromosome 22; a complex folding of the 3D chromatin structure of this region. (c) The topology of three sequences in the BCR–ABL fusion gene region in bone marrow cell nuclei from leukemia patients determined by confocal SPDM. *Left*: The relative mean positions and distances (small white dots) obtained by experimental three-color confocal SPDM of two BCR regions and one ABL region in *t*(9;22) Philadelphia chromosome territory regions of bone marrow cells of CML patients according to (a), measured in 75 nuclei. (From Esa, A. et al., *J. Microsc.*, 199, 96, 2000. With permission.) *Right*: Confocal monospectral virtual microscopy diffraction image calculated for the experimental mean positions shown on the left (vertical axis denotes optical axis of the confocal microscope), using an experimental confocal PSF (Bornfleth et al. 1998). In the case of monospectral labeling, all topological information about the BCR–ABL fusion region has been lost. (From Cremer, C., Far field light microscopy, in *Nature Encyclopedia of the Human Genome*, Nature Publishing Group, London, 2003. *Nat. Publ. Group*, doi: 10.1038/npg.els.0005922, 2003.)

activation, the position of the molecules was determined with precision values significantly better than the FWHM. These microscopic techniques were termed PALM (Betzig et al. 2006), FPALM (Hess et al. 2006), stochastic optical reconstruction microscopy (STORM) (Rust et al. 2006) or PALM with independently running acquisition (PALMIRA) (Geisler et al. 2007, Andresen et al. 2008, Steinhauer et al. 2008). In all these approaches, special fluorochromes were used that can be photo-switched between a "dark" state A and a "bright" state B (Hell 2003, 2009). Photoswitching was also applied for superresolution by the localization of quantum dots using blinking characteristics (Lidke et al. 2005).

In the SPDM concept, the use of appropriate fluorescence lifetimes and stochastic labeling schemes has been proposed as a means of obtaining the optical isolation required for superresolution imaging (Cremer et al. 1996, 2001). Generally, fluorescence lifetime refers to the average time the molecule stays between photon absorption and fluorescence photon emission.

Like lifetimes in radioactivity, on the single molecule level spontaneous fluorescence is connected with a stochastic onset of emission. This stochastic onset of emission may be used as a spectral signature, the target molecules being stochastically labeled by these spectral signatures. From this, a rough estimate may be obtained about the fluorescence lifetime of a fluorophore needed for a specific SPDM imaging problem: Let us assume a simple exponential decay $N(t) = N_0 e^{-t/\tau}$, where $N(t)$ = number of molecules still capable of emission at time t after the start of excitation (i.e., before emission and thus in a "dark" state); τ=lifetime; N_0=number of molecules in a "dark" state at t=0. Then the "decay rate," i.e., the decrease in the number of molecules in the "dark" state at t=0, is $dN/dt=-1/\tau N_0$. Since these molecules left the "dark" state due to photon emission, the number of molecules N_{em} per second with photon emission is the same, $dN_{em}/dt=1/\tau N_0$.

For example, if $N_0=1\times10^5$ molecules labeled with a fluorophore of fluorescence lifetime τ are assumed to be localized in a cellular area $A_0=1\times10^2$ μm^2 (e.g., the nuclear area), then there are N_{obsvol}=31 molecules in the diffraction-limited microscopic observation volume of area $A_{obsvol}=\pi(FWHM/2)^2=\pi(0.1)^2$ $\mu m^2=0.031$ μm^2. In the case of just M=1 fluorescent molecules per second and per A_{obsvol} (condition of "optical isolation"), this corresponds to $dN_{em}/dt=1/\tau N_0=3200/s$ in the area A_0, or $\tau(M=1)=N_0/(dN_{em}/dt)=1\times10^5/3200\sim30$ s. In case M photon-emitting molecules per second in the area A_0 are registered, the lifetime required is reduced to $\tau(M)=1/M\times\tau(M=1)$. For examples, in case M=10 images of photon-emitting molecules are registered per second, the lifetime required would be reduced to τ=3 s only. For other values of N_0, A_0, FWHM, M, etc., the lifetimes that are needed can be estimated in an analogous way.

In case a fluorophore molecule would emit just one fluorescence photon only, there would be no improvement in resolution, due to the spatial probability distribution given by the electromagnetic character of light. However, the general definition of fluorescence lifetime as the time between the absorption and the emission of photons would also include the case that "emission" does not mean just one photon but might mean also a burst of photons, i.e., a number of photons emitted. Under certain conditions, organic molecules have been found to show indeed such a behavior, combined with the long lifetimes (seconds) that are required according to the estimates mentioned earlier.

According to these general principles, we recently applied SPDM on the nanometer resolution scale to conventional fluorochromes that show under special physical conditions "reversible photobleaching." "Reversible photobleaching" has been shown to be a general behavior in several fluorescent proteins, e.g., CFP, Citrine, or eYFP (Patterson and Lippincott-Schwartz 2002, Sinnecker et al. 2005, Hendrix et al. 2008). In such cases, the fluorescence emission of certain types can be described by assuming three different states of the molecule: A fluorescent "bright" state M_{fl}; a reversibly bleached "dark" state M_{rbl} induced by the absorption of photons via a possibly complex mechanism; and an irreversibly bleached "dark" state M_{ibl}. In the M_{rbl} state, the molecule can be excited within a short time (ms range) to a large number of times to the M_{fl} state (fluorescence burst) until it passes into the irreversibly bleached state (i.e., a dark state long compared with the time required for a burst of fluorescence photons); and thus its position can be determined. With transition time constants k_1, k_2, k_3 one can assume the transition scheme (Sinnecker et al. 2005)

$$M_{rbl} \underset{k_1}{\overset{k_2}{\rightleftarrows}} M_{fl} \xrightarrow{k_3} M_{ibl}$$

Following excitation with a short laser pulse, an exponential decay of the fluorescence intensity was observed (Sinnecker et al. 2005). According to the above-mentioned general definition of fluorescence as a process in which the absorption of photons by a molecule triggers the emission of photons with a longer wavelength, in this case the average fluorescence lifetime would refer to the average time (τ) the molecule stays in its excited reversibly "dark" state (M_{rbl}) before the emission of photons via the conversion to the (still excited) "bright" M_{fl} state is triggered. If so, as in the radioactive decay analogy described earlier, for the individual molecules M_1, M_2, M_3,... the individual fluorescence burst onset times t_1, t_2, t_3,.... of the single molecules between photon absorption and photon emission would be distributed stochastically according to $N(t)=N_0 \exp[-t/\tau]$, where $N(t)$ refers to the number of reversibly bleached molecules (dark state) at time t; N_0 refers to the original number of reversibly bleached excited molecules obtained by the illuminating laser pulse; and τ refers to the fluorescence lifetime of the ensemble. If so, the times t_1, t_2, t_3 may be considered as novel spectral signatures to discriminate single molecules from each other, even in the case that their mutual distances are smaller than the $FWHM_{PSF}$ of the optical system used.

We recently showed that using appropriate illumination conditions, reversible photobleaching can be used for effective superresolution imaging ($SPDM_{Phymod}$) of cellular nanostructures labeled with conventional fluorochromes such as Alexa 488 (Reymann et al. 2008) or the Green Fluorescent Protein variant YFP (Lemmer et al. 2008). This novel extension of the SPDM approach is based on the possibility to produce the optical isolation required by allowing only one molecule in a given observation volume and in a given time interval to be in the M_{fl} state. In this specific case, each stochastic time interval $(t_1, t_2, t_3, ..., t_k)$ since the transitions of M_{fl} molecules into the reversibly bleached state (M_{rbl}) and the onset of the fluorescent burst at a transition M_{rbl} to M_{fl} are regarded to represent a specific spectral signature $spec_1$, $spec_2$, $spec_3$, ... $spec_k$ (see previous text for general concept). These spectral signatures, characterized by such a stochastic onset time may be obtained by special physical conditions, such as an exciting illumination intensity in the $10\,kW/cm^2$ to $1\,MW/cm^2$ range (substantially higher than those used by Sinnecker et al. 2005). SPDM imaging approaches based on this method might be specified as $SPDM_{Phymod}$.

Here we provide further technical details about the present state of $SPDM_{Phymod}$ localization microscopy. In addition to some examples taken from previous publications, as original applications, we present first results on nanoimaging of GFP-labeled microtubule in human interphase cells and of GFP-labeled histone 2B molecules in mitotic HeLa cells.

3.4 Approaches to Superresolution of Fluorescence-Labeled Cellular Nanostructures C: Experimental SALM/SPDM with a Large Number of Spectral Signatures

3.4.1 Microscope Setup

The $SPDM_{Phymod}$ experiments were performed with a vertical SMI microscope setup. This allowed to combine SPDM imaging with SI imaging, in this case SMI microscopy. Figure 3.8 shows the basic setup: It provides two laser sources for excitation at $\lambda = 488\,nm$ [Ar488] (2060-6 S, Spectra-Physics/Newport, Darmstadt, Germany) and $\lambda = 568\,nm$ [Kr568] (Lexel 95 K, Lexel Laser, Fermont, CA). Both laser lines are combined by a dichroic mirror [DS1] (F33-492, AHF Analysentechnik, Tübingen, Germany) and after deflection of 90° by a mirror [M1] they pass through an aperture stop [B1]. Using a collimator built up of two achromates with focal lengths of 20 and 100 mm (G322201000 und G322324000, Linos Photonics, Göttingen, Germany) the beams are expanded by a factor of 5 to gain a homogenous illumination.

For the SPDM measurements of molecule positions in the object plane (x, y) reported here, only objective lens O2 was used (for an application of the vertical SMI (SMI Vertico) in 3D nanoimaging using both objective lenses see Lemmer et al. (2008)).

3.4.2 Software for Data Registration and Evaluation

In the following, the basic requirements presently used in our group are described (Kaufmann et al. 2009).

3.4.2.1 General Considerations

For data acquisition, a time stack consisting of typically 2000 frames was recorded with a frame rate between less than 10 and 16 fps. This data was stored on a computer hard disk and evaluated later by algorithms implemented in MATLAB® (7.0.1, The MathWorks, Inc., Natick, MA), a matrix-based programming language for numerical computations.

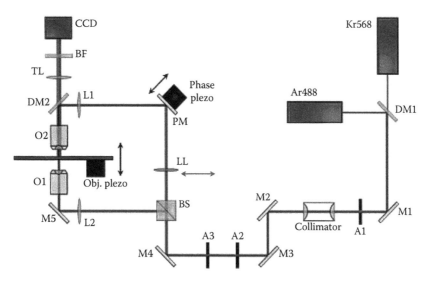

FIGURE 3.8 Microscope setup for a combination of SPDM$_{Phymod}$ and SI. Via mirrors [M1], a collimator arrangement (5× expansion of beam diameter), aperture stops [A1–A3] and mirrors [M1–M4] (10Z40ER.2, Newport) the beam enters the vertical SMI microscope at an appropriate height and is then split into two beams by a nonpolarizing 50:50 beam splitter [BS] (NT32-505, Edmund Optics, Karlsruhe, Germany). The coherent counterpropagating beams are focused by two lenses [L1, L2] into the back focal planes of two opposing oil immersion objective lenses [O1, O2] (HCX PL APO, 63×, NA = 0, 7 – 1, 4, Leica, Bensheim, Germany). The interference of both beams results in a standing wave field along the optical axes with a cos^2-shaped intensity distribution. One laser beam is guided by a dichroic mirror [DM2] (F52-489, AHF Analysentechnik) and a mirror [PM] (10Z40ER.2, Newport), which is mounted on a piezoelectrical stage in the upper objective lens. This allows to shift the relative phase of the standing wave field. The beam in the lower part of the microscope is deflected by a mirror [M5] into the objective. The sample is placed between the objective lenses and can precisely (accuracy ± 5 nm) be moved along the optical axis by a piezoelectrical stage (P-621.ZCD, Physik Instrumente, Karlsruhe, Germany). To achieve the high laser intensity, which is necessary for the localization mode, a lens [LL] can be mounted in the optical pathway. This leads to a more focused spot in the object region. For detection, a conventional epi-fluorescence setup is used. The emitted fluorescent light passes the dichroic mirror [DM2] and is focused by a tube lens [TL] (11020 515 073 010, 1.0×, f = 200 mm, Leica) onto the CCD chip of a highly sensitive 12 bit black-and-white camera [CCD] (SensiCam QE, PCO Imaging, Kehlheim, Germany). For a maximal reduction of the background signal of the excitation laser, a blocking filter [BF] (F73-491, AHF Analysentechnik) is mounted in front of the CCD chip. (From Kaufmann, R. et al., *Proc. SPIE*, 7185, 71850J-1, 2009. With permission.)

In principle, the algorithms were supposed to be applicable for SPDM$_{Phymod}$ data with high background noise and large bleaching gradients (e.g., dense labeling, large observation/activation volume) as well as for low noise scenario with negligible bleaching (e.g., photoactivable molecules in PALM/FPALM and/or small excitation volumes). In particular, the main program was designed to consider the following parameters essential for a correct SPDM evaluation:

1. The position of the fluorescence signal in the object plane/in the object space
2. Estimates for the localization accuracy on the single molecule level
3. The characteristic parameters of the model function used to determine single molecule positions
4. Estimates for the number of detected photons per molecule as a decisive parameter of localization microscopy
5. The position of the detected signal (individual molecule) in the data stack to analyze the influence of time and to extract relevant characteristics of the fluorochromes used

3.4.2.2 Estimate of Photon Numbers

In a first step, the count numbers measured for each pixel of the CCD camera is changed to the corresponding photon numbers. This is required for an estimate of the underlying noise (nonlinear Poisson statistics of photon emission as a dominant feature), essential for a correct interpretation of the single molecule localization.

To translate the count numbers into the incident photon numbers, the count numbers of the CCD camera are multiplied by a factor dependent on the quantum efficiency of the CCD sensor (ratio of the number of detected photons divided by the number of incident photons) and on the conversion rate (number of detected photons/photoelectrons needed for one count):

$$S_{Ph}(x, y, t) = \frac{1}{c_{eff}(\overline{\lambda})} \cdot c_{conv} \cdot S_{Co}(x, y, t)$$

where

$S_{Ph}(x, y, t)$ is the number of incident photons as a function of the pixel position (x, y) and of time $(t_{1,2,3,\ldots,k,\ldots,n})$ denoting the x, y (t_k) images obtained by the CCD sensor at times $t_k, k = 1, 2, 3\ldots)$

$S_{Co}(x, y, t)$ is the number of counts (raw data) as a function of the pixel position and time

c_{conv} is the conversion factor between the number of counts and the number of photoelectrons: $c_{conv} = N_{photoelectrons}/S_{Co}$; In the CCD sensor used, two photoelectrons were required to produce one count in the low light mode, and four photoelectrons were required to produce one count in the standard modus, i.e., $c_{conv} = 2/\text{count}$ and $4/\text{count}$, respectively

$c_{eff}(\overline{\lambda})$ denotes the efficiency of the CCD sensor to change the number $N_{inc} = S_{ph}$ of photons incident on the CCD sensor into the number $N_{photoelectrons}$ of photoelectrons, assuming a mean emission wavelength $\overline{\lambda}$ of the incident light:

$$c_{eff}(\overline{\lambda}) = N_{photoelectrons}/S_{ph}$$

$$S_{Ph}(x, y, t) = \frac{1}{c_{eff}(\overline{\lambda})} \cdot c_{conv} \cdot S_{Co}(x, y, t)$$

In this way, an estimate of the number of photons contributing to the SPDM image is obtained. For example, using $c_{conv} = 2/\text{count}$ (low light mode) and c_{eff} (490 nm $< \lambda <$ 560 nm) $= 0.64$ (manufacturer's information)

$$S_{Ph}(x, y, t) = 1/0.64 \cdot 2 \cdot S_{Co}(x, y, t) = 3.13\ S_{Co}(x, y, t) \approx 3\ S_{Co}(x, y, t)$$

is estimated.

3.4.2.3 Data Segmentation

Following signal detection and registration, in a second step an optional preprocessing is performed. While signals with high signal-to-noise ratio (low background) can be used as raw data for the following segmentation, in the case of high background and photobleaching effects active during several succeeding image frames, an additional computing step is required to segment signals originating from single molecules only. For this a differential photon stack (D_{Ph})

$$D_{Ph}(x, y, t') = S_{Ph}(x, y, t_{k+1}) - S_{Ph}(x, y, t_k)$$

between a succeeding (registered at t_{k+1}) and a preceding image frame (registered at t_k) is calculated. The error σ in photon number produced by the Poisson statistics of the incident photons and the noise σ_{CCD} of the CCD sensor detection (approx. 4 counts per pixel) were estimated by the Gaussian law of error propagation

$$\sigma[D_{Ph}(x, y, t')] = [S_{Ph}(x, y, t_{k+1})^2 + S_{Ph}(x, y, t_k)^2 + 2\sigma_{CCD}^2]^{1/2}$$

3.4.2.4 Position Determination

The data stacks $S_{Ph}(x, y, t)$ (in case of low background) and $D_{Ph}(x, y, t')$ (in case of high background) were then used for high precision localization (x, y) of single molecules by adapting either Gaussian model functions or calculating the fluorescence intensity bary (="gravity") center. To reduce the computing efforts, regions of interest (ROIs) of typically 8×8 pixel were used, containing the signal. Prior to thresholding, a bandpass filtering (BP) was applied

$$BP(x, y, z) = S_{Ph}(x, y, t) \otimes G_{fine}(M_{Sys}) - S_{Ph}(x, y, t) \otimes G_{broad}(M_{Sys})$$

where
$BP(x, y, z)$ is the bandpass-filtered data stack
$G_{fine}(M_{Sys})$, $G_{broad}(M_{Sys})$ are Gaussian kernels with large and small sigma values

The sigma values were chosen in such a way that at the magnification factor M_{Sys} of the microscope a Nyquist conform signal treatment was guaranteed.

The threshold was adapted to allow a maximum of ROIs containing single molecule signals. To eliminate false-positive signals, a number of Boole filters were used. The threshold and fitting parameters were tested by using simulated SPDM data stacks.

The ROIs obtained by thresholding were used to calculate the corresponding molecule positions. In the "barycenter" localization procedure, the molecule positions (x, y) were determined by

$$\vec{r}_s(i) = \frac{1}{\sum_k A_k(i)} \cdot \sum_k \vec{r}_k(i) A_k(i) + \vec{r}_{ROI}(i)$$

where
$\vec{r}_s(i)$ is the localization vector (x, y) of a molecule in the image
A_k are the signals (gray value) of the individual pixels k of the corresponding ROIs (i), and
$\vec{r}_{ROI}(i)$ is the localization vector of the ROI (i).

This procedure of localization allows the reconstruction of a SPDM image with a large number of molecule positions on a one-core processor and in real time.

For this, a two-dimensional Gaussian

$$f(x, y) = p_1 \exp\left(-\frac{(x - p_2)^2 + (y - p_3)^2}{2p_4^2}\right) + p_5 + p_6(x - p_2) + p_7(y - p_3)$$

was fitted to the signals of a thresholded ROI.

The parameters p_1–p_6 were used as start parameters, which were then optimized by the application of a Levenberg–Marquardt algorithm. The additions to the pure Gaussian function serve to describe the contribution of the background.

As start parameters,

$$p_1 = \max_{ROI} - \min_{ROI}$$

$$p_2 = x(\max_{ROI})$$

$$p_3 = y(\max_{ROI})$$

$$p_4 = \text{round}((\lambda / 2NA) \cdot (1 / \text{Pix}))$$

$$p_5 = \min_{ROI}$$

$$p_6 = p_7 = 0$$

were used.

The values $\max_{ROI}$ and $\min_{ROI}$ are the maxima and the minima, respectively, of the ROI analyzed; NA is the numerical aperture of the objective lens used; λ is the wavelength of the laser beam used for the excitation of fluorescence; and Pix is the effective pixel size in the image (corresponding to 102 nm in the object plane). The final results are analyzed for parameter plausibility by additional filters.

Figure 3.9 shows an example on how the $\text{SPDM}_{\text{Phymod}}$ evaluation was performed. Figure 3.9a shows a frame of a human cell nucleus with a large number of individual H2B-emGFP molecules detected (for labeling procedure see following text). The excitation wavelength was 488 nm, at a power (at laser exit) of 100 mW, using an objective lens with NA = 1.4. The entire SPDM was obtained by the combination of about 2000 individual frames at a total registration time of about 300 s. Seven of these individual frames, each of them containing the positions of different H2B-emGFP molecules, are presented in Figure 3.9b. As an example, Figure 3.9c shows a small area of the upper frame with the (x, y) positions of 4 H2B molecules localized. In Figure 3.9d, the intensity (counts) registered for this insert is plotted as a function of the (x, y) coordinates (object plane); in addition, the Gaussian approximations for the four molecules are shown that were determined with the algorithm described earlier. Figure 3.9e gives the (x, y) molecule positions determined from the maximum of the Gaussians. Due to the low depth of field of view for the high NA objective lens used, the Levenberg–Marquardt algorithm eliminated molecule positions with "out-of-focus" axial positions larger than a few hundred nm. Therefore, the image obtained corresponds to an optical section with an estimated thickness in the range of 600 nm.

3.4.3 Specimen Preparation

3.4.3.1 Cells

HeLa cells were used which had been stably transfected with GFP-tagged histone H2B (kindly obtained from the German Cancer Research Institute/DKFZ, Heidelberg; T.A. Knoch, "Approaching the 3D organization of the human cell nucleus," PhD thesis, University Heidelberg, Heidelberg, 2002).

The cells were cultivated in RPMI medium supplemented with 10% fetal calf serum (FCS), 1% L-glutamin and 1% penicillin/streptomycin, at 5% CO_2.

VH7 diploid human fibroblast cells (kindly provided by Prof. Dr. Beauchamp from DKFZ, Heidelberg) were cultivated in Dulbecco's modified eagle medium (DMEM) supplemented with 10% FCS, 1% L-glutamine, 1% penicillin/streptomycin and 1% sodium-pyruvate, at 5% CO_2.

3.4.3.2 Labeling of Tubulin

The cells were transfected using the organelle-lights-tubulin-GFPkit of Invitrogen, according to the protocol of the manufacturer. 75,000 transfected fibroblast cells were seeded on 24×24 mm cover slips

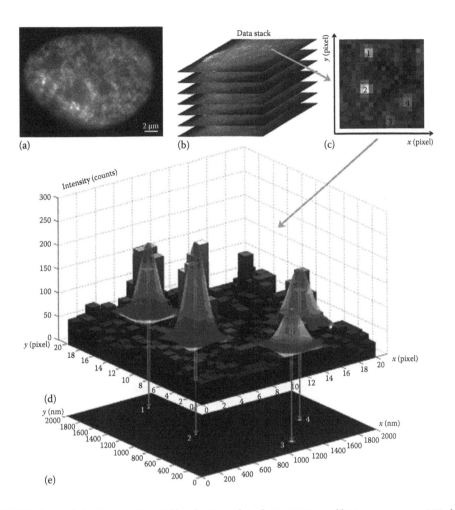

FIGURE 3.9 Example for the experimental localization of single EmGFP tagged histone proteins in VH7 human fibroblast nuclei by SPDM. See text for details. (From Kaufmann, R. et al., *Proc. SPIE*, 7185, 71850J-1, 2009. With permission.)

(medium as above). 20 h after the transfection, tubulin emGFP was expressed. Then the cells were fixed with 4% paraformaldehyde (PFA) in phosphate-buffered saline (PBS) and embedded in ProLong® Gold antifade reagent (Invitrogen).

3.4.3.3 Labeling of Histones in Human Fibroblasts

In fibroblast cells, EmGFP-conjugated histone proteins H2B were expressed using Organelle Lights™ (Invitrogen) according to the manufacturer's protocol. 20 h after transfection, cells were fixed with 4% formaldehyde in PBS and embedded with ProLong® Gold antifade reagent (Kaufmann et al. 2009).

3.4.4 Experimental SPDM$_{Phymod}$ Nanoimaging of the Distribution of emGFP-Tagged Tubulin Molecules in Human Fibroblast Nuclei

In previous reports (Lemmer et al. 2008, Kaufmann et al. 2009), we have shown that SPDM$_{Phymod}$ allows to resolve fluorescent-labeled membrane proteins and histone proteins in human cells at molecule density substantially higher than which can be resolved by conventional fluorescence microscopy. As a first example, here we show preliminary original results for the application to SPDM$_{Phymod}$ nanoimaging of emGFP-tagged tubulin molecules in human fibroblasts in interphase (Figure 3.10). In the entire cell,

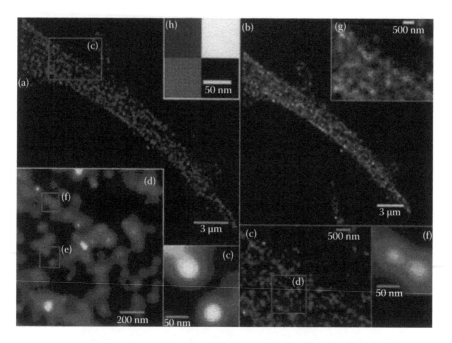

FIGURE 3.10 (See color insert following page 15-30.) SPDM$_{Phymod}$ of emGFP-tagged tubulin molecules in human fibroblasts. (a) A cell protrusion indicating single tubulin molecules, scale bar 3 μm; (b) Conventional epifluorescence image calculated from (a); (c) Insert from (a), scale bar 500 nm; (d) Insert from (c), scale bar 200 nm; (e and f) Inserts from (d), scale bar 50 nm; (g) Conventional epifluorescence image calculated from (c); (h) Conventional epifluorescence image calculated from (e); The distance of the two molecules in (e) was estimated to be 106 nm; The distance of the two molecules in (f) was estimated to be 62 nm. The mean localization accuracy determined was 35±10 nm; total number of tubulin molecules counted in the entire fibroblast cell (optical section of ca. 600 nm thickness): N=42,821, and N=1032 in (a).

the mean localization accuracy was estimated to be 35±10 nm. Figure 3.10a shows a cell protrusion with fluorescent single tubulin molecules (scale bar 3 μm; in Figure 3.10b); a conventional epifluorescence image of a part of (a) is presented. In this case, the molecule density was sufficiently low to locate single molecules even in the conventional (low intensity) mode; even the conventional epifluorescence image of insert (c) presented in (g) still allowed to identify a number of individual molecules but the image appears to be rather "fuzzy." The images in Figure 3.10d–f indicate the superresolution obtainable by the SPDM$_{Phymod}$ mode. Figure 3.10d shows the insert of the SPDM image (a) at higher magnification, with two inserts shown in Figure 3.10e and f at still higher magnification; here individual tubulin molecules at a distance of 62 nm (e) and 106 nm (f) were still well separated from each other. The "size" of the localized molecules corresponds to the standard deviation of the localization accuracy of the molecules.

3.4.5 Experimental SPDM$_{Phymod}$ Nanoimaging of GFP-Labeled Histone Distribution in Human Cell Nuclei

The chromatin nanostructure of mammalian cells is still widely unexplored (Cremer and Cremer 2001, 2006a, b, Misteli 2007). The quantitative analysis of the spatial distribution of histone H2B molecules, a key element of the basic nucleosomes, is expected to contribute to decide between conflicting models of these structures. In previous reports (Kaufmann et al. 2009, Gunkel et al. 2009), we have shown that SPDM$_{Phymod}$ allows highly advanced superresolution imaging of the nuclear distribution of such proteins. In optical sections of human fibroblast nuclei, typically 50,000–100,000 emGFP-tagged H2B molecules were localized, with mean localization accuracies in the 16 nm range, or about 1/30 of the excitation wavelength used; localization accuracies of 5 nm and smaller were achieved, corresponding to an optical resolution potential of about 10 nm. Practically, molecules with distances lower than 20 nm were clearly identified as separate entities (Figure 3.11).

In the present report, we show an original experimental example for SPDM$_{Phymod}$ nanoimaging of GFP-labeled histone H2B molecules in human HeLa cells. After cell fixation with PFA (no metaphase arresting substances, no hypotonic treatment), SPDM$_{Phymod}$ was applied. Figure 3.12 shows the SPDM image of a HeLa nucleus. The mean localization accuracy was 25 ± 7 nm, indicating a mean optical resolution potential of about 60 nm. Since many molecules had a substantially better localization accuracy, locally a considerably better resolution (in terms of the smallest distance resolvable between two single molecules) may be obtained. Figure 12a and b present the SPDM image (a) and the conventional epifluorescence image (b), respectively. The insert in Figure 3.12a is shown at a higher magnification in (c) where already a large number of single H2B molecules can be identified. For each molecule in c–e, the individual localization accuracy obtained from the optimized adaptation of Gaussian functions (compare with Figure 3.9) was given as the "size" of the localized molecules. The insert (d) in (c) and the insert (e) in (d) are presented at still higher magnifications (scale bars of 100 and 30 nm, respectively). Individual molecules with a distance of 70 nm were still well separated from each other. Currently, comparative quantitative analyses of the distribution of histone molecules in HeLa cell nuclei and human diploid cell nuclei are performed to elucidate differences in nuclear genome nanostructure between "normal" cells and cancer cells (Bohn M. et al., manuscript in preparation).

In addition to the superresolution analysis of single molecules in interphase cells, the application of localization microscopy methods to cells in mitosis is highly promising, not only in Bacteria (Biteen et al. 2008) but also in human cells. For example, due to the lack of appropriate superresolution techniques, it still unknown to what extent the differences in histone density believed to play an important role in functional nuclear genome structure are maintained in mitotic chromosomes. Apart from such questions of basic biophysical interest, the "banding" pattern obtained by various dyes as well as by fluorescence in situ hybridization (FISH) methods allows the cytogenetic identification of a large amount of disease-related chromosomal alterations. However, using conventional epifluorescence microscopy, the spatial resolution has been quite limited. For example, assuming a 700 band pattern, each band contains about

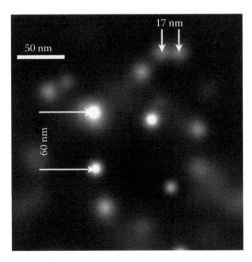

FIGURE 3.11 SPDM$_{Phymod}$ of emGFP-tagged histone 2B molecules in human fibroblast nuclei. Arrows indicate two resolved H2B molecules with an estimated distance of 17 nm. Scale bar: 50 nm. (From Kaufmann, R. et al., *Proc. SPIE*, 7185, 71850J-1, 2009. With permission.)

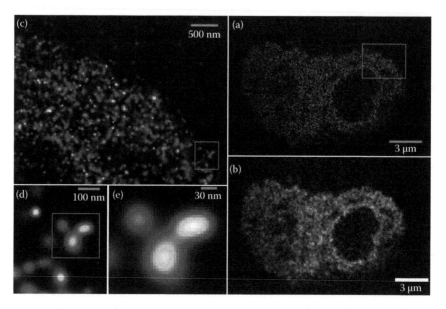

FIGURE 3.12 (See color insert following page 15-30.) SPDM$_{Phymod}$ image of the Histone H2B distribution in a HeLa cell nucleus with stably transfected GFP-tagged H2B. (a and b) Entire nucleus, (scale bar 3 µm): (a) SPDM image; (b) epifluorescence image (calculated from the SPDM image by convolution with a PSF with FWHM = 200 nm; (c) insert from SPDM image (a) at higher magnification (scale bar 500 nm); (d) insert from (c) at higher magnification (scale bar 100 nm); (e) insert from (d) at higher magnification (scale bar 30 nm) featuring three single histone molecules at distances of about 70–80 nm. The "size" of the molecules corresponds to the localization accuracy. Total number of H2B molecules counted: $N = 17,409$ (optical section of ca. 600 nm thickness).

$6300 \times$ Megabase pairs (Mbp)/700 = 9 Mbp of DNA on average. A better spatial resolution might allow a substantial increase in the detection of small aberrations, e.g., small deletions (lack of a specific piece of DNA), or amplifications of genes, such as tumor-related genes.

As a starting point toward such novel perspectives, we here report for the first time an SPDM$_{Phymod}$ image of a mitotic human cell (Figure 3.13). In spite of the lack of metaphase-arresting substances, the lack of hypotonic treatment, and the PFA fixation used in this case, mitotic chromosomes in the aggregations have remained distinguishable. While at visual inspection, the difference between the SPDM image and the conventional epifluorescence image of the entire cell (Figure 3.13a and b) appears to be small, the insert examples with an increasing magnification show the drastic difference between conventional resolution and SPDM. Individual histone H2B molecules with distances around 70 nm were clearly distinguished. Assuming a 30 nm straight chromatin fiber, this would correspond to a genomic distance of about 7000 base pairs (7 kbp). In combination with appropriate identification of the associated DNA sequence, e.g., by SPDM imaging using FISH probes (Weiland et al., manuscript in preparation; Rouquette et al. 2009), this might allow to count small amplifications/deletions in individual mitotic cells at a highly superior spatial resolution.

3.4.6 3D-Nanoimaging by Combination of SPDM and SMI Microscopy

3.4.5.1 Three-Dimensional Nanoimaging

In the SPDM nanoimaging results presented so far, the vertical SMI microscope setup (Figure 3.8) was used for the SPDM imaging of (x, y) positions of molecules in the object plane. For this only objective lens O2 was applied. By first imaging an object suitable for SPDM$_{Phymod}$ at low illumination intensities with both objective lenses O1 and O2 in the SMI mode, and then with lens O2 and higher intensities

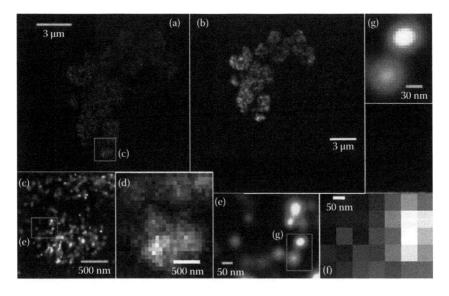

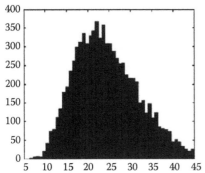

FIGURE 3.13 SPDM nanoimaging of Histone 2B distribution in a mitotic HeLa cell. (a) SPDM image of entire mitotic cell with stably GFP-tagged H2B molecules (no treatment for mitotic arrest; no hypertonic treatment; PFA fixation), scale bar 3 μm; (b) corresponding "conventional" epifluorescence (calculated from a); (c) magnified insert from (a) (scale bar 500 nm); (d) "conventional" epifluorescence image of (c), calculated from (c); (e) magnified insert from (c), scale bar 50 nm; (f) "conventional" epifluorescence image of (e), calculated; (g) magnified insert from (e), scale bar 30 nm; the "size" of the molecules corresponds to the individual localization accuracy. Center-to-center distance between the two molecules in (g): $d=72$ nm. *Below*: Localization accuracy of individual GFP-H2B molecules in the mitotic cell analyzed. "Ordinate": Number of molecules identified with a given localization accuracy (abscissa). Mean localization accuracy: 24 ± 7 nm. Total number of H2B molecules counted in this mitotic cell: 8284.

in the SPDM$_{\text{Phymod}}$ mode, it was possible by a combination of both image stacks to create a 3D image with a 3D effective optical resolution many times better than the Abbe-limit in the lateral (x, y) direction and the conventional confocal resolution limit of about 600 nm in the axial (z) direction. In a first step, the object assumed to have a fluorescent thickness of less than about 130 nm is subjected to the SMI microscopy mode. For this, low illumination intensities are used to allow a continuous excitation of fluorescence emission. From this, the axial (z)-distribution of the fluorescent molecules is obtained with a resolution in the 40 nm range: SMI microscopy (see Section 3.2) allows the measurement of the z-position (z_0) of a subwavelength-sized fluorescence object (e.g., a number of molecules) with an accuracy of a few nm, and its size down to an axial diameter Δz of a few tens of nm. Hence, molecules in such a complex have z-positions $z_0 \pm \Delta z/2$.

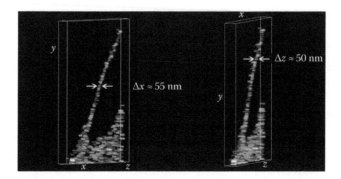

FIGURE 3.14 Combination of SPDM and SMI microscopy. 3D SPDM/SMI reconstruction of a human cancer cell protrusion (human Cal-51 cells). The volume rendering results are shown in two different orientations. (With kind permission of Springer Science & Business Media: Lemmer, P. et al., *Appl. Phys. B*, 93, 1, 2008.)

Immediately after the low illumination intensity SMI imaging, the object is registered at higher illu-mination intensities in the $SPDM_{Phymod}$ mode, yielding the (x, y) positions of the individual molecules. Combining both image informations, a 3D image with a lateral effective optical resolution correspond-ing to the SPDM mode and a lateral resolution corresponding to the SMI mode is obtained. For a detailed description of the method, see Lemmer et al. (2008).

Figure 3.14 shows an example for this 3D nanoimaging application of the vertical SMI setup first using both objective lenses O1 and O2 for SMI imaging and then lens O2 only for SPDM imaging. In this case, YFP molecules localized in the membrane of human Cal-51 cells were imaged.

3.5 Discussion

General ideas about the formation of images have a long tradition. For example, the optics expert and state commissioner of science and technology, J.W. von Goethe, already at the beginning of the nine-teenth century stated that "the bright, the dark, and the colour together constitute what discriminates an object from another object, and the parts of the object from each other. From these three we recon-struct the visible world" (Goethe 1808). Thanks to the progress in modern optical technologies and molecular fluorescence-labeling procedures, in recent years it has become possible to develop such visions into useful nanoimaging approaches. In this review, superresolution methods based on focused/ SI and especially on localization microscopy have been discussed. While in focused/SI approaches, the idea is to excite "color" only in molecules in a given small area at a specific time and to leave the others "in the dark," in localization microscopy approaches the idea is to induce molecules to be "bright" with a given "color" at a specific time in such a way that their next neighbor distance is larger than a mini-mum value given by the conventional optical resolution.

3.5.1 Perspectives for Nanostructure Analysis in Fixed Specimens

Presently, multiple approaches for LOBSTER exist, each of these methods having its range of biological, biophysical, and biomedical applications. Recently, various methods of SALM have been shown to allow the nanoimaging of individual molecules at an optical resolution close to the dimensions of proteins.

In particular, in this report one of these approaches was described more extensively, SPDM with conventional fluorophores under special physical conditions ($SPDM_{Phymod}$). Since biological specimens labeled with such fluorescent proteins are most common, and fluorescence registration can be performed using physiological conditions (our own unpublished results), SPDM methods using such fluorophores have a vast range of applications, including the potential for in vivo measurements.

Here, the current state of $SPDM_{Phymod}$ using fluorescent proteins has been summarized. In addition, here we have presented original evidence that both GFP and the widely used GFP-variant EmGFP allow single molecule nanoimaging. As original examples, single molecule nanoimaging of tubulin proteins in human fibroblast interphase cells was presented along with single molecule nanoimaging of the distribution of Histone H2B molecules in mitotic HeLa cells.

The localization accuracies and molecule densities obtained in the preliminary experiments presented here allowed to clearly identify and separate single molecules at a distance as low as 60 nm, thus demonstrating the potential of the $SPDM_{Phymod}$ method for superresolution imaging. In other applications (Lemmer et al. 2008, Kaufmann et al. 2009, Gunkel et al. 2009), even better localization accuracies (down to a few nm) and higher molecule densities were obtained: Single molecules with a distance below 20 nm were clearly resolved, and the potential of this method for multicolor nanoimaging was demonstrated.

At the effective optical resolution level of about 10–20 nm laterally and 40–50 nm axially presently achieved by $SPDM_{Phymod}$, numerous applications in the structural elucidation of cellular nanostructures are feasible. Examples for such nanostructures are individual gene domains in genetically active and inactive states; environmentally induced changes of chromatin nanostructure; size and nuclear distribution of replication factories and repair complexes; NPC distribution; the arrangement of polyribosomes; or the distribution of ion channels on the cell membrane. An additional important application will be the possibility to count single molecules, e.g., on the cell membrane, or RNA transcripts.

3.5.2 Perspectives for In Vivo Imaging at the Nanometer Scale

To date, laseroptical nanoscopy has typically been performed on fixed cells in order to demonstrate the achievable optical resolution. Although the application of these techniques to live cell imaging is still in its beginning, some preliminary results indicate its feasibility. For example, Westphal et al. (2008) analyzed the movement of fluorescently labeled synaptic vesicles in living cells recorded with STED microscopy at a rate of 28 fps and an optical resolution around 60 nm. In this study, the cross section area of the focal STED spot was reduced about 18-fold below the conventional diffraction limit of about 260 nm. In another application (Nägerl et al. 2008), STED microscopy was used for live cell imaging of dendritic spines to dissect synaptic vesicle movement at video rates.

Due to the scanning mechanism, high speed nanoimaging in STED requires small ROIs (in the few micrometer range). In comparison, the various modifications of SALM have the potential for in vivo nanoimaging of large cellular areas (up to $100 \times 100 \, \mu m^2$). For example, Hess et al. (2007) used FPALM to study the dynamics of hemagglutinin cluster distribution in the membranes of living cells at 40 nm effective optical resolution to discriminate between raft theories; using PALM, Shroff et al. (2008) investigated the nanoscale dynamics within individual adhesion complexes in living cells under physiological conditions. Preliminary experiments have demonstrated that live cell imaging is also possible using $SPDM_{Phymod}$ approaches: In live human cells, the distribution of fluorescent-labeled membrane proteins was registered at an effective optical resolution of a few tens of nm (Weiland et al., unpublished observations). In these experiments, a highly sensitive CCD camera was used allowing to register 15–20 fps. The use of high-speed, back-illuminated CCD cameras allowing to register up to 1000 fps will considerably extend the possibilities of this approach in live cell imaging applications.

3.5.3 Perspectives for Single Molecule Counting

Although the $SPDM_{Phymod}$ procedure so far allows to register only a part of all labeled fluorescent proteins, under conditions where each molecule is registered only once, the numbers obtained are minimum absolute numbers. For example, an SPDM count of 60,000 H2B proteins in an optical section of a human fibroblast nucleus of $10 \times 10 \, \mu m = 100 \, \mu m^2$ results in a molecule density of about $600/\mu m^2$. Even at the best corresponding, conventional optical resolution assumed to be $d = 0.61 \cdot 488\,\text{nm}/1.4 = 212\,\text{nm}$, assuming a diffraction-limited observation area of $0.035\,\mu\text{m}^2$, in the best case about $100/0.035 = 2860$ molecules would be resolvable in a nuclear area of $100\,\mu\text{m}^2$, corresponding to a density of resolved molecules of about 30 molecules/μm^2. Since many protein types in the cells occur in numbers in the order of 10^4–10^5, SPDM and related SALM methods for the first time promise to count the number of fluorescent-labeled molecules in the cell even at such high numbers.

For fluorescent proteins, presently the number of molecules counted in SPDM$_{\text{Phymod}}$ experiments is a minimum number since it is not known to what percentage the molecules can be excited in the way required for SPDM$_{\text{Phymod}}$ (which might be true also for other SALM methods). For other molecule types or at other excitation modes, it might also be possible that a molecule is induced to the emission of several fluorescent bursts during the 30–100 s currently required for a full SPDM$_{\text{Phymod}}$ image. This might result in a higher counting than actually present. It is expected that by appropriate calibration methods, even absolute numbers of individual molecules will become countable. For example, in many cases, the mean total number of proteins/target molecules per cell is known from other methods; consequently, the mean total number of SPDM counting events per cell has to be ideally the same for both methods.

One of the several more direct possibilities might be that the same target molecules are labeled both with "SPDM$_{\text{Phymod}}$" molecules like YFP, GFP, RFP (Gunkel et al. 2009) or Alexa dyes (Reymann et al. 2008, Baddeley et al. 2009), and "STORM" molecules like Cy3, Cy5 (Rust et al. 2006, Heilemann et al. 2008, Steinhauer et al. 2008). A direct solution of such problems will require a better knowledge of the photophysics of such dyes; this would also be of utmost importance for optimization techniques. For example, reliable methods for dense labeling, high localization SPDM$_{\text{Phymod}}$ down to the level of a few nm optical resolution and reliable absolute counting of specific small DNA sections would allow numerous applications in chromatin nanostructure biophysics and molecular genetics. However, many interesting applications using relative single protein numbers for counting are already now feasible, such as high sensitivity gene expression analysis on the single cell level. For example, if in cell No. 1 $N_{\text{1evt}} = 100,000 \pm [100,000]^{1/2} = 100,000 \pm 316$ SPDM "events" have been counted (corresponding to a protein number $N_{\text{1prot}} = \alpha\, N_{\text{1evt}}$ (α calibration factor), and in cell No. 2 under the same SPDM conditions $N_{\text{2evt}} = 98,000 \pm [98,000]^{1/2} = 98,000 \pm 313$ SPDM "events" (corresponding to $N_{\text{2prot}} = \alpha\, N_{\text{2evt}}$), a difference in specific gene expression as low as 2% would become measurable on the single cell level as $[N_{\text{1prot}} - N_{\text{2prot}}]/N_{\text{1prot}} = [N_{\text{1evt}} - N_{\text{2evt}}]/N_{\text{1evt}}$.

In addition to an effective optical resolution in the macromolecular range, single molecule counting will open perspectives for many additional biophotonic applications on the single cell level, such as high precision monitoring of specific DNA amplification gene expression, fast counting of virus particles inside and outside cells, counting of proteins in intracellular normal and pathological aggregates, or counting of drug molecules present in specific nanoregions across the cell membrane.

The strategy of SPDM$_{\text{Phymod}}$ and other modes of localization microscopy by engineering ON ("bright")- and OFF ("dark")-states is not restricted to fluorescent proteins, like emGFP, YFP, and mRFP, but has been shown to work also for Alexa dyes (Reymann et al. 2008, Baddeley et al. 2009); it may essentially be applicable to a large number of synthetic, single molecule compatible fluorophores (Steinhauer et al. 2008).

In case sufficiently photostable fluorochromes with photoconvertable "dark" and "bright" spectral signatures can be used, a further improvement of 3D effective resolution is anticipated. For example, if 5,000–10,000 photons could be registered from a single molecule, under ideal conditions an axial (z_0) localization accuracy around 1 nm would be expected (Albrecht et al. 2001). Eventually, an effective optical 3D resolution in the 2 nm range (including single molecule counting in 3D) might become feasible and make possible far-field light optical structural analyses even of the components of macromolecular complexes in the interior of cells.

To summarize, it is anticipated that SPDM/SALM and other novel developments in laseroptical nanoscopy will eventually bridge the gap in resolution between ultrastructural methods (nm resolution) and visible light far-field microscopy (conventionally hundreds of nm resolution) in such a way that the same cellular structure can be imaged and quantitatively analyzed at molecular resolution. Such a "correlative microscopy" will provide an essential contribution to a direct insight into the "physics of life" on the individual cell level.

Acknowledgments

The authors thank Prof. Markus Sauer (Bielefeld University), Dr. Michael Wassenegger (AIPlanta, Neustadt/Weinstraße), and Prof. Victor Sourjik (ZMBH, University of Heidelberg) for stimulating discussions. The financial support of the University Heidelberg (Frontier project grant to M. Wassenegger), of the State of Baden-Wuerttemberg, the Deutsche Forschungsgemeinschaft (SPP1128), and the European Union (In Vivo Molecular Imaging Consortium, www.molimg.gr) to C. Cremer is gratefully acknowledged. We also thank our colleagues Jürgen Reymann, Wei Jiang, Alexander Brunner, Christoph Hörmann, Johann von Hase, and Margund Bach for great support. Paul Lemmer is a fellow of the Hartmut Hoffmann-Berling International Graduate School of Molecular and Cellular Biology of the University Heidelberg and a member of the Excellence Cluster Cellular Networks of University Heidelberg. Patrick Müller has been supported by a BMBF grant to M. Hausmann.

References

Abbe, E. (1873) Beiträge zur Theorie des Mikroskops und der mikroskopischen Wahrnehmung, *Archiv f. mikroskopische Anatomie* 9, 411–468.

Albrecht, B., Failla, A.V., Heintzmann, R., Cremer, C. (2001) Spatially modulated illumination microscopy: Online visualization of intensity distribution and prediction of nanometer precision of axial distance measurements by computer simulations, *Journal of Biomedical Optics* 6, 292.

Albrecht, B., Failla, A.V., Schweitzer, A., Cremer, C. (2002) Spatially modulated illumination microscopy allows axial distance resolution in the nanometer range, *Applied Optics* 41, 80–87.

Andresen, M., Stiel, A.C., Jonas, F., Wenzel, D., Schönle, A., Egner, A., Eggeling, C. et al. (2008) Photoswitchable fluorescent proteins enable monochromatic multilabel imaging and dual color fluorescence nanoscopy, *Nature Biotechnology* 26, 1035–1040.

Baddeley D., Carl C., Cremer C. (2006) Pi microscopy deconvolution with a variable point-spread function, *Applied Optics* 45, 7056–7064.

Baddeley, D., Batram, C., Weiland, Y., Cremer, C., Birk, U.J. (2007) Nanostructure analysis using spatially modulated illumination microscopy, *Nature Protocols* 2, 2640–2646.

Baddeley, D., Jayasinghe, I.D., Cremer, C., Cannell, M.B., Soeller, C. (2009) Light-induced dark states of organic fluorochromes enable 30 nm resolution imaging in standard media, *Biophysics Journal* 96(2), L22–L24.

Bailey, B., Farkas, D., Taylor, D., Lanni, F. (1993) Enhancement of axial resolution in fluorescence microscopy by standing-wave excitation, *Nature* 366, 44–48.

Betzig, E. (1995) Proposed method for molecular optical imaging, *Optics Letters* 20, 237–239.

Betzig, E., Patterson, G.H., Sougrat, R., Lindwasser, O.W., Olenych, S., Bonifacino, J.S., Davidson, M.W. et al. (2006) Imaging intracellular fluorescent proteins at nanometer resolution, *Science* 313(5793), 1642–1645.

Bewersdorf, J., Bennett, B.T., Knight, K.L. (2006) H2AX chromatin structures and their response to DNA damage revealed by 4Pi microscopy, *PNAS* 103, 18137–18142.

Biteen, J.S., Thompson, M.A., Tselentis, N.K., Bowman, G.R., Shapiro, L., Moerner, W.E. (2008) Single-molecule active-control microscopy (SMACM) with photo-reactivable EYFP for imaging biophysical processes in live cells, *Nature Methods* 5, 947–949.

Bornfleth, H., Sätzler, E.H.K, Eils, R., Cremer, C. (1998) High-precision distance measurements and volume-conserving segmentation of objects near and below the resolution limit in three-dimensional confocal fluorescence microscopy, *Journal of Microscopy* 189, 118–136.

Bradl, J., Hausmann, M., Schneider, B., Rinke, B., Cremer, C. (1994) A versatile 2pi-tilting device for fluorescence microscopes, *Journal of Microscopy* 176, 211–221.

Bradl, J., Rinke, B., Schneider, B., Hausmann, M., Cremer, C. (1996) Improved resolution in 'practical' light microscopy by means of a glass fibre 2pi-tilting device, *Proceedings of SPIE* 2628, 140–146.

Brakenhoff, G.J., Blom, P., Barends, P. (1979) Confocal scanning light microscopy with high aperture immersion lenses, *Journal of Microscopy* 117, 219–232.

Brakenhoff, G.J., van der Voort, H.T.M., van Spronsen, E.A., Linnemanns, W.A.M., Nanninga, N. (1985) Three-dimensional chromatin distribution in neuroblastoma nuclei shown by confocal laser scanning microscopy, *Nature* 317, 748–749.

Burns, D.H., Callis, J.B., Christian, G.D., Davidson, E.R. (1985) Strategies for attaining superresolution using spectroscopic data as constraints, *Applied Optics* 24, 154–161.

Chen, K.R. (2009) Focusing of light beyond the diffraction limit, arxiv.org/pdf/4623v1.

Chen, K.R., Chu, W.H., Fang, H.C., Liu, C.P., Huang, C.H., Chui, H.C., Chuang, C.H. et al. (2009) Beyond-limit focusing in the intermediate zone, arxiv.org/pdf/0901.173.

Cremer, T. (1985) *Von der Zellenlehre zur Chromosomentheorie,* Springer-Verlag, Berlin, Heidelberg. Download: http://humangenetik.bio.lmu.de/service/downloads/buch_tc/index.html.

Cremer, C. (2003) Far field light microscopy, in: *Nature Encyclopedia of the Human Genome.* Nature Publishing Group, London. DOI: 10.1038/npg.els.0005922.

Cremer, C., Cremer, T. (1972) Procedure for the imaging and modification of object details with dimensions below the range of visible wavelengths, German Patent Application No. 2116521 filed April 5, 1971 (in German).

Cremer, C., Cremer, T. (1978) Considerations on a laser-scanning-microscope with high resolution and depth of field, *Microscopica Acta* 81, 31–44.

Cremer, T., Cremer, C. (2001) Chromosome territories, nuclear architecture and gene regulation in mammalian cells, *Nature Reviews Genetics* 2, 292–301.

Cremer, T., Cremer, C. (2003) Chromatin in the cell nucleus: Higher order organization, in: *Nature Encyclopedia of the Human Genome,* Nature Publishing Group, London. DOI: 10.1038/npg.els.0005768.

Cremer, T., Cremer, C. (2006a) Rise, fall and resurrection of chromosome territories: A historical perspective. Part I. The rise of chromosome territories, *European Journal of Histochemistry* 50, 161–176.

Cremer, T., Cremer, C. (2006b) Rise, fall and resurrection of chromosome territories: A historical perspective. Part II. Fall and resurrection of chromosome territories during the 1950s to 1980s. Part III. Chromosome territories and the functional nuclear architecture: Experiments and models from the 1990s to the present, *European Journal of Histochemistry* 50, 223–272.

Cremer, C., Zorn, C., Cremer, T. (1974) An ultraviolet laser microbeam for 257 nm, *Microscopica Acta* 75, 331–337.

Cremer, C., Hausmann, M., Bradl, J., Rinke, B. (1996) Method and device for measuring distances between object structures, German Patent Application No. 196.54.824.1/DE, filed December 23, 1996, European Patent EP 1997953660, 08.04.1999, Japanese Patent JP 1998528237, 23.06.1999, United States Patent 09331644, 25.08.1999.

Cremer, C., Edelmann, P., Bornfleth, H., Kreth, G., Muench, H., Luz, H., Hausmann, M. (1999) Principles of spectral precision distance confocal microscopy for the analysis of molecular nuclear structure, in: *Handbook of Computer Vision and Applications* (B. Jähne, H. Haußecker, P. Geißler, Eds.), Vol. 3, Academic Press, San Diego, CA, pp. 839–857.

Cremer, C., Failla, A.V., Albrecht, B. (2001) Far field light microscopical method, system and computer program product for analysing at least one object having a subwavelength size, US Patent 7,298,461 B2 filed October 9, 2001.

Davidovits, P., Egger, M.D. (1972) Scanning optical microscope, US Patent 3643015, filed February 15, 1972.

Dorn, R., Quabis, S., Leuchs, G. (2003) Sharper focus for a radially polarized light beam, *Physical Review Letters* 91, 233901.

Edelmann, P., Esa, A., Hausmann M., Cremer, C. (1999) Confocal laser scanning microscopy: In situ determination of the confocal point-spread function and the chromatic shifts in intact cell nuclei, *Optik* 110, 194–198.

Egner, A., Jakobs, S., Hell, S.W. (2002) Fast 100-nm resolution three-dimensional microscope reveals structural plasticity of mitochondria in live yeast, *Proceedings of the National Academy of Sciences of the United States of America* 99, 3370–3375.

Egner, A., Geisler, C., von Middendorff, C., Bock, H., Wenzel, D., Medda, R., Andresen, M. et al. (2007) Fluorescence nanoscopy in whole cells by asynchronous localization of photoswitching emitters, *Biophysical Journal* 93, 3285–3290.

Esa, A., Edelmann, P., Trakthenbrot L., Amariglio N., Rechavi G., Hausmann M., Cremer C. (2000) 3D-spectral precision distance microscopy (SPDM) of chromatin nanostructures after triple-colour labeling: A study of the BCR region on chromosome 22 and the Philadelphia chromosome, *Journal of Microscopy* 199, 96–105.

Esa, A., Coleman, A.E., Edelmann, P., Silva, S., Cremer, C., Janz, S. (2001) Conformational differences in the 3D-nanostructure of the immunoglobulin heavy-chain locus, a hotspot of chromosomal translocations in B lymphocytes, *Cancer Genetics and Cytogenetics* 127, 168–173.

Failla, A., Cavallo, A., Cremer, C. (2002a) Subwavelength size determination by spatially modulated illumination virtual microscopy, *Applied Optics* 41, 6651–6659.

Failla, A.V., Cavallo, A., Cremer, C. (2002b) Subwavelength size determination by spatially modulated illumination virtual microscopy, *Applied Optics* 41(31), 6651–6659.

Frohn, J., Knapp, H., Stemmer, A. (2000) True optical resolution beyond the Rayleigh limit achieved by standing wave illumination, *Proceedings of the National Academy of Sciences of United States of America* 97, 7232–7236.

Geisler, C., Schönle, A., von Middendorff, C., Bock, H., Eggeling, C., Egner, A., Hell, S.W. (2007) Resolution of $\lambda/10$ in fluorescence microscopy using fast single molecule photo-switching, *Applied Physics A* 88, 223–226.

Goethe, J.W. (1808) Entwurf einer Farbenlehre—Einleitung, in: *Zur Farbenlehre*, Vol. 1, Cotta Publishing House, Tuebingen, Germany.

Gunkel, M., Erdel F., Rippe, K., Lemmer, P., Kaufmann, R., Hörmann, C., Amberger, R. et al. (2009) Dual color localization microscopy of cellular nanostructures, *Biotechnology Journal* 4, 927–938.

Gustafsson, M. (2005) Nonlinear structured-illumination microscopy: Wide-field fluorescence imaging with theoretically unlimited resolution, *Proceedings of the National Academy of Sciences of United States of America* 102, 13081–13086.

Gustafsson, M., Shao, L., Carlton, P.M., Wang, C.J.R., Golubovskaya, I.N., Cande, W.Z., Agard, D.A. et al. (2008) Three-dimensional resolution doubling in wide-field fluorescence microscopy by structured illumination, *Biophysical Journal* 94, 4957–4970.

Hänninen, P.E., Hell, S.W., Salo, J., Soini, E., Cremer, C. (1995) Two-photon excitation 4Pi confocal microscope: Enhanced axial resolution microscope for biological research, *Applied Physics Letters* 66, 1698–1700.

Heilemann, M., Herten, D.P., Heintzmann, R., Cremer, C., Müller, C., Tinnefeld, P., Weston, K.D. et al. (2002) High-resolution colocalization of single dye molecules by fluorescence lifetime imaging microscopy, *Analytical Chemistry* 74, 3511–3517.

Heilemann, M., van de Linde, S., Schüttpelz, M., Kasper, R., Seefeldt, B., Mukherjee, A., Tinnefeld, P. et al. (2008) Subdiffraction-resolution fluorescence imaging with conventional fluorescent probes, *Angewandte Chemie* 47, 6172–6176.

Heintzmann, R., Cremer, C. (1999) Lateral modulated excitation microscopy: Improvement of resolution by using a diffraction grating, *Proceedings of SPIE* 356, 185–196.

Hell, S.W. (1990) Double confocal microscope, European Patent 0491289.

Hell, S.W. (2003) Toward fluorescence nanoscopy, *Nature Biotechnology* 2, 134–1355.

Hell, S.W. (2007) Far-field optical nanoscopy, *Science* 316, 1153–1158.

Hell, S.W. (2009) Microscopy and its focal switch, *Nature Methods* 6, 24–32.

Hell, S., Stelzer, E.H.K. (1992) Properties of a 4Pi confocal fluorescence microscope, *Journal of the Optical Society of America A* 9, 2159–2166.

Hell, S.W., Wichmann, J. (1994) Breaking the diffraction resolution limit by stimulated emission: Stimulated-emission-depletion fluorescence microscopy, *Optics Letters* 19, 780–782.

Hell, S.W., Lindek, S., Cremer, C., Stelzer, E.H.K. (1994) Measurement of the 4Pi-confocal point spread function proves 75 nm axial resolution, *Applied Physics Letters* 64, 1335–1337.

Hendrix, J., Flors, C., Dedecker, P., Hofkens, J., Engelborghs, Y. (2008) Dark states in monomeric red fluorescent proteins studied by fluorescence correlation and single molecule spectroscopy, *Biophysical Journal* 94, 4103–4113.

Hess, S., Girirajan, T., Mason, M. (2006) Ultra-high resolution imaging by fluorescence photoactivation localization microscopy, *Biophysical Journal* 91, 4258–4272.

Hess, S.T., Gould, T.J., Gudheti, M.V., Maas, S.A., Mills, K.D., Zimmerberg, J. (2007) Dynamic clustered distribution of hemagglutinin resolved at 40 nm in living cell membranes discriminates between raft theories, *Proceedings of the National Academy of Sciences of United States of America* 104, 17370–17375.

Hildenbrand, G., Rapp, A., Spori, U., Wagner, C., Cremer, C., Hausmann, M. (2005) Nano-sizing of specific gene domains in intact human cell nuclei by spatially modulated illumination light microscopy, *Biophysical Journal* 88, 4312–4318.

Hüve, J., Wesselmann, R., Kahms, M., Peters, R. (2008) 4Pi microscopy of the nuclear pore complex, *Biophysical Journal* 95, 877–885.

Kaufmann, R., Lemmer, P., Gunkel, M., Weiland, Y., Müller, P., Hausmann, M., Baddeley, D. et al. (2009) SPDM—single molecule superresolution of cellular nanostructures, *Proceedings of SPIE* 7185: 71850J-1–71850J-19.

Lacoste, T.D., Michalet, X., Pinaud, F., Chemla, D.S., Alivisatos, A.P., Weiss, S. (2000) Ultrahigh-resolution multicolor colocalization of single fluorescent probes, *Proceedings of the National Academy of Sciences of United States of America* 97, 9461–9466.

Lemmer, P., Gunkel, M., Baddeley, D., Kaufmann, R., Urich, A., Weiland, Y., Reymann, J. et al. (2008) SPDM: Light microscopy with single-molecule resolution at the nanoscale, *Applied Physics B* 93, 1–12.

Lewis, A., Isaacson, M., Harootunian, A., Muray, A. (1984) Development of a 500 Å spatial resolution light microscope. I. Light is efficiently transmitted through l/16 diameter apertures, *Ultramicroscopy* 13, 227–231.

Lidke, K.A., Rieger, B., Jovin, T.M., Heintzmann, R. (2005) Superresolution by localization of quantum dots using blinking statistics, *Optics Express* 13, 7052–7062.

Martin, S., Failla, A.V., Spoeri, U., Cremer, C., Pombo, A. (2004) Measuring the size of biological nanostructures with spatially modulated illumination microscopy, *Molecular Biology of the Cell* 15, 2449–2455.

Mathee, H., Baddeley, D., Wotzlaw, C., Cremer, C., Birk, U. (2007) Spatially modulated illumination microscopy using one objective lens, *Optical Engineering* 46, 083603/1–083603/8.

Minsky, M., Microscopy Apparatus, US Patent 3,013,467, filed November 7, 1957.

Misteli, T. (2007) Beyond the sequence: Cellular organization of genome function, *Cell* 128, 787–800.

Nägerl, U.V., Willig, K.I., Hein, B., Hell, S.W., Bonhoeffer, T. (2008) Live-cell imaging of dendritic spines by STED microscopy, *Proceedings of the National Academy of Sciences United States of America* 105, 18982–18987.

Naora, H. (1951) Microspectrophotometry and cytochemical analysis of nucleic acids, *Science* 114, 279–280.

Patterson, G.H., Lippincott-Schwartz, J. (2002) A photoactivatable GFP for selective photolabeling of proteins and cells, *Science* 297, 1873–1877.

Pawley, J.B. (2006) *Handbook of Confocal Microscopy*, 2nd edition. Plenum Press, New York.

Petran, M., Hadravsky, M., Egger, D., Galambos, R. (1968) Tandem-scanning reflected-light microscope, *Journal of the Optical Society of America* 58, 661–664.

Pohl, D.W., Denk, W., Lanz, M. (1984) Optical stethoscopy: Image recording with resolution, *Applied Physics Letters* 44, 651–653.

Rauch, J., Hausmann, M., Solovei, I., Horsthemke, B., Cremer, T., Cremer, C. (2000) Measurement of local chromatin compaction by spectral precision distance microscopy, *Proceedings of SPIE* 4164, 1–9.

Rayleigh, L. (1896) On the theory of optical images, with special reference to the microscope, *Philosophical Magazine* 42, 167–195.

Reymann, J., Baddeley, D., Gunkel, M., Lemmer, P., Stadter, W., Jegou, T., Rippe, K. et al. (2008) High-precision structural analysis of subnuclear complexes in fixed and live cells via spatially modulated illumination (SMI) microscopy, *Chromosome Research* 16, 367–382.

Rouquette, J., Cremer, C., Cremer, T., Fakan, S. (2009) Functional nuclear architecture studied by microscopy: State of research and perspectives, to be submitted.

Rust, M., Bates, M., Zhuang, X. (2006) Sub-diffraction-limit imaging by stochastic optical reconstruction microscopy (STORM), *Nature Methods* 3, 793–795.

Schermelleh, L., Carlton, P.M., Haase, S., Shao, L., Winoto, L., Kner, P., Burke, B. et al. (2008) Subdiffraction multicolor imaging of the nuclear periphery with 3d structured illumination microscopy, *Science* 320, 1332–1336.

Schmidt, M., Nagorni, M., Hell, S.W. (2000) Subresolution axial distance measurements in far-field fluorescence microscopy with precision of 1 nanometer, *Review of Scientific Instruments* 71, 2742–2745.

Schmidt R., Wurm, C.A., Jakobs, S., Engelhardt, J., Egner, A., Hell, S.W. (2008) Spherical nanosized focal spot unravels the interior of cells, *Nature Methods* 5, 539–544.

Schneider, M., Barozzi, S., Testa, I., Faretta, M., Diaspro, A. (2005) Two-photon activation and excitation properties of PA-GFP in the 720–920 nm region, *Biophysical Journal* 89, 1346–1352.

Sheppard, C.J.R., Wilson, T. (1978) Image formation in scanning microscopes with partially coherent source and detector, *Optica Acta* 25, 315–325.

Shroff, H., Galbraith, C.G., Galbraith, J.A., Betzig, E. (2008) Live-cell photoactivated localization microscopy of nanoscale adhesion dynamics, *Nature Methods* 5, 417–423.

Silberberg, Y. (2001), Physics at the attosecond frontier, *Nature* 414, 494–495.

Sinnecker, D., Voigt, P., Hellwig, N., Schaefer, M. (2005) Reversible photobleaching of enhanced green fluorescent proteins, *Biochemistry* 44(18), 7085–7094.

Solovei, I., Kreysing, M., Lanctot, C., Sueleyman, K., Peichl, L., Cremer, T., Guck, J. et al. (2009) Nuclear architecture of rod photoreceptor cells adapts to vision in mammalian evolution, *Cell* 137, 356–368.

Steinhauer, C., Carsten Forthmann, C., Vogelsang, J., Tinnefeld, P. (2008) Superresolution microscopy on the basis of engineered dark states, *Journal of the American Chemical Society* 130, 16840–16841.

Stelzer, E.H., Wacker, I., De Mey, J.R. (1991) Confocal fluorescence microscopy in modern cell biology, *Seminars in Cell Biology* 2, 145–152.

Van Aert, S., Van Dyck, D., den Dekker, A.J. (2006) Resolution of coherent and incoherent imaging systems reconsidered—Classical criteria and a statistical alternative, *Optics Express* 14, 3830–3839.

van Oijen, A.M., Köhler, J., Schmidt, J., Müller, M., Brakenhoff, G.J. (1998) 3-Dimensional super-resolution by spectrally selective imaging, *Chemical Physics Letters* 192(1–2), 182–187.

Weiss, S. (1999) Fluorescence spectroscopy of single biomolecules, *Science* 283, 1676–1683.

Westphal, V., Rizzoli, S.O, Lauterbach, M.A., Kamin, D., Jahn, R., Hell, S.W. (2008) Video-rate far-field optical nanoscopy dissects synaptic vesicle movement, *Science* 320, 246–249.

4

Fluorescence Microscopy with Extended Depth of Field

	4.1	Introduction 4-1
	4.2	Software-Based EDF 4-2
	4.3	Optical Methods 4-3
		Extended Focus PSFs • Excitation-Based Systems • Detection-Based Methods
	4.4	Applications and Future Perspectives 4-14
	References 4-15

Kai Wicker
Rainer Heintzmann

King's College London

4.1 Introduction

Extended depth of field (EDF) techniques yield an image of a thick specimen at a high lateral resolution, in which the whole sample is in focus. Quite often such an extended focus projection of a sample is all that is needed in order to answer questions of biomedical nature. For example, when tracking vesicles performing exocytosis and endocytosis within a cell, it may be sufficient to follow their motion in two dimensions (2D). In an ordinary microscope, this can prove difficult when these vesicles leave the focal plane and become blurred or invisible.

A straightforward way of getting extended focus images is to look at sum projections through a stack of conventional (i.e., non-EDF) microscope images taken at different focal positions. Apart from sum projections, there are other, more sophisticated algorithms for rendering EDF images from 3D data.

While EDF projections are a convenient way of displaying 3D data, being able to directly get such images without the need to first acquire 3D data yields a significant advantage in speed and light exposure, and therefore reduces photo bleaching and toxicity. When two slightly tilted (or sheared) extended focus images are recorded, such data can also serve to rapidly produce pairs of stereo images possessing 3D quality, albeit not being truly volumetric (Hayward et al. 2005). If more images are taken under a variety of tilt angles, the extended focus data can be used for optical projection tomography (Sharpe et al. 2002), resulting in volumetric images with high resolutions in all three spatial dimensions.

Optical methods for acquiring extended focus images are thus very useful. Although there are interesting ways to generate extended focus images with various other microscopy modes (such as bright-field transmission, phase contrast, polarization, dark-field or differential interference contrast), we restrict ourselves in this chapter to the imaging of fluorescently tagged samples.

4.2 Software-Based EDF

Probably the most common way to obtain extended focus images is by acquisition of a 3D data set and successive projection of the data. The projection is usually performed along the axial direction.

There are various ways of performing such projections, which differ in the details. For example, the maximum projection (Figure 4.1a and d), probably the most commonly used projection in microscopy, simply selects the maximum voxel value from each line of projection (e.g., along *z*). It can be seen that a maximum projection yields a good contrast and nicely delineated structures, but in the top left corner of the image it can also be observed (see encircled regions in Figure 4.1d and e), that a maximum projection (as opposed to the sum projection, Figure 4.1b and e) runs the risk of having the dim structures completely masked (and thus made invisible) by brighter structures (or even out-of-focus haze in the widefield case) above or below them. A sum or average projection avoids this problem, but does not eliminate haze, leading to sometimes unacceptably low contrast, especially in the widefield case. The maximum projection in the wavelet domain (Forster et al. 2004, more details can be found on the Web site of the Biomedical Imaging Group (2008) at the Ecole Polytechnique Fédérale de Lausanne, http://bigwww.epfl.ch/demo/edf/) avoids both these problems by keeping the most relevant information of several resolution levels. In the widefield case, this yields particularly nice results (Figure 4.1c). However, in the case of data with a lesser amount of blurring (Figure 4.1f), the loss of information and noise amplification can become a problem and the maximum projection seems to yield better results.

Regarding the sum projection, it is worth noting that this is the method that is most closely related to the optical EDF methods. It can also be performed directly by integrating a widefield or a spinning-disk (Petrán et al. 1968) image on the camera while moving the object through the projection path. This

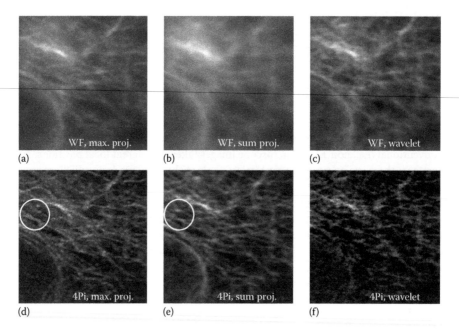

(a)   (b)   (c)

(d)   (e)   (f)

FIGURE 4.1 Extended focus projections using various image processing methods. In the top row (a–c) a simulated widefield image (4Pi data set convolved with a widefield point-spread function) is shown and in the bottom row (d–f) the original 4Pi data was projected. Maximum (a, d) and sum (b, e) projections are shown along with a maximum projection in the wavelet domain (c, f) (Complex Daubechies wavelets at 8 scales with filter length 6 and majority check) (According to Forster, B. et al., *Microsc. Res. Techniq.*, 65, 33, 2004.). The image widths correspond to 11.7 μm.

has been made use of in Hayward's method (Hayward et al. 2005) of stereo projection (see the end of this chapter).

4.3 Optical Methods

Fluorescent light is incoherent, so there is no phase relation between the emitted light and the excitation light. This allows us to treat excitation and emission independently, as the final point-spread function (PSF) is given by the product of the individual PSFs.

There are thus two general approaches to EDF: an excitation-based approach and a detection-based approach. The excitation-based systems are usually scanning systems where the sample is illuminated with an intensity distribution—the illumination point-spread function—which is largely z-independent but still possesses fine enough detail to allow a good lateral resolution. This distribution is then laterally scanned across the sample and all the fluorescence light is detected. Detection-based systems can be either scanning or widefield systems. In either case, the optical systems are designed to have detection PSFs that are largely independent of z.

Some excitation-based and detection-based EDF methods can be combined in order to get an improved performance, e.g., an improved lateral resolution or a reduced haze or background in the image.

4.3.1 Extended Focus PSFs

All methods have one feature in common—they rely on z-independent PSFs. How can such a PSF be obtained?

A distribution of light around the image plane, i.e., the illumination PSF, can be thought of as a superposition of plane waves propagating in different directions. Each plane wave corresponds to a point in the pupil plane: a spherical wave originating from that point is transformed into a plane wave by an objective lens, the plane wave's angle of propagation depending on the point's distance to the optical axis. The resulting interference pattern of all these waves is an inhomogeneous and usually z-dependent distribution of light.

Figure 4.2 exemplifies this mechanism for the situation of two such plane waves, i.e., two points in the pupil plane. In Figure 4.2a, these two points have the same distance to the optical axis, and the two plane waves thus have the same angle to the optical axis. For symmetry reasons, it is obvious that the resulting

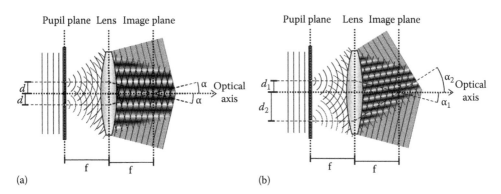

FIGURE 4.2 Light originating from points in the pupil plane will generate plane waves around the image plane. For more than one pupil plane point the corresponding plane waves will interfere in the sample space to generate an intensity distribution, or illumination point-spread function. When the points in the pupil plane all have the same distance d to the optical axis (a) the resulting interference pattern will be oriented parallel to the optical axis and thus be z-independent. When points have different distances d_1, d_2 from the optical axis (b), the interference will no longer be parallel to the optical axis and the z-independence will be destroyed.

interference pattern must be oriented parallel to the optical axis (horizontal fringes) and thus be largely independent of z. In Figure 4.2b, the two points in the pupil plane are located at different distances to the optical axis, resulting in different propagation angles of the plane waves. The resulting interference pattern is tilted at an angle to the optical axis and is therefore no longer independent of z.

As long as they have the same distance to the optical axis, any number of points may be added in the pupil plane without affecting the z-behavior of the PSF. Thus z-independent PSFs could be generated by placing an annular or ring aperture in the pupil plane. The PSF generated by an annular aperture is called a Bessel beam, owing to the fact that its lateral intensity distribution can be expressed as the square of a zeroth-order Bessel function of the first kind.

As such an annular aperture cannot be infinitely narrow, there will be a small range of distances to the optical axis, which will introduce some very coarse z-dependence. But as long as the aperture is kept reasonably narrow, the PSF will be sufficiently z-independent over a large region of interest.

In a more mathematical picture one can describe all plane waves through wave vectors in Fourier space. The length of these k-vectors is $2\pi n/\lambda$, where λ is the wavelength of the light and n is the refractive index of the medium. The direction of the vector in Fourier space corresponds to the waves' directions of propagation in real space. Therefore, the k-vectors of all waves of the same wavelength λ will lie on a spherical shell of radius $2\pi n/\lambda$. This sphere is known as the Ewald sphere (Ewald 1913). McCutchen (1964) showed that any field distribution can be expressed through a generalized aperture, which can be thought of as a spherical aperture that is applied to the Ewald sphere and that defines the strengths and phases of the individual waves' contributions. The final field distribution is the inverse Fourier transform of this generalized aperture function. If a field distribution is generated by illumination from only one side of the sample, at most one half of the Ewald sphere will be available (e.g., $k_z > 0$). The generalized aperture corresponding to an annular aperture can be found by projecting this annular aperture onto that half of the Ewald sphere. The range of spatial frequencies available along the axial direction, Δk_z, defines the axial resolution. A large range of frequencies leads to a high axial resolution, whereas a very small range will lead to an extended depth of field.

Figure 4.3 shows some apertures and the resulting PSFs. While the open aperture with a large range of axial frequencies (Figure 4.3g) results in the conventional widefield PSF, the PSFs of annular apertures (Figure 4.3a–d) have significantly extended depths of field.

For low numerical aperture (NA) widefield systems, the lateral resolution scales approximately linearly with the NA, while the axial resolution scales quadratically. Therefore, low NA systems have a much better lateral than axial resolution, meaning that these systems can also be used for EDF imaging. Note that the small circular aperture (Figure 4.3e) also results in a PSF with an extended depth of field, as it has the same support of the axial frequencies as the annular apertures (Figure 4.3a–d). However, such a low NA will not allow a very good lateral resolution, as only low lateral spatial frequencies are transmitted by the system.

4.3.2 Excitation-Based Systems

4.3.2.1 Bessel Beam Excitation

Because the annular apertures used for the generation of Bessel beams have to be narrow, they are not light efficient. Other ways of generating annular illumination in the pupil plane may be preferable. One possible way is the use of conically shaped lenses, called axicons, which will turn a coherent plane wave illumination into a Bessel beam. When used in combination with a microscope objective, the output of the axicon can be, by means of a single lens, optically Fourier transformed to generate the desired annular illumination in the objective's pupil plane.

Bessel beams are nondiffracting (Durnin 1987; Durnin and Miceli 1987). Their z-independence would make them ideal candidates for excitation-based EDF microscopy. Unfortunately, and in spite of their good FWHM resolution, the Bessel beams' decay in the lateral direction is slow so that at large distances from the PSF center they still have high-intensity tails. When the emission is proportional

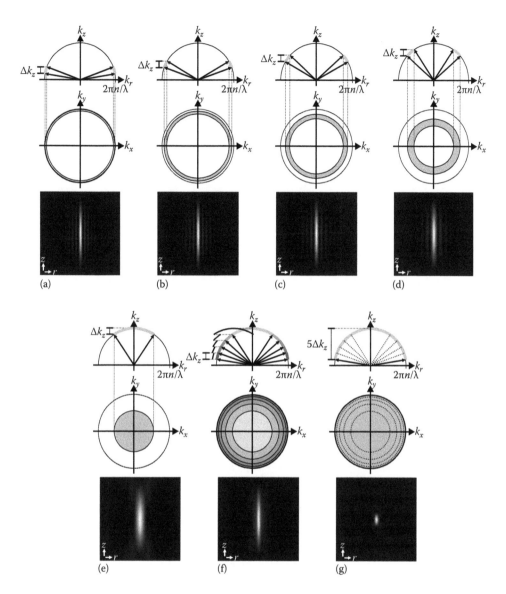

FIGURE 4.3 Different pupil plane apertures (the middle part of each subfigure), their resulting frequency support on the Ewald sphere (the top part) and the resulting point-spread functions (the bottom part). Annular apertures (a–d) as well as low NA circular apertures (e) only transmit those frequency components on the Ewald sphere which lie within a narrow range of axial spatial frequencies, Δk_z. This leads to an extended depth of field compared to the widefield case (g). The widefield case is equivalent to coherently adding the individual apertures (a–e) thus making a larger range of axial frequencies available. Adding the annular apertures incoherently (f) is equivalent to adding the corresponding point-spread functions incoherently. (Modified from Abrahamsson, S. et al., *Proc. SPIE*, 6090, 60900N, 2006.) The resulting point-spread function will also have an extended depth of field while remaining a high lateral resolution.

to the excitation intensity, this method on its own will therefore result in rather blurry images of poor contrast (Figure 4.4b).

4.3.2.1.1 Two-Photon Bessel Beam Excitation

Blurry images of poor contrast can be partially remedied by exploiting any nonlinear relationship between the excitation and the emission intensity. For two-photon excitation, the emission intensity is

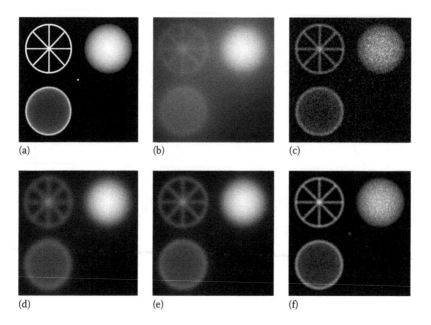

FIGURE 4.4 Extended depth of field simulations showing various excitation and detection schemes. (a) Shows the synthetic sample comprising a solid sphere, a hollow sphere, a "wagon wheel noodle" and a point emitter. The image width corresponds to 6 μm. Integrated detection for one-photon Bessel beam excitation (b) results in hazy images, where a lot of detail is lost. Combining this type of excitation with image-inverting interferometric detection (From Wicker, K. and Heintzmann, R., *Opt. Express*, 15, 12206, 2007. With permission.) improves the resolution and removes much of the haze (c). The haze can also be removed by two-photon Bessel beam excitation (Botcherby et al. 2006), however, at the cost of a lower resolution (d). A combination with the image-inverting interferometric detection (e, f) can recover some of this resolution: using the constructive interferometer channel alone (e) already yields some improvement in resolution, but taking the difference of the constructive and the destructive output (f) further enhances the resolution while also improving the contrast. The noninterferometric images (b, d) were simulated assuming a maximum photon number of 15,000 per pixel. The interferometric images (c, e, f) assumed the same intensity before the interferometer, meaning that the sum of the constructive and the destructive channels would also have a maximum photon number of 15,000 per pixel.

proportional to the square of the local illumination intensity. Using the Bessel beam excitation in a two-photon mode will therefore significantly reduce the unwanted tails in the emission. Another advantage is the deeper penetration depth that can usually be achieved with two-photon excitation, if the two-photon excitation wavelength is in the infrared. This approach has been pursued by Botcherby et al. (2006). They generate the Bessel beam in yet another way: through a circularly symmetrical binary phase grating, parallel light is diffracted into a central zeroth and circular higher orders (see setup Figure 4.5). While all other orders are blocked, the first circular order is focused on a narrow ring in a pupil plane. There, an annular aperture mask further truncates the field. The resulting annular pupil plane illumination yields a Bessel beam like PSF with a greatly extended depth of field. Because of this method's high efficiency of Bessel beam generation, it is possible to achieve the high illumination intensities necessary for two-photon excitation.

4.3.2.1.2 Multiplex Bessel Beam Excitation

Using spatial light modulators (SLM) for the holographic generation of Bessel beams allows higher flexibility in choosing beam parameters and can also facilitate beam scanning. Furthermore, they can be used to easily generate an array of several nondiffracting beams (Tao et al. 2005), which could be used

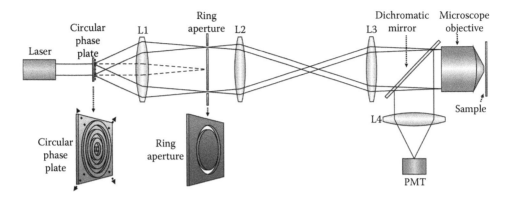

FIGURE 4.5 Schematic representation of the setup used to realize two-photon Bessel beam excitation (Botcherby et al. 2006). A circular binary phase grating diffracts incoming light into one central and circular higher orders. The first order is focused (L1) into a narrow ring which is further truncated by a ring aperture in a secondary pupil plane. Relay lenses (L2, L3) project this annular illumination into the pupil plane of the microscope objective, resulting in a Bessel beam like sample illumination. Because of the high efficiency of this method, light intensities in the sample are high enough for two-photon excitation. (Redrawn from Botcherby, E.J. et al., *Opt. Comm.*, 268, 253, 2006.)

to multiplex and speed up the EDF imaging, similar to spinning-disk confocal systems (Petrán 1968) or the PAM microscope (Heintzmann et al. 2001) in confocal microscopy. However, to do this the detection needs to be able to discriminate between emissions corresponding to each beamlet.

4.3.2.2 Accelerating Airy Beams for Excitation

Bessel beams are not the only class of beams suitable for EDF imaging. Although not truly z-independent, Airy beams (not to be confused with the Airy pattern in PSFs) do fulfill some of the requirements for EDF imaging. Although we are not aware of attempts of the use of the Airy beam excitation for EDF microscopy, we feel that this approach is promising, especially when combined with a wave-front coding EDF detection, which will be described in more detail further on.

While largely retaining its general shape and not becoming blurred, the Airy beam's lateral intensity distribution (Figure 4.7) is shifted during its propagating along z. In fact, its maximum undergoes a constant acceleration so that the Airy beam's trajectory is parabolic or "banana-shaped." But over limited distances, this trajectory is approximately linear and can be used for excitation-based EDF imaging.

For larger distances this beam curvature cannot be neglected. Images taken using this method will therefore not yield linear EDF projections of a sample, but rather images that correspond to a parabolically skewed sample. From an application point of view, this skewing of the sample can be tolerated in most cases. The beam curvature can be seen in Figure 4.6, which shows the propagation of a 1D Airy wave packet in generalized coordinates. While a true Airy wave packet retains its shape it cannot be produced with finite energy and therefore is not physically possible to realize (Figure 4.6a). A finite energy approximation (Figure 4.6b) can be realized but is not truly diffraction-less.

Theoretically predicted by Berry and Balazs (1979) as a nonspreading solution to the Schrödinger equation, Airy beams were demonstrated by Siviloglou et al. (2007). (It should be mentioned however, that Airy beam like PSFs have been used for detection-based EDF imaging for several years, see Section 4.3.3.2 on wave-front coding.) These beams were produced by means of a cubic-phase plate in the pupil plane (Figure 4.8). Such a cubic-phase plate can be any optical element that introduces phase shifts of $\varphi = \alpha(k_x^3 + k_y^3)$ to the wave in the pupil plane, where k_x, k_y represent the pupil plane coordinates and α determines the magnitude of the phase shift. Siviloglou et al. accomplished this by means

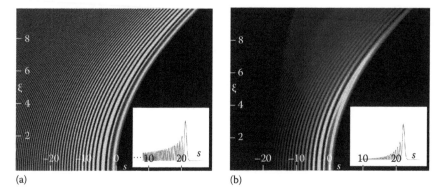

(a)   (b)

FIGURE 4.6 (See color insert following page 15-30.)   Propagation of 1D Airy wave packets in generalized coordinates ξ and s. While true Airy wave packets are diffractionless, this theoretical solution requires infinite energy and therefore cannot be realized physically (a). Finite energy approximations can be realized but are no longer truly diffractionless (b). In both cases, the maximum of the distribution undergoes constant acceleration, resulting in parabolic trajectories. The trajectories can be extended to negative values of ξ by mirroring with respect to the s-axis. (Reprinted figure with permission from Siviloglou, G.A. et al., *Phys. Rev. Lett.*, 99, 213901, 2007. Copyright 2007 by the American Physical Society.)

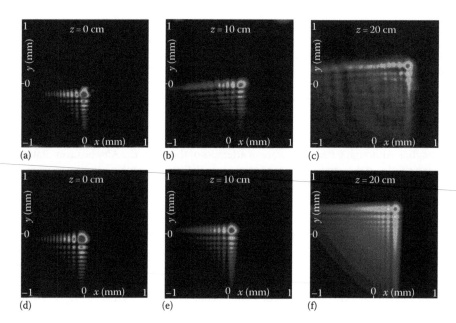

FIGURE 4.7   Experimentally measured (a–c) and theoretically predicted Airy beam (d–f) for different focal positions z. The intensity distribution remains largely unchanged over a long propagation distance. However, the resulting illumination point-spread function is not truly z-independent, as the maximum of the distribution is shifted in dependence on z. (Reprinted figure with permission from Siviloglou, G.A. et al., *Phys. Rev. Lett.*, 99, 213901, 2007. Copyright 2007 by the American Physical Society.)

of a spatial light modulator that enabled a phase shift from -20π to 20π modulo 2π ($\alpha = 10\pi$ for k_x, k_y ranging from -1 to 1). Figure 4.7a–c show the measured wave packet over a propagation distance of 20 cm, and Figure 4.7d–f the predicted theoretical values. These images nicely show how the shape of the distribution remains largely unchanged over a large distance, while the position of the maximum is shifted laterally.

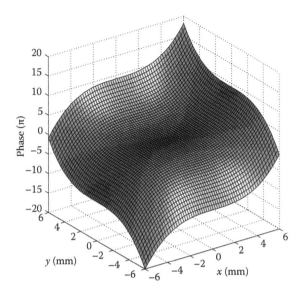

FIGURE 4.8 Representation of a cubic-phase plate used in an aperture plane of an objective for the generation of Airy beams and for wave-front coding techniques. Such cubic-phase plates introduce phase shifts to the incident light, the magnitude of which depends on the pupil plane coordinates k_x and k_y.

As well as not being truly z-independent, the resulting illumination PSF (Figure 4.7a) also is not very suitable for direct microscopy imaging (integrating detection). Its unsymmetrical, patterned shape and slow decay in two directions lead to blurred images similar to that of Figure 4.11b, which shows a related detection-based method with a similar PSF.

However, as will be shown in more detail in Section 4.3.3.2 on wave-front coding, such raw images are well suited for deblurring by 2D deconvolution.

4.3.3 Detection-Based Methods

Just as for the excitation-based methods, the straightforward approach to EDF detection would be through the use of annular apertures. They allow widefield EDF imaging with Bessel-like PSFs. However, the annular apertures' limitations also apply to detection: most of the light is discarded by the narrow annular aperture, making this method usually unacceptably inefficient for fluorescence microscopy. Furthermore, the slow decay of the Bessel function causes the detected images to become blurry. Unlike for the excitation-based methods, this cannot be remedied by tricks as two-photon excitation.

4.3.3.1 Sum of Annular Apertures

Abrahamsson et al. (2006) presented an elegant method that combines the extended depth of field of annular apertures (Figure 4.3a–e) with the light efficiency of using the full numerical aperture, not discarding any light. While coherently adding several different size annular apertures will only yield a larger aperture and thus a PSF where the extended depth of field of the individual rings' PSFs is lost (Figure 4.3g), being able to sum them incoherently would prevent interference between the individual PSFs and thus conserve their EDF properties as well as their individual lateral resolution capabilities (Figure 4.3f). This would allow an efficient EDF detection with a good lateral resolution.

To achieve this, Abrahamsson et al. invented an optical element consisting of a stack of transparent disks of decreasing radii. This circularly symmetric "stair step device" (Figure 4.9) introduces phase delays to the incident light, with a different phase delay corresponding to each annular step or ring. If the phase delay between any two steps is longer than the coherence length of the fluorescence light, the

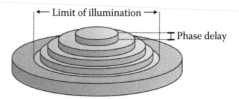

FIGURE 4.9   Placing a "stair step device" as presented in Abrahamsson et al. 2006. In the pupil plane will introduce phase delays to the incident light. If these phase delays are longer than the coherence length of the fluorescence light, light transmitted through the different rings can no longer interfere. The resulting point-spread function will be an incoherent sum of the individual point-spread functions corresponding to different annular apertures and will have an extended depth of field while retaining a high lateral resolution as well as good light efficiency.

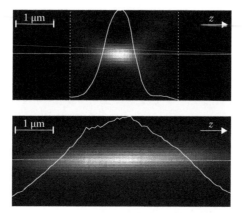

FIGURE 4.10   The point-spread function measured for the setup employing a "stair step device" (b) has a significantly extended depth of field compared to a conventional widefield point-spread function (a). (We would like to thank S. Abrahamsson for providing the data for this image.)

individual rings can no longer interfere and can thus be treated individually. The resulting PSF will be an incoherent sum of the individual PSFs corresponding to the different rings. Abrahamsson et al. used a device with five steps, yielding a measured PSF with an approximately five times longer depth of field than the ordinary widefield PSF (Figure 4.10).

    This technique is a widefield technique and does not require scanning. Nevertheless, it could also be used in combination with a scanning EDF excitation and a pinhole to further improve the lateral resolution.

4.3.3.2 EDF Using Wave-Front Coding

Another widefield method employs cubic-phase plates to generate EDF images (Dowski and Cathey 1995, Bradburn et al. 1997, Marks et al. 1999, Tucker et al. 1999). It is closely related to the excitation-based method of accelerating Airy beams.

    In ordinary widefield imaging, the lateral size of the PSF increases as it blurs with defocus. At the same time, the lateral intensity OTF, which for the in-focus slice is an autocorrelation of the pupil function, is modulated in phase and amplitude with defocus. Assuming a symmetric setup, and therefore a symmetric intensity, the PSF yields a real-valued intensity OTF. This in turn inevitably leads to the existence of ring-shaped zeros where the modulation introduces sign changes within the OTF at

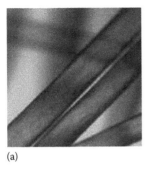 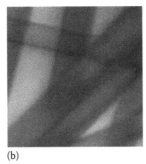 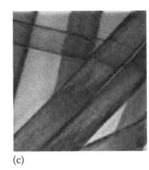

(a)   (b)   (c)

FIGURE 4.11 Whereas in a widefield image (a) out-of-focus information becomes blurred, the wave-front coded EDF image (b) has constant blurring for all focal positions. The remaining blurring is reduced by digital filtering, yielding high resolution EDF images (c). The average width of the hairs in the images is 65 μm. (Images taken from Tucker et al. *Opt. Express*, 4, 467, 1999. With kind permission of the authors.)

defocus. Because the position of the zeros changes with the degree of defocus it becomes extremely difficult to repair the recorded images. Dowski and Cathey (1995) suggested altering an optical system in such a way that the said modulation with defocus will be in phase only rather than in phase and magnitude. That way, no spatial frequencies are lost with defocus. One way this can be achieved is by placing a cubic-phase plate (Figure 4.8) in the pupil plane of the microscope. As was the case for accelerating Airy beams, where a cubic-phase plate was used to generate the said beams, the lateral PSF in wave-front coding retains its general shape over quite an extended field. However, as with Airy beams, when this distribution is "accelerated" over longitudinal position, the resulting 3D PSF is bent parabolically.

Because of the extended, patterned shape of the PSF, unprocessed images taken with the cubic-phase plate EDF method (Figure 4.11b) are quite blurred and do not show an immediate improvement over the widefield images (Figure 4.11a). However, no spatial frequency information is lost because of defocusing, allowing the computational deblurring by 2D deconvolution (Figure 4.11c).

4.3.3.3 Interferometric Image Inversion

4.3.3.3.1 Scanning Interferometric EDF

In 2007, we proposed a method for improving the resolution of scanning EDF systems (Wicker and Heintzmann 2007) by adding an interferometer (e.g., of Mach–Zehnder type) with partial image inversion to the descanned output of a point scanning microscope (Figure 4.12). Between the beam-splitter and the beam-combiner, the interferometer contains optics, which cause the two wave distributions propagating in the individual interferometer arms to be laterally inverted relative to each other. If the point source emitting the light is located on the image inversion axis, the image inversion yields two distributions which are practically identical. Interference between those identical distributions will lead to all the light being collected in the detector of the constructive output of the interferometer (e.g., a photomultiplier tube, PMT), while the destructive output remains dark. If the point source is far away from the optical axis, the inverted field distributions will have hardly any spatial overlap. This means they cannot interfere, leading to half the signal being detected in both the constructive and the destructive output. This results in a constructive-output PSF, which has 100% signal at its center and then drops quickly to 50% signal with an increasing distance from the center (Figure 4.13a). The destructive-output PSF on the other hand starts at 0% signal and then quickly rises to 50% signal. Taking the difference between the constructive and the destructive outputs yields a final detection PSF that has full signal and a FWHM resolution that is about 33% narrower than that of the ordinary widefield PSF. Although the optical-transfer function (OTF) has the same support as an ordinary widefield OTF, it transfers high spatial frequencies more efficiently, leading to the improved resolution (Figure 4.13b). Because the image

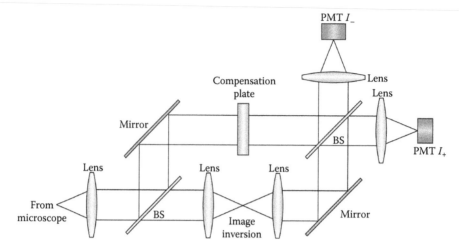

FIGURE 4.12 An interferometer with a partial image inversion can be used for efficient high resolution EDF imaging. This figure shows a Mach–Zehnder type interferometer that has image-inverting optics in one interferometer arm. The difference of signal measured in the constructive (I_+) and the destructive (I_-) channels yields an EDF point-spread function with a lateral resolution that is significantly better than that of an ordinary widefield PSF. (From Wicker, K. and Heintzmann, R., *Opt. Express*, 15, 12206, 2007. With permission.)

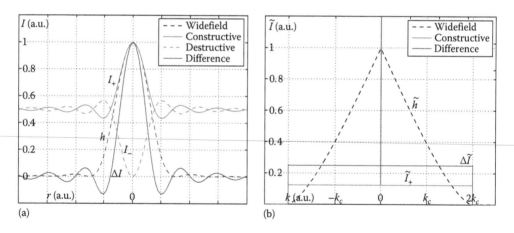

FIGURE 4.13 (See color insert following page 15-30.) (a) From a maximum signal of 100% the constructive channel's point-spread function in interferometric detection quickly approaches a constant background of 50%. Subtracting the destructive channel's signal yields a point-spread function which has a better full-width half-maximum resolution than an equivalent widefield point-spread function. (b) From the corresponding optical-transfer functions it can be seen that compared to the widefield case interferometric detection has a much higher transfer efficiency for high spatial frequencies. The lateral point-spread function and the optical-transfer function are independent of defocus, and are therefore ideally suited for extended depth of field imaging. (From Wicker, K. and Heintzmann, R., *Opt. Express*, 15, 12206, 2007. With permission.)

inversion affects only the lateral components of the field distribution, the resulting PSF will be independent of z and therefore ideally suited for EDF imaging.

In spite of the good resolution, this method alone will generally not yield good-quality images: because the constructive and destructive channels are recorded independently, the 50% background will lead to relatively high noise levels. Only emitters on or close to the scan axis will contribute to the signal, while every emitter in the sample will contribute to the background. As the noise in the two channels is independent, subtracting the two signals will only discard the background signal, but not the noise stemming from it. This means that the signal-to-noise level approaches zero even at relatively small distances from the scan axis.

This technique on its own can therefore only be used for relatively sparse samples or a very limited field of view in order to reduce the number of emitters, or when detecting a very high number of photons.

However, this method can be combined with some of the excitation-based methods mentioned above. In combination with the Bessel beam excitation it would increase the resolution of that method as well as reduce the high Bessel tails by half. The resulting images (Figure 4.4c) show much higher contrast and detail than images acquired with the Bessel excitation only (Figure 4.4b). Combining our method with the two-photon Bessel excitation further enhances the image quality, as the Bessel tails are significantly reduced and therefore hardly contribute to noise (Figure 4.4e and f).

4.3.3.3.2 Widefield Interferometric EDF

When signal-to-noise issues are of no concern (i.e., for sparse samples and high photon numbers), the scanning interferometric method can be converted into a widefield technique. The integrating PMTs in both outputs are replaced by cameras that are placed in the pupil plane of the optical system. In Fourier space, the wave distribution of a point source and its mirrored counterpart will interfere to generate a fringe pattern. The direction and spacing of these fringes depend on the lateral position of the point source but not on its axial position. As all the point sources are incoherent to each other, the recorded image will be an incoherent sum of several such fringe patterns. Fourier transforming the recorded image back into real space yields

$$I_{\pm}(\vec{r}_{xy}) = h(\vec{r})\int 2S(\vec{r}')d^3\vec{r}' \pm \left[S\otimes h\right](\vec{r}) \pm \left[S\otimes h\right](-\vec{r})\Big|_{z=0} \quad (4.1)$$

for the constructive (I_+) and the destructive (I_-) channel respectively, where $S(\vec{r})$ is the sample fluorophore concentration at $\vec{r}=(x,y,z)$, $\otimes$ denotes the convolution operator, and $h(\vec{r})$ is the system's interferometric extended focus PSF. Interestingly, this PSF turns out to be identical to the difference PSF of the scanning interferometric EDF technique, suffering similar signal-to-noise issues. Poisson noise in the actual recorded Fourier-space image translates into evenly distributed noise in the final Fourier transformed real-space image. Every emitter will contribute to that noise background, so in the case of many emitters, the signal may be drowned in noise.

Equation 4.1 does not yield useful EDF images, as in addition to the desired image (second term) it contains an unwanted bright central spot (first term, the integral over the sample accounting for the total number of emitters) and a ghost mirror image (third term). Limiting the sample to one half of the field of view (e.g., through only partial illumination) partly solves this problem, as ghost and true images will no longer overlap. However, one can also get rid of both unwanted terms completely, by taking a second set of images, while introducing a phase shift of $\pi/2$ in one of the two interferometer arms. The resulting real-space images

$$I_{2,\pm}(\vec{r}_{xy}) = h(\vec{r})\int 2S(\vec{r}')d3\vec{r}' \pm i\left[S\otimes h\right](\vec{r}) \mp i\left[S\otimes h\right](-\vec{r})\Big|_{z=0}$$

can be combined with the first two through

$$F_{\pm}(\vec{r}_{xy}) = I_+(\vec{r}_{xy}) - I_-(\vec{r}_{xy}) + iI_{2,+}(\vec{r}_{xy}) - iI_{2,-}(\vec{r}_{xy}) = 4\left[S\otimes h\right](\vec{r})\Big|_{z=0}$$

to produce a final image that has no central spot or ghost image (Wicker and Heintzmann, unpublished results).

4.3.3.4 mRSI: Rotational Shearing Interferometry Coherence Imaging

A very similar method was proposed and experimentally verified by Potuluri et al. (2001). They built a microscope combining a rotational shear interferometer (RSI, of which the image-inverting interferometer is only a special case) and coherence imaging. They also recorded images of fringe patterns generated by the RSI images of individual incoherent point sources. However, as their coherence imaging does not take place in Fourier space, they get varying image magnifications depending on the z-position in the sample. Therefore, their method is not a linear EDF projection but rather a cone projection with an extended depth of field.

4.4 Applications and Future Perspectives

Many of the above mentioned techniques can be used for a number of applications as suggested and exemplified below. Extended focus projections can be performed along different directions (by tilting the sample, or splitting the beam and sending the parts through modified optics). Using two such different projection directions forms the basis of stereoscopic imaging. If the observer's eyes are each presented with an extended focus projection from a slightly different direction, the brain is capable of producing a 3D impression. If the extended focus is generated by a computational projection from a 3D data set, the simplest way to produce the stereo pair is by applying a subpixel shear between the individual slices in positive and negative x directions for the right and left stereo images, accordingly. Hayward et al. (2005) mimicked that scheme for widefield microscopy by moving the sample sideways (e.g., toward the right) while refocusing downward (into the sample) and integrating on the CCD, then sideways in the opposite direction (e.g., the left) while not acquiring an image and then taking the second image of a stereo pair shifting in the first direction again (e.g., the right) while refocusing upward and integrating on the CCD camera. 2D image processing (Fourier filtering) can then remove most of the haze to yield useful stereo pairs. The advantage of such a scheme is that the speed is not limited by the need to take many images per stack but rather by the speed of refocusing.

A recently developed fast refocusing system based on moving a mirror in an image plane at (almost) unit magnification of the sample (Botcherby et al. 2008) should well be applicable to this scheme. To avoid movement of the sample, one could tilt the mirror to a different angle between its refocusing movements, which would tilt the sample space accordingly. This should allow the acquisition of stereo pairs by refocusing at video rate or at higher rates even for weakly fluorescent (or transmitting) samples.

Even though stereo imaging causes a 3D impression for the viewer, it is not able to reconstruct truly volumetric information (3D data sets). From information theory, we know that this cannot be possible with just two ordinary images taken. However, with prior knowledge about the sample (e.g., only a sparse distribution of discrete structures such as points, lines and surfaces being present) current computer vision approaches are able to locate such objects and thus create 3D models of the sample. A particularly interesting possible application of extended focus stereo acquisition would be its application to 3D molecule localization problems as encountered in Pointillism (Lidke et al. 2005), PALM (Betzig et al. 2006), fPALM (Hess et al. 2006), or STORM (Rust et al. 2006, Huang et al. 2008). In these cases, the obtainable axial resolution would scale with the chosen parallax, which in turn would further limit the number of particles that can be simultaneously imaged within a certain depth, due to arising ambiguity problems of particle assignment.

An interesting approach toward true 3D imaging is to obtain more than just one image. In optical projection tomography (OPT) (Sharpe et al. 2002) whole mount mouse embryos of up to 15 mm thickness have been imaged in 3D by acquisition of 400 extended focus images along a series of rotation angles of the sample. OPT typically does not use specific extended focus techniques, but simply makes use of a low numerical aperture system with its greatly extended depth of field (scaling with the inverse square of the NA) at only moderate loss in resolution (scaling linearly with the NA). However, OPT would potentially greatly benefit from employing some of the aforementioned techniques. The main reason for the expected benefit is that the collection efficiency of low NA systems unfortunately also scales with the inverse square of the NA.

A disadvantage of OPT is that it always requires rotation of the sample and thus sequential image acquisition. Schemes adapted from integral photography, attempting to capture the entire "light field" simultaneously, avoid this problem. This approach has recently been applied to microscopy (Levoy et al. 2006). Here a microlens array is placed in an image plane and a CCD with a high pixel number is placed at one focal distance behind it. In this way, an image of the back focal plane is formed behind each microlens. These miniature images of the back focal plane for the specific image plane position of each microlens contain the directional information of the light. Thus composite-extended focus images can be assembled from a particular projection direction (within the range of full NA angles) by always assembling the equivalent corresponding aperture pixels behind each individual microlens into a full image. In this manner, not only can stereo images be generated but even tomographic reconstructions of focus series captured with a single exposure become possible. Unfortunately the price to pay is, as in OPT, a reduced lateral resolution. However, in contrast to OPT, the high-angle photons are not discarded but used in other projections.

References

Abrahamsson, S., Usawab, S., and Gustafsson, M., A new approach to extended focus for high-speed, high-resolution biological microscopy, *Proc. SPIE* **6090**, 60900N (2006).

Berry, M. V. and Balazs, N. L., Nonspreading wave packets, *Am. J. Phys.* **47**, 264 (1979).

Betzig, E., Patterson, G.H., Sougrat, R. et al., Imaging intracellular fluorescent proteins at nanometer resolution, *Science* **313**, 1642–1645 (2006).

Biomedical Imaging Group, Ecole Polytechnique Fédérale de Lausanne, viewed August 14 (2008), http://bigwww.epfl.ch/demo/edf/.

Botcherby, E. J., Juškaitis, R., and Wilson, T., Scanning two photon fluorescence microscopy with extended depth of field, *Opt. Commun.* **268**, 253–260 (2006).

Botcherby, E. J., Juškaitis, R., Booth, M. J., and Wilson, T., An optical technique for refocusing in microscopy, *Opt. Commun.* **281**, 880–887 (2008).

Bradburn, S., Cathey, W. T., and Dowski, Jr., E. R., Realizations of focus invariance in optical/digital systems with wavefront coding, *Appl. Opt.* **36**, 9157–9166 (1997).

Dowski, Jr., E. R. and Cathey, W. T., Extended depth of field through wave-front coding, *Appl. Opt.* **34**, 1859 (1995).

Durnin, J., Exact solution for nondiffracting beams. I. The scalar theory, *J. Opt. Soc. Am. A* **4**, 651–654 (1987).

Durnin, J. and Miceli, Jr., J. J., Diffraction-free beams, *Phys. Rev. Lett.* **58**, 1499–1501 (1987).

Ewald, P. P., Zur Theorie der Interferenzen der Röntgentstrahlen in Kristallen, *Physik. Z.* **14**, 465–472 (1913).

Forster, B., Van De Ville, D., Berent, J., Sage, D., and Unser, M., Complex wavelets for extended depth-of-field: A new method for the fusion of multichannel microscopy images, *Microsc. Res. Tech.* **65**, 33–42 (2004).

Hayward, R., Juškaitis, R., and Wilson, T., High resolution stereo microscopy, *Proc. SPIE* **5860**, 45–50 (2005).

Heintzmann, R., Hanley, Q. S., Arntd-Jovin, D., and Jovin, T. M., A dual path programmable array microscope (PAM): Simultaneous acquisition of conjugate and non-conjugate images, *J. Microsc.* **204**, 119–137 (2001).

Hess, S. T., Girirajan, T. P. K., and Mason, M. D., Ultra-high resolution imaging by fluorescence photoactivation localization microscopy, *Biophys. J.* **91**, 4258–4272 (2006).

Huang, B., Wang, W., Bates, M., and Zhuang, X., Three-dimensional super-resolution imaging by stochastic optical reconstruction microscopy, *Science* **319**, 810–813 (2008).

Levoy, M., Ng, R., Adams, A., Footer, M., and Horowitz, M., Light field microscopy, *ACM Trans. Graph.* **25**, 924–934 (2006).

Lidke, K. A., Rieger, B., Jovin, T. M., and Heintzmann, R., Superresolution by localization of quantum dots using blinking statistics, *Opt. Express* **13**, 7052–7062 (2005).

Marks, D. L., Stack, R. A., Brady, D. J., and van der Gracht, J., Three-dimensional tomography using a cubic-phase plate extended depth-of-field system, *Opt. Lett.* **24**, 253–255 (1999).

McCutchen, C. W., Generalized aperture and three-dimensional diffraction image, *J. Opt. Soc. Am.* **54**, 240–244 (1964).

Petrán, M., Hadravský, M., Egger, M. D., and Galambos, R., Tandem-scanning reflected-light microscope, *J. Opt. Soc. Am.* **58**, 661 (1968).

Potuluri, P., Fetterman, M. R., and Brady, D. J., High depth of field microscopic imaging using an interferometric camera, *Opt. Express* **8**, 624–630 (2001).

Rust, M. J., Bates, M., and Zhuang, X., Sub-diffraction-limit imaging by stochastic optical reconstruction microscopy (STORM), *Nat. Methods* **3**, 793–796 (2006).

Sharpe, J., Ahlgren, U., Perry, P., et al., Optical projection tomography as a tool for 3D microscopy and gene expression studies, *Science* **296**, 541 (2002).

Siviloglou, G. A., Broky, J., Dogariu, A., and Christodoulides, D. N., Observation of accelerating airy beams, *Phys. Rev. Lett.* **99**, 213901 (2007).

Tao, S. H., Yuan, X. C., and Ahluwalia, B. S., The generation of an array of nondiffracting beams by a single composite computer generated hologram, *J. Opt. Soc. Am. A* **7**, 40–46 (2005).

Tucker, S. C., Cathey, W. T., and Dowski, Jr., E. R., Extended depth of field and aberration control for inexpensive digital microscope systems, *Opt. Express* **4**, 467–474 (1999).

Wicker, K. and Heintzmann, R., Interferometric resolution improvement for confocal microscopes, *Opt. Express* **15**, 12206–12216 (2007).

5

Single Particle Tracking

Kevin Braeckmans
Dries Vercauteren
Jo Demeester
Stefaan C. De Smedt

Ghent University

5.1 Introduction 5-1
5.2 SPT Instrumentation 5-2
5.3 Calculating Trajectories of Individual Particles........................ 5-4
    Particle Localization by Image Processing • Building Trajectories
5.4 Trajectory Analysis 5-9
5.5 Applications 5-13
5.6 Conclusions and Perspectives 5-13
References 5-14

5.1 Introduction

Insight into the dynamic behavior of molecules and particles in biomaterials is an important aspect in many research areas. In cell biology, for example, it has become clear that techniques are required for studying the intracellular transport of molecules since the cell organization has proven to be a highly dynamic process (Levi and Gratton 2007). Also in the biomedical and pharmaceutical fields, the mobility of molecules and particles is an important aspect because the successful delivery of (macromolecular) therapeutics, such as proteins and polynucleotides, depends on the mobility of drug molecules during the several phases of the delivery process in the human body (Sanders et al. 2000). To that end, intelligent carrier materials capable of protecting the therapeutic molecules against degradation and facilitating their transport in extracellular and intracellular matrices are being developed (Remaut et al. 2007).

Several complementary advanced fluorescence microscopy techniques have been developed for studying the mobility of molecules and particles on the micro- and nano-scale, such as fluorescence recovery after photobleaching (FRAP), fluorescence correlation spectroscopy (FCS), and single particle tracking (SPT) (De Smedt et al. 2005, Chen et al. 2006, Remaut et al. 2007). FRAP is a well-known fluorescence microscopy technique that was invented in the 1970s (Peters et al. 1974, Axelrod et al. 1976). It has proven to be a very useful and convenient tool for measuring the diffusion of fluorescently labeled molecules at typical imaging concentrations (usually >100 nM) in a micron-sized area (Meyvis et al. 1999, Sprague and McNally 2005). A typical FRAP experiment involves three distinct steps, namely, the registration of the fluorescence before photobleaching, fast photobleaching within a defined area using a high-power laser beam, and subsequent imaging of the fluorescence recovery arising from the diffusional exchange of photobleached molecules by intact ones from the immediate surroundings. It is then possible to extract the diffusion coefficient and a local (im)mobile fraction from the recovery curve by the fitting of a suitable mathematical FRAP model (Braeckmans et al. 2003, 2007, Mazza et al. 2008). FRAP has been used, for example, to study the mobility of molecules in cells (Verkman 2003, Elsner et al. 2003, Braga et al. 2004, Karpova et al. 2004), as well as in extracellular matrices, such as mucus, (tumor) cell interstitium, and vitreous (Remaut et al. 2007). FCS, on the other hand, is based on the temporal measurement of fluorescence intensities in a very small volume (<1 fL). The movement of fluorescently labeled molecules in and out of this detection volume gives rise to fluorescence fluctuations whose duration is directly related to the velocity of the molecules. By autocorrelation analysis, it is possible to calculate the diffusion coefficient from the fluorescence fluctuation trace (Gosch and Rigler 2005). FCS has proven to be a valuable tool for studying the association and dissociation of macromolecular drug complexes and the stability of nucleic acids in vitro and in vivo (De Smedt et al. 2005, Remaut et al. 2007). It can also be used to study the movement of molecules and complexes in living cells, although there are several difficulties related to performing quantitative intracellular correlation analysis, such as photobleaching and autofluorescence (Schwille 2001, Thompson et al. 2002a, Bacia and Schwille 2003, Haustein and Schwille 2004). In addition, just as for FRAP, the mobility parameters calculated from an FCS measurement only provide time-averaged information of the many molecules that have passed the detection volume. FRAP and FCS, therefore, rather provide ensemble average information on the dynamics of molecules on the micrometer scale. SPT, on the other hand, can provide detailed information on the mobility of individual molecules. As the name implies, SPT is a fluorescence microscopy–based method where the movement of individual fluorescently labeled particles or molecules can be visualized in time and space (Saxton and Jacobson 1997, Suh et al. 2005, Levi and Gratton 2007). The particles and molecules in this context are usually smaller than the resolution of the microscope, in which case they are observed in the image plane as a spot of light whose intensity distribution depends on the response of the optical system to the point source, also called the point spread function (Jonkman and Stelzer 2002). The imaging of single particles is possible if the concentration is sufficiently low such that there is less than one particle per diffraction-limited spot. When using a high numerical aperture objective lens, the individual particles should be separated by a distance of at least ~500 nm in order to be detectable as single entities.

In this chapter, the basic principles of SPT will be outlined. First, we will discuss the hardware and instrumentation that is typically required to acquire SPT movies. Next, we will explain how the trajectories of individual particles can be obtained by image processing of the SPT movies. Then, we will explain how to analyze the trajectories quantitatively in order to obtain detailed information on their mobilities. Finally, we will give a brief overview of the different fields to which SPT has been applied.

5.2 SPT Instrumentation

The microscope setup is typically an epi-fluorescence microscope adjusted for widefield laser illumination with a fast and sensitive CCD camera, such as an electron-multiplying CCD camera, to record movies of the particle movement. Laser illumination is preferred over standard light sources for fluorescence microscopy, such as the typical Hg vapor or metal halide lamps, because full power is available at a single wavelength allowing an efficient excitation of selected fluorophores and a suppression of the excitation light scattered into the emission path using suitable "notch" filters. An example of a typical SPT setup is shown in Figure 5.1. Three (solid state) lasers, covering a wide range of the visible spectrum, are used as excitation sources. Using dichroic mirrors (DM), the individual beams are aligned along the same line of propagation. Just as in a confocal laser scanning microscope, an acousto-optical tunable filter (AOTF) can be used to modulate the transmitted intensity of each of the laser beams. The exciting laser beam is subsequently directed toward the microscope to provide widefield illumination of the sample. This can be most easily achieved by expanding the laser beam in front of the microscope using a beam expander and to focus the expanded laser beam onto the back focal plane of the objective lens. As a result, a wide laser beam will exit the objective lens parallel to the optical axis. However, since the laser light is coherent, this type of illumination will create an illumination profile at the sample with many impurities due to light diffraction and interference by dust and other defects in the light path. The quality of the sample illumination can be improved by replacing the beam expander by a diffuser, which scatters the laser light over many angles. It is even possible to accurately control the angular extent over which the laser light is scattered by using a holographic diffuser. In this case, the laser spot on the diffuser can be considered

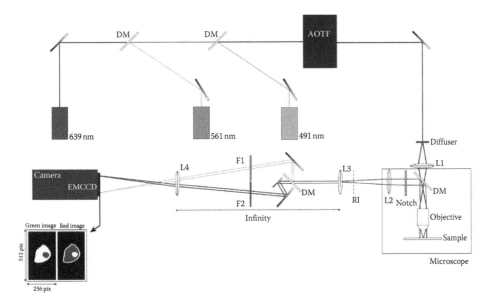

FIGURE 5.1 Schematic overview of a typical SPT setup. Three laser beams are co-aligned using DM and directed through an AOTF, which allows to modulate the transmitted laser intensity. Homogeneous widefield laser illumination is provided at the sample using a holographic diffuser and an achromat lens (L1), which projects an image of the laser spot on the diffuser onto the back focal plane of the objective lens. Fluorescence light from the sample is collected by the objective lens and separated from the excitation beam using a DM. Any reflected or scattered excitation laser light is suppressed by using an additional laser notch filter. A RI of the sample is formed by the microscope tube lens (L2) outside the microscope stand. Additional optics can be used to split up the fluorescence image into two spectral regions for simultaneous dual color SPT. The RI coincides with the focal plane of the achromat lens L3 and the achromat lens L4 projects a (magnified) RI on the CCD chip. Since the light rays between L3 and L4 are parallel, optical components can be inserted with a minimal distortion of the final image. A DM with band-pass filters (F1 and F2) can be used to split up the fluorescence light into two spectral regions (colors). Two images, one for each color, are then formed on both halves of the CCD chip. By synchronizing the AOTF with the CCD camera, the sample illumination can be switched off during the camera read-out phase, thus minimizing photobleaching while recording an SPT movie.

to be the widefield light source, analogous to a lamp filament or the arch of light between the electrodes of a mercury vapor lamp. An achromat lens (L1) is used to create a real image (RI) of the diffuser laser spot at the back focal plane of the objective lens. As a result, parallel rays of light will emerge from the objective lens over many angles, thus creating a smooth and homogeneous illumination of the sample, just as is the case for epi-fluorescence illumination using a standard widefield lamp.

The fluorescence light from the sample that is captured by the objective lens is separated from the excitation light using a suitable DM, which can be conveniently mounted into a standard epi-fluorescence filter cube. Since SPT is about detecting very faint light signals, it is absolutely important to filter out any remaining excitation light at this point. The use of a suitable laser "notch" filter is, therefore, highly recommended, which can be conveniently placed into the exit position of the epi-fluorescence filter cube. Then, a real (magnified) image of the sample will be formed by the microscope's tube lens (L2) at the microscope exit. In case of single-color SPT, one can put the camera directly at this position, possibly with an additional magnification lens to match the image magnification to the pixel size of the EMCCD camera. In case of dual-color SPT, however, the emitted fluorescence light has to be split up into two spectrally different channels. This can be done by placing two achromat lenses (L3 and L4) into the emission path such that the RI emerging from the microscope is at the back focal plane of L3. This creates a parallel optical path between L3 and L4 such that optical elements can be easily put in this region without distorting the final image. Furthermore, if f_3 and f_4 are the focal lengths of L3 and L4, respectively, then an extra magnification of f_4/f_3 is obtained from this lens pair. This again allows to match the final magnification of the image to the pixel size of the CCD camera. A DM and fluorescence band-pass filters (F1 and F2) can be used to split up the fluorescence light into two spectrally different channels. The light in each channel is directed onto the final projection lens (L4) at a small angle with respect to the optical axis such that an image of the sample is formed on the two halves of the CCD chip for each spectral channel. Of course, the field of view in the *x*-direction needs to be limited to avoid spatial overlap of the two images on the CCD chip. This can be done by using a slit diaphragm at the position RI that allows restricting the field of view (in the *x*-direction) to the central part of the image. Such a setup allows for simultaneous dual-color SPT imaging using a single CCD camera. It is clear that a similar dual-color setup can be created using two CCD cameras having the advantage of retaining the full field of view. However, in that case one has to make sure that both cameras are accurately synchronized with each other (Koyama-Honda et al. 2005).

We note that the AOTF can be used to limit the amount of photobleaching during imaging, especially when using very short illumination times (in the order of milliseconds). The response time of an AOTF is typically in the order of microseconds and, therefore, can be used as a very fast (and relatively inexpensive) shutter. When capturing an image with a CCD camera there are essentially two phases: first the illumination phase during which photons are collected by the CCD pixels, second, the read-out phase during which the pixel charges are read out, amplified, and digitized. If the illumination time is short compared to the read-out time, the read-out phase will be the rate-limiting step. In that case, the camera has to wait for a relatively long time before the next illumination sequence can start. During this detector "dead time," the AOTF can be used to switch off the laser illumination in order to avoid unnecessary illumination, and hence photobleaching, of the sample. By synchronizing the AOTF with the CCD camera, the sample illumination can be restricted to when the camera is actually integrating light. Evidently, this becomes less of an issue when using long illumination times, in which case the read-out time eventually becomes negligible.

5.3 Calculating Trajectories of Individual Particles

Having recorded an SPT movie, information on the mobility of individual particles can be obtained from their trajectories. Calculating the trajectories basically involves two steps. First, image processing of the SPT movie needs to be performed in order to find the particles and to calculate their positions in each frame of the movie. Second, knowing the particle coordinates, a suitable algorithm can be employed to decide which positions should be linked between subsequent frames to form the trajectories of individual particles. Both steps will be discussed in the following text.

5.3.1 Particle Localization by Image Processing

The first step to obtain the particle trajectories is identifying the particles in all frames of the SPT movie and to calculate their center locations, possibly with additional parameters, such as the size and intensity distribution. At this point, it should be noted that, although the resolution of the microscope is typically around 250 nm, the position of a particle can be determined with subresolution accuracy by finding the center of the diffraction-limited spot. Various methods have been reported for finding the position of particles, such as cross-correlation, centroid identification, a Gaussian fit, and even a pattern-recognition method (Cheezum et al. 2001, Levi et al. 2006). The accuracy with which the particle position can be calculated is influenced by the number of detected photons, background noise by out-of-focus fluorescence or autofluorescence, read-out noise of the CCD camera, dark current, and pixelation. Theories have been developed to try to estimate the localization accuracy based on such parameters for single molecule imaging (Thompson et al. 2002b, Ober et al. 2004). In practice, however, the position accuracy for a given experimental setup can be determined empirically by recording an SPT movie of stationary particles and calculating their apparent trajectories. The standard deviation of the calculated position of a stationary particle is then used as a measure for the localization accuracy. Reported localization accuracies are usually in the order of tens of nanometers (Schutz et al. 1997, 2000, Sonnleitner et al. 1999, Kubitscheck et al. 2000, Seisenberger et al. 2001, Vrljic et al. 2002, Lakadamyali et al. 2003, Bausinger et al. 2006), although the spatial resolution for a mobile particle will be inherently worse depending on how fast the particle is moving during image acquisition.

The process of identifying particles in images can be automated with suitable image analysis algorithms (Sbalzarini and Koumoutsakos 2005, Anthony et al. 2006), thus allowing to find the positions of many particles simultaneously. To emphasize the principle of tracking many individual particles simultaneously, this technique is also sometimes explicitly referred to as "multiple particle tracking" (Tseng et al. 2002, Suh et al. 2003, 2005). In the following, we will discuss the algorithms we have found in our own research to perform very well for detecting particles or objects in a wide range of image types, also in the presence of complex background features, such as is shown in Figure 5.2A. Our method for particle detection basically involves three steps. First, the nonuniform background is removed as much as possible. This can be most easily done using a so-called unsharp filter, generally defined as

$$F = 1 - f \times B \qquad (5.1)$$

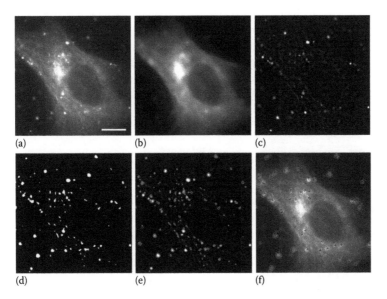

(a)   (b)   (c)

(d)   (e)   (f)

FIGURE 5.2 (See color insert following page 15-30.) Particles can be identified and located by image processing. (a) An SPT image is shown of a RPE cell with EGFP-tubulin and YOYO-1 labeled pDNA-polymer complexes of approximately 100 nm diameter. The scale bar is 10 μm. (b) Using a low pass filter (21×21 median filter), the particles can be removed from the image, resulting in an approximate background image containing only the lower spatial frequencies. (c) By subtracting the background image from the original image (unsharp filtering, Equation 5.1), a filtered image is obtained with the particles on a much more homogeneous background. (d) Intensity thresholding can be used to binarize the filtered image. The pixels above the threshold are assigned a value of 1 (displayed in white), while the pixels below the threshold are set to zero (displayed in black). (e) A contour line is calculated for each object in the binary image. (f) Pixels inside a contour belong to a (potential) object of interest. From the pixel values in the original image, object properties can be calculated, such as mean intensity, size, and center location. Based on user-defined criteria, a final selection is made to filter out unwanted objects and retain the objects of interest.

where

> I is the original image
> B is the background image
> f is a real number between 0 and 1
> F is the resulting filtered image

Since the particles are usually very small (approx. 5 pixels diameter according to the Nyquist criterion), the background image B can be calculated from I using a suitable low-pass filter. We have found that a simple median or mean filter works equally well to more advanced algorithms, such as morphological opening. By choosing a filter kernel that is larger than the objects of interest, they can be effectively removed from the image resulting in the background image B containing only the lower spatial frequencies (see Figure 5.2B). At the same time, the filter kernel should not be too large either in order to retain as much of the background features as possible. Finally, the unsharp filter can be carried out where we usually remove the entire background ($f = 1$ in Equation 5.1). As is shown by the example in Figure 5.2C, this procedure effectively removes the inhomogeneous background and results in a filtered image F where the particles stand out with good contrast against a much more uniform background. A first selection of possible objects can now be made by applying intensity thresholding to the filtered image F. To allow for automation of the entire process and to assure reproducibility, it is preferred to select the intensity threshold based on a suitable algorithm. While there are many algorithms to choose from, we have found the following approach to perform very well. In a typical SPT image, such as the one shown in Figure 5.2C, most of the pixels constitute the background (low intensity), while the pixels belonging to the particles (high intensity) are far less abundant. The intensity histogram of such an SPT image typically looks like the one shown in Figure 5.3. The large Gaussian-like peak originates from the many background pixels and the higher intensities coming from the particles make up the long tail at the right. A suitable threshold, therefore, will be an intensity that separates the Gaussian background peak from the "particle tail." By the fitting of a Gaussian distribution to the image histogram, resulting in a mean value μ and a standard deviation σ, the threshold T can be defined as

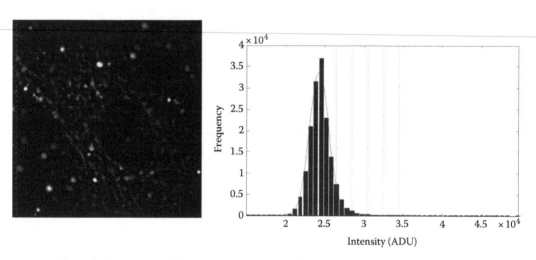

FIGURE 5.3 A typical SPT image mostly consists of dark background pixels with only very few pixels actually belonging to the objects. The corresponding histogram typically has a large Gaussian-like background peak with a tail extending into the high intensity region. A best fit of a Gaussian distribution to the histogram (solid line) yields the mean value μ and the standard deviation σ. A suitable threshold value T can be calculated according to Equation 5.2, in which N is a user-defined parameter that allows to fine-tune the threshold selection for a particular series of experiments. The dotted lines indicate thresholds for $N = 1$–5. The binary image 2D was obtained from $N = 3$.

$$T = \mu + N \times \sigma \tag{5.2}$$

where N is a user-defined parameter (real number) through which the user can fine-tune the threshold selection for a particular series of experiments. Having determined the threshold T, the filtered image F can now be converted to a binary image F_b, where pixels with values below T are set to zero and above T to one, as is shown in Figure 5.2D. Next, the contours of the individual objects in the binary image F_b can be determined (see Figure 5.2E). Objects are then defined as those pixels in the original image I within a particular contour. This allows to calculate the object properties, such as their size, intensity, local background intensity and, most importantly, their center location. We usually calculate the particle position (x_c, y_c) as the intensity-weighted center (similar to a center of mass of a physical object) of the object pixels, also called the "centroid," which is defined as

$$x_c = \frac{\sum_{i \in S} x_i \times (I_i - BG)}{\sum_{i \in S} (I_i - BG)}, \qquad y_c = \frac{\sum_{i \in S} y_i \times (I_i - BG)}{\sum_{i \in S} (I_i - BG)} \tag{5.3}$$

where

S is the set of all pixels with coordinates (x_i, y_i) and intensities I_i belonging to the object
BG is the local background intensity of the object, which we calculate from the mean pixel value along a contour that is drawn close (a few pixels) around the object contour, as is shown in Figure 5.4

Finally, knowing all the object properties, a last selection of "real" particles can be made according to user-defined criteria, such as a range of sizes, a minimum contrast with respect to the local background, sphericity, etc. As is shown in Figure 5.2F, this procedure indeed allows to find multiple particles in even very complex fluorescence images. Since each step of this procedure is a well-defined algorithm, the entire process can be easily automated such that all frames of a movie can be analyzed automatically, and even a batch of movies that are recorded under similar conditions. For each SPT movie, the properties of all particles in each frame are stored in a spreadsheet file, which can then be used for calculating the particle trajectories, as is described in Section 5.3.2. We note that, if one is interested in tracking single molecules, the intensity time trace can be used to verify if a single spot consists of one or more fluorophore molecules. By recording a movie of molecules immobilized on a glass slide or in a silica gel, one can determine the number of fluorescent molecules in a single spot by the discrete steps in the fluorescence time trace due to the photobleaching of single molecules. If there is only a single fluorophore in the spot, the intensity will drop to the background level when the molecule is photobleached. In case there are multiple molecules in a single spot, the fluorescence time trace will show several discrete steps over time (Chirico et al. 2003, Cannone et al. 2003).

Alternative methods for calculating the particle center position have been proposed, which are said to be superior to the centroid calculation as defined earlier (Cheezum et al. 2001, Levi and Gratton 2007). These methods (2-D Gaussian fit, correlation, pattern recognition) rely, however, on a particular shape and intensity distribution of the particles. While these methods perform well for particles that are much smaller than the microscopic resolution, such as single molecules, we have found this to be a very limiting factor for general object-tracking purposes. For example, particles and objects can have different sizes within the same image and the particle shape can be distorted to some extent due to fast movement. The method as outlined earlier, on the other hand, does not rely on any *a priori* assumption on the particle shape or intensity distribution, although a final selection based on particle properties can be made in the end. Also, since we determine the particle contour before calculating the centroid, we have found the centroid calculation to perform much better under low signal-to-noise situations compared to a 2-D Gaussian fit. This is in contrast to what was found by Cheezum et al. We attribute this to the

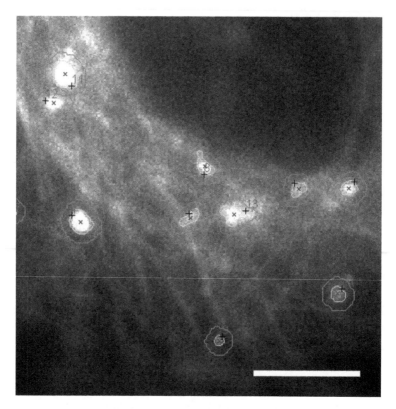

FIGURE 5.4 (See color insert following page 15-30.)   A magnified subregion of image 2F showing yellow contours around particles. The blue × symbols indicate the centroid position. The green contour is drawn at a user-defined distance (here 3 pixels) from the object contour and allows to calculate a local background value for each particle (the average value of the pixels along this contour). The scale bar is 5 μm.

fact that Cheezum et al. did not first determine the particle boundary, but rather applied the different algorithms to large rectangular subareas of the image mostly consisting of background pixels and only a small fraction of pixels belonging to the actual particle. Since all pixels in the subimage were used in the calculation of the centroid, this method was found to be less accurate to the 2-D Gaussian fit. Based on the results by Cheezum et al., the 2-D Gaussian fit has frequently been dubbed to be the best localization algorithm for SPT without considering the particular circumstances from which this conclusion was drawn. We would like to point out that this is not generally the case and that alternative approaches might exist for which the opposite is true.

5.3.2 Building Trajectories

Having found the positions of the particles in all frames of an SPT movie, individual trajectories can be calculated by pairing corresponding features between images. A nearest neighbor algorithm is typically used to construct the trajectories, although similarity in intensity distribution has been used as well to increase the tracking accuracy (Anderson et al. 1992, Sbalzarini and Koumoutsakos 2005). The nearest neighbor algorithm basically connects positions of particles that are closest to each other in subsequent frames. A safeguard has to be included, however, in order not to connect positions that are too far apart. Consider, for example, the following extreme case in which there is only a single particle present in the top left corner of the first image. Imagine that in the next frame this particle disappears from the field of view and that a new particle appears in the lower right corner. Without any restriction on the maximum distance, both positions would be connected to form a trajectory, while clearly this would make no sense. Therefore, the nearest neighbor algorithm has to be applied in consideration of a maximum distance a particle can reasonably move from one frame to another. In case of free diffusion, for example, the maximum step size can be estimated as follows. The distance squared ($\xi = r^2$) that a diffusive particle has covered after a time t follows an exponential distribution (Qian et al. 1991)

$$P(\xi) = \frac{1}{4Dt} e^{-\xi/4Dt} \, d\xi \tag{5.4}$$

having the corresponding cumulative distribution function

$$F(x) = 1 - e^{-x/4Dt} \tag{5.5}$$

which expresses the probability that a particle with diffusion coefficient D has moved a square distance within the interval $[0\ x]$ after a time t. From this, it immediately follows that the particle will not have moved any further than a distance $r = \sqrt{x} = \sqrt{4Dt \ln(1/1 - p)}$ with a probability p. In other words, given a diffusion coefficient D and a time Δt between two subsequent images, the maximum step size can be calculated from this formula for a desired certainly level, e.g., 95%. For example, for $D = 1\,\mu m^2/s$ and $\Delta t = 30\,ms$, there is 95% chance that the particle will not have moved more than 599.6 nm between subsequent images. Increasing the certainty level decreases the chance of losing the particle from one frame to another, but also increases the chance of connecting unrelated positions, for example, when two particles are crossing each other. Therefore, it is important that the particle concentration is sufficiently low so as to minimize such mistakes. In case there is no *a priori* knowledge about the diffusion coefficient, an iterative approach can be followed to find the correct maximum step size. Trajectories can be calculated with a gradually increasing maximum step size until the average diffusion coefficient of the particles as calculated from the trajectories (see the following text) does not significantly increase anymore.

An algorithm to calculate trajectories should also take into account that a particle can (temporarily) disappear from the field of view, for example by photobleaching, blinking, or when moving out of focus. One should also take into account the possibility of new particles appearing during the movie. More advanced algorithms could also take into account the possibility of particles colliding or dissociating.

5.4 Trajectory Analysis

The trajectories can now be further analyzed in terms of mode of motion (diffusive, subdiffusive, active, immobile). Trajectory analysis is usually based on mean square displacement (MSD) analysis. Consider a trajectory consisting of N steps ($N + 1$ locations), as illustrated in Figure 5.5A. Neighboring locations along the trajectory are a time Δt apart, which is the time between subsequent images in the SPT movie. Δt is often referred to as the first time lag. The distance between the first two points (i.e., the first step) is $r_{1,\Delta t}$. The distance between the second and third locations (i.e., the second step) is $r_{2,\Delta t}$, etc. The MSD corresponding to the first time lag is then defined as

$$\left\langle r^2 \right\rangle_{\Delta t} = \frac{\sum_{i=1}^{N} r_{i,\Delta t}^2}{N} \tag{5.6}$$

Next, the distances can be calculated between locations that are separated by a time $2\Delta t$, i.e., between locations 1 and 3, 2 and 4, 3 and 5, etc. Noting these distances as $r_{1,2\Delta t}$, $r_{2,2\Delta t}$ and so on, the MSD corresponding to the second time lag is calculated from

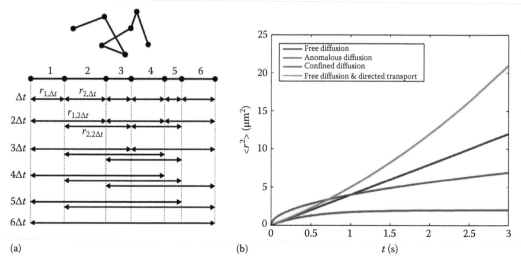

(a)   (b)

FIGURE 5.5 MSD analysis of single trajectories. (a) A trajectory is shown to consist of six steps (seven locations). The MSD analysis principle is most easily shown by putting the six segments next to each other. For the first time lag, six displacements are available for calculating the MSD. For the second time lag, only five displacements are available in the same trajectory for calculating the MSD. For the last time lag, only one pair of points can be found in the track that are separated by a time $6\Delta t$. (b) The MSD versus time-lag curve of a particular trajectory is different depending on the mode of motion, as explained in the main text.

$$\left\langle r^2 \right\rangle_{2\Delta t} = \frac{\sum_{i=1}^{N-1} r_{i,2\Delta t}^2}{N-1} \tag{5.7}$$

In general, the MSD corresponding to the nth time lag is defined as

$$\left\langle r^2 \right\rangle_{n\Delta t} = \frac{\sum_{i=1}^{N-n+1} r_{i,n\Delta t}^2}{N-n+1} \tag{5.8}$$

where $n = 1, 2, \ldots, N$. The type of motion can now be characterized from the MSD versus time-lag curve (Saxton and Jacobson 1997, Suh et al. 2005, Levi and Gratton 2007):

- Free 2-D diffusion:

$$\left\langle r^2 \right\rangle = 4Dn\Delta t \tag{5.9}$$

- Continuous directed motion with free 2-D diffusion:

$$\left\langle r^2 \right\rangle = 4Dn\Delta t + (vn\Delta t)^2 \tag{5.10}$$

- Anomalous diffusion:

$$\left\langle r^2 \right\rangle = 4D(n\Delta t)^\alpha, \quad 0 \le \alpha \le 1 \tag{5.11}$$

As illustrated in Figure 5.5B, a linear relation is expected from the MSD versus the time lag in case of free diffusion, while a parabola is expected if there is (additional) continuous unidirectional motion with velocity v (e.g., flow). If the diffusing particle encounters obstacles or undergoes binding and unbinding events, the particle will slow down as the effect of the barriers becomes dominant. This type of motion is called anomalous subdiffusion and is characterized by the anomalous exponent α. The more α deviates from 1, the more anomalous the motion is said to be ($\alpha = 1$ corresponds to free diffusion). We also note that a diffusive particle may be (temporarily) moving into a subregion of the sample having a certain volume closed off by walls. In that case, the particle will seem to diffuse freely on a short time scale (diffusive movement within the subregion), but the MSD versus the time lag will become less than linear at longer time scales as the walls start to prevent the particle from moving further on. This type of motion is termed as confined motion and is characterized by the asymptotic behavior of the MSD at long time lags, also shown in Figure 5.5B. From the asymptotic value of the MSD, the dimensions of the subvolume can be derived, as is explained in more detail elsewhere (Saxton and Jacobson 1997). Finally, a particle can be classified as being immobile if the MSD remains constant for all time lags and is equal to 4 times the position accuracy squared (as, e.g., determined from comparable stationary particles at the same imaging conditions). Indeed, it can be shown that the limited position accuracy gives rise to an additional term of $4\sigma^2$ in the MSD models mentioned earlier, e.g., for 2-D free diffusion (Crocker and Grier 1996):

$$\left\langle r^2 \right\rangle = 4Dn\Delta t + 4\sigma^2 \tag{5.12}$$

which means that an entirely immobile particle ($D = 0$) would have an apparent MSD of $4\sigma^2$. It follows that the lower limit of the diffusion coefficient that can be measured at a particular time lag is

$$D_{\min} = \frac{\sigma^2}{n\Delta t} \tag{5.13}$$

For example, considering a typical position accuracy of $\sigma = 30\,\text{nm}$ and a time of $\Delta t = 30\,\text{ms}$ between subsequent frames in the movie, $D_{\min} = 0.03\,\mu\text{m}^2/\text{s}$ for the first time lag. In other words, if a particle diffuses more slowly than this, its movement cannot be detected at this time interval because of the limited localization accuracy.

Several methods for classifying trajectories based on their MSD versus time-lag curve have been suggested. A first method classifies the trajectory according to the model that provides the best fit to the experimental data (Anderson et al. 1992, Wilson et al. 1996). Others have classified trajectories by comparing the MSD behavior at short and long time lags (Kusumi et al. 1993). A linear fit to the MSD at short time lags is extrapolated to longer time lags, so that the deviation of the experimental MSD points from the extrapolated vales at longer time lags is a measure for the deviation of the trajectory from free diffusion. In yet another method, a best fit is performed of the anomalous diffusion model to the MSD versus time-lag plot and the trajectories are classified according to the anomalous exponent α. At this point it is important to note that, while in total N MSDs can be calculated from a trajectory with N steps (one for each time lag), the accuracy of the MSD decreases with increasing time lag since correspondingly less distances are available from which the MSD is calculated (see Figure 5.5A). In fact, for diffusive motion one can show that the variance $\sigma_{n\Delta t}^2$ of the MSD corresponding to the nth time lag is (Qian et al. 1991, Saxton 1997)

$$\sigma_{n\Delta t}^2 = F \left\langle r^2 \right\rangle_{n\Delta t}^2 \tag{5.14}$$

where $\sigma_{n\Delta t}$ is the standard deviation on the set of square displacements at time lag $n\Delta t$ from which the MSD $\langle r^2 \rangle_{n\Delta t}$ is calculated. Let $K = N - n + 1$ be the number of steps from which the MSD is calculated, then $F = (4n^2 K + 2K + n - n^2)/(6nK^2)$ if $K \geq n$ and $F = 1 + (K^2 - 4nK^2 + 4n - k)/(6n^2 K)$ if $K \leq n$. As illustrated in Figure 5.6 for a free diffusion trajectory of 100 steps, the standard deviation on the MSD increases rapidly for large time lags, as expected. It has, therefore, been suggested that the MSDs should not be analyzed any further than 25% of the longest time lag, at which point the relative error on the MSD has grown already to approximately 45%. In any case, when calculating the diffusion coefficient by a least-squares fit to the MSD versus time-lag curve, the proper weights $1/\sigma_{n\Delta t}^2$ should be used for each time lag (Saxton 1997). Finally, let it be noted that the precision with which trajectories can be analyzed increases with their length N. Consider for example the first time lag ($n = 1$). Then $F = N$ and the relative error on the MSD according to Equation 5.14 becomes $\sigma_{n\Delta t}/\langle r^2 \rangle = 1/\sqrt{N}$. The more steps in a trajectory, the better the precision with which the diffusion coefficient can be calculated. This is also one of the reasons why it is interesting to follow particles in three dimensions. If particle motion is relatively slow, this can be achieved by acquiring 3-D image stacks over time using a confocal microscope. An alternative approach has been developed, however, allowing fast 3-D tracking of single particles. Rather than acquiring full images, a laser beam is quickly scanned in a circular manner around a single particle thus allowing to sense the particle movement and to adjust the laser position to where the particle is moving (Levi et al. 2005). While this technique provides extremely detailed information, it can only follow one or two particles at the same time and, therefore, might be less convenient for obtaining a good statistical view on how particles behave at a given time. This is, for example, important when studying the uptake and transport of drug complexes in living cells, in which case the drug complexes are all applied at once to the cells.

For the trajectory analysis, some have argued that more detailed information can be obtained by using the entire distribution of (square) displacements for each time lag, rather than just the mean value (Anderson et al. 1992, Schutz et al. 1997, Sonnleitner et al. 1999, Apgar et al. 2000, Vrljic et al. 2002, Hellriegel et al. 2004, 2005). This type of analysis has proven to be a valuable method for studying multicomponent mobilities within a single trajectory (Hellriegel et al. 2004). For example, a molecule may diffuse within a compartment for a certain period of time before "hopping" to the next compartment.

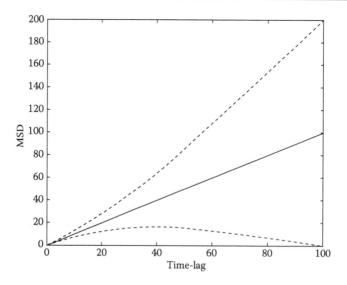

FIGURE 5.6 Because fewer steps are available within a single trajectory for larger time lags, the accuracy of the corresponding MSD becomes increasingly worse. The solid line represents the MSD versus time-lag curve for a free diffusion trajectory of 100 steps, while the dashed lines indicate the $+\sigma$ and $-\sigma$ curves.

If sampled at a sufficiently high rate, the trajectory of such a molecule will show restricted diffusion on a small time scale and free diffusion (hopping between compartments) on a longer time scale (Ritchie et al. 2005). This kind of movement has been observed, for example, for membrane incorporated molecules (Fujiwara et al. 2002, Kusumi et al. 2005a). Also the intracellular transport of nonviral nucleic acid complexes may be composed of different kinds of motions (Suh et al. 2003, Bausinger et al. 2006, de Bruin et al. 2007), similar to what has been observed for viruses (Lakadamyali et al. 2003).

5.5 Applications

SPT was originally mainly used for studying cell membrane organization by observing the dynamic behavior of proteins and lipids in living cells (Anderson et al. 1992, Kusumi et al. 1993, Schmidt et al. 1996, Saxton and Jacobson 1997, Schutz et al. 1997, 2000, Sonnleitner et al. 1999, Dietrich et al. 2002, Fujiwara et al. 2002, Vrljic et al. 2002, Kusumi et al. 2005b). Noteworthy, dual color SPT has been used to study the interaction of differently labeled membrane proteins by observing their simultaneous movements (Koyama-Honda et al. 2005). Manley et al. (2008) have recently demonstrated that super-resolution PALM microscopy can be combined with SPT to study the motion of membrane proteins at high molecule densities normally incompatible with SPT, opening up the path to a more detailed insight into the spatiotemporal membrane organization. Meanwhile, SPT has also contributed to other cell biological studies, such as revealing the walking mechanisms of molecular motors (Yildiz et al. 2003, 2004) and gaining insight into nuclear dynamics and organization (Levi et al. 2005). In the last few years, SPT has also been used for studying the intracellular infection pathway of viruses (Pelkmans et al. 2001, Seisenberger et al. 2001, Lakadamyali et al. 2003, Babcock et al. 2004, Lill et al. 2006). Lakadamyali et al., for example, have found that the intracellular transport of influenza virus particles comprises of three distinct stages, starting with a slow movement in the cell periphery, followed by rapid directed transport toward the nucleus where it exhibits bidirectional movement until its envelope fuses with an endosomal vesicle in which it resides. Using cytoskeleton-depolymerizing drugs, the authors could show that the virus is transported along actin filaments in the first stage, while the second and third stages involve active transport along microtubules. In the last decade, SPT also has found its way to the drug delivery field (Ishii et al. 2001, Dawson et al. 2003, Suh et al. 2003, Berezhna et al. 2005, Kulkarni et al. 2005, 2006, Suh et al. 2005, Bausinger et al. 2006, de Bruin et al. 2007). For example, Suh et al. have demonstrated an active transport of polymeric nucleic acid complexes to the perinuclear region where a large fraction of the particles become relatively immobile over time. From such experiments, it was realized that it might be the transnuclear transport, which is the bottleneck for successful transfection, rather than the cytoplasmic transport. Also, Bausinger et al. have shown directly the interaction of endocytosed polyplexes with both the actin and microtubule cytoskeleton by fluorescent labeling of the actin and tubulin networks of living cells. De Bruin et al. have confirmed that actin filaments play a crucial role in the uptake and initial transport of the endocytosed particles, while transport along microtubules ensures rapid translocation to the perinuclear region.

5.6 Conclusions and Perspectives

Although SPT is a relatively new technique, which is still in a developing stage, it has already proven to be an extremely useful tool for studying the dynamics of molecules and particles in biological and pharmaceutical materials. Contrary to the complementary FRAP and FCS techniques, which are ensemble-averaging methods, SPT provides information on a single particle basis. Not only does this allow to study heterogeneous populations of particles, it is even possible to analyze transient heterogeneous behavior of a single particle over time. Such detailed information can provide new insights into the structure and organization of materials, as well as in the interaction of particles and molecules with the material components. As brighter and more photostable fluorophores become available, together with faster and more sensitive CCD cameras, it will be possible to obtain longer trajectories with better accuracy, as well as to study faster processes. Also, the advent of increasingly powerful and fast switchable multicolor LED light sources holds the potential to bring SPT to a larger number of labs since it might replace the relatively complicated laser illumination (Cole and Turner 2008). Theoretical developments, on the other hand, will further increase the information that can be obtained from the trajectories. We expect that SPT will continue to gain in importance and gradually will make the transition to nonspecialist labs in the coming years in pace with instrumental and theoretical developments.

References

Anderson C. M., Georgiou G. N., Morrison I. E. G., Stevenson G. V. W., and Cherry R. J. 1992. Tracking of cell-surface receptors by fluorescence digital imaging microscopy using a charge-coupled device camera—Low-density-lipoprotein and influenza-virus receptor mobility at 4-degrees-C. *Journal of Cell Science* 101: 415–425.

Anthony S., Zhang L. F., and Granick S. 2006. Methods to track single-molecule trajectories. *Langmuir* 22: 5266–5272.

Apgar J., Tseng Y., Fedorov E. et al. 2000. Multiple-particle tracking measurements of heterogeneities in solutions of actin filaments and actin bundles. *Biophysical Journal* 79: 1095–1106.

Axelrod D., Koppel D. E., Schlessinger J., Elson E., and Webb W. W. 1976. Mobility measurement by analysis of fluorescence photobleaching recovery kinetics. *Biophysical Journal* 16: 1055–1069.

Babcock H. P., Chen C., and Zhuang X. W. 2004. Using single-particle tracking to study nuclear trafficking of viral genes. *Biophysical Journal* 87: 2749–2758.

Bacia K. and Schwille P. 2003. A dynamic view of cellular processes by in vivo fluorescence auto- and cross-correlation spectroscopy. *Methods* 29: 74–85.

Bausinger R., von Gersdorff K., Braeckmans K. et al. 2006. The transport of nanosized gene carriers unraveled by live-cell imaging. *Angewandte Chemie—International Edition* 45: 1568–1572.

Berezhna S., Schaefer S., Heintzmann R. et al. 2005. New effects in polynucleotide release from cationic lipid carriers revealed by confocal imaging, fluorescence cross-correlation spectroscopy and single particle tracking. *Biochimica et Biophysica Acta—Biomembranes* 1669: 193–207.

Braeckmans K., Peeters L., Sanders N. N., De Smedt S. C., and Demeester J. 2003. Three-dimensional fluorescence recovery after photobleaching with the confocal scanning laser microscope. *Biophysical Journal* 85: 2240–2252.

Braeckmans K., Remaut K., Vandenbroucke R. E. et al. 2007. Line FRAP with the confocal laser scanning microscope for diffusion measurements in small regions of 3-D samples. *Biophysical Journal* 92: 2172–2183.

Braga J., Desterro J. M. P., and Carmo-Fonseca M. 2004. Intracellular macromolecular mobility measured by fluorescence recovery after photobleaching with confocal laser scanning microscopes. *Molecular Biology of the Cell* 15: 4749–4760.

Cannone F., Chirico G., and Diaspro A. 2003. Two-photon interactions at single fluorescent molecule level. *Journal of Biomedical Optics* 8: 391–395.

Cheezum M. K., Walker W. F., and Guilford W. H. 2001. Quantitative comparison of algorithms for tracking single fluorescent particles. *Biophysical Journal* 81: 2378–2388.

Chen Y., Lagerholm B. C., Yang B., and Jacobson K. 2006. Methods to measure the lateral diffusion of membrane lipids and proteins. *Methods* 39: 147–153.

Chirico G., Cannone F., and Diaspro A. 2003. Single molecule photodynamics by means of one- and two-photon approach. *Journal of Physics D—Applied Physics* 36: 1682–1688 (July 21).

Cole R. W. and Turner J. N. 2008. Light-emitting diodes are better illumination sources for biological microscopy than conventional sources. *Microscopy and Microanalysis* 14: 243–250.

Crocker J. C. and Grier D. G. 1996. Methods of digital video microscopy for colloidal studies. *Journal of Colloid and Interface Science* 179: 298–310.

Dawson M., Wirtz D., and Hanes J. 2003. Enhanced viscoelasticity of human cystic fibrotic sputum correlates with increasing microheterogeneity in particle transport. *Journal of Biological Chemistry* 278: 50393–50401.

de Bruin K., Ruthardt N., von Gersdorff K. et al. 2007. Cellular dynamics of EGF receptor-targeted synthetic viruses. *Molecular Therapy* 15: 1297–1305.

De Smedt S. C., Remaut K., Lucas B. et al. 2005. Studying biophysical barriers to DNA delivery by advanced light microscopy. *Advanced Drug Delivery Reviews* 57: 191–210.

Dietrich C., Yang B., Fujiwara T., Kusumi A., and Jacobson K. 2002. Relationship of lipid rafts to transient confinement zones detected by single particle tracking. *Biophysical Journal* 82: 274–284.

Elsner M., Hashimoto H., Simpson J. C. et al. 2003. Spatiotemporal dynamics of the COPI vesicle machinery. *Embo Reports* 4: 1000–1005.

Fujiwara T., Ritchie K., Murakoshi H., Jacobson K., and Kusumi A. 2002. Phospholipids undergo hop diffusion in compartmentalized cell membrane. *Journal of Cell Biology* 157: 1071–1081.

Gosch M. and Rigler R. 2005. Fluorescence correlation spectroscopy of molecular motions and kinetics. *Advanced Drug Delivery Reviews* 57: 169–190.

Haustein E. and Schwille P. 2004. Single-molecule spectroscopic methods. *Current Opinion in Structural Biology* 14: 531–540.

Hellriegel C., Kirstein J., Brauchle C. et al. 2004. Diffusion of single streptocyanine molecules in the nanoporous network of sol-gel glasses. *Journal of Physical Chemistry B* 108: 14699–14709.

Hellriegel C., Kirstein J., and Brauchle C. 2005. Tracking of single molecules as a powerful method to characterize diffusivity of organic species in mesoporous materials. *New Journal of Physics* 7: 1–14.

Ishii T., Okahata Y., and Sato T. 2001. Mechanism of cell transfection with plasmid/chitosan complexes. *Biochimica et Biophysica Acta—Biomembranes* 1514: 51–64.

Jonkman J. E. N. and Stelzer E. H. K. 2002. Resolution and contrast in confocal and two-photon microscopy. In *Confocal and Two-Photon Microscopy*, ed. A. Diaspro, pp. 101–125. New York: Wiley-Liss, Inc.

Karpova T. S., Chen T. Y., Sprague B. L., and McNally J. G. 2004. Dynamic interactions of a transcription factor with DNA are accelerated by a chromatin remodeller. *Embo Reports* 5: 1064–1070.

Koyama-Honda I., Ritchie K., Fujiwara T. et al. 2005. Fluorescence imaging for monitoring the colocalization of two single molecules in living cells. *Biophysical Journal* 88: 2126–2136.

Kubitscheck U., Kuckmann O., Kues T., and Peters R. 2000. Imaging and tracking of single GFP molecules in solution. *Biophysical Journal* 78: 2170–2179.

Kulkarni R. P., Wu D. D., Davis M. E., and Fraser S. E. 2005. Quantitating intracellular transport of polyplexes by spatio-temporal image correlation spectroscopy. *Proceedings of the National Academy of Sciences of the United States of America* 102: 7523–7528.

Kulkarni R. P., Castelino K., Majumdar A., and Fraser S. E. 2006. Intracellular transport dynamics of endosomes containing DNA polyplexes along the microtubule network. *Biophysical Journal* 90: L42–L44.

Kusumi A., Sako Y., and Yamamoto M. 1993. Confined lateral diffusion of membrane-receptors as studied by single-particle tracking (nanovid microscopy)—Effects of calcium-induced differentiation in cultured epithelial-cells. *Biophysical Journal* 65: 2021–2040.

Kusumi A., Ike H., Nakada C., Murase K., and Fujiwara T. 2005a. Single-molecule tracking of membrane molecules: Plasma membrane compartmentalization and dynamic assembly of raft-philic signaling molecules. *Seminars in Immunology* 17: 3–21.

Kusumi A., Nakada C., Ritchie K. et al. 2005b. Paradigm shift of the plasma membrane concept from the two-dimensional continuum fluid to the partitioned fluid: High-speed single-molecule tracking of membrane molecules. *Annual Review of Biophysics and Biomolecular Structure* 34: 351–U54.

Lakadamyali M., Rust M. J., Babcock H. P., and Zhuang X. W. 2003. Visualizing infection of individual influenza viruses. *Proceedings of the National Academy of Sciences of the United States of America* 100: 9280–9285.

Levi V. and Gratton E. 2007. Exploring dynamics in living cells by tracking single particles. *Cell Biochemistry and Biophysics* 48: 1–15.

Levi V., Ruan Q. Q., and Gratton E. 2005. D particle tracking in a two-photon microscope: Application to the study of molecular dynamics in cells. *Biophysical Journal* 88: 2919–2928.

Levi V., Serpinskaya A. S., Gratton E., and Gelfand V. 2006. Organelle transport along microtubules in *Xenopus melanophores*: Evidence for cooperation between multiple motors. *Biophysical Journal* 90: 318–327.

Lill Y., Lill M. A., Fahrenkrog B. et al. 2006. Single hepatitis-B virus core capsid binding to individual nuclear pore complexes in HeLa cells. *Biophysical Journal* 91: 3123–3130.

Manley S., Gillette J. M., Patterson G. H. et al. 2008. High-density mapping of single-molecule trajectories with photoactivated localization microscopy. *Nature Methods* 5: 155–157.

Mazza D., Braeckmans K., Cella F. et al. 2008. A new FRAP/FRAPa method for three-dimensional diffusion measurements based on multiphoton excitation microscopy. *Biophysical Journal* 95: 3457–3469.

Meyvis T. K. L., De Smedt S. C., Van Oostveldt P., and Demeester J. 1999. Fluorescence recovery after photobleaching: A versatile tool for mobility and interaction measurements in pharmaceutical research. *Pharmaceutical Research* 16: 1153–1162.

Ober R. J., Ram S., and Ward E. S. 2004. Localization accuracy in single-molecule microscopy. *Biophysical Journal* 86: 1185–1200.

Pelkmans L., Kartenbeck J., and Helenius A. 2001. Caveolar endocytosis of simian virus 40 reveals a new two-step vesicular-transport pathway to the ER. *Nature Cell Biology* 3: 473–483.

Peters R., Peters J., Tews K. H., and Bahr W. 1974. Microfluorimetric study of translational diffusion in erythrocyte-membranes. *Biochimica Et Biophysica Acta* 367: 282–294.

Qian H., Sheetz M. P., and Elson E. L. 1991. Single-particle tracking—Analysis of diffusion and flow in 2-dimensional systems. *Biophysical Journal* 60: 910–921.

Remaut K., Sanders N. N., De Geest B. G. et al. 2007. Nucleic acid delivery: Where material sciences and bio-sciences meet. *Materials Science & Engineering R-Reports* 58: 117–161.

Ritchie K., Shan X. Y., Kondo J. et al. 2005. Detection of non-Brownian diffusion in the cell membrane in single molecule tracking. *Biophysical Journal* 88: 2266–2277.

Sanders N. N., De Smedt S. C., and Demeester J. 2000. The physical properties of biogels and their permeability for macromolecular drugs and colloidal drug carriers. *Journal of Pharmaceutical Sciences* 89: 835–849.

Saxton M. J. 1997. Single-particle tracking: The distribution of diffusion coefficients. *Biophysical Journal* 72: 1744–1753.

Saxton M. J. and Jacobson K. 1997. Single-particle tracking: Applications to membrane dynamics. *Annual Review of Biophysics and Biomolecular Structure* 26: 373–399.

Sbalzarini I. F. and Koumoutsakos P. 2005. Feature point tracking and trajectory analysis for video imaging in cell biology. *Journal of Structural Biology* 151: 182–195.

Schmidt T., Schutz G. J., Baumgartner W., Gruber H. J., and Schindler H. 1996. Imaging of single molecule diffusion. *Proceedings of the National Academy of Sciences of the United States of America* 93: 2926–2929.

Schutz G. J., Schindler H., and Schmidt T. 1997. Single-molecule microscopy on model membranes reveals anomalous diffusion. *Biophysical Journal* 73: 1073–1080.

Schutz G. J., Kada G., Pastushenko V. P., and Schindler H. 2000. Properties of lipid microdomains in a muscle cell membrane visualized by single molecule microscopy. *Embo Journal* 19: 892–901.

Schwille P. 2001. Fluorescence correlation spectroscopy and its potential for intracellular applications. *Cell Biochemistry and Biophysics* 34: 383–408.

Seisenberger G., Ried M. U., Endress T. et al. 2001. Real-time single-molecule imaging of the infection pathway of an adeno-associated virus. *Science* 294: 1929–1932.

Sonnleitner A., Schutz G. J., and Schmidt T. 1999. Free Brownian motion of individual lipid molecules in biomembranes. *Biophysical Journal* 77: 2638–2642.

Sprague B. L. and McNally J. G. 2005. FRAP analysis of binding: Proper and fitting. *Trends in Cell Biology* 15: 84–91.

Suh J., Wirtz D., and Hanes J. 2003. Efficient active transport of gene nanocarriers to the cell nucleus. *Proceedings of the National Academy of Sciences of the United States of America* 100: 3878–3882.

Suh J., Dawson M., and Hanes J. 2005. Real-time multiple-particle tracking: Applications to drug and gene delivery. *Advanced Drug Delivery Reviews* 57: 63–78.

Thompson N. L., Lieto A. M., and Allen N. W. 2002a. Recent advances in fluorescence correlation spectroscopy. *Current Opinion in Structural Biology* 12: 634–641.

Thompson R. E., Larson D. R., and Webb W. W. 2002b. Precise nanometer localization analysis for individual fluorescent probes. *Biophysical Journal* 82: 2775–2783.

Tseng Y., Kole T. P., and Wirtz D. 2002. Micromechanical mapping of live cells by multiple-particle-tracking microrheology. *Biophysical Journal* 83: 3162–3176.

Verkman A. S. 2003. Diffusion in cells measured by fluorescence recovery after photobleaching. In *Biophotonics, Part A: Methods in Enzymology*, eds. G. Marriott and I. Parker, pp. 635–648. New York: Academic Press.

Vrljic M., Nishimura S. Y., Brasselet S., Moerner W. E., and McConnell H. M. 2002. Translational diffusion of individual class II MHC membrane proteins in cells. *Biophysical Journal* 83: 2681–2692.

Wilson K. M., Morrison I. E. G., Smith P. R., Fernandez N., and Cherry R. J. 1996. Single particle tracking of cell-surface HLA-DR molecules using R-phycoerythrin labeled monoclonal antibodies and fluorescence digital imaging. *Journal of Cell Science* 109: 2101–2109.

Yildiz A., Forkey J. N., McKinney S. A. et al. 2003. Myosin V walks hand-over-hand: Single fluorophore imaging with 1.5-nm localization. *Science* 300: 2061–2065.

Yildiz A., Tomishige M., Vale R. D., and Selvin P. R. 2004. Kinesin walks hand-over-hand. *Science* 303: 676–678.

6

Fluorescence Correlation Spectroscopy

6.1	Introduction	**6**-1
6.2	Background	**6**-2
	What Is a Correlation? • How Are Correlation Functions Calculated from Experiments? • Correlator Schemes	
6.3	Instrumentation	**6**-15
6.4	State of the Art	**6**-18
	FCS with Confocal Illumination • Fluorescence Correlation Microscopy • Two-Photon Excitation Fluorescence Correlation Spectroscopy • Total Internal Reflection Fluorescence Correlation Spectroscopy • Fluorescence Cross-Correlation Spectroscopy • Scanning Fluorescence Correlation Spectroscopy • CCD-Based Fluorescence Correlation Spectroscopy • Computational Advances	
6.5	Summary	6-25
6.6	Future Perspectives	6-26
	Acknowledgments..	6-26
	References	**6**-26

Xianke Shi
Thorsten Wohland

National University of Singapore

6.1 Introduction

The end of the twentieth and the beginning of the twenty-first century saw many exciting developments in science and the emergence of novel questions. In particular, the advances in modern biology and nanotechnology brought with them the need to understand processes on a molecular basis. The advent of widely available lasers and computers and the renaissance of light microscopy with the introduction of the first commercial confocal microscope provided an ideal platform for highly sensitive detection coupled with good temporal and spatial resolutions. This formed not only the basis for the investigation of molecular events but in addition was accessible to a wide range of researchers due to the low cost of the new instrumentation. Consequently, the field of light microscopy and fluorescence spectroscopy grew at an accelerated pace and is still growing strongly with an ever-increasing number of new techniques and methods being published (Thompson et al., 2002; Haustein and Schwille, 2007; Hwang and Wohland, 2007; Kolin and Wiseman, 2007; Liu et al., 2008). The first single molecule detection, achieved by fluorescence microscopy in 1976 (Hirschfeld, 1976), led the way for the development of techniques that could detect single molecules on a routine basis.

This was an important event since the access to the behavior of single molecules gave scientists access to the molecular basis of parameters and their distributions for the first time. An ensemble measurement determines only the average value and the standard deviation but cannot determine the underlying distribution of parameters giving rise to these experimental values. Thus, questions whether there is one or multiple molecular species in a sample, e.g., proteins of multiple conformations, cannot be answered. Single molecule experiments, on the contrary, give access to the full distribution of the fluorescent parameters and can thus reconstruct the distribution of different molecular characteristics (Ambrose et al., 1994; Nie et al., 1994; Plakhotnik et al., 1997; Xie and Lu, 1999).

However, although single molecule spectroscopy techniques provide the ultimate resolution for molecular processes, they also suffer from disadvantages. Single molecule observations are time consuming since one has to record many single molecule events to obtain good statistics and often require complicated data evaluation procedures. One solution to this problem was invented by Madge et al. (Magde et al., 1972) who introduced fluorescence correlation spectroscopy (FCS), a technique with single molecule sensitivity but at the same time based on the fast statistical treatment of the recorded data. In FCS, the fluorescence intensity from a very small observation volume ($\sim 10^{-15}$ L) is recorded with high temporal resolution. Any process that causes variations in the fluorescence intensity and happens on a timescale slower than the recording speed of FCS will leave characteristic fluctuations in the intensity trace. These fluctuations can be identified by either a Fourier transform or an auto-correlation analysis.

In this chapter, we introduce the basic idea of FCS, review the state of the art, discuss the advantages, and give an outlook as to where the next frontiers for this technique could lie.

6.2 Background

6.2.1 What Is a Correlation?

A correlation between two variables a and b describes the dependence between these two variables. In practical terms, this means that if we know the value of a or b we can make some prediction of the value of b or a, respectively. While a correlation is thus a weaker criterion than causation (a correlation is not a proof of causation, but causation requires correlation), it is a very useful concept since it allows some predictions about a second variable from the observation of the first.

As an example, we provide three instances of variables a and b in Table 6.1 in which the two are either correlated, anticorrelated, or are uncorrelated. The correlation between the two variables has some measurable effects. Since the values of two correlated variables change with a similar pattern, we expect that a multiplication of the two variables with each other will reinforce this pattern. If we calculate now the average values of a, b, and the product ab, we expect (Berne and Pecora, 2000)

$$\langle a \cdot b \rangle \neq \langle a \rangle \langle b \rangle \tag{6.1}$$

As a matter of fact, $\langle a \cdot b \rangle$ can be larger, smaller, or equal to $\langle a \rangle \langle b \rangle$, depending on whether the variables a and b are correlated, anticorrelated, or have no correlation, respectively (see Table 6.2, last column).

We can simplify the description of correlations by normalizing $\langle a \cdot b \rangle$ with $\langle a \rangle \langle b \rangle$:

$$g = \frac{\langle a \cdot b \rangle}{\langle a \rangle \langle b \rangle} \tag{6.2}$$

The value of g is now 1 for uncorrelated variables, >1 for correlated, and <1 for anticorrelated variables.

Up to now we dealt with what is commonly called a cross-correlation function (CCF) between two variables a and b. It is interesting to ask whether a variable, which is observed over a duration of time, can be correlated with itself and thus if a variable is autocorrelated. This can give us an insight into the time course of the variable and thus the underlying process causing this signal. If we compare the variable $a(t)$ at the same point in time, this is of course trivially the case. But if we compare the values of a variable at different points in time, e.g., $a(t)$ and $a(t + \tau)$, where τ describes the difference in time, a

TABLE 6.1 Examples of Correlated, Anticorrelated, and Uncorrelated Variables

(1) Correlated variables

a 1 1 0 1 0 0 1 1 1 0 1 0 1 0 1 1 0 0 0 0 $\langle a \cdot b \rangle > \langle a \rangle \langle b \rangle$

b 1 1 0 1 0 0 1 1 1 0 1 0 1 0 1 1 0 0 0 0 $\langle a \cdot b \rangle = \frac{1}{2};\ \langle a \rangle \langle b \rangle = \frac{1}{2} \cdot \frac{1}{2} = \frac{1}{4}$

(2) Anticorrelated variables

a 1 1 0 0 1 0 1 0 1 1 1 0 1 0 0 0 1 0 1 0 $\langle a \cdot b \rangle < \langle a \rangle \langle b \rangle$

b 0 0 1 1 0 1 0 1 0 0 0 1 0 1 1 1 0 1 0 1 $\langle a \cdot b \rangle = 0;\ \langle a \rangle \langle b \rangle = \frac{1}{2} \cdot \frac{1}{2} = \frac{1}{4}$

(3) Uncorrelated variables

a 1 0 1 0 1 0 1 0 1 0 1 0 1 0 1 0 1 0 1 0 $\langle a \cdot b \rangle = \langle a \rangle \langle b \rangle$

b 1 1 1 1 1 1 1 1 1 1 1 1 1 1 1 1 1 1 1 1 $\langle a \cdot b \rangle = \frac{1}{2};\ \langle a \rangle \langle b \rangle = \frac{1}{2} \cdot 1 = \frac{1}{2}$

Note: (1) Correlated variables: Values a and b have always the same values and thus are perfectly correlated. The knowledge of either a or b allows an exact prediction of the other variable. (2) Anticorrelated variables: The case is similar to 1, only here the variable b has always exactly the opposite value of a. Again a perfect prediction of b from a and vice versa is possible. (3) Uncorrelated variables: In this case there is no relation of a and b, and the value of a is not predictive of the values of b.

TABLE 6.2 Autocorrelation and Cross-Correlation Functions at Different Times

(1) Autocorrelation function at different times

$a(t)$ 1 1 1 0 0 1 1 1 0 0 0 0 0 1 1 1 0 0 1 0 $\langle a(t) \cdot b(t) \rangle > \langle a(t) \rangle \langle b(t) \rangle$

$a(t)$ 1 1 1 0 0 1 1 1 0 0 0 0 0 1 1 1 0 0 1 0 $\langle a(t) \cdot a(t) \rangle = \frac{1}{2};\ \langle a(t) \rangle \langle a(t) \rangle = \frac{1}{2} \cdot \frac{1}{2} = \frac{1}{4}$

$a(t)$ 0 1 1 1 0 0 1 1 1 0 0 0 0 0 1 1 1 0 0 1 $\langle a(t) \cdot a(t+\tau) \rangle > \langle a(t) \rangle \langle a(t+\tau) \rangle$

$a(t+\tau)$ 1 1 1 0 0 1 1 1 0 0 0 0 0 1 1 1 0 0 1 0 $\langle a(t) \cdot a(t+\tau) \rangle = 0.3;\ \langle a(t) \rangle \langle a(t+\tau) \rangle = \frac{1}{2} \cdot \frac{1}{2} = \frac{1}{4}$

(2) Cross-correlation function at different times

$a(t)$ 1 0 0 1 0 0 0 0 1 0 0 1 0 0 0 1 0 0 0 0 $\langle a(t) \cdot b(t) \rangle > \langle a(t) \rangle \langle b(t) \rangle$

$b(t)$ 0 1 0 0 1 0 0 0 0 1 0 0 1 0 0 0 1 0 0 0 $\langle a(t) \cdot b(t) \rangle = 0;\ \langle a(t) \rangle \langle b(t) \rangle = \frac{1}{2} \cdot \frac{1}{2} = \frac{1}{16}$

$a(t)$ 1 0 0 1 0 0 0 0 1 0 0 1 0 0 0 1 0 0 0 0 $\langle a(t) \cdot b(t+\tau) \rangle > \langle a(t) \rangle \langle b(t+\tau) \rangle$

$b(t+\tau)$ 1 0 0 1 0 0 0 0 1 0 0 1 0 0 0 1 0 0 0 0 $\langle a(t) \cdot b(t+\tau) \rangle = \frac{1}{4};\ \langle a(t) \rangle \langle b(t+\tau) \rangle = \frac{1}{4} \cdot \frac{1}{4} = \frac{1}{16}$

Note: (1) Autocorrelation function at different times: The autocorrelation at $\tau=0$ is at a maximum and decreases for $\tau>0$. (2) Cross-correlation function at different times: The cross-correlation has a peak at $\tau>0$ since the signal b is delayed by τ compared to signal a.

correlation will be found only if the value of a variable persists longer than the time τ, i.e., $a(t)$ changes on a timescale larger than τ. The autocorrelation function (ACF) is said to describe the self-similarity of a signal (values of the variable) in time. Therefore, the ACF is an indication in how far we can make predictions into the future of the values of a variable. If the variable has a long correlation in time we can make predictions far ahead, if the correlation time is short we can predict the future only over short time spans. If the correlation is strong we can make good predictions, if the correlation is weak our predictions will be less accurate.

In the following, we give two examples of an ACF and of a CCF to point out the difference between the two. The ACF of the signal $a(t)$ is largest if compared at the same time, i.e., the time shift $\tau=0$, and for shifts in time $\tau > 0$ is either smaller or can reach at maximum equal values as the initial value (e.g., for oscillating signals):

$$\left\langle a(t) \cdot a(t) \right\rangle \ge \left\langle a(t) \cdot a(t+\tau) \right\rangle \tag{6.3}$$

The CCF for signals $a(t)$ and $b(t)$ does not have the same restrictions and the CCF ($\langle a(t) \cdot b(t+\tau)\rangle$) can be larger at time shifts $\tau > 0$ than for $\tau=0$. This can be easily understood considering a signal $b(t)$, which is delayed in time by τ_p compared to the signal $a(t)$. In that case, the maximum value of the CCF is found at $\langle a(t) \cdot b(t+\tau_p)\rangle$. In Figure 6.1, we give some examples of ACFs and CCFs and the signals from which they were calculated. The first example (Figure 6.1A) shows a signal consisting of random noise. Since this signal is random, the signals at different time points have no correlation and the ACF is flat. In Figure 6.1B, we show a signal that contains an oscillation. The oscillating signal can hardly be seen within the noisy signal. However, the ACF clearly can distinguish the signal from the noise and leads to an oscillating ACF with the same frequency as the signal. In Figure 6.1C, we have added a transient peak on top of a noisy signal. The ACF picks up the signal above the random background and represents how the signal in the peak develops. (Note, that the ACF does not say anything about when a signal peak occurs. It only predicts that if a peak occurs it will develop on average in a particular way as represented by the correlation function.) Figure 6.1D shows two different signals in which a peak occurs at the same time. The CCF shows that the signals occur at the same time, and how the two signals develop in dependence of each other. In the last example, Figure 6.1E, we show two signals in which peaks occur. But now the second signal is delayed with respect to the first. Consequently, the CCF shows a maximum just at this delay between the two signals.

After these preliminary explanations about correlation functions, we can now proceed to FCS. In FCS, we are interested in the fluctuations in the fluorescence signal stemming from a sample containing some fluorescent probes (a fluctuation here means a transient deviation of the fluorescence from its average value). The fluorescent probes can undergo a range of different processes within the sample, which change the fluorescence emission from the probe, e.g., chemical reactions, enzymatic reactions, translational and rotational diffusions, and photophysical transitions. Any molecule that is within the observation volume and undergoes one of these processes will create fluctuation in the fluorescence. These fluctuations contain information about the characteristic time of the process and of the frequency of occurrence. By measuring the fluorescence signal and subjecting it to an autocorrelation analysis, we can find out in how far a signal that is found a time τ in the future is related to the signal at present. For times smaller than the characteristic duration of a molecular process, τ_p, one expects that the fluctuations are dependent on each other. But for times much longer than τ_p there should be no dependence anymore between the fluctuations. For times similar to τ_p, there will be a transition from correlation to noncorrelation. We thus measure the fluorescence signal from the observation volume as a function of time, $F(t)$, and define the ACF in analogy to Equation 6.2 mentioned earlier, which describes the correlation between the fluorescence signals that occur at time τ apart (Figure 6.2):

$$G(\tau) = \frac{\left\langle F(t+\tau)F(t) \right\rangle}{\left\langle F(t+\tau) \right\rangle \left\langle F(t) \right\rangle} = \frac{\left\langle F(t+\tau)F(t) \right\rangle}{\left\langle F(t) \right\rangle^2} \tag{6.4}$$

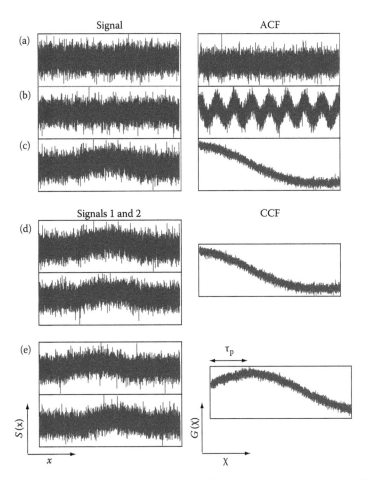

FIGURE 6.1 The figure shows signals in the left column and the corresponding correlations in the right column. The *y*-axis for the left column represents the signal strength $S(x)$ in arbitrary units. The *x*-axis in the left column represents the variable *x* against which the signal is measured (e.g., time, space etc.). The *y*-axis in the right column represents the correlation $G(\chi)$. The *x*-axis in the right column represents then the correlation variable χ (i.e., the difference in *x* for which the correlation is calculated). (A) A random signal leads to a flat ACF. (B) If as in this case a periodic signal is hidden in a very noisy signal, the ACF will reveal this. (C) If peaks are found within the signal, the ACF will give information about the peaks and their average development in time. (D) For two different signals, the CCF represents the correlation between the two signals. In this case, while the noise in the two channels is different, they both register the same peak. Therefore, the CCF is very similar to the ACF. (E) The peak in the two signals is now delayed by τ_p. Therefore, the CCF has its maximum at τ_p. It should be noted that this maximum will only be found if this delay τ_p holds on average true for all peaks detected in the two signals. If the delay between the peaks is random, the CCF will be flat.

For the derivation of the right-hand side, we assume that the processes under investigation are stationary, i.e., the statistical properties of the processes are invariant to a shift in time and

$$\left\langle F\left(t+\tau\right)\right\rangle = \left\langle F\left(t\right)\right\rangle \tag{6.5}$$

Assuming, without loss of generality, that the processes under investigation have a finite duration and thus that the fluorescence signal is independent of itself for long times (compared to the duration of the process), we can immediately distinguish three cases:

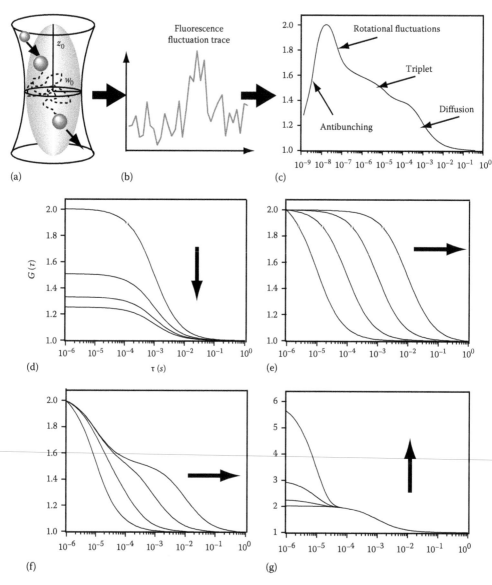

FIGURE 6.2 Characteristics of fluorescence correlation functions. (a) Particles diffuse through the confocal volume with half axis w_0 and z_0 gives rise to (b) fluorescence intensity fluctuations. (c) This graph shows a typical fluorescence correlation function with four different parts: antibunching representing an anticorrelation since a molecule can emit only one photon at a time; rotational motion of fluorophores since the rotation of the fluorescence transition dipoles will lead to fluctuations if the excitation light is linearly polarized; transitions to the triplet state, which represents a forbidden transition and thus has a long lifetime typically in the microsecond range; diffusion of fluorescent molecules through the focal volume. (d) The change in an ACF due to changes in concentration; the arrow indicates increasing concentration. (e) Changes in an ACF for different diffusion coefficients; the arrow indicates increasing diffusion time. (f) ACFs if two different particles are present in equal amounts, and the diffusion time of first particle is set at a lower value; the arrow indicates increasing diffusion time of the second particle. (g) Influence of the number of particles found in the triplet state on an ACF; the arrow indicates increasing fraction of triplet state.

$$1.\ \tau = 0:\ G(0) = \frac{\left\langle F(t)^2 \right\rangle}{\left\langle F(t) \right\rangle^2} \tag{6.6a}$$

$$2.\ \tau = \infty:\ G(\infty) = 1 \tag{6.6b}$$

$$3.\ 0 < \tau < \infty:\ G(\tau) = \frac{\left\langle F(t+\tau)F(t) \right\rangle}{\left\langle F(t) \right\rangle^2} \tag{6.6c}$$

Note that in literature sometimes the fluorescence *fluctuation* correlation function is used instead of the above-defined fluorescence *signal* correlation function. The fluctuations of the fluorescence signal, $\delta F(t)$, are defined as the deviations from the average fluorescence signal $\langle F(t) \rangle$:

$$\delta F(t) = F(t) - \langle F \rangle \tag{6.7}$$

The fluorescence fluctuation correlation function can be derived from the fluorescence signal correlation function:

$$G(\tau) = \frac{\left\langle F(t+\tau)F(t) \right\rangle}{\left\langle F(t) \right\rangle^2} = \frac{\left\langle \left(\delta F(t+\tau) + \langle F \rangle \right)\left(\delta F(t) + \langle F \rangle \right) \right\rangle}{\left\langle \left(\delta F(t) + \langle F \rangle \right) \right\rangle^2}$$

$$= \frac{\left\langle \delta F(t+\tau)\delta F(t) \right\rangle + \langle F \rangle^2}{\langle F \rangle^2} = \frac{\left\langle \delta F(t+\tau)\delta F(t) \right\rangle}{\left\langle F(t) \right\rangle^2} + 1 \tag{6.8}$$

which shows that the two differ only by an additive factor of 1. Here we have used the property that the average of the fluctuations over time is 0, i.e., $\langle \delta F(t) \rangle = 0$.

We can define now the fluorescence signal $F(t)$ and its fluctuations $\delta F(t)$ of a sample measured with an instrument over space ($\vec{r} = (x, y, z)$)

$$F(t) = \kappa \int I(\vec{r}) \cdot \mathrm{CEF}(\vec{r}) \cdot S(\vec{r}) \cdot C(\vec{r}, t)\, d\vec{r} \tag{6.9a}$$

$$\delta F(t) = \kappa \int I(\vec{r}) \cdot \mathrm{CEF}(\vec{r}) \cdot S(\vec{r}) \cdot \delta C(\vec{r}, t)\, d\vec{r} \tag{6.9b}$$

where

κ is a product of the fluorophore absorption cross section, its quantum yield, and the overall system detection efficiency

$I(\vec{r})$ is the illumination intensity profile

$\mathrm{CEF}(\vec{r})$ is the normalized collection efficiency function of the system for different points in the sample

$\mathrm{CEF}(\vec{r})$ is given by the transmission function of the pinhole $T(\vec{r})$ and the point-spread function of the microscope $\mathrm{PSF}(\vec{r})$ (Rigler et al., 1993):

$$\text{CEF}(\vec{r}) = \int T(\vec{r}) \cdot \text{PSF}(\vec{r}) d\vec{r} \qquad (6.10)$$

where

$S(\vec{r})$ is a function describing the extension of the sample

$C(\vec{r},t)$ and $\delta C(\vec{r},t)$ are functions describing the concentration of particles and their fluctuations, respectively, within the sample

By inserting Equation 6.9 into Equation 6.8, the correlation function can then be written as

$$G(\tau) = \frac{\iint I(\vec{r}) I(\vec{r}') \cdot \text{CEF}(\vec{r}) \cdot \text{CEF}(\vec{r}') \cdot S(\vec{r}) \cdot S(\vec{r}') \cdot \langle \delta C(\vec{r}', t + \tau) \delta C(\vec{r}, t) \rangle d\vec{r} \, d\vec{r}'}{\langle C \rangle^2 \left(\int I(\vec{r}) \cdot \text{CEF}(\vec{r}) d\vec{r} \right)^2} + 1 \qquad (6.11)$$

This equation can be analytically or numerically solved for different illumination profiles, collection efficiency functions, and functions describing the fluctuation in fluorescent particles (e.g., diffusion, flow, fluorophore blinking, or chemical reactions).

We will show here the most commonly used case of diffusion when using a confocal setup in an infinite sample, i.e., $S(\vec{r}) = 1$ and thus poses no constraints on the measurement. In a confocal setup, the illumination profile is given by a laser beam, which is focused into a small diffraction-limited spot and is given by a Gaussian laser beam profile (Saleh and Teich, 1991). The pinhole removes out-of-focus light by reducing the collection efficiency along the z-axis the further the emission happens from the focal plane. The two effects together can be approximated by a simple rotationally symmetric function, which is Gaussian in all three dimensions:

$$W(\vec{r}) = I(\vec{r}) \cdot \text{CEF}(\vec{r}) \cdot S(\vec{r}) = \frac{2P}{\pi w_0^2} e^{-2x^2/w_0^2} e^{-2y^2/w_0^2} e^{-2z^2/z_0^2} \qquad (6.12)$$

Here w_0 and z_0 are the distances at which the laser beam has decreased to $1/e^2$ of the intensity maximum at the center of the beam. It can be shown that the correlation of concentration fluctuations is described by the so-called diffusion propagator (Weidemann et al., 2002):

$$\langle \delta C(\vec{r}', t + \tau) \delta C(\vec{r}, t) \rangle = \langle C \rangle \frac{e^{-\left((r-r')^2/4D\tau\right)}}{8(\pi D\tau)^{3/2}} \qquad (6.13)$$

Putting this information together we get

$$G(\tau) = \frac{\iint W(\vec{r}) W(\vec{r}') \cdot \langle \delta C(\vec{r}', t + \tau) \delta C(\vec{r}, t) \rangle d\vec{r} \, d\vec{r}'}{\left(\int \langle C \rangle W(\vec{r}) d\vec{r} \right)^2} + 1$$

$$= \frac{\iint e^{-2(x^2 - x'^2)/w_0^2} e^{-2(y^2 - y'^2)/w_0^2} e^{-2(z^2 - z'^2)/z_0^2} \cdot \frac{e^{-\left((r-r')^2/4D\tau\right)}}{8(\pi D\tau)^{3/2}} d\vec{r} \, d\vec{r}'}{\langle C \rangle \left(\int e^{-2x^2/w_0^2} e^{-2y^2/w_0^2} e^{-2z^2/z_0^2} d\vec{r} \right)^2} + 1 \qquad (6.14)$$

The integration gives the following solution

$$G(\tau) = \frac{1}{\langle C \rangle \pi^{3/2} w_0^2 z_0} \left(1 + \frac{4D\tau}{w_0^2}\right)^{-1/2} \left(1 + \frac{4D\tau}{w_0^2}\right)^{-1/2} \left(1 + \frac{4D\tau}{z_0^2}\right)^{-1/2} + 1 \qquad (6.15)$$

where $\pi^{3/2} w_0^2 z_0$ is valid for the 3D Gaussian profile given in Equation 6.12 and is in general a constant depending on the actual observation volume, $W(\vec{r})$. For practical purposes the function is often written as

$$G(\tau) = \frac{1}{N}\left(1 + \frac{4D\tau}{w_0^2}\right)^{-1}\left(1 + \frac{4D\tau}{z_0^2}\right)^{-1/2} + G_\infty \qquad (6.16)$$

With N defined as the apparent number of particles in the observation volume.[*] The relation between N and actual concentration depends crucially on the function $W(\vec{r})$. G_∞, the convergence value, is introduced as a free parameter. Its value, which is 1 for an infinite measurement time, can differ slightly from 1 for finite measurement times (Equation 6.8). Therefore introducing G_∞ as a free parameter usually improves the quality of fits. In general, the parameter G_∞ will deviate from 1 by less than 1% in solution measurements. If G_∞ deviates strongly from 1, it can be an indication of photobleaching (Dittrich and Schwille, 2001), i.e., the fluorescence signal decays exponentially during the measurement owing to the destruction of fluorescent molecules. Other problems in its value can originate from instabilities of the setup or a moving sample. Examples of correlation functions for different situations are given in Table 6.3.

6.2.2 How Are Correlation Functions Calculated from Experiments?

ACFs have to be calculated from the measurements of intensities or photon counts from a sample. Two important timescales have to be determined for the autocorrelations in advance. Firstly, the minimum sample time $\Delta\tau$, i.e., the minimal time interval over which emitted photons are counted. This time determines the fastest measurable process since as a rule of thumb one has to measure about 10 times faster than the fastest process one wants to observe. In addition, one has to keep in mind that the shorter the $\Delta\tau$, the lower the number of detected photons in each time interval and the larger the standard deviation of the measurement. The second timescale is the total measurement time T. The standard deviation of the ACF is proportional to $\sqrt{T}$ and thus longer measurement times will improve the quality of the measurement, e.g., a four times longer measurement will decrease the standard deviation by a factor 2. This limited effect of T on the standard deviation makes an increase in measurement time not always a good option. The number of points available to calculate the correlation are then given by $M = T/\Delta\tau$. The correlation time τ is given as multiples of the minimum sampling time $\Delta\tau$, i.e., $\tau = m\Delta\tau$. Thus, the ACF for different correlation times $m\Delta\tau$ ($0 \leq m \leq M$) can be calculated as follows (note that this is the fluorescence *signal* correlation function as shown in the second term in Equation 6.8):

$$G(m\Delta\tau) = \frac{(1/M - m)\sum_{k=1}^{M-m} n(k\Delta\tau) n(k\Delta\tau + m\Delta\tau)}{\left((1/M)\sum_{k=1}^{M} n(k\Delta\tau)\right)^2} = \frac{M^2}{M-m} \cdot \frac{\sum_{k=1}^{M-m} n(k\Delta\tau) n(k\Delta\tau + m\Delta\tau)}{\left(\sum_{k=1}^{M-m} n(k\Delta\tau)\right)^2} \qquad (6.17)$$

---

[*] Please note that N is sometimes defined slightly differently by using a correction factor in front of the ACF by writing γ/N instead of $1/N$ (see Thompson, 1991 for an explanation). This would allow the determination of absolute concentrations. We did not follow this approach here since in most cases the absolute concentration of samples are not calculated from the theoretical observation volume size but are rather determined by using a standard dye solution of a fixed known concentration against which all other concentrations are calibrated. In this case an N_{calib} is determined for the standard solution of concentration c_{calib} and the concentration of any other sample x is then $c_x = (N_x/N_{calib}) \cdot c_{calib}$.

TABLE 6.3 Commonly Used Models for ACF Fitting

Type	$G(\tau)$	References
2-D translational diffusion	$\dfrac{1}{N} g_{2D} + G_\infty$	Elson and Magde (1974)
3-D translational diffusion	$\dfrac{1}{N} g_{3D} + G_\infty$	Aragon and Pecora (1976)
3-D translational diffusion with multiple components	$\dfrac{1}{N} \dfrac{\sum_i \beta_i^2 F_i g_{3D_i}}{\left(\sum_i \beta_i F_i\right)^2} + G_\infty$ β_i is the fluorescence yield of particle i compared to the fluorescence yield of particle 1. F_i is the mole fraction of component i in the overall sample	Thompson (1991) and Yu et al. (2005)
3-D translational diffusion with finite-sized particles	$\dfrac{1}{N}\left(\dfrac{1}{1+\left(R/\omega_0\right)^2}\right) g_{3D} + G_\infty$ R is the particle radius and $1 > R/\omega_0 > 0.2$	Starchev et al. (1998) and Wu et al. (2008)
Confined diffusion in z-direction	$\dfrac{1}{N} g_{2D} \cdot g_{z^*} + G_\infty$ For different forms of g_{z^*} see the reference	Gennerich and Schild (2000)
Anomalous diffusion	$\dfrac{1}{N}\left[1+\left(\dfrac{4D\tau}{\omega_0^2}\right)^\alpha\right]^{-1}\left[1+\dfrac{1}{K^2}\left(\dfrac{4D\tau}{\omega_0^2}\right)^\alpha\right]^{-1/2} + G_\infty$ α is the anomalous diffusion factor	Schwille et al. (1999b)
Flow or line-scan Scanning FCS (SFCS)	$\dfrac{1}{N}\exp\left[-\left(\tau V/\omega_0\right)^2\right] + G_\infty$	Magde et al. (1978) and Koppel et al. (1994)
Single flow with free diffusion	$\dfrac{1}{N} g_{3D} \cdot \exp\left[-\left(\dfrac{\tau V}{\omega_0}\right)^2 \cdot g_{2D}\right] + G_\infty$	Magde et al. (1978)
Circle-scan SFCS	$\dfrac{1}{N} g_{3D} \cdot \exp\left[-\dfrac{C^2 \sin^2(\pi f \tau)}{\omega_0^2 + D\tau}\right] + G_\infty$	Berland et al. (1996)

Total internal reflection FCS (TIR-FCS)

f is the frequency of the scan in circular scanning FCS

C is the radius of the scanned circle in circular scanning FCS

$$\frac{1}{2N} g_{2D} \left[\left(1 - \frac{2D\tau}{z_0^2}\right) \cdot w\left(i\sqrt{\frac{D\tau}{z_0^2}}\right) + \sqrt{\frac{4D\tau}{\pi z_0^2}} \right] + G_\infty$$

Starr and Thompson (2001)

CCD-based TIR-FCS

where $w(x) = \exp(-x^2)\mathrm{erf}(-ix)$

$$\frac{1}{N}\left[\frac{2\sqrt{\omega_0^2 + D\tau}}{a\sqrt{\pi}}\left[\exp\left(-\frac{a^2}{4(\omega_0^2 + D\tau)}\right) - 1\right] + \frac{1}{a}\,\mathrm{erf}\left(\frac{a}{2\sqrt{\omega_0^2 + D\tau}}\right)\right]^2 + G_\infty$$

Guo et al. (2008) and Ries et al. (2008)

a is the size of the binned pixel in the CCD

Two foci cross-correlation for flow

$$\frac{1}{N}\exp\left\{-\left(\frac{B}{\omega_0}\right)^2\left[\left(\frac{\tau V}{B}\right)^2 + 1 - 2\left(\frac{\tau V}{B}\right)\cos\theta\right]\right\} + G_\infty$$

Brinkmeier et al. (1999)

θ is the angle between the flow and the two foci vector

V is the speed of the particles or the scanned laser line

Remarks: The basic term for the correlation function is a hyperbolic function for each dimension, if the illumination profile is Gaussian:

$$g_{1D} = \left(1 + \frac{4D\tau}{\omega^2}\right)^{-1/2}$$

where ω is the distance at which the effective intensity of the laser focus has decayed to $1/e^2$ of its maximum value at the center. In most setups, the beam is rotationally symmetric and we can define the 2D correlation term (e.g., for diffusion in a membrane) with ω_0 as the radius of the observation volume in xy-direction as defined earlier:

$$g_{2D} = \left(1 + \frac{4D\tau}{\omega_0^2}\right)^{-1}$$

For the 3D case, the focal volume is elongated along the z-axis with a radius z_0. This leads to the following function:

$$g_{3D} = \left(1 + \frac{4D\tau}{\omega_0^2}\right)^{-1}\left(1 + \frac{4D\tau}{z_0^2}\right)^{-1/2} = \left(1 + \frac{4D\tau}{\omega_0^2}\right)^{-1}\left(1 + \frac{4D\tau}{K^2\omega_0^2}\right)^{-1/2}$$

where $K = z_0/\omega_0$ represents the ratio of the radius of the focal volume in z-direction over the radius in the xy-direction.

(continued)

TABLE 6.3 (continued)   Commonly Used Models for ACF Fitting

Type	$G(\tau)$	References

If in addition to diffusion other processes are inducing fluctuations, e.g., blinking due to transition to the triplet state (also called fluorescence bunching), they have to be introduced by multiplying the basic correlation functions with the functions given in the following:

Type	$f(\tau)$	References
Triplet-state	$\left[1+\left(\dfrac{F_{trip}}{1-F_{trip}}\right)\cdot\exp(-\tau/\tau_{trip})\right]$	Widengren et al. (1994)
	F_{trip} is the fraction of the particles that have entered the triplet state.	
Rotational diffusion	$A\left(1-\exp(-\tau/\tau_r)\right)$	Kask et al. (1989)
	A is the coefficients for rotational correlations and τ_r, the rotational correlation time.	
Antibunching	$1-\exp\left(-|\tau|/\tau_{antibunching}\right)$	Thompson (1991)
Two foci cross-correlation for flow	$\dfrac{1}{N}\exp\left\{-\left[\left(\dfrac{B}{\omega_0}\right)^2\left(\dfrac{\tau V}{B}\right)^2+1-2\left(\dfrac{\tau V}{B}\right)\cos\theta\right]\right\}+G_\infty$	Brinkmeier et al. (1999)
	B is the distance between two foci	
	V is the constant speed of the nontranslating particles or the scanned laser line	

For instance, 3D diffusion of a fluorophore with a triplet component would result in

$$G(\tau)=\frac{1}{N}\cdot g_{3D}\cdot f_{trip}(\tau)+G_\infty=\frac{1}{N}\left(1+\frac{4D\tau}{\omega_0^2}\right)^{-1}\left(1+\frac{4D\tau}{K^2\omega_0^2}\right)^{-1/2}\left[1+\left(\frac{F_{trip}}{1-F_{trip}}\right)\cdot\exp\left(-\tau/\tau_{trip}\right)\right]+G_\infty$$

For infinite measurement times this would be the correct procedure. But at finite measurement times the denominator is calculated over a wider range (1 to M) than the terms in the numerator (1 to $M-m$, or m to M). This leads to problems especially at long correlation times where the introduced error of the asymmetry is greater than the error of the correlation. It has been shown that a symmetric normalization removes this problem (Schätzel et al., 1988; Schätzel, 1990). For the symmetric normalization, the averages in the denominator are taken over the same range as the range of the two terms $n(k\Delta\tau)$ and $n(k\Delta\tau + m\Delta\tau)$ in the numerator. This leads to the following corrected equation:

$$
\begin{aligned}
G\left(m\Delta\tau\right) &= \frac{(1/M-m)\sum_{k=1}^{M-m} n\left(k\Delta\tau\right) n\left(k\Delta\tau + m\Delta\tau\right)}{\left[(1/M-m)\sum_{k=1}^{M-m} n\left(k\Delta\tau\right)\right]\left[(1/M-m)\sum_{k=1}^{M-m} n\left(k\Delta\tau + m\Delta\tau\right)\right]} \\[2mm]
&= (M-m)\cdot \frac{\sum_{k=1}^{M-m} n\left(k\Delta\tau\right) n\left(k\Delta\tau + m\Delta\tau\right)}{\left(\sum_{k=1}^{M-m} n\left(k\Delta\tau\right)\right)\left(\sum_{k=1}^{M-m} n\left(k\Delta\tau + m\Delta\tau\right)\right)}
\end{aligned}
\tag{6.18}
$$

6.2.3 Correlator Schemes

At the moment there are two correlator schemes commonly used. Either, the correlations are calculated on a linear correlation timescale, i.e., the correlation function will contain as many points as the measurement, M. However, for large m and thus large correlation times, $m\Delta\tau$, statistics is low and the standard deviation too high to be of practical use. In a second scheme, a semi-logarithmic timescale for the correlation time as proposed by Schäetzel (1988), statistics is improved for longer measurement times by adapting longer sample times. In the most common scheme, there is a group of 16 channels that have the length, $\Delta\tau$, followed by the next group of 8 channels with, $2\Delta\tau$, followed by another group of 8 channels with, $4\Delta\tau$, etc. (Meseth et al., 1999). The correlation time of each channel (sometimes referred to as lag time), i.e., the time τ of the point it represents in $G(\tau)$ is just the sum of all the sampling times of all the previous channels. This means that the larger the sampling time of a channel the longer is the correlation time (see Figure 6.3). For the normalization in the denominator of Equation 6.18, so-called monitor channels are introduced for each channel. One monitor channel, called direct monitor, for $F(t)$, i.e., $\tau=0$, and one delayed monitor for $F(t+\tau)$. These monitor channels sum up all the counts that were recorded in a particular channel. The monitor channels can then be used in the denominator to calculate the normalized ACF. This configuration has been preferentially used since it reduces the number of points in the correlation function tremendously, reduces the computation time, and thus makes online autocorrelations possible.

This scheme is found in several hardware as well as software correlators commonly used. While hardware correlators are in general faster than their software counterparts, software correlators with time resolutions down to 25 ns have been created (Magatti and Ferri, 2003). This time resolution can capture most processes with the exception of antibunching (fluorophore lifetimes are on the order of 1–10 ns) and rotational diffusion of molecules (the rotational diffusion time of, for instance, green fluorescent protein in solution is on the order of 16–20 ns (Swaminathan et al., 1997)).

We depict the channel width and the correlation times associated with each channel in Figure 6.3. The semi-logarithmic scheme covers a much wider range of times with much fewer points. The computational advantage can be seen as follows. In the semi-logarithmic correlator, the first group of 16 channels (in this example, the sampling time is $\Delta\tau=12.5$ ns) can be combined to give one group of 8 channels with double the sampling time ($\Delta\tau=25$ ns). If the first 16 channels of 12.5 ns are filled for the second time, they give another 8 channels of 25 ns sampling time. Now we have 16 channels of 25 ns, which can be combined to give 8 channels of 50 ns. If the first 16 channels have been filled four times, enough data

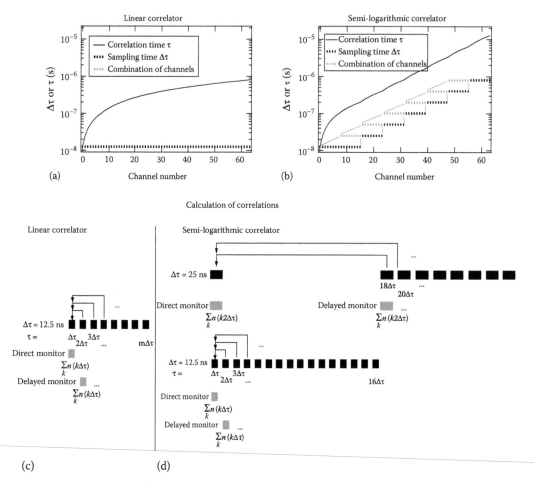

FIGURE 6.3 Correlator schemes and calculation of correlation functions. (a) and (b) show the channel width $\Delta\tau$ and the correlation time τ associated with the different channels in a linear correlator and a semi-logarithmic correlator, respectively. In a linear correlator, all $\Delta\tau$ are equal and the correlation times covered is proportional to the number of channels. In the semi-logarithmic correlator scheme, a wider range of correlation times can be covered with the same number of channels as in the linear correlator since the channel time is doubled every eight channels. Here the basic channel width $\Delta\tau = 12.5$ ns. The correlation time of the 64th channel is 0.8 µs for the linear correlator but 25.6 µs for the semi-logarithmic correlator. (c) and (d) depict the channel structure including the direct and delayed monitors, and how correlation functions are calculated in the two correlator schemes. Arrows between channels indicate multiplication for the calculation of ACFs.

exists to create 8 channels of 100 ns and so on. For the calculation of the correlation function, there are then only half as many calculations to do for the 8 channels of 25 ns compared to the calculations for the 16 channels of 12.5 ns. For the channels of 50 ns, there is only one-quarter of the calculations necessary compared to the 12.5 ns channels, etc. This means that the combined calculation time for all higher channels amounts to just the same amount of time for the calculations, which have to be performed on the first 16 channels. This makes the semi-logarithmic scheme very effective and has the additional advantage to limit the noise for the longer channels.

6.3 Instrumentation

The instrumentation for FCS is nowadays widely commercially available. Nevertheless, new setups are created in fast succession to allow researchers to customize their instruments to particular scientific questions or to allow the performance of novel experiments not possible with the less flexible commercial instruments (Bacia and Schwille, 2007; Hwang and Wohland, 2007; Liu et al., 2008). For this discussion, we divide the FCS setups into essentially four parts, comprising the light source, the excitation scheme, the detection scheme, and the evaluation scheme. The evaluation scheme has already been discussed under the chapter describing correlations and correlators. The other three schemes are depicted in Figure 6.4 and are described in the following paragraphs.

Light source: The main light sources for FCS are lasers. Since FCS usually requires relatively little power (~100 µW or less for a confocal setup, or <100 mW for TIRF setups) almost all commercially available laser sources can be used and the choice of laser is mainly governed by required wavelength,

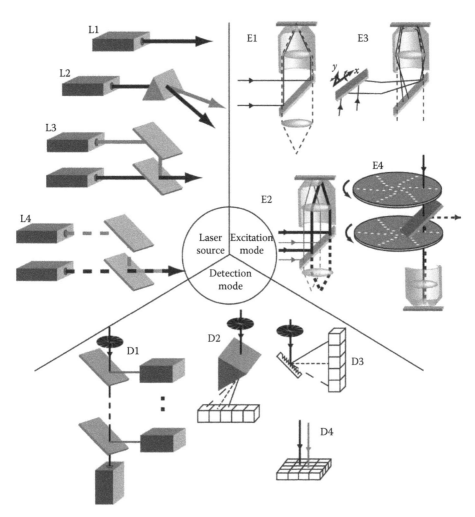

FIGURE 6.4 FCS setup components. Typical FCS setups can be divided into four parts. The light source and coupling of the source into the microscope (L1–L4), the excitation scheme (E1–E4), e.g., confocal illumination, total internal reflection, etc., the detection scheme (D1–D4), and the correlator scheme. The first three are depicted here, and details are given in the text. The correlator scheme is described in the text and Figure 6.3.

beam quality, and cost. We have obtained reasonable correlation curves even with laser pointers in our lab. For two-photon excitation (TPE)-pulsed infrared lasers are necessary. The most common type is the Ti-Sapphire laser (Schwille et al., 1999a). These lasers are more costly but have the advantage of using IR light, which allows much deeper penetration into tissues than lasers with wavelength in the visible range (Helmchen and Denk, 2005). If one laser with one wavelength is coupled into the microscope, scheme L1 can be used. If a multiline laser is used, either laser line selection filters using L1 can be used or a prism can be used to select the desired laser line (L2, Kassies et al., 2005). Often several lasers have to be coupled simultaneously into the microscope. This can be achieved with a combination of mirrors and dichroic mirrors (L3). An interesting scheme for the reduction of cross talk between the detection channels is alternating or interleaved excitation in which the different lasers are alternatingly switched on and off to selectively excite and record the fluorescence of the different fluorescent labels (L4, Kapanidis et al., 2004; Muller et al., 2005).

Excitation mode: The excitation scheme E1 is the classical confocal case where a laser is coupled into a microscope and focused into the sample space. It is versatile, gives 3D resolution, a small confocal volume element (~fL), and, when scanning the laser beam, can be used to image the sample and subsequently park the laser at any spot to collect correlation functions at any desired point (Pan et al., 2007a). This scheme can be expanded to E2 where two or more different lasers are coupled into the microscope to create several focal spots simultaneously. If the spots are sufficiently separated to limit cross talk, correlation functions can be collected at several points simultaneously, considerably increasing the number of measurements achievable (Gosch et al., 2004; Kannan et al., 2007). E3 is an excitation scheme using TIR at the surface of the mounting coverslip (prism TIRF (Thompson et al., 1981) with external excitation can also be used but is not shown here). Only molecules in the sample close to the surface are illuminated by an evanescent wave, which penetrates about 100–300 nm (Axelrod, 1981) into the sample. This illumination scheme has an intrinsic 3D resolution due to the thin slice of sample on the surface illuminated. This has the advantage of better background reduction since most of the sample is not illuminated. The scheme has been used either with a pinhole and point detector (Starr and Thompson, 2001; Ries et al., 2008) or with a CCD camera allowing the spatial or temporal correlation of all points in an image (Bachir et al., 2006; Kannan et al., 2007). The last illumination scheme reported here is the use of a Nipkow disk in a spinning disk confocal microscope for the illumination and collection of fluorescence intensities at multiple spots simultaneously. If the spinning disk microscope is synchronized with the detector (e.g., a CCD camera) the correlation functions can be calculated in a whole image (Sisan et al., 2006) anywhere in the sample. This scheme does not suffer from the problem encountered in the TIRF system, which is confined to the surface of a sample.

Detection mode: On the detection side, we differentiate between 4 different possibilities. The simplest and most common possibility is D1 in which the excitation light is sequentially separated according to wavelength by dichroic mirrors. While this is easy to construct, it suffers from two major disadvantages. The more dichroic mirrors are used the more of the emitted light is lost due to reflection and absorption, and thus if more than 2 labels are to be used this can limit the light detection efficiency significantly. This is especially a problem in biological samples when several labels are used. The excitation power has to be kept low so that the sample is not damaged, and often labels emit low amounts of photons (e.g., FPs often emit 10 times less light compared to small organic dyes). Another disadvantage is the limited spectral resolution achievable by dichroic mirrors. Therefore the detection schemes D2 and D3, in which either a prism or a grating is used to separate the wavelength, are preferable solutions if several labels have to be distinguished with good spectral resolution (Burkhardt et al., 2005; Hwang et al., 2006b; Previte et al., 2008). For the detection of the different wavelengths, either a detector array or an optical fiber bundle connected to separate detectors can be used. The three detection schemes D1–D3 are usually combined with E1 or E2 using a pinhole to achieve rejection of out-of-focus light and 3D spatial resolution. The last detection scheme D4 uses an array of detection elements where the size of the detection elements is on the order of the point-spread function of the microscope or smaller. This scheme has been used with the excitation schemes E2–E4. In this case, no pinhole is used in the detection scheme but the detection elements are put directly in the image plane of the microscope and thus each detection element functions as its own pinhole. Detection elements (e.g., pixels on a CCD chip) can be combined to create detection areas, which correspond to pinholes of different sizes. An important advantage here is that the combination of detection elements can even be done after recording the data so that one recorded data set contains all the information for the correlation functions of all possible pinhole sizes (Kannan et al., 2007). By autocorrelating simultaneously the different spatially separated detection elements one can simultaneously record several ACFs (Gosch et al., 2004), or when using TIR-FCS one can create ACF images.

Typical detectors used for FCS are photomultiplier tubes (PMT), avalanche photodiodes (APD), complementary metal–oxide–semiconductor (CMOS) detectors, and electron multiplying charge-coupled device cameras (EMCCD). While PMTs can have faster response times than APDs, APDs have the higher quantum efficiency. The same holds for CMOSs and EMCCDs, respectively. EMCCDs have the highest quantum efficiency of the mentioned detectors but are by far the slowest detectors with time resolution, at the moment, of not better than 0.2 ms when using a 2D array of pixels. While this is sufficient to measure dynamics in biological membranes, it is just on the limit for solution measurements. Single line resolution for EMCCDs can be as low as 0.02 ms making solution measurements possible but lacking the capability for high multiplexing and imaging. In contrast, the time resolution of PMTs and APDs is in the nanosecond range. Some PMT and APD arrays have become commercially available recently but they contain several orders of magnitude less detection elements compared to an EMCCD.

A typical setup using L1, E1, and D1 is shown in Figure 6.5. But any of the schemes in Figure 6.4 can be combined to deliver a range of different setups with different capabilities. For instance, L1, E1, and D3 have been combined for spectral correlations (Burkhardt et al., 2005; Previte et al., 2008) and L1, E2, and D4 for multi-spot correlations (Gosch et al., 2004), or L1, E3, D4 for imaging TIR-FCS (Kannan et al., 2007; Guo et al., 2008).

Fluorophores: FCS is a single molecule sensitive technique and as such works best when using fluorophores with a high yield of photon counts per particle per second (*cps*; also referred to as molecular brightness) and good photostability. This can be easily achieved in *in vitro* studies. Here a wide range of

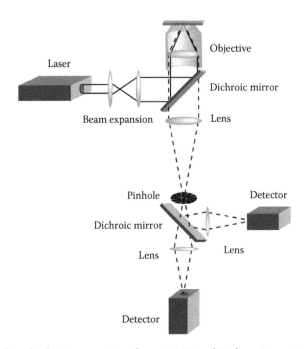

FIGURE 6.5 A typical confocal FCS setup using schemes L1, E1, and D1 from Figure 6.4.

small organic fluorophores with large extinction coefficients and high quantum yields are available. In this case, the choice of fluorophores is mainly guided by the availability of coupling chemistry for the fluorophore to the molecular system under investigation and by the condition that the fluorescent label should not interact or disturb the observed system (Wohland et al., 1999; Kim et al., 2007). Some interesting labels have recently found wider application. The first of these labels are quantum dots, which can be all excited at a single wavelength but exhibit a narrow and size dependent emission (Alivisatos, 1996; Heuff et al., 2007). The second class of dyes are so-called tandem dyes, which consist of a donor molecule for the absorption of light and an acceptor molecule to which the energy from the donor is transferred by FRET (Glazer and Stryer, 1983; Kronick, 1986). The third class of dyes are Megastokes dyes, small organic fluorophores with large Stokes shifts (Hwang et al., 2006a). All three classes of dyes exhibit a variety of Stokes shifts, which makes them very good candidates for Fluorescence Cross-correlation experiments (Hwang and Wohland, 2004; Hwang et al., 2006a).

In biology a range of FPs is available for the genetic tagging of proteins (Tsien, 1998; Campbell et al., 2002; Shaner et al., 2004). However, FPs have in general a lower molecular brightness compared to small organic dyes and thus reduce the signal-to-noise ratio for FCS. Nevertheless, the efficiency of coupling, the controlled labeling ratio, and the specificity of the labeling process often outweigh the disadvantage of the less optimal fluorescent properties of these molecules. Recently, a range of improved FPs have been reported including photoactivatable FPs (Patterson and Lippincott-Schwartz, 2002), and FPs with exceptionally large Stokes shifts (Kogure et al., 2006; Shcherbo et al., 2007), mirroring the development of the Megastokes dyes.

Extrinsic labels can be attached to proteins (see review, Prummer and Vogel, 2008) either by using self-labeling proteins (George et al., 2004) or by so-called FlAsH (Fluorescein Arsenical Helix binder) probes (Griffin et al., 1998). Self-labeling proteins are small enzymes, which can be genetically tagged to any protein of interest (similar to GFP and with only about half the molecular weight or less). These enzymes will label themselves when exposed to their substrates carrying the desired label. FlAsH probes need a small 6 amino acid motif. FlAsH, which is not fluorescent, will recognize the motif that it will bind and subsequently increases its molecular brightness. Another approach useful *in vivo* and *in vitro* is nitrilotriacetate (NTA) labeling, which binds reversibly to hexahistidine tags on the protein of interest (Guignet et al., 2004).

Ultimately, the choice of fluorophore will be governed by the ease of coupling and the availability of the label. FCS with its high sensitivity can work with many labels even when their fluorescence properties are not optimal and especially when only qualitative answers are required (e.g., does a protein bind, does a molecule aggregate, etc.). But quantitative information, i.e., the accuracy and precision of the measurement parameters extractable by FCS, will depend on the photostability and the brightness of the fluorophore.

6.4 State of the Art

6.4.1 FCS with Confocal Illumination

Although confocal systems were already used earlier (e.g., Ricka and Binkert, 1989), recent advances of FCS and the widespread use of confocal FCS systems start with the work of Rigler et al. (1993). Previous FCS measurements suffered from poor signal-to-noise ratio due to technical limitations. In Rigler et al.'s study, an epi-illuminated microscope configuration is used together with a strong focused laser beam, a small pinhole and an APD as detector (L1, E1, and D1 of Figure 6.4). The excitation laser beam is reflected by a long-pass dichroic mirror and coupled into a high numerical aperture (NA) objective, which generates a tightly focused laser beam inside the sample. The fluorescence emission is collected by the same objective and is spatially filtered by a pinhole, rejecting out-of-focus light, and generating the typical observation volume of FCS (Equation 6.12). The fluorescence signal is then detected by the APD and processed by a correlator. The tightly focused laser beam in this setup ($\sim 10^{-15}$ L) ensures that a minimum number of molecules are detected, and thus the fluctuations caused by single molecules can be easily distinguished. This is important to guarantee a good signal-to-noise ratio, and with increasing concentrations the signal-to-noise ratio in FCS will decrease. In typical biological samples, concentrations range from 1 nM to 1 μM that results in about 1–1000 particles in the observation volume, a range just measurable by FCS. It was the application of confocal illumination, which pushed the sensitivity of FCS to the single-molecule level. In addition, the use of high quantum yield fluorescent dyes and high quantum efficiency detectors further increased the signal-to-noise ratio and shortened the data collection time. The improved sensitivity also extends the application of FCS and allows one to probe, e.g., the conformational fluctuations of DNA molecules (Wennmalm et al., 1997) or the photodynamics of fluorescent dyes (Widengren and Rigler, 1995). Currently, the confocal geometry is the standard setup of FCS.

6.4.2 Fluorescence Correlation Microscopy

The concept of fluorescence correlation microscopy (FCM) was first introduced in 1995 by Terry et al. (Terry et al., 1995) to describe the combination of imaging techniques with FCS. This combination allows the user to obtain an image of the sample first before identifying a position on the image where subsequent FCS measurements could be performed. This is especially useful in intracellular applications (Brock et al., 1998). The typical volume of a eukaryotic cell is 10^{-12} L, which is three orders of magnitude larger than the observation volume of FCS. Using FCM, protein dynamics could be specifically investigated in subcellular compartments. The early FCM prototype used a charge couple device (CCD) camera to obtain an image and guide FCS positioning. The optical pathway of imaging and FCS in this case should be properly aligned each time, especially in samples as heterogeneous as living cells. The next generation FCM utilized the confocal laser scanning microscope (CLSM), which achieves three-dimensional images due to its sectioning capability (Wachsmuth et al., 2003). Modifications of commercial CLSM instruments with custom-built FCS attachment and single pinhole were also reported (Brock and Jovin, 1998; Pan et al., 2007a). These technical improvements made it possible to obtain high-resolution cellular images and position the FCS observation volume accurately within cells and organisms (Pan et al., 2007b). Several commercial FCM instruments as well as FCS upgrades for confocal microscopes are available nowadays.

6.4.3 Two-Photon Excitation Fluorescence Correlation Spectroscopy

TPE describes a nonlinear process in which two photons are absorbed simultaneously ($<10^{-15}$ s) by a molecule. The molecule is promoted to an excited state and then follows the normal fluorescence emission pathway (Denk et al., 1990). High photon flux densities are required to achieve the simultaneity and the probability of TPE is proportional to the square of the illumination intensity. Thus in a confocal illumination scheme, where high intensities are found only in the focus, fluorescence excitation is intrinsically confined to a very small volume near the focal plane, eliminating the need for a pinhole. The first TPE-FCS was reported by Berland et al. in 1995 (Berland et al., 1995) to investigate molecular dynamics in living cells. A femtosecond-pulsed Ti-Sapphire infrared laser was used in a confocal microscope to drive two-photon absorption and a short pass dichroic mirror to separate emission from excitation light; no pinhole was used. TPE-FCS has other inherent advantages when applied in living cells. Due to the intrinsically confined excitation volume, TPE reduces bulk photobleaching and phototoxicity, allowing long-term and multiple FCS measurements in living cells. TPE also reduces excitation light scattering and out-of-focus absorption, allowing longer working distance of FCS in tissue samples. Several technical discussions of TPE-FCS (Schwille et al., 1999a; Nagy et al., 2005; Marrocco, 2008; Petrasek and Schwille, 2008a) and biological applications (Kohler et al., 2000; Chirico et al., 2001; Lippitz et al., 2002; Garcia-Marcos et al., 2008) have been reported.

6.4.4 Total Internal Reflection Fluorescence Correlation Spectroscopy

TIR describes an optical phenomenon, which occurs at a medium boundary, from higher refractive index to lower. If an incident light beam strikes at an angle larger than the critical angle ($\theta_{crit} = \sin^{-1}(n_t/n_i)$), the light beam is totally reflected back internally. TIR induces an evanescent wave across the lower refractive index medium whose intensity decays exponentially with the distance from the boundary surface. The decay length is between 100 and 300 nm and can be controlled by adjusting the angle of incidence. The superior axial resolution makes TIR-related techniques suitable for surface and membrane-associated studies (Hansen and Harris, 1998; Lieto et al., 2003; Ohsugi et al., 2006). For FCS, by limiting the area from which the fluorescence is collected, an even smaller observation volume is generated than in the confocal illumination scheme (1.5 μm in axial extension). The first experimental realization of TIR-FCS was reported by Thompson et al. (Thompson et al., 1981) where a prism-based scheme was used for generating TIR. In general, the prism is mounted on top of the inverted microscope stage, followed by a refractive-index-matched medium, a planar substrate of interest, an aqueous solution, and a coverglass. The laser beam is focused through the prism on to the sample and the evanescently excited fluorescence is collected by a high NA objective below the coverglass. A circular pinhole is placed before the detector to restrict the observation volume, and the resulting signal is processed for autocorrelation. Another TIR-FCS setup was reported using an objective-based scheme (Stout and Axelrod, 1989). It uses epi-illumination through the periphery of a high NA oil-immersion objective to generate TIR (E3 in Figure 6.4). By changing the angle of a tilting mirror, the incident laser beam is adjusted to satisfy the condition for TIR on the surface between the coverglass and the aqueous sample solution. The evanescently excited fluorescence is then collected by the same objective, and a dichroic mirror separates the laser beam from the fluorescence signal. The objective-based TIR-FCS benefits from the availability of commercial TIR attachments for inverted microscopes, which can be easily adapted for FCS. For further reading, the theoretical expression of TIR-FCS ACF (Starr and Thompson, 2001) and a protocol for detailed TIR-FCS application (Thompson and Steele, 2007) have been reported in literature.

6.4.5 Fluorescence Cross-Correlation Spectroscopy

Fluorescence cross-correlation spectroscopy (FCCS) is an FCS variant that uses dual-color labeling to determine molecular interactions in solutions and cells (Figure 6.6). Single-color FCS has been shown previously to measure receptor–ligand interactions (Rauer et al., 1996; Van Craenenbroeck and Engelborghs, 1999; Wohland et al., 1999; Zemanova et al., 2004; Wruss et al., 2007). It is based on the theory that relative changes in mass upon binding lead to a reduction in the diffusion coefficient. However, in order to distinguish two components (before and after binding) in FCS, their diffusion coefficients must differ by at least a factor of 1.6 (Meseth et al., 1999). Based on the Stokes–Einstein relation ($D^{-1} \sim M^{1/3}$), the mass must differ by at least a factor of 4. Dimerization is therefore difficult to resolve. In addition, single-color FCS cannot resolve specific binding in a multicomponent system, and protein–protein interactions in living cells are generally not assessable due to the complex environment and a large number of potentially interacting components. Dual-color FCCS labels both binding partners of interest with distinct fluorescent dyes; thus, molecular binding can be monitored specifically. The fluorescence signals from both binding partners are detected in separate channels. Aside from the autocorrelation of signals from each channel, the signals from both channels are cross-correlated:

$$G(\tau) = \frac{\langle F_a(t+\tau) F_b(t) \rangle}{\langle F_a(t) \rangle \langle F_b(t) \rangle}$$

As described previously in ***background***, a $G(\tau) > 1$ therefore indicates that two particles are correlated (binding in this case). FCCS determines molecular binding based on the amplitude of the

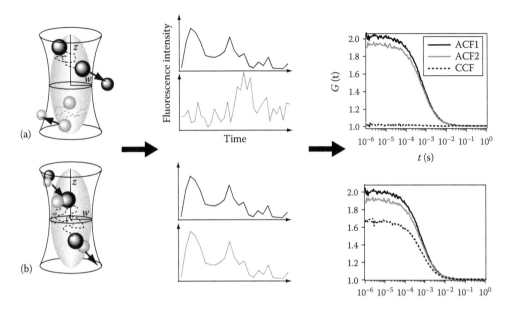

FIGURE 6.6 Fluorescence cross-correlation spectroscopy (FCCS). A typical FCCS measurement yields three correlation functions, two ACFs, one for each channel, and one CCF. (a) If the two differently labeled particles are not interacting, then the ACFs give the concentrations and diffusion coefficients of the two species, but the CCF is flat. (b) If the two differently labeled particles do interact, then the ACFs give the concentration of the two species and the CCF gives the concentration of interaction particles. In addition, the ACFs will contain the diffusion coefficient of free and interactions particles, while the CCF will give the diffusion coefficient of the interacting particles.

cross-correlation curve independent of mass. Since this technique is independent of distance between and orientation of the fluorophores, FCCS represents an attractive alternative to Förster resonance energy transfer (FRET) measurements, which are typically used to study molecular interactions (Liu et al., 2008). Recent advances in FCCS are discussed in the following paragraphs.

6.4.5.1 Dual Laser Excitation Fluorescence Cross-Correlation Spectroscopy

Although the concept of FCCS using multiple colors has been proposed in 1994 (Eigen and Rigler, 1994), the experimental realization was first reported by Schwille et al. in 1997 (Schwille et al., 1997). In their report, two different, wavelength-separated fluorescent dyes, rhodamine green and Cy-5, were used to label two complementary DNA strands. The confocal illumination is carried out by aligning two laser beams (488 and 647 nm) to the same focal volume. The fluorescence signal is separated by dichroic mirrors and filters, and captured by two detectors (L3, E1, and D1 of Figure 6.4). The hybridization of differently labeled DNA strands produces positive cross-correlation signals and the amplitude of the cross-correlation curve increases upon time as more double strand DNAs are formed. The cross-correlation curve easily distinguishes reaction products from free reactants, and quantification is possible. The potential of FCCS to effectively measure biomolecular interactions has been demonstrated both *in vitro* (Kettling et al., 1998; Foldes-Papp and Rigler, 2001; Korn et al., 2003; Camacho et al., 2004) an *in vivo* (Bacia et al., 2002; Saito et al., 2004; Baudendistel et al., 2005; Muto et al., 2006). Dual laser excitation FCCS (DLE-FCCS) has high *cps* as each fluorophore is excited with a laser matching its absorption maximum. Fluorophores can also be selected to have widely separated emission maxima to minimize cross talk. One limitation of DLE-FCCS is the experimental difficulty of aligning and maintaining two laser beams to the same focal volume. This is complicated by chromatic aberrations and the differences of the focal spot size resulting from different wavelengths. Laser misalignment will cause reduced cross-correlation amplitudes and a shift toward slower decays (Weidemann et al., 2002).

6.4.5.2 Two-Photon Excitation Fluorescence Cross-Correlation Spectroscopy

TPE is a nonlinear process, which is theoretically symmetry forbidden, and exhibits different selection rules. The TPE excitation spectra of common fluorophores differ considerably from their one-photon excitation (OPE) counterparts, although the emission spectra remain the same. A previous study shows that TPE spectra of different fluorescent dyes exhibit large overlaps (Xu et al., 1996); thus, simultaneous excitation of distinct dyes is possible. TPE-FCCS borrowed this concept to simplify the instrumentation and experimental procedures (Heinze et al., 2000). As one laser line is used to excite two fluorophores, system alignment is simplified and instrument stability is dramatically improved. TPE-FCCS immediately found several applications in the following years (Dittrich and Schwille, 2002; Kim et al., 2004, 2005; Collini et al., 2005; Swift et al., 2007; Ruan and Tetin, 2008), and a triple-color coincidence analysis was also reported recently (Heinze et al., 2004). Despite the advantages of TPE, the *cps* achieved in TPE-FCS/TPE-FCCS is generally low, because of photobleaching (Dittrich and Schwille, 2001; Petrasek and Schwille, 2008a) and saturation (Berland and Shen, 2003). In addition, the high cost of suitable Ti:sapphire laser systems hamper the wider usage of TPE-FCCS.

6.4.5.3 Single Wavelength Fluorescence Cross-Correlation Spectroscopy

Single wavelength FCCS (SW-FCCS) was firstly reported by Hwang and Wohland in 2004 (Hwang and Wohland, 2004). The SW-FCCS theory is essentially the same as that of TPE-FCCS, except OPE is used instead of TPE (L1, E1, and D1 in Figure 6.4). Ricka and Binkert have reported previously the excitation of fluorescent and scattering polystyrene beads with single wavelength to obtain autocorrelations and cross-correlations (Ricka and Binkert, 1989). The experiment is only feasible with fluorophores that possess similar excitation maxima but largely different Stokes shifts. In Hwang and Wohland's report, a ligand–receptor interaction was successfully monitored using OPE SW-FCCS, with particles labeled either with fluorescein or tandem dyes/quantum dots. The fluorophore pairs of fluorescein and QDs excited by 488 nm laser generate reasonable *cps* for FCCS analysis. The resolution of SW-FCCS was investigated with spectrally similar dyes (Hwang and Wohland, 2005) and contributions of sample concentration, impurities, labeling ratios, and spectral cross talk to the results were discussed. Recently, *in vivo* SW-FCCS applications were presented in cell cultures (Liu et al., 2007; Sudhaharan et al., 2009) and living embryos (Shi et al., 2009) using FP pairs of EGFP and mRFP (Campbell et al., 2002), and triple-color detection was achieved *in vitro* to probe higher order interactions (Hwang et al., 2006a). SW-FCCS simplifies the setup of DLE-FCCS with affordable and commercially available instruments, and eliminates the potential artifacts that arise from laser misalignment. But SW-FCCS does not have the benefit of the background reduction inherent in the TPE scheme. In general, DLE-FCCS achieves highest *cps* and offers the widest selection of fluorophore pairs; TPE-FCCS possesses a simplified experimental procedures and deep penetration depth but is more costly; SW-FCCS has simplified experimental procedures, good *cps*, and is readily available but relies on the availability of suitable fluorophores. Depending on the sample and available instrumentation, either technique can be advantageous.

6.4.5.4 Pulsed Interleaved Excitation Fluorescence Cross-Correlation Spectroscopy

Intracellular FCCS applications mostly rely on FPs for target labeling. Aside from the low photostability, FPs also suffer from spectral cross talk due to their long-tailed emissions. Spectral cross talk produces false-positive cross-correlation amplitudes. Pulsed interleaved excitation FCCS (PIE-FCCS) utilizes two excitation lasers that are pulsed alternatively in the range of nanoseconds such that the fluorescence emission generated from one pulse is complete before the next excitation pulse arrives (Thews et al., 2005). Photons are collected by one detector using time-correlated single photon counting (TCSPC) and their arrival times are recorded with respect to their excitation pulse. Off-line cross-correlation analysis of photons collected after each pulse produce cross-correlation functions that are free of cross talk. With PIE-FCCS, complexes that undergo FRET can still be analyzed quantitatively, and the FRET efficiency can be determined directly (Muller et al., 2005). The major problem in PIE-FCCS is reduced signal-to-noise ratio because of light scattering and auto-fluorescence in biological samples. PIE also decreases molecular brightness due to limited excitation cycles. However, recent reports indicate that pulsed excitation when optimized can suppress triplet population buildup and increase the molecular brightness and the FCS performance (Donnert et al., 2007; Persson et al., 2008; De and Goswami, 2009).

6.4.5.5 Multiple Focal Spot Excitation Fluorescence Cross-Correlation Spectroscopy

Two-beam FCCS was first introduced by Brinkmeier et al. to determine the flow velocity in micro-structured channels (Brinkmeier et al., 1999). For the setup, two polarized beam splitters are used to generate two parallel laser beams that are slightly displaced (this can be achieved using L3, E3, and D4 in Figure 6.4). The two parallel beams are coupled into the back aperture of the microscope objective to create two spatially separated focal spots. The fluorescence signal from each volume is collected by the same objective and focused onto two closely placed optic fibers for cross-correlation. Another simplified scheme was proposed using two different pinholes to generate two separate observation volumes (Jaffiol et al., 2006), and even the detection of different parts of one pinhole by two detectors was used (Pan et al., 2007a). Conventional one-beam FCS is sufficient to measure the mobility of single molecules. However, by introducing another measurement volume, it is now possible to separate isotropic and anisotropic dynamics, thus allowing one to monitor flow directions and discriminate against photodynamics. Two-beam FCCS has been mainly used to determine flow parameters in microfluidic systems (Dittrich and Schwille, 2002; Jung and Van Orden, 2005; Jung and Van Orden, 2006). Diffractive optical elements have also been used to create multifocal spot of 2×2 (Blom et al., 2002b; Gosch et al., 2005) and 4×1 (Blom et al., 2002a) arrays for fluorescence detection and autocorrelations or cross-correlations (D2 and D3 in Figure 6.4). This simultaneous multi-foci excitation can be used directly in biochip microarrays and multiplexed detection schemes.

6.4.5.6 Total Internal Reflection Fluorescence Cross-Correlation Spectroscopy

A dual-color TIR-FCCS has been reported by Leutenegger et al. using objective-based TIR (Leutenegger et al., 2006). Its high fluorescence collection efficiency generates a twofold to threefold increase in *cps* compared to conventional confocal FCCS. TIR-FCCS could be potentially used to probe weak molecular interactions on cell or model membranes.

6.4.6 Scanning Fluorescence Correlation Spectroscopy

The concept of Scanning Fluorescence Correlation Spectroscopy (SFCS) was brought up by Weissman and Petersen to study molecular weight and lateral diffusion of immobile or slowly diffusing molecules (Weissman et al., 1976; Petersen, 1986; Petersen et al., 1986). In general, the observation volume is scanned across the sample in a controlled way while the fluorescence signal is collected. The scanning trajectory can be linear or circular depending on the applications, and the correlation curves can be calculated from all points along the trajectory or from selected ones only. There are several advantages when scanning is introduced in FCS. Firstly, SFCS allows simultaneous measurements of autocorrelations at different locations. Diffusion that is slower than the scanning speed can be efficiently measured along the scanning trajectory (Ruan et al., 2004). In addition, one can compensate for sample movement by selecting the appropriate sequences from the recorded scans. Secondly, by synchronizing scanning and data acquisition, correlations can be performed in both space and time. Additional spatial information can be obtained besides high temporal resolution. SFCS has been used to determine the direction and speed of the flow of mobile and the position of immobile particles (Skinner et al., 2005), and the blood flow direction in living zebra fish embryos has been measured (Pan et al., 2007b, 2009). Thirdly, by controlling the scanning radius in circle-scan SFCS to the size comparable to the observation volume, diffusion coefficients could be precisely measured without knowing the exact size of the observation volume (Petrasek and Schwille, 2008b). This is advantageous for *in vivo* applications as the heterogeneous cytoplasm tends to affect the shape and size of the observation volume. Fourthly, SFCS distributes the laser power over a larger area and thus reduces fluorophore depletion in the observation volume allowing the accurate observation of slowly diffusing particles. The reduced photobleaching effect is especially useful when TPE is combined with SFCS (Berland et al., 1996; Petrasek and Schwille, 2008a). SFCS has also been adapted for dual-color cross-correlation (Amediek et al., 2002) and two foci cross-correlation (Ries and Schwille, 2006) analysis. For a recent review see Petrasek and Schwille (2008c).

6.4.7 CCD-Based Fluorescence Correlation Spectroscopy

CCD-based FCS is one step beyond multi-foci FCS. In this case, a CCD camera is used for photon collection instead of detector arrays (D4 in Figure 6.4). This potentially allows mapping the flow of a whole channel or studying the protein diffusion of an entire cell membrane. However, conventional CCD cameras do not have sufficient sensitivity and speed for FCS. Recently, electron-multiplying CCD (EMCCD) cameras were introduced with single-photon sensitivity, over 90% quantum efficiency and read-out speeds in the microsecond to millisecond range, features that make these cameras suitable for FCS applications. EMCCD-based FCS has been experimentally proved using focused laser excitation (Burkhardt and Schwille, 2006; Kannan et al., 2006). There is limited time resolution at full frame acquisition, but faster readout could be achieved by using a sub-region of the EMCCD. A time resolution of 4 ms for 20 lines of 512 pixels and 20 µs for one-line measurements has been realized, which is sufficient to resolve protein diffusion on a plasma membrane or even diffusion in solution, respectively. EMCCD provides an excellent platform for multiplexing FCS detection. Nevertheless, a high-speed excitation scheme should also be developed to match the detection. Sisan et al. reported spatially resolved EMCCD-FCS using a spinning disk confocal microscope (Sisan et al., 2006). A rotating disk containing thousands of microlenses, together with a corotating disk of pinholes, produces an array of diffraction-limited observation volumes that could rapidly scan the sample (E4 in Figure 6.4). This technique spatially resolved hindered diffusion of fluorescent microspheres in a collagen matrix. Another report by Kannan et al. used TIR-based excitation (Kannan et al., 2007). As discussed before, the TIR-based excitation excites a 100–300 nm layer along the glass–water interface, and the whole excitation area can be directly captured by the EMCCD. This system was shown to be able to resolve diffusion on bilayers and give a full FCS image of the cell membrane close to the coverglass, with up to 3500 ACFs measured simultaneously. Up to now, the time resolution has been restricted by the limited read-out rate of EMCCD cameras, but newer models already provide read-out times of less than 0.3 ms for regions of interest as large as 20 lines of a chip. EMCCD-based FCS is a promising tool for high-throughput, and imaging FCS applications.

6.4.8 Computational Advances

The calculations of correlation functions are simple and essentially include only sums and products of integer numbers (counts of photons, see Equation 6.17). Nevertheless, fitting and evaluation of the final correlation functions is mathematically complex. There are at least four areas on which researchers have focused to improve data acquisition and treatment. Firstly, the correlator scheme and the calculation of the correlations have been optimized. Klaus Schätzel has shown in a series of articles that a semi-logarithmic timescale for the correlator reduces computation time to allow online calculations of ACFs, while reducing the noise for long correlation times (Schätzel et al., 1988). In addition, he introduced the symmetric normalization removing the bias for long correlation times (Schätzel et al., 1988, 1990). Since these schemes require great speed they have been mainly realized as hardware correlators. Recently, with the advance of computer technology several authors have devised schemes, which allow the calculation of correlations in batches of data and that can be performed by software correlators in real time (Eid et al., 2000; Magatti and Ferri, 2003; Wahl et al., 2003; Culbertson and Burden, 2007). However, these approaches suffer from the problem that they reduce the originally acquired intensity data (10^6–10^9 data points) to a limited number of points in a correlation function (~10^3 data points) and thus represent a loss of information. New recording schemes, acquiring the time point of photon arrivals at the detector instead of the number of photons collected in a finite interval, have been used (Eid et al., 2000). This leads to a reduction in the stored data if the number of photons per second recorded is smaller than the number of time intervals per second (e.g., for $\Delta\tau = 10^{-6}$ s, and a count rate of 100,000 s^{-1}, 10^6 values have to be stored in the first case but only 10^5 values in the second case; even if the bit width for the second case is four times larger it would still lead to a reduction in file size). This recording scheme then contains the maximum information, and correlations as well as other parameters (e.g., photon-counting histograms (Chen et al., 1999) and fluorescence distribution analysis (Kask et al., 1999)) can be calculated from the data.

Secondly, new fit models in FCS are constantly derived for a range of different situations giving researchers a much wider base with which to test their experimental curves. A very interesting approach has been followed by Culbertson et al (Culbertson et al., 2007) who have measured the actual observation volume for their instrument. They then simulate ACF curves, taking account of the experimental observation volume, with variable parameters to find the best fit to their data. This allows them to take account of any aberrations in their system and get better FCS fits.

Thirdly, an important topic in data fitting is taking account of the experimental noise when fitting data. By weighting data points according to their noise, one can decide by statistical means (e.g., *F*-test) on the proper fitting model (Meseth et al., 1999). The paper on the error in FCS by Koppel et al. (Koppel, 1974) gave a first approximation of the standard deviation for each point in an ACF. The signal-to-noise ratio in dependence of observation profiles and the different fit parameters were studied by Qian (Qian, 1990) and Kask et al. (Kask et al., 1997), and the influence of the pinhole was studied by Rigler (Rigler et al., 1993). Meseth et al. made empirical tests using Koppel's standard deviation to derive limits for FCS data evaluation (Meseth et al., 1999), and Wohland et al. improved on Koppel's formula by directly calculating the standard deviation from the experimental data (Wohland et al., 2001). The dependence of the noise on measurement time, fluorescence intensity, and number of particles were considered empirically by Starchev et al. (Starchev et al., 2001). Saffarian then gave the description of the standard deviation for FCS experiments (Saffarian and Elson, 2003). The influence of the focal volume and resulting experimental artifacts was investigated by Hess et al. (Hess and Webb, 2002), and more recently FCS artifacts were described concerning optical saturation (Cianci et al., 2004; Marrocco, 2004; Nishimura and Kinjo, 2004; Enderlein et al., 2005). These advances in the understanding of the standard deviation made tests of quality of fitting and decisions on the proper fit model more reliable.

Fourthly, the process of fitting has been improved over the years giving better and more stable parameter evaluations. New fitting algorithms using a maximum entropy method for heterogeneous systems allowed the evaluation of particle distributions (Sengupta et al., 2003; Modos et al., 2004). Global Fitting has been used for better parameter estimation (Skakun et al., 2005), and recently a new fitting algorithm, which improves data fitting and parameter accuracy, has been introduced (Rao et al., 2006).

Together these advances have made FCS a more reliable and well-understood tool for the measurement of molecular characteristics and interactions.

6.5 Summary

In this chapter, we explained the basics of FCS experiments and their mathematical background, reviewed the different instrumental implementations, and gave an overview of the state of the art in FCS. This chapter is aimed at researchers in different disciplines who would like to apply FCS and need a first glimpse at the basics.

For further reading we suggest a range of recent reviews giving more details on the historical development of FCS (Krichevsky and Bonnet, 2002), new developments in FCS and FCCS (Haustein and Schwille, 2007; Hwang and Wohland, 2007), the application of FCS to receptor proteins (Hovius et al., 2000; Briddon and Hill, 2007), comparisons of FCS with other fluorescence techniques (Liu et al., 2008), the application of FCS to living cells (Kim et al., 2007), the protocol for FCCS and TIR-FCS applications (Bacia and Schwille, 2007; Thompson and Steele, 2007), and the comprehensive book by Rigler and Elson (Rigler and Elson, 2001).

The capability of FCS and related techniques to measure molecular dynamics with single molecule sensitivity, recent advances in fluorescent labeling, data treatment, and instrumentation, and the demonstration that it can be used not only in living cells but as well in more physiologically relevant small organisms (e.g., nematodes (*Caenorhabditis elegans*), fruit flies (*Drosophila melanogaster*), and zebra fish (*Danio rerio*)) makes FCS a powerful tool in the advancement of our understanding of the molecular basis of the processes of life.

6.6 Future Perspectives

FCS and related methods are developing fast with a number of interesting trends. We expect to see some important new developments and applications. Firstly, FCS will facilitate more quantitative analysis in live cells, e.g., fluorescence lifetime correlation spectroscopy (FLCS) allows the separation of ACFs of different signal components quantitatively (Kapusta et al., 2007) and FCCS can be used to determine dissociation constants of protein–protein interactions (Maeder et al., 2007; Sudhaharan et al., 2009). Secondly, FCS will be extended to measurements in small living animals making the questions of developmental biology accessible with FCS and providing a better physiological environment for molecular studies (Pan et al., 2007b, 2009; Korzh et al., 2008; Petrasek et al., 2008). Thirdly, FCS multiplexing has reached a stage where FCS images can now be taken of whole cells (Sisan et al., 2006; Kannan et al., 2007). This advance also facilitates the combination of spatial and temporal correlations, which tremendously increases the information accessible from experiments (Kolin and Wiseman, 2007). Fourthly, new FCS modalities will broaden the applicability of the technique. Scanning beams will allow corrections of sample movement and will extend measurements to very slow moving particles, an area up to now neglected by classifying molecules as immobile (Petrasek and Schwille, 2008c). Pulsed excitations will lead to increase in signal-to-noise ratios possibly allowing the use of a wider range of fluorophores (Muller et al., 2005). The creation of even smaller observation volumes using stimulated emission depletion (STED, Eggeling et al., 2009) or increased detection capabilities with 4Pi microscopy (Arkhipov et al., 2007) will allow the study of molecular dynamics at the nanoscale and give access to even higher fluorophore concentrations. And new correlation methods using absorption signals will alleviate the problems of poor fluorophore photophysics (Octeau et al., 2009). Fifthly, new approaches to calculate correlation functions using more accurate illumination profiles (Culbertson et al., 2007) and using better fitting algorithms (Skakun et al., 2005; Rao et al., 2006) will lead to better data evaluation. And lastly, FCS is and will be more often used in combination with other complementary spectroscopic techniques to create powerful systems customized for the solution of particular problems (e.g., PCH (Chen et al., 1999), FLIM (Breusegem et al., 2006)).

Acknowledgments

We would like to thank Ping Liu for helping us with the figures. Xianke Shi received a scholarship from the National University of Singapore. Both authors were funded by a grant from the Biomedical Research Council in Singapore (07/1/21/19/488, R-143-000-351-305).

References

Alivisatos AP. 1996. Semiconductor clusters, nanocrystals, and quantum dots. *Science* 271:933.
Ambrose WP, Goodwin PM, Martin JC, and Keller RA. 1994. Single molecule detection and photochemistry on a surface using near-field optical excitation. *Phys Rev Lett* 72:160–163.

Amediek A, Haustein E, Scherfeld D, and Schwille P. 2002. Scanning dual-color cross-correlation analysis for dynamic co-localization studies of immobile molecules. *Single Mol* 3:201–210.

Aragon SR and Pecora R. 1976. Fluorescence correlation spectroscopy as a probe of molecular dynamics. *J Chem Phys* 64:1791.

Arkhipov A, Huve J, Kahms M, Peters R, and Schulten K. 2007. Continuous fluorescence microphotolysis and correlation spectroscopy using 4Pi microscopy. *Biophys J* 93:4006–4017.

Axelrod D. 1981. Cell-substrate contacts illuminated by total internal reflection fluorescence. *J Cell Biol* 89:141–145.

Bachir AI, Durisic N, Hebert B, Gruter P, and Wiseman PW. 2006. Characterization of blinking dynamics in quantum dot ensembles using image correlation spectroscopy. *J Appl Phys* 99:064503.

Bacia K and Schwille P. 2007. Practical guidelines for dual-color fluorescence cross-correlation spectroscopy. *Nat Protoc* 2:2842–2856.

Bacia K, Majoul IV, and Schwille P. 2002. Probing the endocytic pathway in live cells using dual-color fluorescence cross-correlation analysis. *Biophys J* 83:1184–1193.

Baudendistel N, Muller G, Waldeck W, Angel P, and Langowski J. 2005. Two-hybrid fluorescence cross-correlation spectroscopy detects protein–protein interactions in vivo. *Chemphyschem* 6:984–990.

Berland K and Shen G. 2003. Excitation saturation in two-photon fluorescence correlation spectroscopy. *Appl Opt* 42:5566–5576.

Berland KM, So PT, Chen Y, Mantulin WW, and Gratton E. 1996. Scanning two-photon fluctuation correlation spectroscopy: Particle counting measurements for detection of molecular aggregation. *Biophys J* 71:410–420.

Berland KM, So PT, and Gratton E. 1995. Two-photon fluorescence correlation spectroscopy: Method and application to the intracellular environment. *Biophys J* 68:694–701.

Berne BJ and Pecora R. 2000. *Dynamic Light Scattering: With Applications to Chemistry, Biology, and Physics.* Mineola, NY: Courier Dover Publications.

Blom H, Johansson M, Gosch M, Sigmundsson T, Holm J, Hard S, and Rigler R. 2002a. Parallel flow measurements in microstructures by use of a multifocal 4 × 1 diffractive optical fan-out element. *Appl Opt* 41:6614–6620.

Blom H, Johansson M, Hedman AS, Lundberg L, Hanning A, Hard S, and Rigler R. 2002b. Parallel fluorescence detection of single biomolecules in microarrays by a diffractive-optical-designed 2 × 2 fan-out element. *Appl Opt* 41:3336–3342.

Breusegem SY, Levi M, and Barry NP. 2006. Fluorescence correlation spectroscopy and fluorescence lifetime imaging microscopy. *Nephron Exp Nephrol* 103:e41–49.

Briddon SJ and Hill SJ. 2007. Pharmacology under the microscope: The use of fluorescence correlation spectroscopy to determine the properties of ligand–receptor complexes. *Trends Pharmacol Sci* 28:637–645.

Brinkmeier M, Dorre K, Stephan J, and Eigen M. 1999. Two beam cross correlation: A method to characterize transport phenomena in micrometer-sized structures. *Anal Chem* 71:609–616.

Brock R and Jovin TM. 1998. Fluorescence correlation microscopy (FCM)-fluorescence correlation spectroscopy (FCS) taken into the cell. *Cell Mol Biol (Noisy-le-grand)* 44:847–856.

Brock R, Hink MA, and Jovin TM. 1998. Fluorescence correlation microscopy of cells in the presence of autofluorescence. *Biophys J* 75:2547–2557.

Burkhardt M, Heinze KG, and Schwille P. 2005. Four-color fluorescence correlation spectroscopy realized in a grating-based detection platform. *Opt Lett* 30:2266–2268.

Burkhardt M and Schwille P. 2006. Electron multiplying CCD based detection for spatially resolved fluorescence correlation spectroscopy. *Opt Express* 14:5013–5020.

Camacho A, Korn K, Damond M, Cajot JF, Litborn E, Liao B, Thyberg P, Winter H, Honegger A, Gardellin P, and Rigler R. 2004. Direct quantification of mRNA expression levels using single molecule detection. *J Biotechnol* 107:107–114.

Campbell RE, Tour O, Palmer AE, Steinbach PA, Baird GS, Zacharias DA, and Tsien RY. 2002. A monomeric red fluorescent protein. *Proc Natl Acad Sci USA* 99:7877–7882.

Chen Y, Muller JD, So PT, and Gratton E. 1999. The photon counting histogram in fluorescence fluctuation spectroscopy. *Biophys J* 77:553–567.

Chirico G, Bettati S, Mozzarelli A, Chen Y, Muller JD, and Gratton E. 2001. Molecular heterogeneity of *O*-acetylserine sulfhydrylase by two-photon excited fluorescence fluctuation spectroscopy. *Biophys J* 80:1973–1985.

Cianci GC, Wu J, and Berland KM. 2004. Saturation modified point spread functions in two-photon microscopy. *Microsc Res Tech* 64:135–141.

Collini M, Caccia M, Chirico G, Barone F, Dogliotti E, and Mazzei F. 2005. Two-photon fluorescence cross-correlation spectroscopy as a potential tool for high-throughput screening of DNA repair activity. *Nucleic Acids Res* 33:e165.

Culbertson MJ and Burden DL. 2007. A distributed algorithm for multi-tau autocorrelation. *Rev Sci Instrum* 78:044102.

Culbertson MJ, Williams JT, Cheng WW, Stults DA, Wiebracht ER, Kasianowicz JJ, and Burden DL. 2007. Numerical fluorescence correlation spectroscopy for the analysis of molecular dynamics under nonstandard conditions. *Anal Chem* 79:4031–4039.

De AK and Goswami D. 2009. Adding new dimensions to laser-scanning fluorescence microscopy. *J Microsc* 233:320–325.

Denk W, Strickler JH, and Webb WW. 1990. Two-photon laser scanning fluorescence microscopy. *Science* 248:73–76.

Dittrich PS and Schwille P. 2001. Photobleaching and stabilization of fluorophores used for single-molecule analysis with one-and two-photon excitation. *Appl Phys B* 73:829–837.

Dittrich PS and Schwille P. 2002. Spatial two-photon fluorescence cross-correlation spectroscopy for controlling molecular transport in microfluidic structures. *Anal Chem* 74:4472–4479.

Donnert G, Eggeling C, and Hell SW. 2007. Major signal increase in fluorescence microscopy through dark-state relaxation. *Nat Methods* 4:81–86.

Eggeling C, Ringemann C, Medda R, Schwarzmann G, Sandhoff K, Polyakova S, Belov VN et al. 2009. Direct observation of the nanoscale dynamics of membrane lipids in a living cell. *Nature* 457:1159–1162.

Eid JS, Muller JD, and Gratton E. 2000. Data acquisition card for fluctuation correlation spectroscopy allowing full access to the detected photon sequence. *Rev Sci Instrum* 71:361.

Eigen M and Rigler R. 1994. Sorting single molecules: Application to diagnostics and evolutionary biotechnology. *Proc Natl Acad Sci USA* 91:5740–5747.

Elson EL and Magde D. 1974. Fluorescence correlation spectroscopy. I. Conceptual basis and theory. *Biopolymers* 13:1–27.

Enderlein J, Gregor I, Patra D, Dertinger T, and Kaupp UB. 2005. Performance of fluorescence correlation spectroscopy for measuring diffusion and concentration. *Chemphyschem* 6:2324–2336.

Foldes-Papp Z and Rigler R. 2001. Quantitative two-color fluorescence cross-correlation spectroscopy in the analysis of polymerase chain reaction. *Biol Chem* 382:473–478.

Garcia-Marcos A, Sanchez SA, Parada P, Eid J, Jameson DM, Remacha M, Gratton E, and Ballesta JP. 2008. Yeast ribosomal stalk heterogeneity in vivo shown by two-photon FCS and molecular brightness analysis. *Biophys J* 94:2884–2890.

Gennerich A and Schild D. 2000. Fluorescence correlation spectroscopy in small cytosolic compartments depends critically on the diffusion model used. *Biophys J* 79:3294–3306.

George N, Pick H, Vogel H, Johnsson N, and Johnsson K. 2004. Specific labeling of cell surface proteins with chemically diverse compounds. *J Am Chem Soc* 126:8896–8897.

Glazer AN and Stryer L. 1983. Fluorescent tandem phycobiliprotein conjugates. Emission wavelength shifting by energy transfer. *Biophys J* 43:383–386.

Gosch M, Serov A, Anhut T, Lasser T, Rochas A, Besse PA, Popovic RS, Blom H, and Rigler R. 2004. Parallel single molecule detection with a fully integrated single-photon 2×2 CMOS detector array. *J Biomed Opt* 9:913–921.

Gosch M, Blom H, Anderegg S, Korn K, Thyberg P, Wells M, Lasser T, Rigler R, Magnusson A, and Hard S. 2005. Parallel dual-color fluorescence cross-correlation spectroscopy using diffractive optical elements. *J Biomed Opt* 10:054008.

Griffin BA, Adams SR, and Tsien RY. 1998. Specific covalent labeling of recombinant protein molecules inside live cells. *Science* 281:269–272.

Guignet EG, Hovius R, and Vogel H. 2004. Reversible site-selective labeling of membrane proteins in live cells. *Nat Biotechnol* 22:440–444.

Guo L, Har JY, Sankaran J, Hong Y, Kannan B, and Wohland T. 2008. Molecular diffusion measurement in lipid bilayers over wide concentration ranges: A comparative study. *Chemphyschem* 9:721–728.

Hansen RL and Harris JM. 1998. Measuring reversible adsorption kinetics of small molecules at solid/liquid interfaces by total internal reflection fluorescence correlation spectroscopy. *Anal Chem* 70:4247–4256.

Haustein E and Schwille P. 2007. Fluorescence correlation spectroscopy: Novel variations of an established technique. *Annu Rev Biophys Biomol Struct* 36:151–169.

Heinze KG, Koltermann A, and Schwille P. 2000. Simultaneous two-photon excitation of distinct labels for dual-color fluorescence crosscorrelation analysis. *Proc Natl Acad Sci USA* 97:10377–10382.

Heinze KG, Jahnz M, and Schwille P. 2004. Triple-color coincidence analysis: One step further in following higher order molecular complex formation. *Biophys J* 86:506–516.

Helmchen F and Denk W. 2005. Deep tissue two-photon microscopy. *Nat Methods* 2:932–940.

Hess ST and Webb WW. 2002. Focal volume optics and experimental artifacts in confocal fluorescence correlation spectroscopy. *Biophys J* 83:2300–2317.

Heuff RF, Swift JL, and Cramb DT. 2007. Fluorescence correlation spectroscopy using quantum dots: Advances, challenges and opportunities. *Phys Chem Chem Phys* 9:1870–1880.

Hirschfeld T. 1976. Optical microscopic observation of single small molecules. *Appl Opt* 15:2965–2966.

Hovius R, Vallotton P, Wohland T, and Vogel H. 2000. Fluorescence techniques: Shedding light on ligand-receptor interaction. *TiPS* 21:266–274.

Hwang LC and Wohland T. 2004. Dual-color fluorescence cross-correlation spectroscopy using single laser wavelength excitation. *Chemphyschem* 5:549–551.

Hwang LC and Wohland T. 2005. Single wavelength excitation fluorescence cross-correlation spectroscopy with spectrally similar fluorophores: Resolution for binding studies. *J Chem Phys* 122:114708.

Hwang LC and Wohland T. 2007. Recent advances in fluorescence cross-correlation spectroscopy. *Cell Biochem Biophys* 49:1–13.

Hwang LC, Gosch M, Lasser T, and Wohland T. 2006a. Simultaneous multicolor fluorescence cross-correlation spectroscopy to detect higher order molecular interactions using single wavelength laser excitation. *Biophys J* 91:715–727.

Hwang LC, Leutenegger M, Gosch M, Lasser T, Rigler P, Meier W, and Wohland T. 2006b. Prism-based multicolor fluorescence correlation spectrometer. *Opt Lett* 31:1310–1312.

Jaffiol R, Blancquaert Y, Delon A, and Derouard J. 2006. Spatial fluorescence cross-correlation spectroscopy. *Appl Opt* 45:1225–1235.

Jung J and Van Orden A. 2005. Folding and unfolding kinetics of DNA hairpins in flowing solution by multiparameter fluorescence correlation spectroscopy. *J Phys Chem B* 109:3648–3657.

Jung J and Van Orden A. 2006. A three-state mechanism for DNA hairpin folding characterized by multiparameter fluorescence fluctuation spectroscopy. *J Am Chem Soc* 128:1240–1249.

Kannan B, Har JY, Liu P, Maruyama I, Ding JL, and Wohland T. 2006. Electron multiplying charge-coupled device camera based fluorescence correlation spectroscopy. *Anal Chem* 78:3444–3451.

Kannan B, Guo L, Sudhaharan T, Ahmed S, Maruyama I, and Wohland T. 2007. Spatially resolved total internal reflection fluorescence correlation microscopy using an electron multiplying charge-coupled device camera. *Anal Chem* 79:4463–4470.

Kapanidis AN, Lee NK, Laurence TA, Doose S, Margeat E, and Weiss S. 2004. Fluorescence-aided molecule sorting: Analysis of structure and interactions by alternating-laser excitation of single molecules. *Proc Natl Acad Sci USA* 101:8936–8941.

Kapusta P, Wahl M, Benda A, Hof M, and Enderlein J. 2007. Fluorescence lifetime correlation spectroscopy. *J Fluoresc* 17:43–48.

Kask P, Piksarv P, Pooga M, Mets U, and Lippmaa E. 1989. Separation of the rotational contribution in fluorescence correlation experiments. *Biophys J* 55:213.

Kask P, Gunther R, and Axhausen P. 1997. Statistical accuracy in fluorescence fluctuation experiments. *Eur Biophys J* 25:163–169.

Kask P, Palo K, Ullmann D, and Gall K. 1999. Fluorescence-intensity distribution analysis and its application in biomolecular detection technology. *Proc Natl Acad Sci USA* 96:13756–13761.

Kassies R, Lenferink A, Segers-Nolten I, and Otto C. 2005. Prism-based excitation wavelength selection for multicolor fluorescence coincidence measurements. *Appl Opt* 44:893–897.

Kettling U, Koltermann A, Schwille P, and Eigen M. 1998. Real-time enzyme kinetics monitored by dual-color fluorescence cross-correlation spectroscopy. *Proc Natl Acad Sci USA* 95:1416–1420.

Kim SA, Heinze KG, Waxham MN, and Schwille P. 2004. Intracellular calmodulin availability accessed with two-photon cross-correlation. *Proc Natl Acad Sci U S A* 101:105–110.

Kim SA, Heinze KG, Bacia K, Waxham MN, and Schwille P. 2005. Two-photon cross-correlation analysis of intracellular reactions with variable stoichiometry. *Biophys J* 88:4319–4336.

Kim SA, Heinze KG, and Schwille P. 2007. Fluorescence correlation spectroscopy in living cells. *Nat Methods* 4:963–973.

Kogure T, Karasawa S, Araki T, Saito K, Kinjo M, and Miyawaki A. 2006. A fluorescent variant of a protein from the stony coral Montipora facilitates dual-color single-laser fluorescence cross-correlation spectroscopy. *Nat Biotechnol* 24:577–581.

Kohler RH, Schwille P, Webb WW, and Hanson MR. 2000. Active protein transport through plastid tubules: Velocity quantified by fluorescence correlation spectroscopy. *J Cell Sci* 113 (Pt 22):3921–3930.

Kolin DL and Wiseman PW. 2007. Advances in image correlation spectroscopy: Measuring number densities, aggregation states, and dynamics of fluorescently labeled macromolecules in cells. *Cell Biochem Biophys* 49:141–164.

Koppel DE. 1974. Statistical accuracy in fluorescence correlation spectroscopy. *Phys Rev A* 10:1938–1945.

Koppel DE, Morgan F, Cowan AE, and Carson JH. 1994. Scanning concentration correlation spectroscopy using the confocal laser microscope. *Biophys J* 66:502–507.

Korn K, Gardellin P, Liao B, Amacker M, Bergstrom A, Bjorkman H, Camacho A et al. 2003. Gene expression analysis using single molecule detection. *Nucleic Acids Res* 31:e89.

Korzh S, Pan X, Garcia-Lecea M, Winata CL, Wohland T, Korzh V, and Gong Z. 2008. Requirement of vasculogenesis and blood circulation in late stages of liver growth in zebrafish. *BMC Dev Biol* 8:84.

Krichevsky O and Bonnet G. 2002. Fluorescence correlation spectroscopy: The technique and its applications. *Rep Prog Phys* 65:251–297.

Kronick MN. 1986. The use of phycobiliproteins as fluorescent labels in immunoassay. *J Immunol Methods* 92:1–13.

Leutenegger M, Blom H, Widengren J, Eggeling C, Gosch M, Leitgeb RA, and Lasser T. 2006. Dual-color total internal reflection fluorescence cross-correlation spectroscopy. *J Biomed Opt* 11:040502.

Lieto AM, Cush RC, and Thompson NL. 2003. Ligand-receptor kinetics measured by total internal reflection with fluorescence correlation spectroscopy. *Biophys J* 85:3294–3302.

Lippitz M, Erker W, Decker H, van Holde KE, and Basche T. 2002. Two-photon excitation microscopy of tryptophan-containing proteins. *Proc Natl Acad Sci USA* 99:2772–2777.

Liu P, Sudhaharan T, Koh RM, Hwang LC, Ahmed S, Maruyama IN, and Wohland T. 2007. Investigation of the dimerization of proteins from the epidermal growth factor receptor family by single wavelength fluorescence cross-correlation spectroscopy. *Biophys J* 93:684–698.

Liu P, Ahmed S, and Wohland T. 2008. The *F*-techniques: Advances in receptor protein studies. *Trends Endocrinol Metab* 19:181–190.

Maeder CI, Hink MA, Kinkhabwala A, Mayr R, Bastiaens PI, Knop M. 2007. Spatial regulation of Fus3 MAP kinase activity through a reaction-diffusion mechanism in yeast pheromone signalling. *Nat Cell Biol* 9:1319–1326.

Magatti D and Ferri F. 2003. 25 ns software correlator for photon and fluorescence correlation spectroscopy. *Rev Sci Instrum* 74:1135.

Magde D, Elson E, and Webb WW. 1972. Thermodynamic fluctuations in a reacting system-measurement by fluorescence correlation spectroscopy. *Phys Rev Lett* 29:705–708.

Magde D, Webb WW, and Elson EL. 1978. Fluorescence correlation spectroscopy. III. Uniform translation and laminar flow. *Biopolymers* 17:361–376.

Marrocco M. 2004. Fluorescence correlation spectroscopy: Incorporation of probe volume effects into the three-dimensional Gaussian approximation. *Appl Opt* 43:5251–5262.

Marrocco M. 2008. Two-photon excitation fluorescence correlation spectroscopy of diffusion for Gaussian–Lorentzian volumes. *J Phys Chem A* 112:3831–3836.

Meseth U, Wohland T, Rigler R, and Vogel H. 1999. Resolution of fluorescence correlation measurements. *Biophys J* 76:1619–1631.

Modos K, Galantai R, Bardos-Nagy I, Wachsmuth M, Toth K, Fidy J, and Langowski J. 2004. Maximum-entropy decomposition of fluorescence correlation spectroscopy data: Application to liposome-human serum albumin association. *Eur Biophys J* 33:59–67.

Muller BK, Zaychikov E, Brauchle C, and Lamb DC. 2005. Pulsed interleaved excitation. *Biophys J* 89:3508–3522.

Muto H, Nagao I, Demura T, Fukuda H, Kinjo M, and Yamamoto KT. 2006. Fluorescence cross-correlation analyses of the molecular interaction between an Aux/IAA protein, MSG2/IAA19, and protein–protein interaction domains of auxin response factors of arabidopsis expressed in HeLa cells. *Plant Cell Physiol* 47:1095–1101.

Nagy A, Wu J, and Berland KM. 2005. Observation volumes and {gamma}-factors in two-photon fluorescence fluctuation spectroscopy. *Biophys J* 89:2077–2090.

Nie S, Chiu DT, and Zare RN. 1994. Probing individual molecules with confocal fluorescence microscopy. *Science* 266:1018–1021.

Nishimura G and Kinjo M. 2004. Systematic error in fluorescence correlation measurements identified by a simple saturation model of fluorescence. *Anal Chem* 76:1963–1970.

Octeau V, Cognet L, Duchesne L, Lasne D, Schaeffer N, Fernig DG, and Lounis B. 2009. Photothermal absorption correlation spectroscopy. *ACS Nano* 3:345–350.

Ohsugi Y, Saito K, Tamura M, and Kinjo M. 2006. Lateral mobility of membrane-binding proteins in living cells measured by total internal reflection fluorescence correlation spectroscopy. *Biophys J* 91:3456–3464.

Pan X, Foo W, Lim W, Fok MH, Liu P, Yu H, Maruyama I, and Wohland T. 2007a. Multifunctional fluorescence correlation microscope for intracellular and microfluidic measurements. *Rev Sci Instrum* 78:053711.

Pan X, Yu H, Shi X, Korzh V, and Wohland T. 2007b. Characterization of flow direction in microchannels and zebrafish blood vessels by scanning fluorescence correlation spectroscopy. *J Biomed Opt* 12:014034.

Pan X, Shi X, Korzh V, Yu H, and Wohland T. 2009. Line scan fluorescence correlation spectroscopy for 3D microfluidic flow velocity measurements. *J Biomed Opt* 14:024049.

Patterson GH and Lippincott-Schwartz J. 2002. A photoactivatable GFP for selective photolabeling of proteins and cells. *Science* 297:1873–1877.

Persson G, Thyberg P, and Widengren J. 2008. Modulated fluorescence correlation spectroscopy with complete time range information. *Biophys J* 94:977–985.

Petersen NO. 1986. Scanning fluorescence correlation spectroscopy. I. Theory and simulation of aggregation measurements. *Biophys J* 49:809–815.

Petersen NO, Johnson DC, and Schlesinger MJ. 1986. Scanning fluorescence correlation spectroscopy. II. Application to virus glycoprotein aggregation. *Biophys J* 49:817–820.

Petrasek Z, Hoege C, Hyman A, and Schwille P. 2008. Two-photon fluorescence imaging and correlation analysis applied to protein dynamics in *C. elegans* embryo. *Proceedings of SPIE* 6860:68601L.

Petrasek Z and Schwille P. 2008a. Photobleaching in two-photon scanning fluorescence correlation spectroscopy. *Chemphyschem* 9:147–158.

Petrasek Z and Schwille P. 2008b. Precise measurement of diffusion coefficients using scanning fluorescence correlation spectroscopy. *Biophys J* 94:1437–1448.

Petrasek Z and Schwille P. 2008c. Scanning fluorescence correlation spectroscopy. In: Rigler R and Vogel H, eds., *Single Molecules and Nanotechnology*. Berlin: Springer. pp. 83–105.

Plakhotnik T, Donley EA, and Wild UP. 1997. Single-molecule spectroscopy. *Annu Rev Phys Chem* 48:181–212.

Previte MJ, Pelet S, Kim KH, Buehler C, and So PT. 2008. Spectrally resolved fluorescence correlation spectroscopy based on global analysis. *Anal Chem* 80:3277–3284.

Prummer M and Vogel H. 2008. Mobility and signaling of single receptor proteins. In: Rigler R and Vogel H, eds., *Single Molecules and Nanotechnology*. Berlin: Springer. pp. 131–162.

Qian H. 1990. On the statistics of fluorescence correlation spectroscopy. *Biophys Chem* 38:49–57.

Rao R, Langoju R, Gosch M, Rigler P, Serov A, and Lasser T. 2006. Stochastic approach to data analysis in fluorescence correlation spectroscopy. *J Phys Chem A* 110:10674–10682.

Rauer B, Neumann E, Widengren J, and Rigler R. 1996. Fluorescence correlation spectrometry of the interaction kinetics of tetramethylrhodamin alpha-bungarotoxin with *Torpedo californica* acetylcholine receptor. *Biophys Chem* 58:3–12.

Ricka J and Binkert T. 1989. Direct measurement of a distinct correlation function by fluorescence cross correlation. *Phys Rev A* 39:2646–2652.

Ries J, Petrov EP, and Schwille P. 2008. Total internal reflection fluorescence correlation spectroscopy: Effects of lateral diffusion and surface generated fluorescence. *Biophys J* 95:390–399.

Ries J and Schwille P. 2006. Studying slow membrane dynamics with continuous wave scanning fluorescence correlation spectroscopy. *Biophys J* 91:1915–1924.

Rigler R and Elson ES. 2001. *Fluorescence Correlation Spectroscopy: Theory and Applications*. Springer, Berlin.

Rigler R, Mets, Widengren J, and Kask P. 1993. Fluorescence correlation spectroscopy with high count rate and low background: Analysis of translational diffusion. *Eur Biophys J* 22:169–175.

Ruan Q and Tetin SY. 2008. Applications of dual-color fluorescence cross-correlation spectroscopy in antibody binding studies. *Anal Biochem* 374:182–195.

Ruan Q, Cheng MA, Levi M, Gratton E, and Mantulin WW. 2004. Spatial-temporal studies of membrane dynamics: Scanning fluorescence correlation spectroscopy (SFCS). *Biophys J* 87:1260–1267.

Saffarian S and Elson EL. 2003. Statistical analysis of fluorescence correlation spectroscopy: The standard deviation and bias. *Biophys J* 84:2030–2042.

Saito K, Wada I, Tamura M, and Kinjo M. 2004. Direct detection of caspase-3 activation in single live cells by cross-correlation analysis. *Biochem Biophys Res Commun* 324:849–854.

Saleh BEA and Teich MC. 1991. *Fundamentals of Photonics*. New York: Wiley-Interscience.

Schätzel K. 1990. Noise on photon correlation data: I. Autocorrelation functions. *Quantum Opt* 2:287–305.

Schätzel K, Drewel M, and Stimac S. 1988. Photon correlation measurements at large lag times: Improving statistical accuracy. *J Mod Opt* 35:711–718.

Schwille P, Meyer-Almes FJ, and Rigler R. 1997. Dual-color fluorescence cross-correlation spectroscopy for multicomponent diffusional analysis in solution. *Biophys J* 72:1878–1886.

Schwille P, Haupts U, Maiti S, and Webb WW. 1999a. Molecular dynamics in living cells observed by fluorescence correlation spectroscopy with one- and two-photon excitation. *Biophys J* 77:2251–2265.

Schwille P, Korlach J, and Webb WW. 1999b. Fluorescence correlation spectroscopy with single-molecule sensitivity on cell and model membranes. *Cytometry* 36:176–182.

Sengupta P, Garai K, Balaji J, Periasamy N, and Maiti S. 2003. Measuring size distribution in highly heterogeneous systems with fluorescence correlation spectroscopy. *Biophys J* 84:1977–1984.

Shaner NC, Campbell RE, Steinbach PA, Giepmans BN, Palmer AE, and Tsien RY. 2004. Improved monomeric red, orange and yellow fluorescent proteins derived from *Discosoma* sp. red fluorescent protein. *Nat Biotechnol* 22:1567–1572.

Shcherbo D, Merzlyak EM, Chepurnykh TV, Fradkov AF, Ermakova GV, Solovieva EA, Lukyanov KA et al. 2007. Bright far-red fluorescent protein for whole-body imaging. *Nat Methods* 4:741–746.

Shi X, Foo YH, Sudhaharan T, Chong SW, Korzh V, Ahmed S, Wohland T. 2009. Determination of dissociation constants in living zebrafish embryos with single wavelength fluorescence cross-correlation spectroscopy. *Biophys J* 97:678–686.

Sisan DR, Arevalo R, Graves C, McAllister R, and Urbach JS. 2006. Spatially resolved fluorescence correlation spectroscopy using a spinning disk confocal microscope. *Biophys J* 91:4241–4252.

Skakun VV, Hink MA, Digris AV, Engel R, Novikov EG, Apanasovich VV, and Visser AJ. 2005. Global analysis of fluorescence fluctuation data. *Eur Biophys J* 34:323–334.

Skinner JP, Chen Y, and Muller JD. 2005. Position-sensitive scanning fluorescence correlation spectroscopy. *Biophys J* 89:1288–1301.

Starchev K, Zhang J, and Buffle J. 1998. Applications of fluorescence correlation spectroscopy - Particle size effect. *J Colloid Interface Sci* 203:189–196.

Starchev K, Ricka J, and Buffle J. 2001. Noise on fluorescence correlation spectroscopy. *J Colloid Interface Sci* 233:50–55.

Starr TE and Thompson NL. 2001. Total internal reflection with fluorescence correlation spectroscopy: Combined surface reaction and solution diffusion. *Biophys J* 80:1575–1584.

Stout AL and Axelrod D. 1989. Evanescent field excitation of fluorescence by epi-illumination microscopy. *Appl Opt* 28:5237–5242.

Sudhaharan T, Liu P, Foo YH, Bu W, Lim KB, Wohland T, and Ahmed S. 2009. Determination of in vivo dissociation constant, K_d, of CDC42-effector complexes in live mammalian cells using single wavelength fluorescence cross-correlation spectroscopy (SW-FCCS). *J Biol Chem* 284 (20):13602–13609.

Swaminathan R, Hoang CP, and Verkman AS. 1997. Photobleaching recovery and anisotropy decay of green fluorescent protein GFP-S65T in solution and cells: Cytoplasmic viscosity probed by green fluorescent protein translational and rotational diffusion. *Biophys J* 72:1900–1907.

Swift JL, Burger MC, Massotte D, Dahms TE, and Cramb DT. 2007. Two-photon excitation fluorescence cross-correlation assay for ligand-receptor binding: Cell membrane nanopatches containing the human micro-opioid receptor. *Anal Chem* 79:6783–6791.

Terry BR, Matthews EK, and Haseloff J. 1995. Molecular characterisation of recombinant green fluorescent protein by fluorescence correlation microscopy. *Biochem Biophys Res Commun* 217:21–27.

Thews E, Gerken M, Eckert R, Zapfel J, Tietz C, and Wrachtrup J. 2005. Cross talk free fluorescence cross correlation spectroscopy in live cells. *Biophys J* 89:2069–2076.

Thompson NL. 1991. Fluorescence correlation spectroscopy. In: Lakowicz, JR, ed., *Topics in Fluorescence Spectroscopy*, Vol. 1: *Techniques*. New York: Plenum Press. pp. 337–378.

Thompson NL and Steele BL. 2007. Total internal reflection with fluorescence correlation spectroscopy. *Nat Protoc* 2:878–890.

Thompson NL, Burghardt TP, and Axelrod D. 1981. Measuring surface dynamics of biomolecules by total internal reflection fluorescence with photobleaching recovery or correlation spectroscopy. *Biophys J* 33:435–454.

Thompson NL, Lieto AM, and Allen NW. 2002. Recent advances in fluorescence correlation spectroscopy. *Curr Opin Struct Biol* 12:634–641.

Tsien RY. 1998. The green fluorescent protein. *Annu Rev Biochem* 67:509–544.

Van Craenenbroeck E and Engelborghs Y. 1999. Quantitative characterization of the binding of fluorescently labeled colchicine to tubulin in vitro using fluorescence correlation spectroscopy. *Biochemistry* 38:5082–5088.

Wachsmuth M, Weidemann T, Muller G, Hoffmann-Rohrer UW, Knoch TA, Waldeck W, and Langowski J. 2003. Analyzing intracellular binding and diffusion with continuous fluorescence photobleaching. *Biophys J* 84:3353–3363.

Wahl M, Gregor I, Patting M, and Enderlein J. 2003. Fast calculation of fluorescence correlation data with asynchronous time-correlated single-photon counting. *Opt Express* 11:3583–3591.

Weidemann T, Wachsmuth M, Tewes M, Rippe K, and Langowski J. 2002. Analysis of ligand binding by two-colour fluorescence cross-correlation spectroscopy. *Single Mol* 3:49–61.

Weissman M, Schindler H, and Feher G. 1976. Determination of molecular weights by fluctuation spectroscopy: Application to DNA. *Proc Natl Acad Sci USA* 73:2776–2780.

Wennmalm S, Edman L, and Rigler R. 1997. Conformational fluctuations in single DNA molecules. *Proc Natl Acad Sci USA* 94:10641–10646.

Widengren J and Rigler R. 1995. Fluorescence correlation spectroscopy of triplet states in solution: A theoretical and experimental study. *J Phys Chem* 99:13368–13379.

Widengren J, Rigler R, and Mets U. 1994. Triplet-state monitoring by fluorescence correlation spectroscopy. *J Fluoresc* 4:255–258.

Wohland T, Friedrich K, Hovius R, and Vogel H. 1999. Study of ligand–receptor interactions by fluorescence correlation spectroscopy with different fluorophores: Evidence that the homopentameric 5-hydroxytryptamine type 3As receptor binds only one ligand. *Biochemistry* 38:8671–8681.

Wohland T, Rigler R, and Vogel H. 2001. The standard deviation in fluorescence correlation spectroscopy. *Biophys J* 80:2987–2999.

Wruss J, Runzler D, Steiger C, Chiba P, Kohler G, and Blaas D. 2007. Attachment of VLDL receptors to an icosahedral virus along the 5-fold symmetry axis: Multiple binding modes evidenced by fluorescence correlation spectroscopy. *Biochemistry* 46:6331–6339.

Wu B, Chen Y, and Muller JD. 2008. Fluorescence correlation spectroscopy of finite-sized particles. *Biophys J* 94:2800–2808.

Xie XS and Lu HP. 1999. Single-molecule enzymology. *J Biol Chem* 274:15967–15970.

Xu C, Zipfel W, Shear JB, Williams RM, and Webb WW. 1996. Multiphoton fluorescence excitation: New spectral windows for biological nonlinear microscopy. *Proc Natl Acad Sci USA* 93:10763–10768.

Yu L, Tan M, Hob B, Ding JL, and Wohland T. 2005. Determination of critical micelle concentrations and aggregation numbers by fluorescence correlation spectroscopy: Aggregation of a lipopolysaccharide. *Anal Chim Acta* 556 (1):216–225.

Zemanova L, Schenk A, Hunt N, Nienhaus GU, and Heilker R. 2004. Endothelin receptor in virus-like particles: Ligand binding observed by fluorescence fluctuation spectroscopy. *Biochemistry* 43:9021–9028.

7

Two-Photon Excitation Microscopy: A Superb Wizard for Fluorescence Imaging

7.1 Introduction 7-1
7.2 Basic Principles 7-1
    Two-Photon Excitation Process • Two-Photon Point-Spread Function
7.3 Probes and Architecture Considerations 7-5
    2PE Fluorescent Probes • Design Considerations of a 2PE Set Up
7.4 Applications and Advantages 7-8
    Applications • Advantages due to Excitation Localization
7.5 Conclusions 7-10
References 7-11

Francesca Cella
University of Genoa

Alberto Diaspro
*University of Genoa
and Italian Institute
of Technology*

7.1 Introduction

Fluorescence optical microscopy is one of the most important tools to investigate biological samples and to observe structures in living cells. In particular, the development of confocal (Sheppard and Wilson, 1978) and two-photon excitation (2PE) microscopy (Denk et al., 1990) allowed to perform optical sectioning and the imaging of three-dimensional structures of biological samples (Diaspro, 2001). In recent years, such techniques have gained great advantages from the development of green fluorescent proteins (GFPs) as probes (Chalfie et al., 1994; Tsien, 1998) and variants (Patterson and Lippincott-Schwarz, 2002).

The first theoretical prediction of the simultaneous absorption of a photon couple was made in 1931 by Maria Göppert-Mayer (Göppert-Mayer, 1931), but the first experimental demonstration of 2PE microscopy on biological cells was realized in 1990 (Denk et al., 1990). The development of commercially available mode-locked lasers made 2PE microscopy a popular tool for the imaging of biological samples. Furthermore, its intrinsic optical sectioning capabilities and the reduced damage of the sample allowed to perform high-resolution and long-term imaging on living cells (Diaspro et al., 2005).

7.2 Basic Principles

7.2.1 Two-Photon Excitation Process

Fluorescence microscopy techniques based on a single-photon excitation (1PE) make use of photons in the visible range to match the energy gap between the ground state and the excited one (Lakowicz, 1999).

On the other hand, two-photon excitation microscopy takes advantage of a nonlinear process based on the simultaneous absorption of two photons in the IR range whose total energy has to be sufficient to trigger the transition of the fluorescent molecule to the excited state (Callis, 1997).

The relationship between the energy, E, and the wavelength, λ, of the exciting photons is given by

$$E = \frac{h \cdot c}{\lambda} \tag{7.1}$$

where
   c is the light speed
   h is Planck's constant

Since the sum of the energy of the photons required in a multiphoton absorption event has to be equal to the energy needed for the single-photon absorption, the wavelengths λ_1 and λ_2 of the photons used for a two-photon excitation can be chosen under the constraint:

$$\frac{1}{\lambda_{1P}} \cong \left(\frac{1}{\lambda_1} + \frac{1}{\lambda_2} \right) \tag{7.2}$$

where λ_{1P} is the wavelength used in the single-photon excitation process. For practical reasons, the easiest choice is to set $\lambda_1 = \lambda_2$ that leads to the experimental condition $\lambda_1 \approx 2\lambda_{1P}$ (Denk et al., 1990; Diaspro, 2001), as shown in Figure 7.1. One of the main advantages of 2PE is given by the high localization of the excitation process due to the fact that the probability of two-photon absorption is considerable only in the region close to the focal plane. Since an out-of-focus excitation is avoided, the reduction of the overall photobleaching and the phototoxicity of the thick samples (Patterson and Piston, 2000) represents another immediate advantage of two-photon excitation microscopy over confocal microscopy (Diaspro, 2001). In two-photon excitation microscopy, all the emitted photons are detected as opposed to the case in confocal microscopy in which the out-of-focus fluorescent signal is spatially rejected by a pinhole. Furthermore, the shift to the IR range of the excitation wavelength allows the application of 2PE to the thick scattering samples since the use of the higher wavelengths reduces the light scattering processes. The probability for the n-photon excitation event is proportional to the n-power of the instantaneous intensity of the excitation beam and to the molecular cross section, δ_n, for the multiphoton absorption. Hence, the fluorescence emission can be referred to as the function of the temporal characteristics of

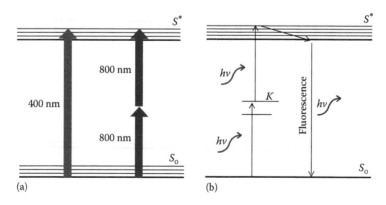

FIGURE 7.1 (a) Simplified scheme of a single-photon and a two-photon absorption. (b) Simplified Perrin–Jablonsky diagram for a 2PE. The molecule is brought to the excited state by the simultaneous absorption of two photons and the fluorescence emission is the same obtained in the 1PE.

the excitation light to the molecular cross section. High light fluxes are required in order to obtain 2PE since the value of the molecular cross section, δ_2, for 2PE is typically small (1 GM = 10^{-58} m^4 s^{-1}) and high photon flux densities can be reached by a tight focusing of the excitation laser beam. The excitation can be obtained by means of a high power continuous wave (CW) laser (Booth and Hell, 1998) but this will produce a substantial thermal effect on the sample. The best choice is to combine the spatial and the temporal focusing using short-pulsed lasers for performing the excitation. In this way, short pulses allow to reduce the average power delivered and the thermal effects on the sample, thus providing a considerable reduction of the damage on the sample induced by the incident light.

Short-pulsed lasers (~140 fs) allow a selective excitation of a single electronic transition since the order of the magnitude of the timescale of the 2PE process is

$$\tau_{2PE} = 10^{-16} - 10^{-17} \, s \tag{7.3}$$

In order to produce a two-photon fluorescence emission, the pulse duration (Koenig, 2000) has to be short enough (10^{-13} s) in order to prevent the relaxation to the ground state during the pulse allowing the simultaneous absorption of two photons. Moreover, in order to enable the collection of the emitted photons, the typical repetition rate (100 MHz) has to be chosen at least one order slower than the typical fluorescence lifetime (10^{-9} s).

The 2PE fluorescence intensity is strictly related to the probability that a fluorophore in the focus absorbs two photons in the same quantum event and is proportional to the square of the excitation intensity $I(t)$, to the molecular cross section, δ_2, and to the quantum yield η:

$$I_f(t) \propto \delta_{2PE} \cdot \eta \cdot I(t)^2 \tag{7.4}$$

The dependence between the excitation intensity $I(t)$ (expressed in photons cm^2 s^{-1}) and the instantaneous power $P(t)$ delivered on the illumination area A is

$$I(t) = \frac{\lambda \cdot P(t)}{h \cdot c \cdot A} \tag{7.5}$$

considering $A \cong \pi[\lambda_{exc}/(2 \cdot NA]^2$ (Born and Wolf, 1993)
where
   λ_{exc} is the wavelength of the excitation radiation
   NA is the numerical aperture of the objective
   c is the speed of light in the vacuum
   h is Planck's constant

The fluorescence intensity obtained under the 2PE conditions is

$$I_f(t) = k \cdot \pi \cdot \delta_{2PE} \cdot \eta \cdot P(t)^2 \cdot \left(\frac{NA^2}{h \cdot c \cdot \lambda_{exc}} \right)^2 \tag{7.6}$$

In case of the CW excitation the relation between the time averaged in the 2PE molecular fluorescence intensity and the average laser power P_{ave} is

$$\langle I_{f,CW} \rangle \cong k \cdot \pi \cdot \delta_{2PE} \cdot \eta \cdot \frac{1}{T} \int_0^T P^2(t) \cdot \left(\frac{NA^2}{h \cdot c \cdot \lambda_{exc}} \right)^2 \cong k \cdot \pi \cdot \delta_{2PE} \cdot \eta \cdot P^2_{ave} \cdot \left(\frac{NA^2}{h \cdot c \cdot \lambda_{exc}} \right)^2 \tag{7.7}$$

When a pulsed laser is used, the relationship between the collectable fluorescence signal and the average excitation power becomes (So et al., 2001)

$$\langle I_{f,pulsed} \rangle \cong \delta_{2PE} \cdot \eta \cdot \frac{P^2_{ave}}{\tau^2_P \cdot f^2_P} \cdot \left(\frac{NA^2}{h \cdot c \cdot \lambda_{exc}} \right)^2 \frac{1}{T} \int_0^{\tau_P} dt \cong \delta_{2PE} \cdot \eta \cdot \frac{P^2_{ave}}{\tau_P \cdot f_P} \cdot \left(\frac{NA^2}{h \cdot c \cdot \lambda_{exc}} \right)^2 \qquad (7.8)$$

where the dependence of the peak power and the average excitation power is $P_{peak}(t) = P_{ave}/\tau_P \cdot f_P$ where τ_P and f_P are respectively the pulse width and the pulse repetition rate (Figure 7.2).

7.2.2 Two-Photon Point-Spread Function

The intensity distribution of the excitation light of the wavelength λ_{exc} in the focus of a lens with a numerical aperture NA can be described in the paraxial approximation as (Sheppard and Gu, 1990; Born and Wolf, 1999)

$$I(u, v) = \left| 2 \cdot \int_0^1 J_0(v \cdot \rho) e^{-(1/2)u\rho^2} \rho \cdot d\rho \right|^2 \qquad (7.9)$$

where
  J_0 is the zeroth order Bessel function
  ρ represents the radial coordinate
  u, v are defined as the optical coordinate

$$u = \frac{8\pi \sin^2(\alpha/2) \cdot z}{\lambda}, \quad v = (2\pi \sin(\alpha) \cdot r)/\lambda$$

in which α is the semiangular aperture of the objective lens.

The single-photon fluorescence intensity is linearly proportional to the spatial distribution of the fluorophore $C(r, z)$ inside the sample and to the illumination intensity distribution. In a 2PE regime, even if the increased excitation wavelength, due to the use of the near IR radiation, produces a worsening of the optical resolution, the quadratic dependence of the fluorescent signal from the illumination intensity would lead to a localization of the excited region (cfr. Figure 7.3) to subfemtoliter (fm) volumes (Agard, 1984).

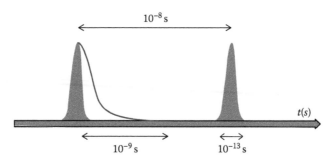

FIGURE 7.2 Pulse duration, pulse repetition rate, and fluorescence lifetime (not to scale).

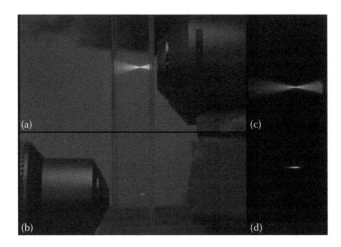

FIGURE 7.3 Acquisition of excitation volumes has been performed focusing laser beams into the rhodamine solution: The 1PE was provided by a 543 nm laser (a, c) and the 2PE was provided by a modelocked pulsed laser ($\lambda = 800$ nm) (b, d).

Hence, the square dependence of the emitted fluorescent light from the excitation power induces a signal axial falling off as the fourth power of the distance from the focal plane (Xu, 2002). Due to this fact, multiphoton excitation techniques provide an intrinsic axial sectioning capability and allow the performance of 3D imaging with an increased signal-to-noise ratio (SNR) when compared to confocal microscopy techniques. Indeed, in confocal microscopy all the molecules within the illuminated region are excited and the optical sectioning is reached by means of a pinhole, which allows the performance of a spatial selection of the in-focus contributions rejecting the out-of-focus fluorescent signal. Differently, in the 2PE regime, the optical sectioning is provided as only the molecules close to the focal region are excited and all the emitted photons can be collected.

7.3 Probes and Architecture Considerations

7.3.1 2PE Fluorescent Probes

The probability of the absorption of two photons in the same quantum event is strongly related to the molecular cross section of the molecule for 2PE, and the popularity of two-photon excitation microscopy is partially due to the fact that a great part of the common fluorescent molecules also own the 2PE cross section (Xu, 2001). The molecular cross section is expressed in GM units (1 GM $= 10^{-50}$ cm^4 s^{-1}) and its value has been measured (Xu and Webb, 1996) for most popular probes. Table 7.1 reports the properties under 2PE for the most commonly used dyes.

The 2PE cross section can be measured for a wide range of wavelengths and Figure 7.4 summarize the cross section value of some fluorophores within the wavelength range provided by the most popular ultrafast lasers (Girkin and Wokosin, 2001) used for performing 2PE. A fluorophore should have a cross section local maximum corresponding to double the wavelength used for a single-photon excitation, but this rule is fulfilled only for symmetrical fluorescent molecules and it is no longer valid for molecules with a more complex structure like some green fluorescent protein (GFP) mutants (Blab et al., 2001).

The cross section values (cfr. Table 7.1) prove that the use of a single wavelength allows the excitation of efficiently different probes under 2PE conditions, thus providing a powerful tool for co-localization studies and easily allowing the emission spectral separation of different dyes. This fact resulted in the recent development of new organic fluorophores able to provide a larger 2PE cross section.

TABLE 7.1   Summary of the Properties of the Most Common Dyes

Fluorophores	λ (nm)	$\eta\delta_2$	δ_2
	Extrinsic Fluorophores		
Bis-MSB	691/700	6.0±1.8	6.3±1.8
Bodipy	920	17±4.9	—
Calcium green	740–990	—	~80
Calcofluor	780/820	—	—
Cascade blue	750–880	2.1±0.6	~3
Coumarin 307	776.700–800	19±5.5	~20
CY2	780/800	—	—
CY3	780	—	—
CY5	780/820	—	—
DAPI (free)	700/720	0.16±0.05	~3.5
Dansyl	700	1	—
Dansyl hydrazine	700	0.72±0.2	—
DiI	700	95±28	—
Filipin	720	—	—
FITC	740–820	—	~25–38
Fluorescein (pH~11)	780	—	38±9.7
Fura-2 (free)	700	11	—
Fura-2 (high Ca)	700	12	—
Hoechst	780/820	—	—
Indo-1 (free)	700	4.5±1.3	12±4
Indo-1 (high Ca)	590/700	1.2±0.4	2.1±0.6
Lucifer yellow	840–860	0.95±0.3	~2
Nile red	810	—	—
Oregon green Bapta 1	800	—	—
Rhodamine B	840	—	210±55
Rhodamine 123	780–860	—	—
Syto 13	810	—	—
Texas red	780	—	—
Trile probe (Dapi, FITC, and Rhodamine)	720/740	—	—

Source: After Diaspro, A. et al., *Quart. Rev. Biophys.*, 38, 97, 2005.

7.3.2 Design Considerations of a 2PE Set Up

The commercial availability of pulsed lasers sources promoted the development of 2PE microscopy systems. In principle, a 2PE set up can be obtained by coupling a conventional confocal scanning system with a pulsed laser source (Diaspro, 2001). Concerning the detection system, two different approaches can be followed, namely, the descanned and the non-descanned configurations (Sandison and Webb, 1994).

The former approach can be obtained by directly coupling the IR laser source with a commercial confocal microscope scanning head utilizing the conventional acquisition pathway. An example of a descanned 2PE set up (Mazza et al., 2008) is presented in Figure 7.5, where a modelocked Ti:Sapphire Laser (Chameleon-XR Coherent) is externally coupled to a scanning confocal microscope (Leica TCS SP5).

The latter requires that the classical scanning system be modified in order to minimize optical elements in the detection path (Girkin et al., 2001). In this configuration, the photodetector is placed

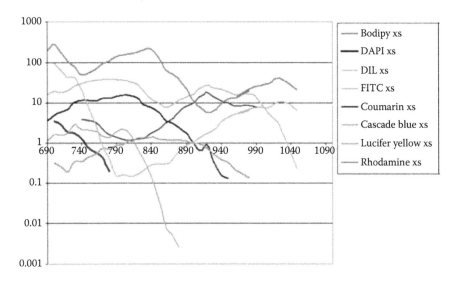

FIGURE 7.4 (See color insert following page 15-30.)   2PE cross sections for common dyes in the IR range. (After Diaspro, A. et al., *Quart. Rev. Biophys.*, 38, 97, 2005.)

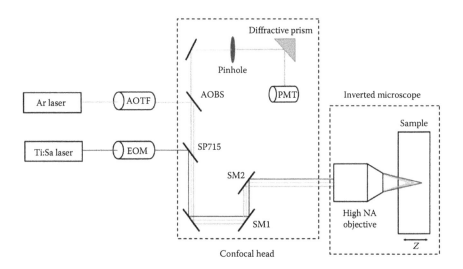

FIGURE 7.5   Schematic representation of a 2PE set up. The power of the IR laser is controlled by an EOM (electro-optic modulator). A short-pass (SP 715 nm) dichroic mirror prevents the reflected IR light from reaching the detector. Scanning is accomplished with conventional scanning mirrors SM1 and SM2. The fluorescent light coming from the sample is discriminated from the excitation light by the AOBS (acousto optical beam splitter) and brought to the detector after passing through a diffractive element which provides the selection of the detected wavelength range. An emitted fluorescence signal is acquired by a photo-multiplier tube. (After Mazza, D. et al., *Biophys. J.*, 95, 3457, 2008.)

directly beneath the microscope objective thus providing a better detection sensitivity. Imaging in the 2PE regime has to be performed using high numerical aperture objectives and the fluorescence radiation has to be collected by high sensitivity detectors, such as photomultiplier tubes (PMT), avalanche photodiodes, or charge-coupled device cameras (CCD). Low cost and good performance in terms of sensitivity makes PMTs the most common choice for the detection of the 2PE fluorescence signal.

7.4 Applications and Advantages

7.4.1 Applications

Despite the early theoretical prediction of the possibility of performing a two-photon absorption in the same quantum event (Göppert-Mayer, 1931), the first applications to biological samples of 2PE were performed after the development of pulsed lasers (Denk et al., 1990; Denk, 1996). In 2PE, the out-of-focus fluorescent contributions are intrinsically selected without using spatial filtering (pinhole) because the absorption probability is restricted to the focal region. For this reason, the high sectioning capabilities and the high SNR gained with nonlinear excitation can be exploited to perform the 3D imaging (cfr. Figure 7.6) and to minimize the photo-bleaching and the photo-damage effects on the sample.

Since a 2PE typically uses near IR wavelengths, it permits a deeper penetration capability (Theer and Denk, 2006) in scattering samples in comparison with a single-photon excitation, while at the same time allowing the performance of a high-resolution imaging in the turbid media.

For this reason, 2PE techniques are suitable for several *in vivo* and *in vitro* applications in the study of brain tissues (Christie et al., 2001), where the real-time analysis of intact neuronal networks and the study of the signaling mechanism at the synaptic level in brain slices can be performed. The reduced photobleaching due to the nonlinear excitation makes 2PE one of the most suitable tools for the long-term imaging of the dynamic processes and the structural reorganization in an intact brain or in brain slices (cfr. Figure 7.7).

In such a research area, the 2PE optical sectioning capabilities can also be exploited to perform imaging *in vivo* of the 3D structure in a mouse brain (cfr. Figure 7.8).

7.4.2 Advantages due to Excitation Localization

The 2PE high spatial confinement is mainly due to the quadratic dependence between the fluorescence intensity and the excitation laser power. In the single-photon excitation case, the whole sample region confined within the illumination double cone is involved in the fluorescence excitation process, while

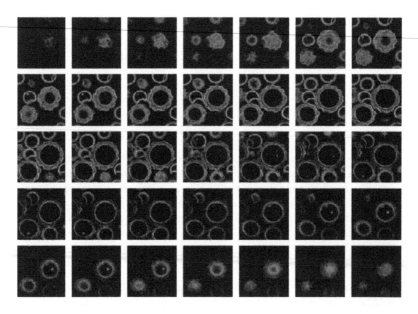

FIGURE 7.6 Example of optical sectioning (Diaspro, 2005) by means of a 2PE (780 nm). Detection of an auto-fluorescence signal from Colpoda's cistis allows the observation of the 3D structure of the membrane. (Courtesy of P. Ramoino.)

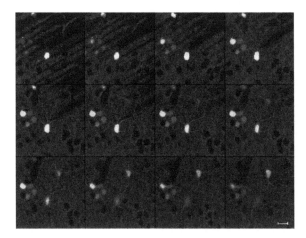

FIGURE 7.7   GAD67/GFP expression transgenic mice, imaging of brain slices. 2PE imaging allows the observation of the neuronal 3D structures (z sampling for the optical sectioning is 0.6 µm). The excitation wavelength used is 890 nm, Leica HCX APO L 63X Water Immersion U-V-I objective; NA 0.90. Scale bar 15 µm. (Courtesy of P. Bianchini and A. Vercelli.)

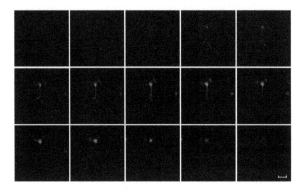

FIGURE 7.8   RFP expression in CAG-mRFP1 transgenic mice, *in vivo* imaging in brain. Optical sectioning (z sampling 1.713 µm) by means of 2PE allows a visualization of the brain structural conformation. The excitation wavelength used is 890 nm, Leica HCX APO L 63X Water Immersion U-V-I objective; NA 0.90. Scale bar 50 µm. (Courtesy of P. Bianchini and A. Vercelli. With permission.)

in the 2PE excitation regime the excited fluorescent molecules are strictly localized in a sub femtoliter volume close to the focal region (cfr. Figure 7.9c). This characteristic can be exploited in different advanced fluorescence techniques such as in fluorescence recovery after photobleaching (FRAP) and photoactivation (PA). FRAP and photoactivation belong to the category of the perturbation techniques and find applications in the study of dynamic processes within living cells. The basic concept behind these techniques is to permanently change the local properties of a restricted population of molecules inside the sample in order to follow the kinetics and the redistribution of the "signed" molecules. In FRAP experiments, the perturbation is performed by means of photobleaching, a process described as a first-order reaction that brings the molecules into a permanent "dark" state, while photoactivation consists of a light induced photoconversion process able to induce a permanent change in the spectral properties of the fluorescent protein.

As shown in Figure 7.9a and b the higher localization of the 2PE confines the photobleaching process to a region close to the focal plane while the 1PE produces photobleaching of an extended volume within the illumination cone. In the quantitative analysis of an FRAP experiment, the analytical description

FIGURE 7.9   Measurements of the effective photobleached volume has been performed on an immobile sample. The immobile fluorescent sample employed is a polyelectrolyte gel made by an ionic cross-linking of PolyAllylamineHydrochloride (PAH, Sigma-Aldrich), labeled with Alexa555, and sodium phosphate (Mazza et al., 2007). (a) Radial and axial extension of the photo-bleached volume in the 1PE regime ($\lambda_{exc}=543\,nm$). (b) Radial and axial extension of the photo-bleached volume when the excitation is performed in the 2PE regime ($\lambda_{exc}=800\,nm$). Photobleaching occurs in the overall illumination cone if the bleaching is performed using a single-photon excitation while it is well confined in the 2PE case. (c) Schematic representation of the excitation intensity distribution in the 1PE and in the 2PE case. (Modified from Diaspro, A. et al., *Biol. Eng. Online*, 5, 36, 2006.)

of the initial condition produced by the perturbation represents a basic aspect. Recent works (Mazza et al., 2007) show that the effective bleached volume is better analytically described if the bleaching is performed in the 2PE regime than in the single photon one. This work demonstrates that the localization of the excited volume provided by 2PE makes the quantitative analysis of the FRAP experiments more accurate and allows a reduction of the errors made in the estimation of the diffusion coefficient.

In the very same way, the use of 2PE allows a better confinement of the photoactivation process within a subcellular region (Testa et al., 2008). A photoactivation of HeLa cells expressing H2B-paGFP was performed employing 2PE (cfr. Figure 7.10a) and the 1PE (cfr. Figure 7.10b). The experiment shows that the axial localization of the photoactivated volume obtained inducing the photoconversion by a 2PE is larger than the one obtained using UV light. This suggests that 2PE microscopy is an important tool to photoactivate selective cellular compartments and to observe kinetic dynamics *in vivo* (cfr. Figure 7.10c). HeLa cells were transfected with PaGFP-Rac and the fluorescence recovery after the photoactivation (FRAPa) experiment (Mazza, 2008) demonstrated that Rac moves from the intracellular vesicles to the plasma membrane (Palamidessi et al., 2008).

7.5 Conclusions

Confocal and 2PE microscopy represent an important evolution in fluorescence microscopy since it couples the typical advantages of the conventional optical microscopy with a higher resolution power and an increased optical sectioning capability. This kind of microscopy also enables the development of advanced fluorescence microscopy techniques such as fluorescence recovery after photobleaching (FRAP), Forster resonance energy transfer (FRET), and fluorescence lifetime imaging (FLIM). In this chapter, the basic principles of 2PE have been discussed, and the advantages gained with this technique have been underlined. Despite the lower resolution when compared to confocal microscopy, the reduction of the sample photobleaching and the improved signal to the background (SNR) suggests that 2PE microscopy is the best tool for *in vivo* imaging and applications. An immediate advantage is due to the

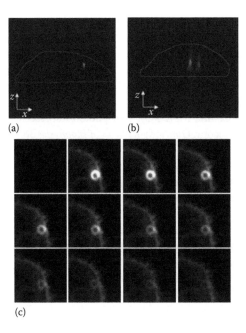

FIGURE 7.10 Axial dimension of the photoactivated volume has been measured on HeLa cells expressing H2B-paGFP. A small region inside the cell was photoactivated employing a 750 nm wavelength (a) and UV light (405 nm) (b). The full-width half-maximum (FWHM) of the activation axial profile is approximately 2 μm if the activation is performed in the IR range and 6 μm if the activation is performed using UV light. (c) HeLa cells were cotransfected with Rab5 and paGFP-Rac. Photoactivation of PaGFP inside vesicles has been performed employing the 2PE excitation (λ=790 nm). The redistribution in the plasma membrane of PaGFP-Rac is observed with a 488 nm light. Diffusion of paGFP-Rac has been observed acquiring a frame every 160 s. (Modified from Testa, I. et al., *J. Microsc.*, 230, 48, 2008.)

longer excitation wavelength used in 2PE, which provides an improvement of the imaging performances in the scattering and the turbid media. The advantages over confocal microscopy are also given from the high localization of the excitation process. Indeed, the excitation spatial confinement provides a strong selectivity that can be exploited in photoactivation and in FRAP experiments as outlined in this chapter.

References

Agard, D. A. (1984). Optical sectioning microscopy: Cellular architecture in three dimensions. *Annual Review of Biophysics* **13**, 191–219.

Blab, G. A., Lommerse, P. H. M., Cognet, L., Harms, G. S., and Schmidt, T. (2001). Two-photon excitation action cross-sections of the autofluorescent proteins. *Chemical Physics Letters* **350**, 71–77.

Booth, M. J. and Hell, S. W. (1998). Continuous wave excitation two-photon fluorescence microscopy exemplified with the 647-nm ArKr laser line. *Journal of Microscopy* **190**, 298–304.

Born, M. and Wolf, E. (1999). *Principles of Optics*, 7th edn. Cambridge: Cambridge University Press.

Callis, P. R. (1997). Two-photon-induced fluorescence. *Annual Review of Physical Chemistry* **48**, 271–297.

Chalfie et. al. (1994) Green fluorescent protein as a marker for gene expression. *Science* **263**:802–805.

Christie, R. H., Backsai, B. J., Zipfel, W. R., Williams, R. M., Kajdasz, S. T., Webb, W. W., and Hyman, B. T. (2001). Growth arrest of individual senile plaques in a model of Alzheimer's disease observed by in vivo multiphoton microscopy. *Journal of Neuroscience* **21**, 858–864.

Denk, W. (1996). Two-photon excitation in functional biological imaging. *Journal of Biomedical Optics* **1**, 296–304.

Denk, W., Strickler, J. H., and Webb, W. W. (1990). Two-photon laser scanning fluorescence microscopy. *Science* **248**, 73–76.

Diaspro, A. (ed.) (2001). *Confocal and Two-photon Microscopy: Foundations, Applications, and Advances.* New York: Wiley-Liss, Inc.

Diaspro, A., Chirico, G., and Collini, M. (2005). Two-photon fluorescence excitation and related techniques in biological microscopy. *Quarterly Reviews of Biophysics* **38**, 97–166.

Diaspro, A., Bianchini, P., Vicidomini, G., Faretta, M., Ramoino, P., and Usai, C. (2006). Multi-photon excitation microscopy. *Biomedical Engineering Online* **5**, 36.

Girkin, J. M. and Wokosin, D. (2001). Practical multiphoton microscopy. In *Confocal and Two-Photon Microscopy: Foundations, Applications and Advances* (ed. A. Diaspro), pp. 207–236. New York: Wiley-Liss, Inc.

Göppert-Mayer, M. (1931). Über Elementarakte mit zwei Quantensprüngen. *Annals of Physics* **9**, 273–295.

Koenig, K. (2000). Multiphoton microscopy in life science. *Journal of Microscopy* **200**, 83–104.

Lakowicz, J. R. (1999). *Principles of Fluorescence Microscopy.* New York: Plenum Press.

Mazza, D., Braeckmans, K., Cella, F., Testa, I., Vercauteren, D., Demeester, J., De Smedt, SS., and Diaspro, A. (2008). A new FRAP/FRAPa method for three-dimensional diffusion measurements based on multiphoton excitation microscopy. *Biophys J.* **95**(7): 3457–3469. Epub Jul 11, 2008.

Mazza, D., Cella, F., Vicidomini, G., Krol, S., and Diaspro, A. (2007). Role of three dimensional bleach distribution in confocal and two-photon fluorescence recovery after photobleaching experiments. *Applied Optics* **46**, 7401–7411.

Palamidessi, A., Frittoli, E., Garré, M., Faretta, M., Mione, M., Testa, I., Diaspro, A., Lanzetti, L., Scita, G., and Di Fiore, P. (2008). Endocytic trafficking of Rac is required for the spatial restriction of signaling in cell migration. *Cell*, **134** (1), 135–147.

Patterson, G. H. and Lippincott-Schwarz, J. (2002). A photoactivatable GFP for selective photolabeling of proteins and cells. *Science* **297**, 1873–1877.

Patterson, G. H. and Piston, D. W. (2000). Photobleaching in two-photon excitation microscopy. *Biophysical Journal* **78**, 2159–2162.

Sandison, D. R. and Webb, W. W. (1994). Background rejection and signal-to-noise optimization in the confocal and alternative fluorescence microscopes. *Applied Optics* **33**, 603–615.

Sheppard, C. J. and Gu, M. (1990). Image formation in two photon fluorescence microscopy. *Optik* **86**, 104–106.

Sheppard, C. J. R. and Wilson, T. (1978). Image formation in scanning microscopes with partially coherent source and detector. *Optica Acta*, **25**, 315–325.

Testa, I., Parazzoli, D., Barozzi, S., Garrè, M., Faretta, M., and Diaspro, A. (2008). Spatial control of pa-GFP photoactivation in living cells. *Journal of Microscopy* **230**, 48–60.

Theer, P. and Denk, W. (2006). On the fundamental imaging-depth limit in two-photon microscopy. *Journal Optical Society of America*, **23** (12), 3139–3149.

Tsien, R. Y. (1998). The green fluorescent protein. *Annual Review of Biochemistry* **67**, 509–544.

Xu, C. (2002). Cross-sections of fluorescence molecules in multiphoton microscopy. In *Confocal and Two-Photon Microscopy: Foundations, Applications and Advances* (ed. A. Diaspro), pp. 75–99. New York: Wiley-Liss, Inc.

Xu, C. and Webb, W. W. (1996). Measurement of two-photon excitation cross-sections of molecular fluorophores with data from 690 nm to 1050 nm. *Journal of the Optical Society of America B* **13**, 481–491.

8

Photobleaching Minimization in Single- and Multi-Photon Fluorescence Imaging

Partha Pratim Mondal
Massachusetts Institute of Technology

Paolo Bianchini
Zeno Lavagnino
University of Genoa

Alberto Diaspro
University of Genoa and Italian Institute of Technology

8.1   Introduction 8-1
8.2   Background 8-2
8.3   Single- and Multi-Photon Excitation 8-3
        Zero-Order Coefficient • First-Order Coefficient • Second-Order
        Coefficient • Cross Section for Single- and Multi-Photon Excitation
8.4   Photobleaching: The Mechanism..8-7
8.5   Photobleaching Minimization Techniques..............................8-11
        Triplet State Depletion Method for Single- and Multi-
        Photon Process • Controlled Light Exposure Microscopy
        (CLEM) • Photobleaching Reduction by Dark State
        Relaxation • Quantum Light Microscopy
Appendix...8-21
References 8-23

8.1 Introduction

The emerging microscopic techniques based on both near- and far-field excitation play a prominent role in bioimaging. While near field is strictly surface-based techniques, far-field high-resolution techniques, such as stimulated emission depletion (STED), 4PI, stochastic optical reconstruction method (STORM), and others, have shown promising applications in biological imaging [1–3]. These far-field techniques having optical sectioning capability and the ability of essay based on fluorescent-tagged physiological study have accelerated our understanding of biological processes. Multi-photon-based fluorescence imaging techniques have the additional advantage of excitation of biomolecules, such as tryptophan, which is excitable in the deep-ultraviolet (UV) range [4,5]. While UV lasers are now available to excite UV-range fluorophores, the issue of photobleaching (both single- and multi-photon based) is important and needs to be addressed for most of the quantitative and qualitative analyses in bioimaging. Furthermore, the vast and emerging nature of the field ranging from photonics to biophysics makes it impossible to write a complete review. Hence, we focus our attention on photobleaching property and its minimization for fluorescence-based imaging. We also mention the most important contributions in this field for the sake of completeness.

The ability to produce optical sectioning of biological specimens makes confocal and two-photon excitation very attractive enabling both in vivo and in vitro imaging. Since confocal imaging and, specially, two-photon imaging have the ability to generate focal plane information, they can be stacked together to obtain the three-dimensional reconstruction of the biological specimen [6,7]. To be able to image fluorescent molecules at $<10^{-9}$ M concentration and specifically at single-molecule level purely depends upon its characteristics such as high quantum yield and photostability. In a confocal imaging and more so in multi-photon imaging, most of the detected fluorescent signals must possess high signal-to-noise ratio (SNR) to form a quantitative image. The fluorescent signal falls off substantially with time at reasonable scan rate. This phenomenon is attributed to photobleaching. Confocal imaging has an additional effect of off-focus photobleaching that severely compromises the SNR. While multi-photon imaging does not have off-focus photobleaching effect, high peak power of the laser pulses cause photodamage and photobleaching of fluorescent molecules in the focal volume [8–12]. Thus, in general, loss of fluorescence signal due to photobleaching effect degrades the image and compromises on image quality [13].

The constructive side of photobleaching is its ability to produce valuable information about biological system dynamics such as diffusion since the 1970s [14–25]. Two most common ways of harnessing photobleaching as an effective tool are fluorescence recovery after photobleaching (FRAP) and fluorescence loss in photobleaching (FLIP) [26,27]. Multi-photon imaging is used as a basic mechanism for confining photobleaching to a region of interest. Although no dyes have been developed that do not bleach, we need to rely on these techniques for keeping photobleaching to a minimum in single- and multi-photon-based fluorescence imaging. Photobleaching and photodamage in fluorescence microscopy are mainly due to reaction with oxygen [28,29]. Well below the saturation level, photobleaching is linearly proportional to the excitation power. Obviously, an easy way to reduce photobleaching would be to reduce excitation power but this also reduces the fluorescence signal resulting in a low SNR. Hence, a proper balance between SNR and photobleaching is necessary. In this chapter, we attempt to do so by describing a few techniques (both theoretical and experimental) developed very recently. We lay our foundation on single- and multi-photon fluorescence imaging. Furthermore, we restrict ourselves to those specific applications that are related to photophysics of the fluorescent molecules in fluorescence-based imaging.

8.2 Background

The foundations of modern light microscopy were established by Ernst Abbe as early as 1873 [30,31]. He demonstrated how the diffraction of light and the objective lens are fundamental in determining the resolution of the imaging system, i.e.,

$$r \approx \frac{\lambda}{2n \sin \theta} \tag{8.1}$$

where
  r is the full width at half maximum of the lateral intensity spot
  λ is the wavelength of the light (in vacuum) used for illuminating the sample

Here, NA $= n \sin \theta$ is the numerical aperture of the objective lens with aperture angle θ, and n is the refractive index of the objective immersion liquid. The unwanted light that expands the apparent depth of field is exactly what confocal imaging eliminates. Thus, we can view only those fluorescent and light-scattering objects that lie within the depth that is given by the axial resolution of the microscope and attain the desired shallow depth of field.

Until 1970, Abbe's diffraction limit was taken as the ultimate resolution that is attainable by an imaging system as far as farfield is concerned. During the early 1970s, Egger and coworkers developed a laser-illuminated confocal microscope in which the objective lens was oscillated in order to scan the beam over the specimen [32]. The theoretical analysis on various modes of confocal and laser-scanning microscopy was provided by Sheppard and Choudhury [33]. The following year, Sheppard et al. [34] and Wilson et al. [35] described an epi-illuminating confocal microscope of the stage-scanning type, equipped with a laser source and a photomultiplier tube (PMT) as the detector. In addition, Cremer and Cremer of Heidelberg designed a specimen-scanning laser-illuminated confocal microscope [36]. Several other laboratories by Brakenhoff et al. [37,38], Wijnaendts van Resandt et al. [39], and Carlsson et al. [40] followed suit. These investigators, respectively, developed the stage-scanning confocal microscope further, verified the theory of confocal imaging, and expanded its application into cell biology. These circumstances culminated in the development of the confocal laser-scanning microscope for biological applications [41–44].

Multi-photon molecular excitation during a single quantum event was first predicted more than 70 years ago and consists of the simultaneous absorption of multiple photons that combine their energies to cause the transition to the excited state of the fluorophore. The theoretical foundation including the quantum description of multi-photon microscopy was laid by Maria-Goppert Mayer in 1930 [45]. This remained in the literature until the development of pulsed lasers and sensitive detectors. The first experimental demonstration of two photon excitation (TPE) was reported by Kaiser and Garret in 1961 on $CaF_2:Eu^{2+}$ [46]. This was followed by the experiment of Singh and Bradley in 1964, who were able to estimate the three-photon absorption cross section for naphthalene crystals [47]. In 1970, three-photon absorption was observed in organic dyes by Rentjepis et al. [48]. Then came the first TPE fluorescence in the chromosomes of living cells [49]. In 1990, the first experimental demonstration of multi-photon microscopy for biological imaging was demonstrated [50]. This was followed by huge volumes of research articles and new research initiatives in multi-photon fluorescence microscopy. On the other hand, the first confocal system was realized by Minsky in 1957 [51], but the concept was later developed extensively by Davidovits et al. [52], Sheppard et al. [33], and Brakenhoff et al. [37].

This is followed by some exciting developments in high-resolution microscopy for both single- and multi-photon excitation. Moreover, noninvasive 3D imaging is made possible in cells and other biological samples by light microscopes developed in the last three decades (1970–present) [33–35,38,53–56]. But two-photon excitation fluorescence microscopy is probably the most exciting development in the field of microscopy in the last two decades and most importantly, perhaps, after the application on living cells by Denk et al. [50]. This chapter concerns mainly with single and multiphoton excitation microscopy from a photophysical point of view.

The advantages of TPE over confocal single-photon microscopy are many. It is important to note that, for 3D imaging, a substantial volume needs to be exposed to illumination light, further resulting in photobleaching. This is particularly severe when exciting molecules in the UV range. Particular advantage with TPE microscopy is that it uses higher wavelength often in infrared range for excitation, that particularly reduces the photodamage. Moreover, infrared light scatters less than visible light and so and TPE may become handy when UV-excited fluorescent molecules are used. Another advantage is the slicing property of TPE process so as to be able to excite localized volume. This is because of the high photon density at the objective focus, which substantially increases the probability of two-photon absorption at the focus only. Because of the broad TPE spectra, several fluorescent molecules can be excited simultaneously by the same infrared beam, which further reduces colocalization errors. TPE substantially reduces the background noise because of the nonoverlap of the absorption and emission spectra, thereby making the optical setup conceptually simple.

8.3 Single- and Multi-Photon Excitation

Although the two-photon excitation process is discussed in this section for the sake of completeness, more stress will be given to the photophysical prospective. Single photon excitation was well known for centuries but the excitation using two photons was first conceptualized by Maria-Goppert Mayer [45]. The quantum mechanical description predicted the occurrence of such an event even though the probability of occurence is very low. Thus, the observation of such a rare event is possible only at very high photon flux. This needed a wait of 50 odd years before it could be experimentally verified with the advent of lasers. The first experimental application in biophysics was from the group of Watt Webb [50].

Initially, the system is in the ground state ($|\phi_i\rangle$). An incoming radiation field perturbs the perturbation-free time-independent Hamiltonian. The resulting Hamiltonian consists of time-dependent perturbation and perturbation-free time-independent Hamiltonian, i.e., $H' = H_0 + V$. The resulting state can be decomposed in stationary state basis set as

$$|\psi, t\rangle_I = \sum_n C_n(t)|\phi_n\rangle \tag{8.2}$$

where $C_n(t) = \langle \phi_n|U_I(t)|\phi_i\rangle$ is the time-evolution operator (see Appendix), $V(t) = A_1 e^{-i\omega t}$ and $A_1 = \frac{e}{mc}\vec{p}.\hat{q}A_0 e^{i\vec{k}.\vec{r}}$.

8.3.1 Zero-Order Coefficient

The first coefficient, $C_n^{(0)}$, defines the probability that the system remains in the initial ground state $|\phi_i\rangle$. Since $|\phi_i\rangle$ and $|\phi_n\rangle$ are orthogonal functions, the corresponding probability becomes

$$p_n^{(0)} = |C_n^{(0)}|^2 = \delta_{ni} \tag{8.3}$$

8.3.2 First-Order Coefficient

Now, the second term, $C_n^{(1)}$, determines the probability that the system makes a transition from $|\phi_i\rangle$ to excited state $|\phi_n\rangle$ by the absorption of a single photon of appropriate energy for making such a transition. The corresponding transition probability is given by

$$p^{(1)}\left(|\phi_i\rangle \to |\phi_n\rangle\right) = \left|C_n^{(1)}\right|^2$$

$$= \frac{2\pi}{\hbar}\left|\langle \phi_n|V|\phi_i\rangle\right|^2 \delta(E_{ni} - \hbar\omega) \tag{8.4}$$

where $E_{ni} = (E_n - E_i)$. This is the transition probability associated with single-photon excitation of fluorescent molecules. Detailed derivation can be found in the Appendix.

8.3.3 Second-Order Coefficient

The third term, $C_n^{(2)}$, is the second–most dominant term for exciting electrons to the excited state $|\phi_n\rangle$ by the simultaneous absorption of two photons of approximately half the excitation energy. The excitation is via an intermediate virtual state $|\phi_m\rangle$. The corresponding transition probability due to two-photon excitation is given by

$$p^{(2)}\left(|\phi_i\rangle \to |\phi_n\rangle\right) = |C_n^{(2)}|^2$$

$$= \frac{2\pi}{\hbar^3}\frac{\left|\langle \phi_n|V|\phi_m\rangle\right|^2 \left|\langle \phi_m|V|\phi_i\rangle\right|^2}{|\omega_{mi} - \omega|^2}\delta\left(E_{nm} + E_{mi} - 2\hbar\frac{\omega}{2}\right) \tag{8.5}$$

assuming $\omega_{nm} = \omega_{mi}$.

Considering the photophysical prospective of single- and multi-photon excitation microscopy, probably the most disheartening reality is the occurrence of photobleaching and photodamage for both in vivo and in vitro imaging. Photobleaching is such a sensitive issue that most of the reported and interpreted results in literature are approximately correct.

8.3.4 Cross Section for Single- and Multi-Photon Excitation

Single photon cross section probability can be rewritten in terms of photon number ($\frac{I}{\hbar\omega}$),

$$p^{(1)}\left(\left|\phi_i\right\rangle \to \left|\phi_n\right\rangle\right) = \left(\frac{I}{\hbar\omega}\right)\sigma_S \tag{8.6}$$

where

$$\sigma_S = \frac{2\pi e^2}{m^2 c^2 \omega}\left|\left\langle\phi_n\left|\hat{\varepsilon}.\vec{p}e^{i\frac{\omega}{c}\hat{n}.\vec{r}}\right|\phi_i\right\rangle\right|^2 \delta\left(E_{ni} - \hbar\omega\right) \tag{8.7}$$

$$= \frac{4\pi^2 e^2}{c}\frac{\omega_{ni}^2}{\omega}\left|\left\langle\phi_n\left|x\right|\phi_i\right\rangle\right|^2 \delta\left(E_{ni} - \hbar\omega\right) \tag{8.8}$$

is the single photon cross section. Here, the perturbing electric field felt by the molecule is $V = Ce^{-i\omega t}$ where the Coulomb gauge $C = \hat{\varepsilon}\vec{P}e^{i\frac{\omega}{c}\hat{n}.\vec{r}}$ and $\hat{\varepsilon}$ is the polarization vector.

Similarly, two-photon cross section probability can be rewritten in terms of photon number ($\frac{I}{\hbar\omega}$),

$$p^{(2)}\left(\left|\phi_i\right\rangle \to \left|\phi_n\right\rangle\right) = \left(\frac{I}{\hbar\omega}\right)^2 \sigma_{2PE} \tag{8.9}$$

where

$$\sigma_{2PE}\left(\left|\phi_i\right\rangle \to \left|\phi_n\right\rangle\right) = \frac{8\pi^3 e^4}{\hbar^2 c^2}\frac{\omega_{nm}^2 \omega_{mi}^2}{\omega^2(\omega_{ni} - \omega)^2}$$

$$\times \sum_{\left|\phi_m\right\rangle}\left|\left\langle\phi_n\left|x\right|\phi_m\right\rangle\right|^2\left|\left\langle\phi_m\left|x\right|\phi_i\right\rangle\right|^2 \delta\left(E_{nm} + E_{mi} - 2\hbar\frac{\omega}{2}\right) \tag{8.10}$$

is the two-photon cross section. Additionally, we should be reminded that for two-photon excitation we use light of wavelength 2λ/frequency $\omega/2$.

Although the multi-photon excitation is an interesting prospect from the theoretical point of view, it is experimentally difficult to realize due to very small cross section. Another disadvantage is that multi-photon microscopy loses out in resolution because of the longer illumination wavelength used for excitation. Nevertheless, it is useful to consider the multi-photon cross section. Generalization takes us to the following multi-photon (q-photon absorption) cross section probability:

$$p^{(q)}\left(\left|i\right\rangle \to \left|n\right\rangle\right) = \left(\frac{I}{\hbar\omega}\right)^q \sigma_{MPE} \tag{8.11}$$

where
$\sigma_{MPE}\left(\left|\phi_i\right\rangle \to \left|\phi_n\right\rangle\right)$ is the multi-photon cross section
$p^{(q)}$ is the probability that the molecule absorbs q-number of photons to cause an equivalent transition to that caused by single high-energy photon

The energy equivalence is given by

$$\hbar\omega = \hbar\omega_1 + \hbar\omega_2 + \cdots + \hbar\omega_q \tag{8.12}$$

Multi-photon excitation processes require two or more photons to interact simultaneously with the molecule. The cross-sectional area (A) of a molecule is estimated by its dipole transition length (for a transition length of 0.1–1 Å, $A = 10^{-17}$–10^{-16} cm^2). The lifetime of virtual states is around 10^{-16} s. The timescale for photon coincidence is determined by the lifetimes of the virtual intermediate states. Thus, the expected single- and two-cross sections are approximately given by A and $A^2\Delta\tau$, respectively. The cross section for n-photon absorption is $\sigma_n = A^n \Delta\tau^{n-1}$. Hence, the expected single-, two- and three-photon cross sections are approximately 10^{-16} cm^2 (s/photon), 10^{-49} cm^4 (s/photon), and 10^{-82} cm^6 (s/photon)^2, respectively. The small cross section values are one of the main obstacles for realizing multi-photon microscopy. Typical values for one of the widely used fluorescent molecules, 4,6-diamidino-2-phenolindole (DAPI), are 1.3×10^{-16} cm^2 at 345 nm for single-photon, 1.6×10^{-49} cm^4 at 700 nm for two-photon, and 2.5×10^{-84} cm^6 at 1000 nm for three-photon excitations. There are several published collections of two-photon excitation spectra and cross sections that provide guidance on the compatibility of dyes and probes with excitation sources [57,58].

Figure 8.1 shows the profile of excitation PSF of both single- and multi-photon excitation microscopy for three different fluorophores. The photobleaching nature of single-photon excitation is clearly evident as compared to two-photon excitation. In two-photon excitation, photobleaching is limited strictly to focal volume whereas for single-photon excitation, photobleaching is spread over the top and bottom layers as well. Figure 8.2 shows single- and two-photon excitation volumes of fixed polyelectrolyte samples labeled with Alexa 555. For single-photon excitation, the axial bleach pattern is obtained by bleaching a circular region with a radius of 1 μm (see Figure 8.2A). The bleach pattern was produced by delivering 1 mW at 543 nm on the sample with scanning speed $v = 0.384$ μm/s. On the other hand, two-photon pattern is obtained at an average power of 3.8 mW (Figure 8.2B). It is clear that because of intensity-squared dependence of two-photon excitation cross-section as compared to single photon cross-section, the excitation occurs only at the focus where the photon flux is high.

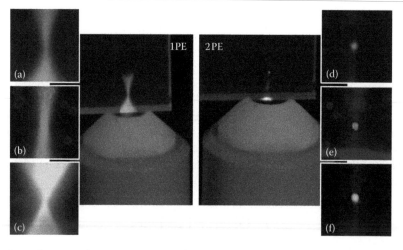

FIGURE 8.1 (See color insert following page 15-30.) Experimental excitation profile of three different fluorophores along the axial axis for single- and two-photon excitation.

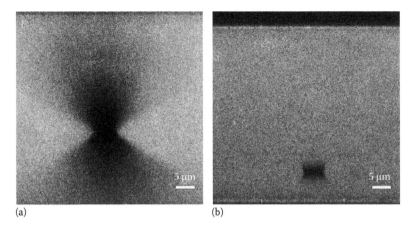

(a)   (b)

FIGURE 8.2 Immobile fluorescent sample built by ionic cross-linking of poly(allylamine hydrochloride) (PAH, Sigma-Aldrich), a positively charged polyelectrolyte, covalently labeled with Alexa Fluor 555 (Invitrogen). (a) Single-photon excitation and (b) two-photon excitation.

8.4 Photobleaching: The Mechanism

The resolution of the fluorescence-based imaging systems such as confocal microscopy, two-photon microscopy, 4PI microscopy, and others play a significant role in photobleaching studies [35, 50,59–64]. Most of these high-resolution fluorescence-based microscopic techniques have the ability to produce optical section of desired biological specimens both in vitro and in vivo. This reduces photobleaching from the upper and bottom layers unlike single photon microscopy. Single photon–based techniques have high SNR but multi-photon imaging suffers from low SNR in a reasonable scan time. Under these imaging conditions, the fluorescence emitted is often observed to decrease substantially with time because of photobleaching effects thereby further reducing the SNR. In most of the single- and multi-photon excitation, the high peak power of the laser pulses causes photodamage to the fluorescent probes [8,65]. Since photobleaching and phototoxicity limit the available signal, we feel that photobleaching is one of the most important factors restraining future developments. Especially, in multi-photon imaging, fluorescence signal is very low, and due to both linear and nonlinear photobleaching effects the process becomes complex, ruling out any analytical solution. Although we are not yet in the position of being able to develop dyes that do not bleach, the other approach is to develop molecule-specific techniques based on the spectroscopic signature of molecules to undo the photobleaching effect. This chapter is directed toward this effort.

To address photobleaching effect due to the involvement of metastable triplet state, we take on the simplest model involving ground singlet state S_0, excited singlet state S_1, and triplet state T_1, as shown in Figure 8.3. The corresponding set of differential rate equations is given by

$$\frac{\partial}{\partial t} S_0 = -f_{\mathrm{exc}} A S_0 + k_2 S_1 + k_4 T_1$$

$$\frac{\partial}{\partial t} S_1 = f_{\mathrm{exc}} A S_0 - (k_2 + k_3) S_1$$

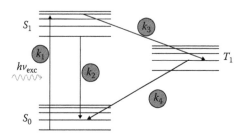

FIGURE 8.3   Jablonsky diagram showing various state transitions in a three-level system.

$$\frac{\partial}{\partial t}T_1 = k_3 S_1 - k_4 T_1$$

The steady-state condition is obtained when the population does not undergo any further change. The steady-state solution of the photobleaching process is given by

$$S_0(t) = k_4(k_2 + k_3)/C_1$$

$$S_1(t) = f_{exc}Af_4/C_1 \qquad\qquad (8.13)$$

$$T_1(t) = f_{exc}Ak_3/C_1$$

where $C_1 = f_{exc} A(k_3 + k_4) + k_4 (k_2 + k_3)$.

To demonstrate the validity of the simplistic photobleaching model, we computationally study two frequently used fluorescent molecules, i.e., green fluorescent protein (GFP) and fluorescein. In the computational study, we have photophysical factors such as singlet and triplet state lifetime and intersystem crossing rate. Figure 8.4 shows the population dynamics of all the three states at various excitation intensities. Initially, the molecules are predominantly in the ground state obeying Boltzmann distribution law. As the excitation intensity is increased, the population of ground state decreases and the triplet state population builds up. Higher triplet state population as compared to excited state (see Figure 8.4) indicates the lower output fluorescence signal from both GFP and fluorescein molecules. It should be reminded that triplet states are metastable states and hence the transition probability of $T_1 \rightarrow S_0$

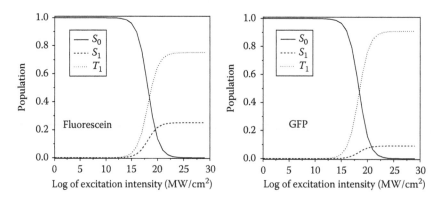

FIGURE 8.4   Population dynamics of S_0, S_1, and T_1 state with varying excitation intensity.

transition is very small. Thus, once a molecule makes a transition to triplet state, it reacts with oxygen resulting in strong photobleaching.

Fluorescence emitted by almost all fluorescent dyes fades during observation. This phenomenon is called photobleaching or dye photolysis and involves a photochemical modification of the dye resulting in the irreversible loss of its ability to fluoresce. Even though, in some cases, a fluorescent molecule can be switched on again after an apparent loss of emission ability or from a natural initial dark state (photocycling) or is able to switch on and off by itself (blinking) within a short timescale, we refer to photobleaching as an irreversible phenomenon. This point is particularly relevant in FRAP and FLIP because these techniques are rigorously based on the idea that fluorescence recovers only because fluorescent molecules diffuse into the bleached sample volume. In this case, it is even more important to be clear that the photobleached state cannot be reversed.

When fluorescent molecules are illuminated at a wavelength for which they exhibit a good cross section [66–69], possibly close to the cross section maximum, there is a shift from the singlet ground energy level (S_0) to the singlet excited energy level (S_1). Such a temporary excess of energy can be dissipated by fluorescence emission or in radiationless processes such as internal conversion and intersystem crossing to the triplet state (T_1). The decay times from S_1 and T_1 to S_0 are different according to the selection rules, and are of the order of 1–10 ns and 10^{-3}–10^{-6} s, respectively. Many factors, such as the molecular environment and the intensity of excitation light, may affect the mechanism, and thus the reaction order and the rates of photobleaching [70]. Here, we will concentrate on the photobleaching due to triplet state involvement only, leaving out the molecular environment effect.

The main causes seem to involve photodynamic interactions between excited fluorophores and molecular oxygen (O_2) in its triplet ground state and dissolved in the sample media. If the dye has a relatively high quantum yield for intersystem crossing, a significant number of dye molecules may cross from a singlet excited state S_1 to the long-lived triplet state T_1, a process that permits these molecules to interact with their environment for a much longer time (milliseconds instead of nanoseconds). Interactions between O_2 and dye triplets may generate singlet oxygen. Singlet oxygen has a longer lifetime than the excited triplet states of the dyes. Moreover, several types of damaging oxygen free radicals can be created when it decays. A fluorophore in the excited triplet state is also highly reactive and may undergo irreversible chemical reactions involving other intracellular organic molecules. All these chemical reactions depend both on the intracellular singlet oxygen concentration and on the distance between the dye and the intracellular components such as proteins, lipids, etc. Therefore, the number of photons emitted before a dye molecule is destroyed depends both on the nature of the dye molecule itself and on its environment. Some fluorophores have a very short lifetime, fading after the emission of only a few hundred photons, whereas other molecules in other surroundings can emit a very large number of photons (tens of millions) before being bleached. If a dye is protected from reaction with environmental molecular oxygen, the observed rate of photobleaching is lower [70]. This occurs naturally in GFP [18].

In addition, either multi-photon events or the absorption of a second photon by a molecule already in an excited state may also be involved in the photobleaching of some dyes, as shown in Figure 8.5. Because of the low absorption cross section and its quadratic dependence on the incident power, multi-photon excitation requires an excitation intensity several orders of magnitude higher than that needed for one-photon excitation. Unfortunately, this increases photobleaching because of the high probability of photochemical degradation in the long-lived triplet state and the interplay of multi-photon ionization processes [11,71]. In single-molecule detection conditions, evidence of a two-step photolysis has been reported for several coumarin and rhodamine derivatives [72]. These authors suggest that excited triplet states T_n, higher in energy than T_1, can occur when two distinct, successive photon absorptions occur at high irradiance. The mechanism is shown in Figure 8.5. Excitation at low irradiance yields longer survival times because, at any instant, fewer molecules are in the T_1 state to be excited to T_n [29,72].

At excitation intensities of 10^{23}–10^{24} photons/cm^2 s, approximately corresponding to 1 mW of power incident on a spot of 0.25 μm radius at 488 nm wavelength [73], fluorescein bleaches with a

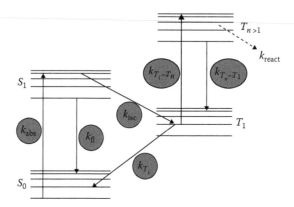

FIGURE 8.5   Perrin–Jablonski diagram of the relevant states showing photobleaching. Absorption (k_{abs}) from S_0 to S_1 may lead to fluorescence (k_{fl}), or intersystem crossing (k_{isc}) to T_1. From T_1, the molecule may relax back to S_0 (k_{T_1}), or absorb a second photon ($k_{T_1 \to T_n}$) and go to T_n. From there, the molecule may either relax back to T_1 or react (k_{react}) leading to irreversible photobleaching.

quantum efficiency, $Q_b = 3 \times 10^{-5}$. Therefore, during its useful lifetime the average fluorescein molecule dissolved in water emits 30,000–40,000 photons before being permanently bleached [68,74,75]. Recently, Deschenes and vanden Bout [29] reported a detailed single-molecule study of rhodamine 6G molecules embedded in a film of poly(methylacrylate) (PMA) under vacuum and using very low excitation intensities, between 0.1 and 10 kW/cm². The authors reported a remarkable decrease in the photobleaching rate at low excitation intensities and an increase in the total number of photons emitted up to 10^9 photons per molecule before bleaching. They propose a four-state model, sketched in Figure 8.5, that accounts for the observed nonlinear intensity dependence of the photobleaching rate. This model seems to fit most of the single-molecule photobleaching studies published up to now.

The high photon flux used in two-photon excitation microscopy seems to lead to higher-order photon interactions. The dependence of fluorescence intensity and photobleaching rate on excitation power has been studied for Indo-1, nicotinamide adenine dinuclestide (NADH), and aminocoumarin under one- and two-photon excitation [10]. The results of these studies suggest that higher-order photobleaching is common in two-photon excitation microscopy.

The prevention of the fading of fluorescence emission intensity is very important not only for quantitative microscopy but also for obtaining high-quality images. Unfortunately, reducing photobleaching by decreasing the excitation time or by lowering the excitation intensity leads to a reduction in the fluorescence signal. As lower signal increases the effect of Poisson noise and obscures low-contrast features, deciding on the optimal excitation level is always a trade-off in which ideally one reduces the light dose until one can no longer see the details one needs to understand.

Even though two-photon excitation microscopy has the advantage of minimizing out-of-focus photodamage and photobleaching [76] and confined excitation volume compared to one-photon excitation microscopy [77], the high photon flux (in gigawatts) used in two-photon excitation microscopy lead to increased higher-order photon interactions as well in the focal volume. Research has shown that such a multi-photon interaction is significant and may lead to photobleaching and photodamage of the biological samples [61,65,78]. Possible mechanisms that can lead to higher-order dependence of photobleaching are coherent three- and higher-order-photon excitation of upper electronic states and consecutive pumping of higher states via intermediate states, although other physical processes do exist that may result in photodamage and photobleaching [79,80]. Here, we address the photobleaching and photodamage due to higher-order interaction processes in the focal volume of two-photon excitation microscopy.

It is very important to notice the following points from a photobleaching point of view:

1. Due to the direct intensity dependence in single-photon excitation, the bleaching along the optical axis is severe as compared to multi-photon excitation.
2. Intensity square dependence in TPE limits the off-focus photobleaching but causes strong photobleaching in the focal plane when compared to single-photon excitation.

Due to multi-photon excitation in the focal volume, deviation from intensity-squared dependence is observed in many experiments [61,65]. The fluorescence intensity versus single- and two-photon excitation power is shown to increase with a slope of 1 and 2, respectively. But the two-photon photobleaching rate increases with a slope greater than 3, thus indicating the presence of higher-order photon interactions [65]. Experiments on Indo-1, NADH, and aminocoumarin produced similar results and suggest that this higher-order photobleaching is common in two-photon excitation microscopy [65]. Hence, the use of two-photon excitation microscopy to study thin biological samples may be limited by increased photobleaching and photodamage. Previous experimental evidence suggests that photobleaching in the focal volume is more pronounced under two-photon excitation than with conventional microscopy [79,80]. Particularly, NADH suggested that the photobleaching rate depended not only on the square of the intensity, but also on a higher-order relationship [79,80]. Nonlinear photo switching is also reported for *A* and *B* forms of a GFP mutant [81].

At low intensity levels, thermally assisted one-photon excitation could become significant if the wavelength of two-photon absorption band overlaps the one-photon absorption band. This causes a severe departure from the square-law dependence. For example, a significant deviation from square-law for rhodamine B at intensity levels of approximately 10^{24}–10^{25} photons/cm^2 s for $\lambda < 730$ nm is reported in the literature [82].

For Ca^{++}-bound species, 2PE dominates at $\lambda < 830$ nm and 3PE at $\lambda > 900$ nm, while for Ca^{++}-free species, 2PE dominates up to 910 nm and 3PE only at $\lambda > 960$ nm. At intermediate wavelengths, a mixture of two- and three-photon excitation was observed [83]. Slopes in the logarithmic plot of fluorescence output versus intensity from 2.0 to 3.0 indicate that a mixture of 2PE and 3PE fluorescence was generated [83]. Also, the quadratic power dependence of 2PE microscopy does not hold under the saturation condition [84]. Further, it is often difficult to distinguish multi-photon excitation spectra especially two-photon from one- and three-photon excitation spectra because the same initial excited states can be reached via one/two/three photon excitation, without violating any selection rules [57].

In Section 8.5, we describe some recent microscopic techniques and address how these techniques are capable of reducing photobleaching and photodamage.

8.5 Photobleaching Minimization Techniques

8.5.1 Triplet State Depletion Method for Single- and Multi-Photon Process

It should be noted that the absorption rate k_1 depends on the excitation intensity. Since single- and two-photon excitation have different absorption cross sections, the intensity required for the same rate of absorption is different. In general case for n-photon excitation, we need to introduce cross section into the rate equations by using the expression

$$\sigma_{\text{MPE}} = \sigma_n = \sigma_1^n \Delta \tau^{n-1} \qquad (8.14)$$

A possible experimental realization of the proposed approach is schematically sketched in Figure 8.6. One laser is used to pump the electrons from singlet S_0 to excited S_1 state. Once the electrons are in excited singlet state S_1, they can dissipate following two different pathways: (1) fluorescence process ($S_1 \to S_0$); (2) photobleaching process ($S_1 \to T_1 \to S_0$). In case of strong photobleaching, the probability

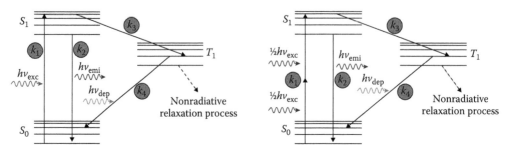

FIGURE 8.6 Molecular energy diagram depicting rate constants for single- and two-photon excitation.

of pathway (2) becomes highly probable and hence a large number of electrons follow this pathway for deexcitation. We propose to deplete the electrons instantaneously by applying another pulse just after the excitation pulse (see Figure 8.6). The corresponding set of differential rate equations manifesting the proposed mechanism is given by

$$\frac{\partial}{\partial t}\begin{pmatrix} S_0 \\ S_1 \\ T_1 \end{pmatrix} = \sigma_1^n \Delta \tau^{n-1} \begin{pmatrix} -f_{exc}A & k_2 & f_{dep}A \\ f_{exc}A & -(k_2+k_3) & 0 \\ 0 & k_3 & -f_{dep}A \end{pmatrix} \begin{pmatrix} S_0 \\ S_1 \\ T_1 \end{pmatrix} \tag{8.15}$$

with the additional constraint on the transition rate that $S_0 + S_1 + T_1 = 1$. $f_{exc} \propto h\nu_{exc}k_1$ and $f_{dep} \propto h\nu_{dep}k_4$ are the photon flux of the excitation and depletion lasers, where ν_{exc} and ν_{dep} are, respectively, the excitation and depletion frequencies.

For single- and two-photon excitation, the excitation cross section $\sigma_1 \approx 10^{-16}$ and $\sigma_2 \approx 10^{-48}$. When the steady state is reached, the population probabilities do not undergo further changes. Hence, we proceed to solve the rate equations for steady state, i.e., $\frac{\partial}{\partial t}\begin{pmatrix} S_0 \\ S_1 \\ T_1 \end{pmatrix} = 0$ where 0 is a null vector. The population density obtained in steady state condition is given by

$$S_0(t) = f_{dep}A(k_2 + k_3)/C_1$$

$$S_1(t) = f_{dep}Af_{exc}A/C_1 \tag{8.16}$$

$$T_1(t) = f_{exc}Ak_3/C_1$$

where $C_1 = (k_2 + k_3)f_{dep}A + (k_3 + f_{dep}A)f_{exc}A$.

Thus, the steady-state solution does not depend on the absorption cross section. This is justifiable because singlet and triplet state population dynamics come to play only when enough molecules are in the excited state. Equation 8.16, respectively, represents the steady state solution of the proposed mechanism. We have chosen to perform the theoretical and computational studies on two popular fluorescent molecules, i.e., fluorescein and GFP that are frequently used in bioimaging applications. The information regarding the transition rates for various molecules are readily available in the literature [27,85,86].

Photobleaching to no-photobleaching is a consequence of transition between occupied-to-unoccupied triplet state. This transition from no-photobleaching to photobleaching state is observed with increasing depletion power, where the crossover point is termed as photobleaching threshold power. Before the photobleaching threshold power, most of the population are in T_1 state and after the transition in the S_0 state. We have noted this interesting transition for two most commonly used fluorescent molecules, i.e., variants of fluorescein and GFP. The plots depicting such transitions are shown in Figure 8.7. Figure 8.7A through D and E through H shows the photobleaching nature of fluorescein and GFP molecules in the presence of depletion pulse. For fluorescein molecule, more than 95% of the molecules are in T_1 at a depletion pulse power of 3 kW/cm². At a power of 50 kW/cm², the population of S_1 state builds up to 22% and as a consequence the triplet state T_1 population falls to 78%. The situation reverses at the depletion power of 1 MW/cm², i.e., the singlet population reaches 78% and triplet population goes down to 22%. The depletion pulse reverse pumps the electron to the ground state S_0. However, it requires a power of around 10 MW/cm² to reduce triplet population (and thereby the photobleaching) to less than 5% (see Figure 8.7D). At lower rate of transition from triplet state to bleached state, the depletion process is not very effective. This requires more power to compete with the rapid bleaching process. Population dynamics similar to that of fluorescein molecule is observed for GFP molecule as well, except that to reduce the T_1 population to less than 2%, a relatively very high power of 3000 MW/cm² is required (see Figure 8.7H). This is because of the different lifetime of triplet states in fluorescein and GFP molecules

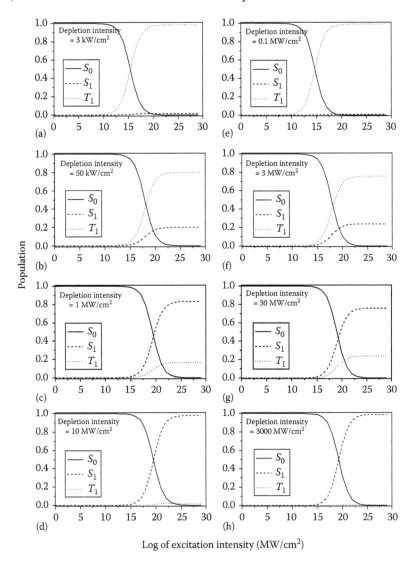

FIGURE 8.7 The normalized population of S_0, S_1, and T_1 as a function of excitation and depletion powers for fluorescein (a through d), and GFP (e through h) molecules.

and intersystem crossing rate. Molecules with longer triplet lifetime are more prone to bleaching as compared to molecules with shorter triplet lifetime. Longer triplet lifetime maximizes the generation of reactive oxygen species and makes the molecule resistant to depletion process, thereby requiring high power for significant depletion of triplet state. On the other hand, molecules with smaller lifetime can be depleted with relatively small power.

8.5.2 Controlled Light Exposure Microscopy (CLEM)

In this section, a spatially controlled light exposure-based microscopic technique is presented [87]. Spatially controlling the light on the sample reduces the photobleaching and photodamage without compromising image quality.

In conventional microscopy imaging, the field of view is illuminated uniformly irrespective of fluorophore distribution in the region. Ideally, regions with high fluorophore distribution require small illumination intensity (low power) for recognizing edges, whereas low fluorophore-distributed regions require relatively high illumination intensity to recognize sharp edges. Hence, local excitation-light dose (product of light intensity and exposure time, $I_{ill} \times \tau_{exposure}$) does not improve SNR, further causing unnecessary photobleaching and photodamage.

The basic principles outlining CLEM system are

1. Nonuniform illumination of the field of view minimizes photobleaching, keeping the image quality as those obtained using non-CLEM.
2. Based on local concentration of fluorescent molecules, the local excitation-light dose is adjusted.
3. The excitation light exposure time (τ) time is reduced in black background and bright fluorescent foreground.

Such a strategy has the potential advantage of reducing not only the photobleaching in the focal volume but also above and below the focal plane. This is a significant improvement considering the immense involvement of photobleaching in single-photon excitation microscopy. The experimental realization of CLEM system is shown in Figure 8.8. In CLEM, the illumination light is spatially controlled and used depending on the distribution of fluorescent molecules. The image of the object is calculated using the illumination light dose and the detection signal for each pixel. Illumination is controlled per pixel by a fast feedback system between the detector and the illuminator. The acoustic-optical modulator (AOM) is placed in the excitation beam path to switch the light ON and OFF depending on the signal from feedback electronics. The feedback electronics is fed by the detected fluorescence

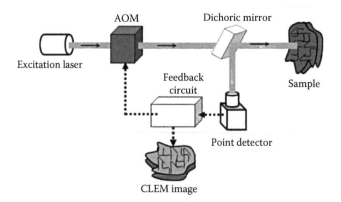

FIGURE 8.8 Schematic diagram depicting CLEM.

signal to determine the excitation light dose required in a pixel. Excitation light dose is manipulated by the duration of excitation (time to exposure) in the pixel. The electronic circuit switches OFF the laser for the rest of pixel dwell time. The signal reflecting the concentration of fluorescent molecule in the pixel is calculated on the basis of light dose and detected fluorescent signal.

The heart of the CLEM technique lies in the feedback circuit. Moreover, it is clear that non-CLEM (widefield and confocal microscope) setup equipped with AOM and feedback circuit is essentially CLEM setup. The feedback circuit performs the following operations:

1. It measures the integrated fluorescence signal during the pixel dwell time.
2. It controls the exposure time for each pixel and hence the local light dose based on the integrated fluorescence signal.
3. An appreciable extrapolation of the fluorescence signal for full-time light exposure.
4. The feedback circuit scans the dark background with minimum dwell exposure time, thereby, saving the scan time. Considering imaging fast events and tracking this is a big advantage.
5. Since this is a pixel-by-pixel technique, it can be easily parallelizable providing fast acquisition of images enabling the observation of live cell dynamics.

For a particular pixel, output fluorescence value determines the optimal light exposure time. If the fluorescence signal is less than a preset threshold value, the pixel is considered to be a part of background and, instantly the feedback circuit switches AOM to the OFF state. In this way, background pixels are illuminated during a minimum light exposure time. When the pixel represents a genuine signal (integrated fluorescense whose value is above the preset threshold), the AOM switches OFF. At this point, the signal is considered to be large enough to obtain a sufficient SNR in that pixel. Most of the bright pixels reduce the light-exposure time because these pixels reach the second threshold level in less than full pixel dwell time. After the pixel dwell time is over, the focus is shifted to next pixel with AOM in the ON state, and a new exposure/decision cycle starts. This process is carried on for all the desired pixels and a CLEM image is formed.

To demonstrate that CLEM reduces photobleaching, a repetitive image of 3D fixed tobacco plant BY-2 cells expressing a microtubule reporter GFP-MAP4 is shown in Figure 8.10. Non-CLEM imaging of nine 3D stacks of confocal optical section reduced fluorescence intensities by 85%, whereas 15% of the initial fluorescence intensity was lost when CLEM was applied. An analysis of the bleaching curves (Figure 8.9) shows that the bleach rate in CLEM is sevenfold slower than the bleach rate in non-CLEM. With respect to SNR, bleached CLEM images are superior to bleached non-CLEM images, which makes the analysis of bleached CLEM images more valuable. This is because in CLEM, most of the emitted photons are used to contribute to image information and not to unnecessary improvement of SNR.

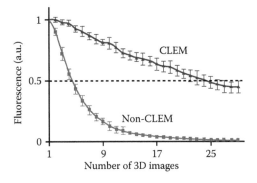

FIGURE 8.9 Analysis of photobleaching curves shows that the bleach rate is sevenfold reduced by CLEM. (Reprinted by permission of Macmillan Publishers: Hoebe, R.A. et al., *Nat. Biotechnol.*, 25, 249, 2007. Copyright 2007.)

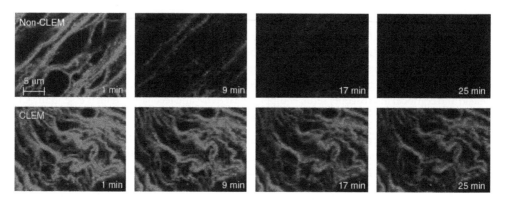

FIGURE 8.10 The difference in photobleaching characteristics between non-CLEM and CLEM of tobacco BY-2 cells expressing microtubule-associated GFP-MAP4. (Reprinted by permission of Macmillan Publishers: Hoebe, R.A. et al., *Nat. Biotechnol.*, 25, 249, 2007. Copyright 2007.)

8.5.3 Photobleaching Reduction by Dark State Relaxation

This section deals with an excitation technique that exploits pulse duration and pulse width optimally to achieve relaxation of dark states between two molecular absorption events [88]. In principle, this is a simple technique, but experimental realization is difficult. This technique, which makes use of dark state relaxation, is an efficient way to reduce photobleaching and photodamage. This further encourages the use of high photon fluxes, which are useful in those applications where low quantum yield fluorescent molecules are used.

To maximize fluorescence emission within a given time span, it is generally desirable to apply high excitation intensities. Besides, high intensities are mandatory in multi-photon microscopy [4]. However, intense excitation results in both enhanced triplet buildup and photobleaching, and thus in losses in the fluorescence signal. It is observed that the total number of photons emitted by a dye that is subject to intense illumination substantially increases when one ensures that fluorophores caught in an absorbing dark state, such as the triplet state, are not further excited.

This technique has a direct implication in single-photon microscopy but this is also applicable in multi-photon microscopy with limited perturbation in experimental setup. High flux illumination densities are desirable for realizing large SNR. This becomes even more demanding in situations where the probe fluorescent molecule has low quantum yield. This can be because of the compatibility of the fluorescent taggers with the biochemical process or to exploit the autofluorescence present in the biological specimen. However, such a high illumination photon flux enhances triplet buildup following a loss in fluorescence signal and also results in higher-order processes such as triplet–triplet transitions. To avoid photobleaching, researchers opt for fast scanning while acquiring images. Unfortunately, this also results in low SNR. So, judicious pulsed excitation ensures minimal higher-order triplet–triplet excitation yielding more photons causing fluorescence.

Before going into the reported experimental realization by Stephen Hell's group, we briefly try to understand the physical process involved. Figure 8.11 shows a simplistic energy diagram of the processes such as excitation, emission, and relaxation for the fluorescent molecule. Assuming that the molecules are in ground state (S_0) in the absence of any excitation light. As soon as the sample (with fluorescent markers) is illuminated by the excitation light, the molecules make a transition to the next higher singlet state (S_1). Once the molecules are in S_1 state, it can follow the following four pathways:

1. Fluorescence emission ($S_1 \rightarrow S_0$): This is the most favorable emission process leading to output fluorescence.
2. Singlet–triplet transition ($S_1 \rightarrow T_1$): This is referred to as singlet to triplet intersystem crossing.

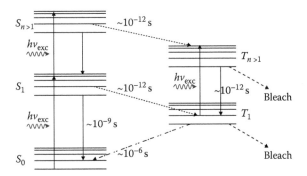

FIGURE 8.11   Molecular energy diagram of a typical fluorescent molecule, showing possible pathways for fluorescence and photobleaching.

3. Triplet–triplet transition ($T_1 \rightarrow T_2$): First-order triplet transition.
4. Order transitions ($T_1 \rightarrow T_n$; $n > 2$ and $S_1 \rightarrow S_n$; $n > 1$): These are referred to as higher-order transition, which becomes prominent at large intensities.

Pathway (1) results in fluorescence, whereas photobleaching occurs due to the activation of pathways (2)–(4). The assumption that we make under normal imaging conditions (low excitation power, high quantum yield of fluorescent molecules) is that only pathway (1) is active and emission occurs via $S_1 \rightarrow S_0$. But at medium power and high power, this assumption breaks down and nonlinear process comes into play resulting in the involvement of pathways (1)–(3). One of the prominent factors that distinguishes the fluorescent pathways from the nonfluorescent ones is the timescale involved in these processes. Thus, one of the many interesting prospects to reduce photobleaching and photodamage is to respect these timescales.

In addition, this observation provides further evidence for dark state absorption because in the multiphoton mode, the pulse intensity is larger than that for its one-photon counterpart by several orders of magnitude. Once a molecule has passed to an absorbing dark state, such as the T_1, the molecule is confronted with a photon flux that leads to a single (and possibly also multi-photon) excitation to the $T_{n>1}$ (see Figure 8.11). These processes are catalyzed by the fact that the cross section for a one-photon absorption of the T_1 state is $\sigma_T \approx 10^{-17}$ cm (Refs. [2,21,57]). Given $I_p > 50$ GW/cm^2 and a photon energy of 2.5×10^{19} J, the excitation rate of a T_1 molecule is ≈ 1 event per ($\tau_p \approx 200$ fs) pulse, that is, the $T_1 - T_{n>1}$ process is nearly saturated. Therefore, once the molecule has crossed to the T_1 state, it is prone to be excited to a more fragile state. Furthermore, the $T_{n>1}$ molecule can absorb another photon from the same pulse, thus setting off a cascade of several consecutive one-photon excitations to increasingly fragile states. This observation explains why the dark-state relaxation illumination modality is so effective.

The saturation of the $S_0 - S_1$ or the $T_1 - T_{n>1}$ excitation eliminates any dependence on pulse peak intensity or energy. Especially the saturation of $T_1 - T_{n>1}$ is consistent with our findings: following a two-photon excitation to the S_1, the molecule crosses to the dark state (e.g., the T_1), which is then subject to a further nearly saturated excitation (e.g., to $T_{n>1}$). The saturation is either provoked by large I_p for short ($\tau_p = 200$ fs) or by repetitive excitation to $T_{n>1}$ within the same long pulse ($\tau_p = 40$ ps) with the lifetime of $T_{n>1}$ being ≈ 200 fs. In fact, other evidence were found for this characteristic when scrutinizing the role of the pulse duration τ_p on the total signal G2p gained by TPE. The data shown in Figure 8.12 confirm that stepping down f leads to a substantial increase in G2p, but changing τ_p by a factor of 200 and hence the pulse energy by 14-fold does not have a considerable effect on G2p. The data shown in Figure 8.5 confirm that stepping down f leads to a substantial increase in G2p, but changing τ_p by $m = 200$ and hence the pulse energy by 14-fold does not have a considerable effect on G2p.

The timescales involved in the pathways (1)–(4) vary substantially for different fluorescent probes. It is clear that the timescale involved in intersystem crossing (10^{-12} s) is approximately 3 orders less than

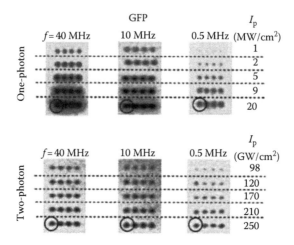

FIGURE 8.12 Photobleaching of GFP and Atto532 decreases with interpulse break and peak power for single- and two-photon excitation. (Reprinted by permission of Macmillan Publishers Ltd: Donnert, G. et al., *Nat. Meth.*, 4, 81, 2007. Copyright 2007.)

fluorescence lifetimes (10^{-9} s), which is 3–4 orders less than dark state lifetimes (10^{-6} s). In the vicinity of these timescales, an interesting prospective for reducing photobleaching and photodamage would be to decrease the pulse repetition rate (f). For example, considering pulse lasers with repitation rate 50–80 MHz, the interpulse duration (12.5–20 ns) is approximately 10 times larger than the typical lifetime ($\approx$1–5 ns) of most organic fluorescent molecules and proteins.

Basically in all imaging modalities, pulse trains with a repetition rate 40–100 MHz are used [2]. This rate is provided by most pulsed lasers, for example, by a mode-locked Ti:sapphire laser operating at f=80 MHz. Moreover, the associated "illumination pause" Δ_t=1/f of 10–25 ns between succeeding pulses is up to 10 times larger than the typical lifetime of the fluorescent state S_1 of organic fluorophores and fluorescent proteins. Shorter Δ_t increases the probability of illuminating molecules that are already excited, whereas longer Δ_t would leave the dye idle. Hence, not surprisingly, increasing Δ_t by decreasing f appeared unattractive.

The role of dark or triplet state relaxation in pulsed mode illumination by gradually reducing the repetition rate f from 40 MHz to 0.5 MHz, which is equivalent to expanding the inter-pulse distance Δ_t from 25 ns to 2 ms is studied by Hell's group. Figure 8.12 shows the bleaching nature of GFP at various pulse peak power and pulse repetition rates. For single-photon excitation, the bleaching increases with pulse repetition rate. This is probably because of triplet–triplet transitions, as schematically shown in Figure 8.11. Similar effects can be seen for two-photon excitation as well.

8.5.4 Quantum Light Microscopy

This section is devoted to quantum mechanical methodology for minimizing photobleaching and photo damage. This is achieved by using nonclassical light for exciting the target fluorescent molecules [90]. These quantum radiation sources emit an even number of photons with nonzero probability and an odd number with zero probability. Such a radiation source is invaluable for multi-photon excitation microscopy. This methodology has the additional advantage of achieving comparable signal with almost three orders of less intensity. The SNR of the output fluorescent signal can be boosted if photobleaching is minimized. Photobleaching occurs prominently because of the involvement of high-order photon interaction in the focal volume. Hence, in TPE microscopy, it becomes essential to isolate two-photon mimicked absorption from that of single- and three-photon mimicked absorption process. Small molecular cross section reduces other higher-order excitation process. Excitation with even coherent state in 2PE microscopy can improve the excitation probability without causing unwanted photodamage and photobleaching.

We propose to employ even coherent state light (a quantum radiation source, for which only an even number of photons have nonzero probability of being observed) for possible isolation of two-photon excitation from one-photon and three-photon excitation. The generation of even coherent state is reported in the literature [89,91–93]. Generally, the macroscopic superposition states (Schrodinger Cat states) are generated by the interaction of two modes of the optical fields [91,92]. It is possible to efficiently generate even coherent state by parametric down conversion [91–93]. Particularly, for generating two-photon state by parametric down conversion process, a nonlinear crystal transforms the part of the incident photon energy to photons with half the energy [94]. A typical nonlinear crystal can spontaneously down convert light with a typical efficiency of around 10^{-7}. Hence, about 10 W of pumping light is needed to generate 1 μW of down-converted power [95]. It is expected that the light source for two-photon excitation experiment requires lower photon flux compared to the flux of random source of light [93].

The state of quantum mechanical systems, such as coherent photon systems, is characterized by the expectation of the moments of bosonic operators $\hat{a}$ and $\hat{a}^\dagger$, which are either in normally ordered form $\langle(\hat{a}^\dagger)^m\,(\hat{a})^n\rangle$, antinormally ordered form $\langle(\hat{a})^n\,(\hat{a}^\dagger)^m\rangle$, or symmetrical form $\langle\{(\hat{a}^\dagger)^m\,(\hat{a})^n\}\rangle$, where $\{\cdots\}$ represents Poisson bracket.

The general form of the *s*-parameterized characteristic function is given by

$$C(\xi, s) = \text{tr}[\hat{\rho}e^{(\xi\hat{a}^\dagger - \xi^*\hat{a} + s\frac{|\xi|^2}{2})}] \tag{8.17}$$

and the density operator $\hat{\rho}$ is the inverse Fourier transform of the characteristic function [96]. The expectation value of operator $\langle\{(\hat{a}^\dagger)^m\,(\hat{a})^n\}\rangle$ can be written as

$$\left\langle\{(\hat{a}^\dagger)^m(\hat{a})^n\}\right\rangle = \int W(\beta,0)\beta^{*m}\beta^n\,d^2\beta \tag{8.18}$$

where the corresponding Wigner function is given by

$$W(\beta, s = 0) = \frac{1}{\pi^2}\int C(\xi, s = 0)e^{(\beta\xi^* - \beta^*\xi)}\,d^2\xi \tag{8.19}$$

Now, the density matrix of the statistical mixture of the states $|\alpha\rangle$ and $|-\alpha\rangle$ is given by

$$\rho_{\text{mix}} = \frac{1}{2}\Big(|\alpha\rangle\langle\alpha| + |-\alpha\rangle\langle-\alpha|\Big) \tag{8.20}$$

and the corresponding Wigner function is given by

$$W(\beta) = (1/\pi)\{e^{-2(x-\alpha)^2 - 2y^2} + e^{-2(x+\alpha)^2 - 2y^2}\} \tag{8.21}$$

The general expression for photon number distribution is given by

$$P_n = \pi\int W(\beta)W_n(\beta)d^2\beta \tag{8.22}$$

where

$$W_n(\beta) = (-1)^n\frac{2}{\pi}e^{-2|\beta|^2}L_n(4|\beta|^2) \tag{8.23}$$

and $\int W_n(\beta)\,d^2\beta = 1$, $L_n(x)$ being the Laguerre polynomial of order n.

Using (8.21) and (8.23) in (8.22), the corresponding number distribution becomes

$$P_n = \frac{\alpha^{2n}}{n!}e^{-\alpha^2} \tag{8.24}$$

This is the well-known Poisson distribution. So, the distribution corresponding to the statistical mixture of the coherent states is Poissonian.

Consider the superposition of two coherent states $|\alpha\rangle$ and $|-\alpha\rangle$ with zero phase for generating even coherent states [97]

$$|\psi\rangle = \frac{1}{[2(1+e^{-2\alpha^2})]^{1/2}}\left(|\alpha\rangle + |-\alpha\rangle\right) \tag{8.25}$$

The Wigner function corresponding to the even coherent state is [97]

$$W(\beta) = W(\beta, s=0) = \frac{1}{\pi(1+e^{-2\alpha^2})}[e^{-2(x-\alpha)^2-2y^2}$$

$$+ e^{-2(x+\alpha)^2-2y^2} + e^{-2x^2-2y^2}\cos(4y\alpha)] \tag{8.26}$$

where $x=\mathrm{re}(\beta)$ and $y=\mathrm{im}(\beta)$. The last term arising from the quantum interference of $|\alpha\rangle$ and $|-\alpha\rangle$ is responsible for the nonclassical behavior of even coherent state light.

Using (8.26) and (8.23) in (8.22) gives the photon number distribution as [97]

$$P_n = \frac{2e^{-\alpha^2}}{1+e^{-2\alpha^2}}\frac{\alpha^{2n}}{n!}, \quad \text{if } n=2m,\ m=1,2,\dots$$

$$P_n = 0, \quad \text{if } n=2m+1,\ m=0,1,\dots \tag{8.27}$$

The quantum interference between the states $|\alpha\rangle$ and $|-\alpha\rangle$ generates an oscillatory behavior in the photon number distribution. This photon number distribution represents even coherent state because only a even number of photons has a nonzero probability of being observed. This is similar to bunching $2n$-photons together. It can be shown that even coherent state light has super-Poissonian photon statistics for any value of intensity α^2. The corresponding Mandel Q parameter is positive, i.e., $Q = \dfrac{4\alpha^2 e^{-2\alpha^2}}{1-e^{-4\alpha^2}} > 0$.

Now, the fluorophores present in the focal volume can undergo multi-mode excitation depending upon the molecular cross section of the sample. The fluorescent signal emerging from the elemental volume element ΔV undergoing all possible modes of excitation processes can be written as

$$I_{fT}(\Delta V) = I_{f(1)}(\Delta V) + I_{f(2)}(\Delta V) + I_{f(3)}(\Delta V) + \cdots$$

$$= \delta_i P_i N_{\Delta V} + \sum_{j\neq i}\delta_j P_j N_{\Delta V} \tag{8.28}$$

where

$I_{f(1)}(\Delta V)$ represents ith order process
P_i is the probability for i-photon excitation
$N_{\Delta V}$ is the number of fluorescent molecules in the focal volume ΔV
δ_i is the molecular cross section corresponding to ith order excitation

The probability $P(n)$ of finding n-photons in a coherent state $|\alpha\rangle$ with complex amplitude α is given by Poisson distribution, $P(n) = |\langle n|\alpha\rangle|^2 = \frac{e^{-\langle n\rangle}\langle n\rangle^n}{n!}$, where $\langle n\rangle = |\alpha|^2$ is the average number of photon equal to the classical intensity of the coherent light. On the other hand, the probability of finding two photons $(n=2)$ in such a mixed superposition of coherent states $|\alpha\rangle$ and $|-\alpha\rangle$ exhibiting even coherent states is given by (Equation 8.6)

$$P_2 = \frac{2e^{-\alpha^2}}{1+e^{-2\alpha^2}}\frac{\alpha^4}{2!} = \alpha^4 \operatorname{sech}(\alpha^2) \tag{8.29}$$

where it should be noted that $P_n \neq 0$, for $n=2m$, $m > 1$.

For small excitation intensity $I = \langle n\rangle = \alpha^2$, practically true for two-photon microscopy [84,98], $\operatorname{sech}(\alpha^2) \approx 1$, and thus, $P_2 \propto \alpha^4 = C_I I^2$ where C_I is a proportionality constant. The explicit expression for the total fluorescent signal for a pulsed laser beam with pulse width τ_p, repetition rate f_p, average power P_{ave}, and the effective power $\left(P(t) = \frac{P_{ave}}{\tau_p f_p}, \text{ for } 0 < t < \tau_p\right)$ [50,84,98] is given by

$$I_{f(2)}(\Delta V, t) = \delta_2 \frac{P_{ave}^2}{\tau_p f_p^2}\left[\frac{\pi(NA)^2}{hc\lambda}\right]^2 N_{\Delta V} \tag{8.30}$$

where NA is the numerical aperture of the objective lens.

It should be noted that for four-photon absorption, the molecular cross section for four-photon excitation is of the order of 10^{-119} considering $\delta_n \cong \delta_1^n \Delta t^{n-1}$ [57], where the time interval Δt is the timescale of molecular energy fluctuations at photon energy scales, as determined by the Heisenberg uncertainty principle, $\Delta t \cong 10^{-17}$. The approximate value of δ_1 is 10^{-17} cm^2.

Since excitation quantum radiation source (even coherent state light) emits photons with probability strength $P_2 >> P_4 >> \cdots >> P_{2m}$ (see 8.27), thus, $I_{f(2)} >> I_{f(4)} >> \cdots >> I_{f(2m)}$ and $I_{f(1)} = I_{f(3)} = \cdots = I_{f(2m+1)} = 0$, the total fluorescent signal in the elemental volume is essentially due to two-photon excitation, i.e.,

$$I_{fT}(\Delta V) \approx \delta_2 P_2 N_{\Delta V} \approx \delta_2 \frac{P_{ave}^2}{\tau_p f_p^2}\left[\frac{\pi(NA)^2}{hc\lambda}\right]^2 N_{\Delta V} \tag{8.31}$$

This approach shows that photobleaching may be reduced due to one and multi-photon excitations in two-photon excitation microscopy. To achieve this, we use even coherent state light. Such a light has the advantage that the probability of observing odd coherent state is zero and there exists nonzero probability for observing even photons.

Appendix

Consider a system initially in an eigen state ϕ_i of the Hamiltonian H_0 obeying the system equation

$$H_0\phi_i = E_i\phi_i \tag{8.32}$$

Now, consider a perturbation of the Hamiltonian by an external potential $V(t)$. The resultant Hamiltonian is $H' = H_0 + V(t)$. The corresponding Schrodinger equation for the perturbed system is given by

$$i\hbar\frac{\partial}{\partial t}|\psi, t\rangle_s = (H_0 + V)|\psi, t\rangle_s \tag{8.33}$$

For better understanding, we switch to interaction picture rather than Schrodinger picture. Changing from Schrodinger notation to interaction picture requires the following transformation:

$$\left|\psi,t\right\rangle_s = e^{-iH_0t/\hbar}\left|\psi,t\right\rangle_I \tag{8.34}$$

Substituting the system function $\left|\psi,t\right\rangle_s$ in Equation 8.34 and further algebraic simplifications produce

$$i\hbar\frac{\partial}{\partial t}\left|\psi,t\right\rangle_I = V_I\left|\psi,t\right\rangle_I \tag{8.35}$$

where $V_I = e^{iH_0t/\hbar}Ve^{-iH_0t/\hbar}$.

Since the system function evolves with time, which is determined by the evolution operator, $\left|\phi,t\right\rangle_I = U_I\left|\phi,0\right\rangle_I$. Substituting this in Equation 8.35 produces

$$i\hbar\frac{\partial}{\partial t}U_I\left|\psi,0\right\rangle_I = V_IU_I\left|\psi,0\right\rangle_I \tag{8.36}$$

From Equation 8.36 the operator equivalence is given by

$$i\hbar\frac{\partial}{\partial t}U_I = V_IU_I \tag{8.37}$$

Integration and substituting initial condition $U_I(t,t_0)_{t=t0}=I$, we get

$$U_I(t) = 1 - \frac{1}{\hbar}\int_{t_0}^{t}V_I(t')U_I(t')dt' \tag{8.38}$$

Integrating and substituting $U_I(t)$ in the above expression and iterating the process n time gives

$$U_I(t) = 1 - \frac{1}{\hbar}\int_{t_0}^{t}V_I(t')dt'$$
$$+ \left(-\frac{1}{\hbar}\right)^2\int_{t_0}^{t}dt'\int_{t_0}^{t'}dt''V_I(t')V_I(t'') + \cdots + n\text{th term} \tag{8.39}$$

Since $\left|\psi,t\right\rangle_I = U_I(t)\left|\psi,0\right\rangle_I$, one can represent in terms of perturbation free Hamiltonian (H_0) states ψ_i states as

$$\left|\psi,t\right\rangle_I = \sum_n\left\langle n|U_I(t)|i\right\rangle|n\rangle \tag{8.40}$$

In general, one can write $\left|\psi,t\right\rangle_I$ in stationary state basis set (complete) $\{|\phi_n\rangle\}$ as

$$\left|\psi, t\right\rangle_{\text{I}} = \sum_{n} C_n(t)\left|\phi_n\right\rangle \tag{8.41}$$

where

$$C_n(t) = \left\langle\phi_n\left|U_I(t)\right|\phi_i\right\rangle \tag{8.42}$$

Substituting $U_I(t)$ in Equation 8.42 gives the coefficient representing transition probability. The zeroth-, first-, and second-order transition probabilities are given by

$$C_n^{(0)}(t) = \left\langle\phi_n\left|I\right|\phi_i\right\rangle \tag{8.43}$$

$$C_n^{(1)}(t) = \frac{-i}{\hbar}\int_{t_0}^{t}\mathrm{d}t'\left\langle\phi_n\left|V_I(t')\right|\phi_i\right\rangle \tag{8.44}$$

$$C_n^{(2)}(t) = \left(\frac{-i}{\hbar}\right)^2\int_{t_0}^{t}\mathrm{d}t'\int_{t_0}^{t'}\mathrm{d}t''\left\langle\phi_n\left|V_I(t')V_I(t'')\right|\phi_i\right\rangle \tag{8.45}$$

It is however very important to realize that the probability for the transition from state $|i\rangle$ to state $|n\rangle$, irrespective of the number of photons absorbed for causing the transition, is $p\left(|i\rangle\to|n\rangle\right)=|C_n^{(1)}+C_n^{(2)}+\cdots+C_n^{(n)}|^2$.

Assume harmonic perturbation of the form $V(t)=Ve^{-i\omega t}$. The probability of $|i\rangle\to|n\rangle$ transition is $|C_n^{(1)}(t)|^2$ and $|C_n^{(2)}(t)|^2$ for single- and two-photon excitation. Carrying out the integration on the transition probability and imposing the condition for large time $(t\to\infty)$ we get

$$\lim_{t\to\infty}|C_n^{(1)}|^2 = \frac{2\pi}{\hbar^2}\delta(\omega_{ni}-\omega)|V_{ni}|^2 \tag{8.46}$$

and

$$\lim_{t\to\infty}C_n^{(2)}(t) = \frac{4\pi^2}{\hbar^4}\frac{|V_{nm}|^2|V_{mi}|^2}{|\omega_{nm}-\omega||\omega_{mi}-\omega|}\delta(\omega_{nm}+\omega_{mi}-\omega) \tag{8.47}$$

References

1. Hell, S.W., 2007. Far-field optical nanoscopy, *Science* 316:1153–1158 (2007).
2. Betzig, E., Patterson, G.H., Sougrat, R., Lindwasser, O.W., Olenych, S.W., Bonifacino, J.S., Davidson, M.W., Schwartz, J.L., and Hess, H.F., 2006, Imaging intracellular fluorescent proteins at nanometer resolution, *Science* 313:1642–1645.
3. Rust, M.J., Bates, M., and Zhuang, X., 2006, Sub-diffraction-limit imaging by stochastic optical reconstruction microscopy (STORM), *Nat. Meth.* 3:793–796.
4. Mertz, J., Xu, C., and Webb, W.W., 1995. Single-molecule detection by two-photon-excited fluorescence, *Opt. Lett.* 20:2532–2534.
5. Lippitz, M., Erker, W., Decker, H., van Holde, K.E., and Basch, T., 2002, Two-photon excitation microscopy of tryptophan-containing proteins, *Proc. Nat. Acad. Sci. USA* 99:2772–2777.
6. Kriete, A., 1992, *Visualization in Biomedical Microscopies*, VCH, Weinheim, Germany.

7. Bonetto, P., Boccacci, P., Scarito, M., Davolio, M., Epifani, M., Vicidomini, G., Tacchetti, C., Ramoino, P., Usai, C., and Diaspro, A., 2004, Three dimensional microscopy migrates to the Web with PowerUp Your Microscope, *Microsc. Res. Tech.* 64:196–203.

8. Brakenhoff, G.J., Muller, M., and Ghauharali, R.I., 1996, Analysis of efficiency of two-photon versus single-photon absorption of fluorescence generation in biological objects, *J. Microsc.* 183:140–144.

9. Mertz, J., 1998, Molecular photodynamics involved in multi-photon excitation fluorescence, *Eur. Phys. J.* D3:53–66.

10. Patterson, G.H. and Piston, D.W., 2000, Photobleaching in two-photon excitation microscopy, *Biophys. J.* 78:2159–2162.

11. Dittrich, P.S. and Schwille, P., 2001, Photobleaching and stabilization of fluorophores used for single molecule analysis with one- and two-photon excitation, *Appl. Phys. B* 73:829–837.

12. Knig, K. and Tirlapur, U.K., 2002, Cellular and subcellular perturbations during multiphoton microscopy. In: *Confocal and Two-Photon Microscopy: Foundations, Applications and Advances* (A. Diaspro, ed.), Wiley-Liss, New York, pp. 191–205.

13. Pawley, J. B., 1995h, Fundamental limits in confocal microscopy. In: *Handbook of Biological Confocal Microscopy* (J. B. Pawley, ed.), Plenum Press, New York, pp. 19–37.

14. Peters, R., Peters, J., Tews, K.H., and Bahr, W., 1974, A microfluorimetric study of translational diffusion in erythrocyte membranes, *Biochem. Biophys. Acta* 367:282–294.

15. Poo, M. and Cone, R.A., 1974, Lateral diffusion of rhodopsin in the photoreceptor membrane, *Nature* 247:438–441.

16. Edidin, M., Zagyansky, Y., and Lardner, T.J., 1976, Measurement of membrane protein lateral diffusion in single cells, *Science* 191:466–468.

17. Axelrod, D., Koppel, D.E., Schlessinger, J., Elson, E., and Webb, W.W., 1976, Mobility measurement by analysis of fluorescence photobleaching recovery kinetics, *Biophys. J.* 16:1055–1069.

18. Tsien, R.Y., 1998, The green fluorescent protein, *Annu. Rev. Biochem.* 67:509–544.

19. Cole, N.B., Smith, C.L., Sciaky, N., Terasaki, M., Edidin, M., and Lippincott-Schwartz, J., 1996, Diffusional mobility of Golgi proteins in membranes of living cells, *Science* 273:797–801.

20. White, J. and Stelzer, E., 1999, Photobleaching GFP reveals protein dynamics inside live cells, *Trends Cell Biol.* 9:61–65.

21. Reits, E.A. and Neefjes, J.J., 2001, From fixed to FRAP: Measuring protein mobility and activity in living cells, *Nat. Cell Biol.* 3:E145–E147.

22. Davis, S.K. and Bardeen, C.J., 2002, Using two-photon standing waves and patterned photobleaching to measure diffusion from nanometers to microns in biological systems, *Rev. Sci. Instrum.* 73:2128–2135.

23. Braeckmans, K., De Smedt, S.C., Roelant, C., Leblans, M., Pauwels, R., and Demeester, J., 2003, Encoding microcarriers by spatial selective photobleaching, *Nat. Mater.* 2:169–173.

24. Braga, J., Desterro, J.M.P., and Carmo-Fonseca, M., 2004, Intracellular macromolecular mobility measured by fluorescence recovery after photobleaching with confocal laser scanning microscopes, *Mol. Biol. Cell* 15:4749–4760.

25. Delon, A., Usson, Y., Derouard, J., Biben, T., and Souchier, C., 2004, Photobleaching, mobility and compartimentalisation: Inferences in fluorescence correlation spectroscopy, *J. Fluoresc.* 14:255–267.

26. McNally, J.G. and Smith, C.L., 2002, Photobleaching by confocal microscopy, In: *Confocal and Two-Photon Microscopy: Foundations, Applications, and Advances* (A. Diaspro, ed.), Wiley-Liss, Inc., New York, pp. 525–538.

27. Song, L., Hennink, E.J., Young, I.T., and Tanke, H.J., 1995, Photobleaching kinetics of fluorescein in quantitative fluorescence microscopy, *Biophys. J.* 68:2588–2600.

28. Mang, T.S., Dougherty, T.J., Potter, W.R., Boyle, D.G., Somer, S., and Moan J., 1987, *Photochem. Photobiol.* 45:501–506.

29. Deschenes, L.A. and Bout, D.A.V., 2002, Avoiding two-photon photochemistry to extend the life of single molecules, *Chem. Phys. Lett.* 365:387–395.

30. Abbe, E., 1873, Beitrge zur Theorie des Mikroskops und der mikroskopischen Wahrnehmung, *Schultzes Arc. f. Mikr. Anat.* 9:413–468.

31. Abbe, E., 1884, Note on the proper definition of the amplifying power of a lens or a lens-system, *J. Royal Microsc. Soc.* 4:348–351.

32. Egger, M.D., 1989, The development of confocal microscopy, *Trends Neurosci.* 12:11.

33. Sheppard, C.J.R. and Choudhury, A., 1977. Image formation in the scanning microscope, *Opt. Acta* 24:1051–1073.

34. Sheppard, C.J.R., Gannaway, J.N., Walsh, D., and Wilson, T., 1978, *Scanning Optical Microscope for the Inspection of Electronic Devices*, Microcircuit Engineering Conference, Cambridge, UK.

35. Wilson, T., Gannaway, J.N., and Johnson, P., 1980, A scanning optical microscope for the inspection of semiconductor materials and devices, *J. Microsc.* 118:309–314.

36. Cremer, C. and Cremer, T., 1978, Considerations on a laser-scanning microscope with high resolution and depth of field, *Microsc. Acta* 81:31–44.

37. Brakenhoff, G.J., Blom, P., and Barends, P., 1979, Confocal scanning light microscopy with high aperture immersion lenses, *J. Microsc.* 117:219–232.

38. Brakenhoff, G.J., van der Voort, H.T.M., van Spronsen, E.A., Linnemans, W.A.M., and Nanninga, N., 1985, Three dimensional chromatin distribution in neuroblastoma nuclei shown by confocal scanning laser microscopy, *Nature* 317:748–749.

39. Wijnaendts van Resandt, R.W., Marsman, H.J.B., Kaplan, R., Davoust, J., Stelzer, E.H.K., and Strickler, R., 1985, Optical fluorescence microscopy in three dimensions: Microtomoscopy, *J. Microsc.* 138:29–34.

40. Carlsson, K., Danielsson, P., Lenz, R., Liljeborg, A., Majlof, L., and Slund, N., 1985, Three-dimensional microscopy using a confocal laser scanning microscope, *Opt. Lett.* 10:53–55.

41. Slund, N., Carlsson, K., Liljeborg, A., and Majlof, L., 1983, PHOIBOS, a microscope scanner designed for micro-fluorometric applications, using laser induced fluorescence. In: *Proceedings of the Third Scandinavian Conference on Image Analysis,* Studentliteratur, Lund, p. 338.

42. Slund, N., Liljeborg, A., Forsgren, P.-O., and Wahlsten, S., 1987, Three dimensional digital microscopy using the PHOIBOS scanner, *Scanning* 9:227–235.

43. Amos, W.B., White, J.G., and Fordham, M., 1987, Use of confocal imaging in the study of biological structures, *Appl. Opt.* 26:3239–3243.

44. White, J.G., Amos, W.B., and Fordham, M., 1987, An evaluation of confocal versus conventional imaging of biological structures by fluorescence light microscopy, *J. Cell Biol.* 105:41–48.

45. Goppert-Mayer, M., 1931, Elementry acts with two quantum jumps, *Ann. Phys.,* 9:273–294.

46. Kaiser, W. and Garret, C.G.B., 1961. Two-photon excitation in $CaF_2:Eu^{2+}$, *Phys. Rev. Lett.* 7: 229–231.

47. Singh, S. and Bradley, L.T., 1964. Three-photon absorption in naphtalene crystals by laser excitation, *Phys. Rev. Lett.* 12:162–164.

48. Rentjepis, P.M., Mitschele, C.J., and Saxman, A.C., 1970. Measurement of ultrashort laser pulses by three photon fluorescence, *Appl. Phys. Lett.* 17:122–124.

49. Berns, M.W., 1976. A possible two-photon effect in vitro using a focused laser beam, *Biophys. J.* 16:973–977.

50. Denk, W., Strickler, J.H., and Webb W.W., 1990, Two-photon laser scanning fluorescence microscopy, *Science* 248:73–76.

51. Minsky, M. 1998, Memoir of inventing the confocal scanning microscope, *Scanning* 10:128–138.

52. Davidovits, P.D. and Egger, M.D., 1971, Scanning laser microscope for biological investigations, *Appl. Opt.* 10:1615–1619.

53. Wilson, T. and Sheppard, C.J.R., 1984, *Theory and Practice of Scanning Optical Microscopy*, Academic Press, London, UK.

54. Wilson, T., 1990, *Confocal Microscopy*, Academic Press, London, UK.

55. Sekar, R.B. and Periasamy, A., 2003, Fluorescence resonance energy transfer (FRET) microscopy imaging of live cell protein localizations, *J. Cell Biol.* 160:629–633.

56. Chirico, G., Cannone, F., Baldini, G., and Diaspro, A., 2003, Two-photon thermal bleaching of single fluorescent molecules, *Biophys. J.* 84:588–598.

57. Xu, C., Zipfel, W., Shear, J.B., Williams, R.M., and Webb, W.W., 1996, *Proc. Natl. Acad. Sci. USA* 93: 10763–10768.

58. Dickinson, M.E., Simbuerger, E., Zimmermann, B., Waters, C.W., and Fraser, S.E., 2003, Multiphoton excitation spectra in biological samples, *J. Biomed. Opt.* 8:329–338.

59. Pawley, J.B., 1995a, *Handbook of Biological Confocal Microscopy*, Plenum Press, New York.

60. Diaspro, A., 2002, *Confocal and Two-Photon Microscopy: Foundations, Applications, and Advances*, Wiley-Liss, New York.

61. Diaspro, A., 2004, Confocal and multiphoton microscopy. In: *Lasers and Current Optical Techniques in Biology* (G. Palumbo and R. Pratesi, eds.), RsC-Royal Society of Chemistry, Cambridge, UK, pp. 429–478.

62. Matsumoto, B., 2002, *Cell Biological Applications of Confocal Microscopy*, 2nd edn, Academic Press, San Diego, CA.

63. Amos, W.B. and White, J.G., 2003, How the confocal laser scanning microscope entered biological research, *Biol. Cell* 95:335–342.

64. Mondal, P.P., 2008, Minimizing photobleaching in fluorescence microscopy by depleting triplet states, *Appl. Phys. Lett.* 92:013902.

65. Patterson G.H. and Piston, D.W., 2000, Photobleaching in two-photon excitation microscopy, *Biophys. J.* 78:2159–2162.

66. Weber, G. and Teale, F.W.J., 1958, Fluorescence excitation spectrum of organic compounds in solution, *Trans. Faraday Soc.* 54:640–648.

67. Chen, R.F. and Scott, C.H., 1985, Atlas of fluorescence spectra and lifetimes of dyes attached to protein, *Anal. Lett.* 18:393–421.

68. Tsien, R.Y. and Waggoner, A.S., 1995, Fluorophores for confocal microscopy. Photophysics and photochemistry. In: *Handbook of Biological Confocal Microscopy* (J.B. Pawley, ed.), Plenum Press, New York, pp. 267–279.

69. Xu, C., 2002, Cross-section of fluorescent molecules in multiphoton microscopy. In: *Confocal and Two-Photon Microscopy: Foundations, Applications, and Advances* (A. Diaspro, ed.), Wiley-Liss, New York, pp. 75–99.

70. Bernas, T., Zarebski, M., Cook, P.R., and Dobrucki, J.W., 2004, Minimizing photobleaching during confocal microscopy of fluorescent probes bound to chromatin: Role of anoxia and photon flux, *J. Microsc.* 215:281–296.

71. Michalet, X., Kapanidis, A.N., Laurence, T., Pinaud, F., Soeren Doose, S., Pflughoefft, M., and Weiss, S., 2003, The power and prospects of fluorescence microscopies and spectroscopies, *Annu. Rev. Biophys. Biomol. Struct.* 32:161–182.

72. Eggeling, C., Widengren, J., Rigler, R., and Seidel, C.A.M., 1998, Photobleaching of fluorescent dyes under conditions used for single molecule detection: Evidence of two step photolysis, *Anal. Chem.* 70:2651–2659.

73. Schneider, M.B. and Webb, W.W., 1981, Measurements of submicron laser beam radii, *Appl. Opt.* 20:1382–1388.

74. Hirschfeld, T., 1976, Quantum efficiency independence of the time integrated emission from a fluorescent molecule, *Appl. Opt.* 15:3135–3139.

75. Mathies, R.A. and Stryer, L., 1986, Single-molecule fluorescence detection: A feasibility study using phycoerythrin. In: *Applications of Fluorescence in the Biomedical Sciences* (D.L. Taylor, A.S. Waggoner, R.F. Murphy, F. Lanni, and R.R. Birge, eds.), Alan R. Liss, New York, pp. 129–140.

76. Konig, K., So, P.T.C., Mantulin, W.W., Tromberg, B.J., and Gratton, E., 1996, Two-photon excited lifetime imaging of autofluorescence in cells during UVA and NIR photostress, *J. Microsc.* 183:197–204.

77. Stelzer, E.H.K., Hell, S., Linder, S., Pick, R., Storz, C., Stricker, R., Ritter, G., and Salmon, N., 1994, Nonlinear absorption extends confocal fluorescence microscopy into the ultra-violet regime and confines the observation volume, *Opt. Commun.* 104:223–228.

78. Chen, T.S., Zeng, S.Q., Luo, Q.M., Zhang, Z.H., and Zhou, W., 2002, High-order photobleaching of green fluorescent protein inside live cells in two-photon excitation microscopy, *Biochem. Biophys. Res. Commun.* 291:1272–1275.

79. Schwille, P., Kummer, S., Moerner, W.E., and Webb, W.W., 1999, Fluorescence correlation spectros-copy of different GFP mutants reveals fast light driven intramolecular dynamics, *Biophys. J.*, 76: A260.

80. Sanchez, E.J., Novotny, L., Holtom, G.R., and Xie, X.S., 1997, Room-temperature fluores-cence imaging and spectroscopy of single molecule by two-photon excitation, *J. Phys. Chem. A* 101:7019–7023.

81. Chirico, G., Cannone, F., Diaspro, A., Bologna, S., Pellegrini, V., Nifosi, R., and Beltram, F., 2004, Multiphoton switching dynamics of single green fluorescent protein, *Phys. Rev. E* 70:030901(R).

82. Hermann, S.J.P. and Ducuing, J., 1972, Dispersion of the two-photon cross section in rhodamine dyes, *Opt. Commun.* 6:101–105.

83. Williams, R.M. and Piston, D.W., 1994, Two-photon molecular excitation provides intrinsic 3-dimen-sional resolution for laser-based microscopy and microphotochemistry, *FASEB J.* 8:804–813.

84. Xu, C., 2001. In: *Confocal and Two-Photon Microscopy: Foundations, Applications and Advances*, (A. Diaspro, ed.), Wiley-Liss Inc., New York, p. 75.

85. Garcia-Parajo, M.F., Segers-Nolten, G.M.J., Veerman, J.-A., Greve, J., and van Hulst, N.F., 2000, Real-time light-driven dynamics of the fluorescence emission in single green fluorescent protein molecules, *Proc. Natl. Acad. Sci. USA* 97:7237.

86. Heikal, A.A., Hess, S.T., Baird, G.S., Tsien, R.Y., and Webb, W.W., 2000, Molecular spectroscopy and dynamics of intrinsically fluorescent proteins: Coral red (dsRed) and yellow (Citrine), *Proc. Natl. Acad. Sci. USA* 97:11996–12001.

87. Hoebe, R.A., Van Oven, C.H., Gadella Jr., T.W.J., Dhonukshe, P.B., Van Noorden, C.J.F., and Manders, E.M.M., 2007, Controlled light-exposure microscopy reduces photobleaching and phototoxicity in fluorescence live-cell imaging, *Nat. Biotechnol.* 25:249.

88. Donnert, G., Eggeling, C., and Hell, S.W., 2007, Major signal increase in fluorescence microscopy through dark-state relaxation, *Nat. Meth.* 4:81–86.

89. Gerry, C.C., 1999, Generation of optical macroscopic quantum superposition states via state reduction with a Mach–Zehnder interferometer containing a Kerr medium, *Phys. Rev. A* 59:4095–4098.

90. Mondal, P.P. and Diaspro, A., 2007, Reduction of higher-order photobleaching in two-photon exci-tation microscopy, *Phys. Rev. E* 75:061904.

91. Jeong, H., Kim, M.S., Ralph, T.C., and Ham, B.S., 2004, Generation of macroscopic superposition states with small nonlinearity, *Phys. Rev. A* 70:061801(R).

92 Jeong, H., Lund, A.P., and Ralph, T.C., 2005, Production of superpositions of coherent states in trav-eling optical fields with inefficient photon detection, *Phys. Rev. A* 72:013801.

93. Nasr, M.B., Abouraddy, A.F., Booth, M.C., Saleh, B.E.A., Sergienko, A.V., Teich, M.C., Kempe, M., and Wolleschensky, R., 2002, Biphoton focusing for two-photon excitation, *Phys. Rev. A* 65:023816.

94. Burmhan, D. and Weinberg, D., 1970, Observation of simultaneity in parametric production of opti-cal photon pairs, *Phys. Rev. Lett.* 25:84–87.

95. Javanainen, J. and Gould, P.L., 1990, Linear intensity dependence of a two-photon transition rate, *Phys. Rev. A* 41:5088.

96. Glauber, R.J., 1963, Coherent and incoherent states of the radiation field, *Phys. Rev.* 131:2766–2788.

97. Buzek, V. and Knight, P. L., 1995. In: *Progress in Physics* (E. Wolf, ed.), Elsevier, Amsterdam, the Netherlands, Chap. XXXIV, p. 1.

98. Diaspro, A., Chirico, G., and Collini, M., 2005, Two-photon fluorescence excitation in biological microscopy and related techniques., *Q. Rev. Biophys.* 38:97–166.

9

Second Harmonic Generation Imaging Microscopy: Theory and Applications

Paolo Bianchini
University of Genoa

Alberto Diaspro
*University of Genoa
and Italian Institute
of Technology*

9.1 Introduction 9-1
9.2 Theoretical and Physical Considerations on Second
Harmonic Generation...9-2
9.3 SHG Imaging Modes and Microscope Design 9-4
9.4 Biological Observations of SHG within Tissues........................9-7
9.5 Summary 9-11
References 9-12

9.1 Introduction

The physical concepts at the basis of modern nonlinear optical microscopies have a long history. The idea that two photons can combine their energies to produce effects to a level at the sum of the two was first advanced by Albert Einstein in his famous Nobel Prize paper of 1905 in *Ann Physik* (Einstein, 1905). Along the development route, Maria Goeppert-Mayer reported the quantum mechanical formulation of two-photon molecular excitation (Goeppert-Mayer, 1931), but another 30 years had to pass before lasers bright enough to deliver the necessary photons to a molecule were produced. Today, two-photon excitation and other nonlinear optical approaches have been successfully used within the microscopy framework. In general, they are collectively known as nonlinear microscopy, which includes multi-photon (2P, 3P) excitation fluorescence microscopy (EFM) (Denk et al., 1990; Diaspro et al., 2005), second and third harmonic generation (SHG, THG) microscopy (So et al., 2000; Zipfel et al., 2003), and coherent anti-Stokes Raman scattering (CARS) microscopy (Cheng et al., 2002). Although 2PEFM is the most commonly used technique for thick tissue imaging, SHG was used several years prior to its invention. The first implementation was based on the generation of second harmonic light from surfaces (Hellwarth and Christensen, 1974; Sheppard et al., 1977) or from endogenous tissue structures such as rat tendons (Rotha and Freund, 1979). Owing to difficulties in signal interpretation, at least in biological imaging, SHG microscopy has gone relatively unnoticed until very recently (Guo et al., 1997; Campagnola et al., 1999; Moreaux et al., 2000b). The discovery that exogenous markers (Chemla and Zyss, 1984; Prasad, 1991) can lead to exceptionally high signal levels has been one of the major causes for the revival of SHG microscopy. For example, in 1993, Loew and Lewis (Bouevitch et al., 1993) demonstrated that second harmonic signals could be produced from model and cell membranes labeled with voltage-sensitive styryl dyes. Interestingly, SHG markers, when properly designed and collectively organized, can produce signal intensities comparable with conventional 2PEFs (Moreaux et al., 2000b). Therefore, it follows that several endogenous protein structures give rise to SHG (Campagnola et al., 2002). The implementation of high-resolution SHG imaging on a laser scanning microscope offered combined 2PEF images as well as a data acquisition rate comparable to that of confocal fluorescence imaging. Since this development, two general forms of SHG imaging have emerged in parallel. The first approach has further exploited the interfacial aspect of SHG to study membrane biophysics by using voltage-sensitive dyes (Moreaux et al., 2000a,b, 2001; Mohler et al., 2003). The second approach has used a contrast mechanism to investigate structural protein arrays (e.g., collagen) in tissues at a higher resolution and with a more detailed analysis than previously possible.

SHG has several advantageous features that make it an ideal approach in the microscopy of living organisms. Since SHG signals arise from an induced polarization rather than from absorption, this leads to a substantially reduced photobleaching and phototoxicity relative to fluorescence methods (including multiphoton). Additionally, because the excitation typically uses the same near-infrared wavelengths (800–1000 nm) produced by titanium-sapphire lasers, which are also used for two-photon excited fluorescence (2PEF), this method is well suited for studying intact tissue samples. Moreover, excellent depths of penetration can be obtained. Furthermore, detailed information about the organization of protein matrices at the molecular level can be extracted from SHG imaging data. This is because the SHG signals have well-defined polarizations with respect to the laser polarization and the specimen orientation, which can be used to determine the absolute orientation of the protein molecules in the array, as well as the degree of organization of proteins in tissues.

This chapter is organized as follows. Initially, the basic principles of SHG will be qualitatively described at the molecular level. Thereafter, typical experimental configurations for combined SHG and 2PEF imaging will be briefly described. Finally, various applications of SHG microscopy will be addressed.

9.2 Theoretical and Physical Considerations on Second Harmonic Generation

Molecular harmonic up-conversion originates from the nonlinear dependence of the induced molecular dipole moment μ on a driving optical electric field E. The response of the material medium is specified by means of the polarization P (dipole moment per unit volume) and depends on the amplitude E of the electric field of the applied optical wave. Under the simplest circumstances, this relationship can be expressed in the time domain as

$$\tilde{P}(t) = \varepsilon_0[X^{(1)}\tilde{E}(t) + X^{(2)}\tilde{E}^2(t) + X^{(3)}\tilde{E}^3(t)] \tag{9.1}$$

where the presence of a tilde over a quantity indicates that the quantity is a rapidly varying function of time. Here

$\chi^{(1)}$ is the linear susceptibility
$\chi^{(2)}$ is the second-order susceptibility
$\chi^{(3)}$ is the third-order susceptibility, etc.

SHG occurs as a result of the second-order response described by $\chi^{(2)}$. The situation is a bit more complicated for THG, which can occur either directly as a consequence of the third-order response $\chi^{(3)}$, or indirectly as a two-step process involving two second-order processes. In the latter case, the first step involves SHG involving $\chi^{(2)}$, and the second step involves sum-frequency mixing, also involving $\chi^{(2)}$, of frequencies ω and 2ω to produce 3ω. In well-designed optical systems, the sequential process can be far more efficient than the direct process, although in situations involving biological materials the direct process usually dominates. Similar considerations regarding direct and indirect processes apply to higher-order harmonic generations.

In 2PEF microscopy, the fluorescence signal is determined by the intensity profile alone of the excitation beam. In SHG, however, the phase plays a crucial role and cannot be overlooked. A fundamental difference between a collimated beam and a focused beam is that the phase of a focused beam is not parallel distributed along the propagation axis. In particular, near the focus, the phase fronts do not appear to travel as quickly in a collimated beam. Intuitively, this arises from the fact that a focused beam comprises a cone of illumination directions, most of which are off-axis, thus slowing the overall axial phase. As a result of this phase retardation, a focused beam necessarily incurs a net axial phase lag of π as it travels through its focal center relative to the corresponding phase of an unfocused (collimated) beam. This phase lag is called a Gouy shift (Born and Wolf, 1993). On the scale of the SHG active area it may be approximated as varying linearly about its $\pi/2$ midpoint. The consequences of this Gouy shift on SHGs are dramatic. The SHG from a focused beam, instead of propagating on-axis, now propagates off-axis in two well-defined symmetric lobes (Moreaux et al., 2000a,b). This may be explained by momentum conservation (or, in nonlinear optics parlance, phase-matching). We emphasize that the SHG angular pattern is specific to a case where the radiating molecules are well ordered and uniformly distributed, at least on the scale of the SHG wavelength, for instance, when molecules are distributed in a membrane. In the case of more complicated molecular distributions, the molecular distribution could possess an axial periodicity in the vicinity of $\lambda/4$ (where λ is the wavelength in the sample medium), in which case up to 25% of the SHG power is radiated in the backward-directed lobes (Mertz and Moreaux, 2001). Alternatively, the molecules could be distributed tightly clustered around the focal center, in which case the SHG radiation is dipolar in nature, meaning that it is equally distributed in the forward and backward directions (we neglect chemical interactions that might occur between the molecules at such small separations). Momentum conservation arguments can again be invoked to explain these patterns. When the molecular distribution possesses axial inhomogeneities, then in effect, these impart their own pseudomomenta to the SHG emission. In the examples above, the pseudomomenta impart backward-directed "kicks" to the SHG of a magnitude sufficient to provoke a partially backward-directed emission. The total power contained in the SHG radiation, whether the excitation is focused or unfocused, or whether the molecular distribution is homogeneous or inhomogeneous, can be calculated by integrating the SHG intensity over all emission angles (Moreaux et al., 2000a,b).

Now, we can define the SHG cross-section in the same units as the corresponding MPEF cross-section of the same order (Xu and Webb, 1996). This parameter directly provides the multi-harmonic scattered power, ω being the radiation oscillation frequency, from a single molecule for a given driving field intensity, according to

$$\sigma_{SHG} = \frac{4h\omega^5}{3\pi\varepsilon_0^3 c^5}|X^2|^2 \tag{9.2}$$

Then the totally radiated SHG power can be derived by integrating the radiation profile over all the emission directions. As earlier reported by Moreaux (Mertz and Moreaux, 2001) one has:

$$P_{SHG} = \frac{1}{2}\Theta_2 N^2 \sigma_{SHG} \tag{9.3}$$

The factor 1/2 is conventionally used when all the radiated powers are defined in units of photons, as opposed to watts. σ_{SHG} can then be defined in Göppert–Mayer units, allowing it to be directly compared to a 2PEF cross-section. More precisely, the fluorescence power emitted by a dipole undergoing a two-photon excitation can be expressed similarly as:

$$P_{2PEF} = \frac{1}{2}\Theta_2 N \sigma_{2PEF} \tag{9.4}$$

where
   I_0 is the driving field intensity at the focal coordinates
   N is the effective number of radiating molecules
   Θ_2 takes into account the fact that the radiated fields from these N molecules are not necessarily in phase with one another since the molecules are distributed over an area whose dimensions are at least of the order of the beam waist (Diaspro et al., 2005)

In practice, σ_{SHG} is usually much smaller than σ_{2PEF} for a single molecule. By exciting a molecule near resonance, however, it is possible to enhance SHG, typically up to two orders in magnitude. Moreover, SHG scales quadratically with N, whereas fluorescence scales only linearly with N. In the case of a homogeneous molecular distribution and a moderate excitation beam focusing, Θ is very roughly given by $0.01\ \lambda^2/\omega_0^2$. As an example, for an excitation wavelength of 800 nm and a focal spot radius of $\omega_0 = 0.5$ μm in water, $\Theta \approx 0.01$, meaning that the arrangement of the N molecules into a uniform distribution has indeed had a very deleterious effect on their ability to efficiently produce SHG. Nevertheless, a sufficient labeling density (i.e., N) can often overcome this problem.

9.3 SHG Imaging Modes and Microscope Design

As reported in the previous paragraph, biological objects extended along the optic axis (z-direction) for a distance of the order of the emission wavelength exhibit linear, forward-directed scattering rules SHG (Moreaux et al., 2001; Diaspro et al., 2002; Zipfel et al., 2003), and other nonlinear generated signals (Volkmer et al., 2001). This is due to the fact that such extended scatterers prevalently emit a phase-matched signal in the forward direction. Backward phase matching only occurs under exceptional circumstances, but the general rule still holds true (Moreaux et al., 2000b). Thus, the SH emission directions could bring additional information to the bulk image. This means that both the backward and the forward directions (BSHG, FSHG) can be usefully exploited. However, a normal confocal microscope is usually not designed and optimized to perform SHG experiments. Some small changes should be done to the optical setup and the best starting point is the possibility of adapting a spectral confocal microscope since effective spectral separation is one of the keys to the success of SHG imaging. The SHG wavelength is always half the illumination wavelength because it is a nonlinear coherent scattering process that conserves energy. Figure 9.1

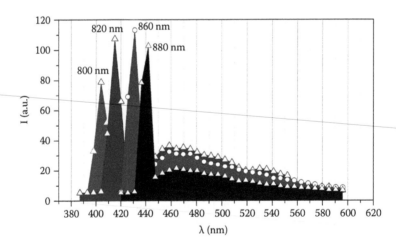

FIGURE 9.1 This figure provides the effective experimental key for analyzing the SHG collected signal. As an example, here it is reported that the emission spectral signature obtained from a mouse Achilles tendon is obtained with an 800, 820, 860, and 880 nm excitation. Each spectrum is obtained by averaging five different images obtained at different x–y sample locations. For all of the excitation wavelengths used, the emission spectra reveal strong SHG signals manifested by a narrow peak at half the excitation wavelength and a bandwidth (full width at half-maximum) in accordance with the excitation laser spectral width. The emission spectra also show a broad feature from 410 to 600 nm that corresponds to a two-photon excitation autofluorescence. Spectral data have been obtained for slightly different values of excitation power, P (namely, 54.2 mW for λ_{ex} 1/4 800 nm, 57.3 mW for λ_{ex} 1/4 820 nm, 60.1 mW for λ_{ex} 1/4 860 nm, and 60.3 mW for λ_{ex} 1/4 880 nm). The spectra have been corrected for the spectral dependence of the microscope objective transmission, the spectrograph grating efficiency, and the transmission of the optical components (filters, dichroic mirrors) in the optical pathway (see also Figure 9.2).

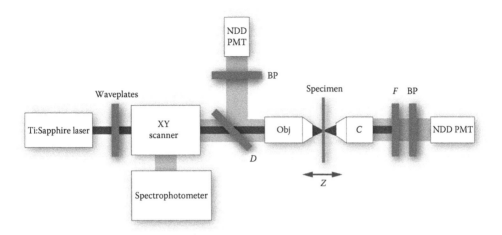

FIGURE 9.2 Simplified optical scheme of the SHG/2PEF microscope at www.lambs.it, based on a Leica SP5 AOBS resonant scanner spectral confocal scanning head (Leica Microsystems, Germany). The SHG is collected both in back-scattering and in a transmitted light configuration. The BSHG spectral component of interest is selected by means of the Leica built-in prism-based spectrophotometer or eventually by a nondescanned PMT. The FSHG signal is isolated by means of optical filters. The polarization tunable section can occasionally be improved for a polarization anisotropy analysis both in the backward and the forward pathways. (Similar work may be found in Bianchini, P. and Diaspro, A., *Multiphoton Microscopy in the Biomedical Sciences IX* SPIE, San Jose, CA, 2009.)

shows a SHG spectra from a mouse tendon illuminated at several different wavelengths. Here, the spectral width scales as $1/\sqrt{2}$ of the illumination spectral width (as expected from the sum-frequency generation across the fundamental's spectrum). In order to isolate the FSHG signal (Volkmer et al., 2001), the transmission channel has to be endowed with stop filters to avoid propagation of the excitation radiation to the photon sensitive elements and to cut fluorescence or autofluorescence contributions (Figure 9.2). In order to be able to assign the collected signal to the SHG or other components, there are four key experiments that can be performed, according to the following physical evidences:

1. Emitted intensity should be proportional to the second power of the incident light, as reported in Figure 9.3.
2. Photobleaching should not occur.
3. SHG spectral components follow the excitation wavelength by proper scaling.
4. A lifetime measurement should be almost zero. Polarization experiments could also be useful in making the origin of the signal more evident.

In Figure 9.2, we have shown a special setup that can be easily generalized. More specifically, Figure 9.2 shows the optical scheme of the second harmonic microscope used to produce most of the images reported in this chapter. It is composed of a Leica DMI6000B-CS inverted microscope, coupled to a Leica TCS-SP5 AOBS spectral confocal scanning head (Leica Microsystems, Germany). The laser system is a Ti:Sapphire Chameleon-XR (Coherent Inc., Santa Clara, CA, United States), tunable between 705 and 980 nm and characterized by a pulse width of 140 fs delivered at a repetition rate of 90 MHz by means of a home-built set-up. The FSHG signal is collected exploiting the transmission pathway of the microscope. This mode allows the collection of most part of the generated SH signal at the cost of some drawbacks. The first is related to the fact that an SH generated radiation cannot pass through a thick sample (greater than 500 mm) to reach the detector. The second is that the SHG light comes together with a transmitted excitation radiation and a potentially primed fluorescence. While the former cannot be optically overcome, the latter can be solved by inserting two band pass filters: an infrared light stop and a SHG wavelength filter. A high numerical aperture condenser is used to

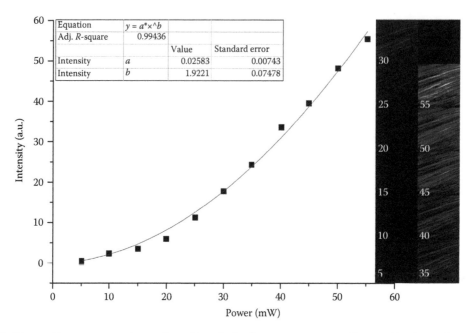

FIGURE 9.3 SHG intensity scales quadratically with the illumination power as illustrated by the graph. On the right side, the related images obtained at the different excitation power conditions, are shown. The specimen is a fixed piece of a mouse's Achilles tendon excited at 860 nm.

collect as much light as possible, since the phenomenon is localized within the focal volume. Usually, a common microscopy photomultiplier is appropriate to obtain high quality images at an appreciable signal-to-noise ratio. The BSHG is acquired using the confocal laser scanning head: the microscope has five different epichannels. The wavelength of the light collected by each channel is selected by a spectrophotometer, avoiding the use of filters and enabling spectral validation. In fact, a distinction between a 2PEF and an SHG signal is occasionally necessary to perform a synchronous spectrum. For example, in the case of a rat tendon sample, the Ti:Sapphire laser was tuned to different excitation wavelengths, λ_{ex} (800, 820, 860, and 880 nm). For each λ_{ex}, a spectra series was acquired and reported after correction for the spectral dependence of the microscope objective transmission, the spectrograph grating efficiency, and the transmission of the optical components (filters, dichroic mirrors, etc.) in the optical path of the two-photon system, in Figure 9.1. For all of the excitation wavelengths being used, the emission spectra reveal strong SHG signals manifested by a narrow peak at half the excitation wavelength and a bandwidth (full width at half maximum) in agreement with the excitation laser spectral width $1/\sqrt{2}$.

The SHG signal strength from biophotonic structures varies according to the relative orientation between the beam and the organized structure. This allows one to determine the absolute orientation of the protein molecules in the array, as well as the degree of organization of proteins in tissues. Thus, polarization information can lead to a third mode, PSHG (Stoller et al., 2002; Gao et al., 2006). The insertion of a half wave plate and a polarizer in the excitation pathway is one of the ways of performing a detailed analysis of polarization dependences (Bianchini and Diaspro, 2009) (Figure 9.2). The PSHG mode allows the assessment of tissue geometry or their morphological features by rotating the linear polarized laser beam at specific angles obtaining two images for each polarization (respectively BSHG and FSHG images). The stitching of the resulting images, using a different lookup table for each angle, immediately provides information about fiber orientation, as shown in Figure 9.4. Therefore, a quantitative evaluation of the set of images adds important optical properties of the sample namely the nonlinear susceptibility, $\chi^{(2)}$ (Bianchini and Diaspro, 2008).

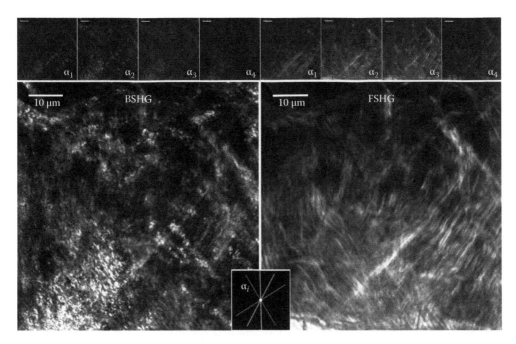

FIGURE 9.4 (See color insert following page 15-30.) Transversal section of a murine Achilles tendon 15 mm thick. The objective used is a 63×1.4 NA oil immersion. The sample is acquired in the PSHG mode. α_i angles are related to the orientation of the incoming beam polarization with respect to the sample. At the top inset row there are the images resulting from the acquisition at different angles false color coded. The full color images come from the stitching of the images in the top row acquired in the BSHG and the FSHG modes, respectively. All the scale bars are 10 mm. (From Bianchini, P. and Diaspro, A., *J. Biophoton.*, 1, 443, 2008. Wiley-VCH Verlag GmbH & Co. KGaA. Reproduced with permission.)

9.4 Biological Observations of SHG within Tissues

The SHG microscopy capability can be demonstrated by SHG imaging. A special strong attribute of the SHG method resides in the property that in tissues, the contrast is produced purely from endogenous species. This, coupled with the physical basis that the SHG signals arise from an induced polarization rather from absorption, leads to reduced photobleaching and phototoxicity relative to fluorescence methods (including multiphoton). Additionally, because the excitation typically uses the same near-infrared wavelengths (800–1000 nm) produced by titanium–sapphire lasers that are also used for a two-photon excited fluorescence (2PEF), this method is well suited for studying intact tissue samples, since excellent depths of penetration can be obtained. The SHG properties have recently been exploited to examine a wide range of structural protein arrays in situ (Campagnola et al., 2002; Deng et al., 2002; Zoumi et al., 2002, 2004; Brown et al., 2003; Cox et al., 2003a,b; Mohler et al., 2003; Dombeck et al., 2003, 2004; Boulesteix et al., 2004; Williams et al., 2005; Yeh et al., 2005; Sun et al., 2006). The profusion of the recent reports has focused on visualizing collagen fibers in natural tissues, including skin, tendons, cartilages, blood vessels, and corneas (Stoller et al., 2001, 2002; Zoumi et al., 2002, 2004; Han et al., 2004; Yeh et al., 2005), acto-myosin complexes in muscles (Campagnola et al., 2002) and microtubule-based structures in live cells (Campagnola et al., 2002; Dombeck et al., 2003).

In the following gallery, biological sample images obtained with different methods are shown. Figure 9.5 shows an SH signal generated by myosin muscular fibers of two different organisms; a rat (Figure 9.5A) and a zebrafish (Figure 9.5B and C). Muscle structure has been studied for a long time by ultrastructural methods, including electron microscopy and x-ray crystallography. A report in 1997 suggested

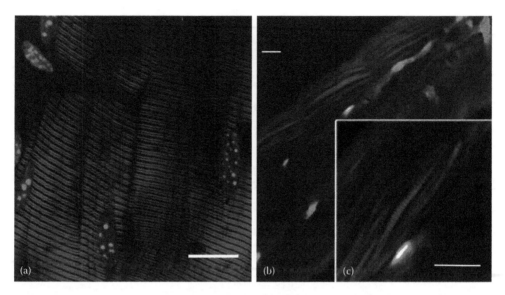

FIGURE 9.5 (See color insert following page 15-30.) (a) Mouse muscle fibers (gastrocnemius muscle) single optical section. Infrared excitation at 800 nm. Nuclei are labeled with Hoechst33342 (green) and acquired in the epichannel in the spectral range 440–480 nm. SHG (magenta) from myosin fibers is acquired in the transmission channel. (Sample courtesy of E. Ralston, NIH, Bethesda, MD.) (b, c) Tail of a whole zebrafish. Epithelial blood vessel cells are genetically modified in order to express EGFP (green). Forward SHG comes from the tail muscles (magenta). Panel (c) is a magnification of the muscle fibers to evidence the contrast achieved by SHG imaging. The scale bar is 20 μm.

that a source of SHG lies within the striated muscle (Guo et al., 1997) then, examining muscle tissue on a high-resolution laser scanning microscope, it was confirmed that myosin fibers exhibit large SHG signal. Actually detailed, high-contrast features could be resolved in SHG optical sections throughout the full ~550 μm thickness of a freshly dissected, unfixed sample of mouse lower leg muscle (Campagnola et al., 2002). The rat muscle, shown in Figure 9.5A comes from an optically transparent slice whereas the zebrafish tail shown in Figure 9.5B and C comes from the whole fixed animal. In this case, the transparency of the samples and their ordered fiber arrangement make the FSHG detectable. A 2PEFM is possible at the very same time. A fluorescence signal comes from the EGFP expressed by the epithelial blood cells. In this way, the blood vessel is clearly visible in fluorescence while the SHG signal allows one to follow the muscle architecture along the zebrafish tail. Therefore, the SHG contrast shows all the muscle cells (Figure 9.5C), and the image is comparable to that observed in standard histological transverse sections, but now it can be acquired within an unfixed, unsectioned tissue and an unstained tissue.

Figure 9.6 shows peritumoral blood vessels acquired on a mouse *in vivo* where the relevance of the SHG signal to follow anatomical details that can provide a useful guidance for the understanding of complementary fluorescence signals becomes evident. Now, since a tendon is prevalently composed of collagen, it can be handled either as a whole fragment or cut in thin histological sections. Although looking at the whole of the Achilles tendon through low magnification (5×) and a low numerical aperture (0.4 NA) objective, it is possible to observe the external macro structure of the tendon by imaging BSHG, FSHG and autofluorescence (Figure 9.7A). A closer look at the tendon structure could be done by moving to a higher numerical aperture objective (63×1.4 NA oil immersion) and a thin (~15 μm) transversal histological section (Figure 9.7B). This latter image is strongly influenced by the scattering of the fibers. In particular, as also reported by Zipfel et al. (2003), the vertically oriented fibrils scatter mostly along the forward direction as shown in Figure 9.8 (B and D frames). Figure 9.8A and C completes the information since laterally oriented fibrils scatter bidirectionally. Moreover, with a focused beam, the

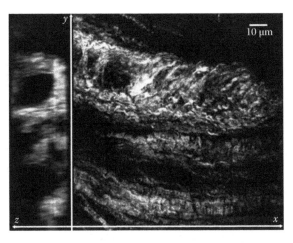

FIGURE 9.6 Peritumoral murine blood vessels acquired in the living animal under general anesthesia (ketamine-xylamine, IM). Access to the tumor surface in the living animal was obtained by dissection of the overlaying skin in a large flap. The backward-directed SHG shows collagen fibers around the vessel. On the left side of the image is a *yz* projection of the 3D stack showing the section of the vessel. (From Bianchini, P. and Diaspro, A., *J. Biophoton.*, 1, 443, 2008. Wiley-VCH Verlag GmbH & Co. KGaA. Reproduced with permission.)

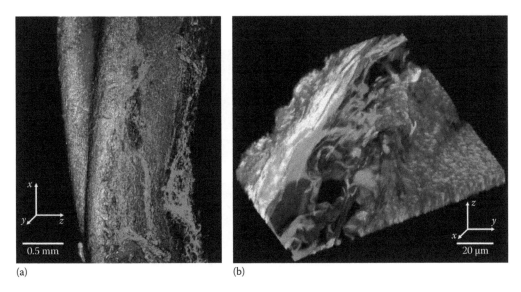

(a)   (b)

FIGURE 9.7 (See color insert following page 15-30.)   A 3D reconstruction of the Achilles tendon. The excitation wavelength is 860 nm. Green is the 2PE autofluorescence, magenta and cyan are the backward and forward SHGs, respectively. (a) Entire tendon viewed with 5×0.4 NA objective. (b) Transverse histological slice of the tendon viewed with 63×1.4 NA oil immersion objective. (From Bianchini, P. and Diaspro, A., *J. Biophotonics*, 1, 443, 2008. Wiley-VCH Verlag GmbH & Co. KGaA. Reproduced with permission.)

forward-directed emission profile is further influenced by the Gouy phase anomaly and by the orientation and distribution of scattering centers (Moreaux et al., 2001). The latter seems to be prevalent in the case of the tendon. In order to elucidate this aspect better, one could acquire an image in the confocal reflection mode (458 nm) obtaining the most probable distribution of the scatterers for the SHG wavelength of interest (Figure 9.9). This preliminary result seems to confirm that the backscattered SHG comes mostly from the scattering centers. This improves the ability to clearly distinguish among fiber

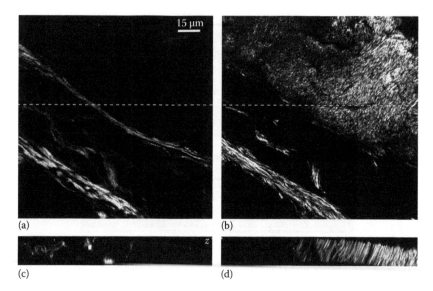

FIGURE 9.8   Transversal section of a murine Achilles tendon 15 mm thick. The objective used is a 63×1.4 NA oil immersion. (a) Backward SHG, (b) SHG in transmission. (c) and (d) are the *yz* projections of (a) and (b) respectively along a centered optical cut. (From Bianchini, P. and Diaspro, A., *J. Biophoton.*, 1, 443, 2008. Wiley-VCH Verlag GmbH & Co. KGaA. Reproduced with permission.)

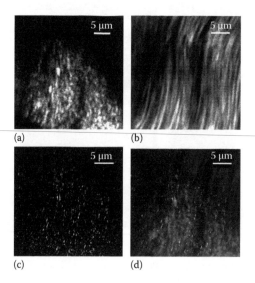

FIGURE 9.9   Longitudinal section of a murine Achilles tendon 100 mm thick. The objective used is a 63×1.4 NA oil immersion. (a–b) are the backward and the forward SHGs excited at 880 nm, (c) A confocal reflected image of the very same section obtained with a 458 nm incident laser line. Panel (d) is the overlay of panels (a), (b), and (c). (Sample courtesy of L. Gallus and P. Ramoino, University of Genoa, Genoa, Italy.)

types and orientations and at the very same time can be used to identify punctuate segmental collagen, characteristic of an ongoing fibrillogenesis or a collagen turnover.

Bone is highly turbid, consisting largely of collagen and minerals. Thus, the same level of penetration is not expected to be achieved as in the case of less-scattering tissues such as tendons or muscles, where the mean penetration depths are ~100 μm. However, here we show that high-contrast SHG images

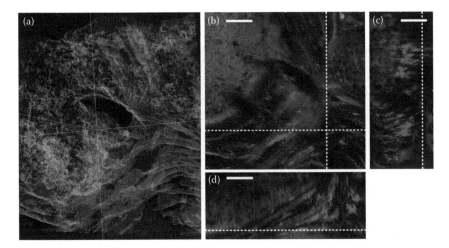

FIGURE 9.10 (See color insert following page 15-30.) A human trabecular bone fragment. Panel (a) is the 3-dimensional reconstruction of the z-stack depicted in panels (b), (c), and (d). The 3D render is obtained by the MicroScoBioJ software package (containing image segmentation, registration, restoration, surface rendering, and surface estimation features) (MicroSCoBioJ, http://imagejdocu.tudor.lu/doku.php?id=plugin:stacks:microscobioj: start). Panels (b), (c), and (d) are the orthogonal view of the acquired optical z-sections. The dashed lines indicate which planes of the stack are shown. In particular, (b) is the xy, (c) is the yz, and (d) is the xz plane. For all the panels, green represents the 2PEF of advanced glycation end products (AGEs), blue represents the backward SHG, and red the forward SHG. All the scale bars are 30 μm.

can still be obtained in the forward and backward collection geometry. Figure 9.10 shows an optical section of a human trabecular bone. While these do not possess the striking regularity of the tendon in Figure 9.10B through D the collagen fibers are clearly identifiable and possess good image contrast. An important question is the depth of penetration that can be achieved. To determine this, a complete 3D stack was acquired and orthogonal views are shown in Figure 9.10C and D. We find that SHG could be collected from depths through approximately ~70 μm. The thicker regions were too opaque to collect a sufficient signal in the transmission mode or, at least, not negligible shading effects disturb the imaging formation. The contrast provided by SHG and 2PEF open the door to the examination of bone pathological conditions such as diseases induced by diabetes or age, where differences in concentration, morphology, and collagen organization may be imaged by this technique.

9.5 Summary

Highly organized nanoperiodic structures in biological samples exhibit strong SHG activity. Many biological structures, such as microfibrils in cell walls, alternating crystalline lamellae in starch granules, crystalline myosin and actin nanofilaments in the myofibrils of skeletal and cardiac muscles, and microtubules in both the cytoskeleton and the mitotic apparatus, and many of the biological birefringent structures like collagen exhibit SHG properties. In contrast to single- and multi-photon absorption, harmonic generation involves only virtual states and does not involve energy deposition. The harmonic signals allow a 3D structural visualization with minimal or no additional preparation of the sample. Meanwhile, 2PF imaging modes can be added to monitor multiple molecular probes in living cells and tissues, such as those composed of transformed cells or taken from transgenic organisms.

Therefore, we have shown that by measuring the signal of SHG which originated along different pathways, i.e., FSHG and BSHG, one can get different information related to the more complex architectural properties of the samples. The SHG in the backscattering pathway is strongly modulated by the positions and dimensions of the scattering centers. Orientation information can be derived by polarization measurements both in the BSHG and the FSHG mode also. The acquisition of the BSHG signal and its interpretation allow the design of experiments for *in vivo* studies and with the capability of differentiating better those molecular species that originate in the SHG signal. We think that some computational effort should be taken in order to separate the contributions in the BSHG signal using as reference the FSHG one complemented by polarization dependent data. This should be done on comparatively simple and known SHG generating samples. Such image contrast mechanisms can be fully integrated in a multidimensional fluorescence approach aiming to provide complementary information based on 4D (x, y, z, t), lifetime, and spectral contrast mechanisms.

References

Bianchini, P. and Diaspro, A. (2008). Three-dimensional (3D) backward and forward second harmonic generation (SHG) microscopy of biological tissues. *J. Biophoton.* **1**(6): 443–450.

Bianchini, P. and Diaspro, A. (2009) SHIM and TPEM: Getting more information from non linear excitation. *Multiphoton Microscopy in the Biomedical Sciences IX*, 1st edn. San Jose, CA, SPIE **7183** (1):718335.

Born, M. and Wolf, E. (1993) *Principles of Optics: Electromagnetic Theory of Propagation, Interference and Diffraction of Light.* Oxford, New York, Pergamon Press.

Bouevitch, O., Lewis, A., Pinevsky, I., Wuskell, J. P., and Loew, L. M. (1993) Probing membrane potential with nonlinear optics. *Biophys J*, 65, 672–679.

Boulesteix, T., Beaurepaire, E., Sauviat, M.-P., and Schanne-Klein, M.-C. (2004) Second-harmonic microscopy of unstained living cardiac myocytes: Measurements of sarcomere length with 20-nm accuracy. *Opt Lett*, 29, 2031–2033.

Brown, E., McKee, T., Ditomaso, E., Pluen, A., Seed, B., Boucher, Y., and Jain, R. K. (2003) Dynamic imaging of collagen and its modulation in tumors in vivo using second-harmonic generation. *Nat Med*, 9, 796–800.

Campagnola, P. J., Wei, M. D., Lewis, A., and Loew, L. M. (1999) High-resolution nonlinear optical imaging of live cells by second harmonic generation. *Biophys J*, 77, 3341–3349.

Campagnola, P. J., Millard, A. C., Terasaki, M., Hoppe, P. E., Malone, C. J., and Mohler, W. A. (2002) Three-dimensional high-resolution second-harmonic generation imaging of endogenous structural proteins in biological tissues. *Biophys J*, 81, 493–508.

Chemla, D. S. and Zyss, J. (1984) *Nonlinear Optical Properties of Organic Molecules and Crystals.* Vol. 1, edited by Chemla, D. and Zyss, J., pp. 23–187, New York, Academic Press.

Cheng, J.-X., Jia, Y. K., Zheng, G., and Xie, X. S. (2002) Laser-scanning coherent anti-stokes Raman scattering microscopy and applications to cell biology. *Biophys J*, 83, 502–509.

Cox, G., Kable, E., Jones, A., Fraser, I., Manconi, F., and Gorrell, M. D. (2003a) 3-Dimensional imaging of collagen using second harmonic generation. *J Struct Biol*, 141, 53–62.

Cox, G. C., Xu, P., Sheppard, C. J. R., and Ramshaw, J. A. (2003b) Characterization of the second harmonic signal from collagen. In Periasamy, A. and So, P. T. C. (Eds.), *Multiphoton Microscopy in the Biomedical Sciences III. Proc SPIE*, 4963, 32–40.

Deng, X., Williams, E. D., Thompson, E. W., Gan, X., and Gu, M. (2002) Second-harmonic generation from biological tissues: Effect of excitation wavelength. *Scanning*, 24, 175–178.

Denk, W., Strickler, J. H., and Webb, W. W. (1990) Two-photon laser scanning fluorescence microscopy. *Science*, 248, 73–76.

Diaspro, A., Fronte, P., Raimondo, M., Fato, M., Deleo, G., Beltrame, F., Cannone, F., Chirico, G., and Ramoino, P. (2002) Functional imaging of living *Paramecium* by means of confocal and two-photon excitation fluorescence microscopy. *Proc SPIE*, 4622, 47–53.

Diaspro, A., Chirico, G., and Collini, M. (2005) Two-photon fluorescence excitation and related techniques in biological microscopy. *Q Rev Biophys*, 38, 97–166.

Dombeck, D. A., Kasischke, K. A., Vishwasrao, H. D., Ingelsson, M., Hyman, B. T., and Webb, W. W. (2003) Uniform polarity microtubule assemblies imaged in native brain tissue by second-harmonic generation microscopy. *Proc Natl Acad Sci USA*, 100, 7081–7086.

Dombeck, D. A., Blanchard-Desce, M., and Webb, W. W. (2004) Optical recording of action potentials with second-harmonic generation microscopy. *J Neurosci*, 24, 999–1003.

Einstein, A. (1905) Creation and conversion of light. *Ann Physik*, 17, 132–148.

Gao, L., Jin, L., Xue, P., Xu, J., Wang, Y., Ma, H., and Chen, D. (2006) Reconstruction of complementary images in second harmonic generation microscopy. *Opt Express*, 14, 4727–4735.

Goeppert-Mayer, M. (1931) Elementary processes with two quantum jumps. *Ann Physik*, 9, 273–294.

Guo, Y., Ho, P. P., Savage, H., Harris, D., Sacks, P., Schantz, S., Liu, F., Zhadin, N., and Alfano, R. R. (1997) Second-harmonic tomography of tissues. *Opt Lett*, 22, 1323–1325.

Han, M., Zickler, L., Giese, G., Walter, M., Loesel, F. H., and Bille, J. F. (2004) Second-harmonic imaging of cornea after intrastromal femtosecond laser ablation. *J Biomed Opt*, 9, 760–766.

Hellwarth, R. and Christensen, P. (1974) Nonlinear optical microscopic examination of structure in poly-crystalline ZnSe. *Opt Commun*, 12, 318–322.

Mertz, J. and Moreaux, L. (2001) Second-harmonic generation by focused excitation of inhomogeneously distributed scatterers. *Opt Commun*, 196, 325–330.

Mohler, W., Millard, A. C., and Campagnola, P. J. (2003) Second harmonic generation imaging of endogenous structural proteins. *Methods*, 29, 97–109.

Moreaux, L., Sandre, O., Blanchard-Desce, M., and Mertz, J. (2000a) Membrane imaging by simultaneous second-harmonic generation and two-photon microscopy. *Opt Lett*, 25, 320–322.

Moreaux, L., Sandre, O., and Mertz, J. (2000b) Membrane imaging by second harmonic generation microscopy. *J Opt Soc Am B*, 17, 1685–1694.

Moreaux, L., Sandre, O., Charpak, S., Blanchard-Desce, M., and Mertz, J. (2001) Coherent scattering in multi-harmonic light microscopy. *Biophys J*, 80, 1568–1574.

Prasad, P. N. (1991) *Introduction to Nonlinear Optical Effects in Molecules and Polymers*. New York, Wiley.

Rotha, S. and Freund, I. (1979) Second harmonic generation in collagen. *JCP*, 70, 1637–1643.

Sheppard, C., Gannaway, J., Kompfner, R., and Walsh, D. (1977) The scanning harmonic optical microscope. *IEEE J Quantum Electron*, 13, 912–912.

So, P. T., Dong, C. Y., Masters, B. R., and Berland, K. M. (2000) Two-photon excitation fluorescence microscopy. *Annu Rev Biomed Eng*, 2, 399–429.

Stoller, P. C., Kim, B. M., Rubenchik, A. M., Reiser, K. M., and Da Silva, L. B. (2001) Measurement of the second-order nonlinear susceptibility of collagen using polarization modulation and phase-sensitive detection. In Haglund, R. F., Neev, J., and Wood, R. F. (Eds.), *Commercial and Biomedical Applications of Ultrashort Pulse Lasers; Laser Plasma Generation and Diagnostics. Proc. SPIE*, 4276, 11–16.

Stoller, P., Reiser, K. M., Celliers, P. M., and Rubenchik, A. M. (2002) Polarization-modulated second harmonic generation in collagen. *Biophys J*, 82, 3330–3342.

Sun, Y., Chen, W.-L., Lin, S.-J., Jee, S.-H., Chen, Y.-F., Lin, L.-C., So, P. T. C., and Dong, C.-Y. (2006) Investigating mechanisms of collagen thermal denaturation by high resolution second-harmonic generation imaging. *Biophys J*, 91, 2620–2625.

Volkmer, A., Cheng, J.-X., and Sunney XIE, X. (2001) Vibrational imaging with high sensitivity via epidetected coherent anti-stokes Raman scattering microscopy. *Phys Rev Lett*, 87, 023901-1–023901-4.

Williams, R. M., Zipfel, W. R., and Webb, W. W. (2005) Interpreting second-harmonic generation images of collagen I fibrils. *Biophys J*, 88, 1377–1386.

Xu, C. and Webb, W. W. (1996) Measurement of two-photon excitation cross sections of molecular fluorophores with data from 690 to 1050 nm. *J Opt Soc Am B*, 13, 481.

Yeh, A. T., Hammer-Wilson, M. J., Sickle, D. C. V., Benton, H. P., Zoumi, A., Tromberg, B. J., and Peavy, G. M. (2005) Nonlinear optical microscopy of articular cartilage. *Osteoarthr Cartil*, 13, 345–352.

Zipfel, W. R., Williams, R. M., Christie, R., Nikitin, A. Y., Hyman, B. T., and Webb, W. W. (2003) Live tissue intrinsic emission microscopy using multiphoton-excited native fluorescence and second harmonic generation. *Proc Natl Acad Sci USA*, 100, 7075–7080.

Zoumi, A., Yeh, A., and Tromberg, B. J. (2002) Imaging cells and extracellular matrix in vivo by using second-harmonic generation and two-photon excited fluorescence. *Proc Natl Acad Sci USA*, 99, 11014–11019.

Zoumi, A., Lu, X., Kassab, G. S., and Tromberg, B. J. (2004) Imaging coronary artery microstructure using second-harmonic and two-photon fluorescence microscopy. *Biophys J*, 87, 2778–2786.

10

Green Fluorescent Proteins as Intracellular pH Indicators

Fabio Beltram
Scuola Normale Superiore, Italian Institute of Technology, and National Council of Research

Ranieri Bizzarri
Scuola Normale Superiore, Italian Institute of Technology, and National Council of Research

Stefano Luin
Scuola Normale Superiore

Michela Serresi
Scuola Normale Superiore and Italian Institute of Technology

10.1 Introduction 10-1
10.2 pH-Dependent Properties of Green Fluorescent Proteins 10-3
  Protein Structure and Folding • The Basis of pH Sensing • Kinetics of GFP Protonation and Resolution of pH_i Imaging Measurements
10.3 GFP-Based pH Indicators 10-9
  Ratiometric Fluorescent Indicators • GFP-Based pH Indicators Applied *In Vivo* • pH_i Measurement: Instrumentation and Methods
10.4 Summary and Future Perspectives 10-17
Acknowledgments 10-18
References 10-18

10.1 Introduction

Intracellular pH (pH_i) is one of the fundamental modulators of cell function (Srivastava et al. 2007). Owing to the high reactivity of H^+ ions with proteins and other biomolecules, pH_i plays a major role in processes as varied as cell metabolism and growth (Wang et al. 1997, Putney and Barber 2003), ionic current flow through membrane channels and cellular excitability, solute movement on membrane transporter proteins (and therefore general ion homeostasis) (Fliegel 2005, Hunte et al. 2005), and muscle cellular contractility (Wakabayashi et al. 2006). Alterations of pH_i strongly affect cell viability. For instance, mitochondria are characterized by mildly alkaline pH and deviations from this pattern may lead to cell apoptosis (Abad et al. 2004). The aberrations of the normal organellar pH homeostasis can lead to an impairment of posttranslational modifications and processing of secreted proteins (Carnell and Moore 1994), to a mislocalization of the biosynthetic cargo (Chanat and Huttner 1991), and to severe defects in the functionality of organelles (Puri et al. 2002). Abnormalities noted in many human tumors (i.e., breast cancer and colorectal cancer) and papillomas have also been attributed to the modification of the pH of the secretory compartment (Kellokumpu et al. 2002).

The pH_i of eukaryotic cells falls normally in the range of 7.1–7.4. pH_i and is regulated by the flux of acid equivalents (H^+, OH^-, or HCO_3^-) across the surface membrane, usually on specific ion-coupled transporters such as the NHE proteins (Hunte et al. 2005), and by acid-generating processes, such as aerobic and anaerobic metabolisms (Sun et al. 2004). Remarkably, pH_i is not spatially uniform and depends strongly on the nature and the function of subcellular domains. For example, in the endocytic pathway, the progressive luminal acidification of endosomes is essential for the distribution and degradation of internalized ligands into lysosomes (Mellman and Warren 2000). Also, differences in pH between early, recycling, and late endosomes within many cell types have been observed (Rybak et al. 1997). The synaptic activity is another biological event regulated by pH in a spatially selective manner. Neurotransmitters stored in presynaptic vesicles are released through the fusion of vesicles with the plasma membrane (exocytosis) (Jahn et al. 2003). The lumen of presynaptic vesicles displays an acidic interior. During exocytosis, however, the synaptic vesicles undergo a pH jump as their interior comes in contact with the extracellular medium, characterized by a neutral pH.

High resolution quantitative fluorescence imaging microscopy has recently become a powerful tool in cell and molecular biology because it permits the measurement of both the spatial and the temporal dynamics of molecules and organelles in living specimens (Lakowicz 1999). The superior sensitivity and spectroscopic capabilities of fluorescence microscopy encouraged the development of advanced fluorescent indicators tailored to many different applications *in vivo*. Owing to its relevance for cell biology, the proton concentration has been one of the first parameters monitored by fluorescent probes in living cells (Rink et al. 1982, Tsien 1989). To date, spatially resolved fluorescence sensing is the only method available for the high-resolution detection of pH_i (Lakowicz 1999, Lin et al. 2003). Among the engineered probes, however, those based on green fluorescent protein mutants certainly play the predominant role.

The green fluorescent protein (wild type GFP, abbreviated wtGFP) is an intrinsically fluorescent protein that absorbs violet-blue light and emits green light with a high quantum efficiency. wtGFP was discovered in 1962 as an accessory protein of the bioluminescence system of the hydroid jellyfish *Aequorea victoria*, encompassing also the protein *Aequorin* (Shimomura et al. 1962, Tsien 1998). Thirty years later, the cloning (Prasher et al. 1992), and the successful heterologous expression of the wtGFP gene (Chalfie et al. 1994), demonstrated clearly that the green fluorescent emission is genetically encoded into the primary sequence of the protein and no jellyfish-specific cofactors are needed for the synthesis of the chromophore. Thus, wtGFP can be used as an intrinsic intracellular reporter of target proteins by a simple genetic fusion and a subsequent gene transfer and expression into cells. The modification of the primary sequence of wtGFP yielded many mutants of different colors characterized by a wide range of photophysical properties (Chudakov et al. 2005). Recently, the color palette of GFPs was expanded by the discovery and engineering of new red-shifted proteins from coral organisms (Miyawaki 2005, Shaner et al. 2005, Chudakov et al. 2005).

The main advantages of fluorescent protein–based indicators over simple organic dyes are that they can be designed to respond to a much greater variety of biological events and signals, targeted to subcellular compartments, introduced in a wider variety of tissues and organisms, and they seldom cause photodynamic toxicity (Zhang et al. 2002). Furthermore, some fluorescent proteins possess intrinsic photophysical properties perfectly tailored to the detection of pH_i.

In this chapter, we shall review the use of fluorescent proteins as fluorescent indicators of pH_i. We shall limit our discussion to mutants belonging to the *Aequorea* family (GFPs), on account of the general poor pH-responsiveness and the more complex photophysics of red fluorescent proteins from *Anthozoa* and other corals that have prevented their use so far (Shaner et al. 2005). The nature and optical characteristics of GFPs will be reviewed in Section 10.2. Then, the engineered GFP-based indicators for pH_i measurement will be described in Section 10.3, together with some details on the typical microimaging setups adopted for these measurements. In the last section, the future perspectives about the development and applications of these remarkable indicators will be debated.

10.2 pH-Dependent Properties of Green Fluorescent Proteins

10.2.1 Protein Structure and Folding

wtGFP is constituted by a single peptide chain of 238 amino acids and a 27 kDa molecular weight (Tsien 1998). X-ray spectroscopy showed that this sequence folds into a compact cylindrical three-dimensional structure (referred to as β-barrel) with a diameter of 24 Å and a height of 42 Å (Figure 10.1) (Ormo et al. 1996). The β-barrel is capped on both ends by short α-helical sections and run through by a α-helix, which contains the three amino acids forming the chromophore. The GFP chromophore is buried at the center of the β-barrel (Figure 10.1), and originates from the spontaneous posttranslational cyclization of three consecutive amino acids: Ser[65]–Tyr[66]–Gly[67] following the formation of the native β-barrel tertiary structure. The formation of the GFP chromophore formation comprises three distinct chemical processes and is triggered by the attainment of the native β-barrel tertiary structure: (1) the cyclization of the tripeptide Ser[65]–Tyr[66]–Gly[67], (2) the oxidation of the cyclic intermediate to yield a more conjugated structure, and (3) the

FIGURE 10.1 (See color insert following page 15-30.) 3D structure of wtGFP as derived from x-ray analysis (Ormo et al. 1996). The frontal part of the barrel is torn open to show the chromophore buried at the protein center.

final dehydration step (Scheme 10.1) (Wachter 2007). The rate-limiting step of the process is the oxidation reaction, which requires at least 30 min to occur (Zhang et al. 2006). The final structure of the chromophore consists of two rings, a six-member aromatic phenol coming from the side chain of Tyr[66], and a five-member imidazolidinone resulting from the cyclization of the backbone (Scheme 10.1).

The presence of the compact and the rigid tertiary structure of wtGFP is thought to be responsible for the existence and the high quantum yield of the fluorescence emission (Webber et al. 2001). Indeed, the isolated chromophore is not fluorescent in an aqueous solution (Niwa et al. 1996) and many of the classical fluorescence quenching agents are almost ineffective on the emission of the folded GFP (Ward

SCHEME 10.1 Formation of the wtGFP chromophore from the Ser[65]–Tyr[66]–Gly[67] amino acid sequence. (Based on Wachter, R.M., *Acc. Chem. Res.*, 40, 120, 2007.)

and Bokman 1982). Notably, the chromophore is surrounded by four entrapped water molecules and several charged and polar residues such as Gln[69], Gln[94], Arg[96], His[148], Thr[203], Ser[205], and Glu[222] (Brejc et al. 1997). These residues may act as a proton donor and an acceptor, and they are thought to participate in a structured hydrogen-bond network responsible for most of the spectral and photophysical properties of the protein (Brejc et al. 1997, Kummer et al. 2000, Wachter et al. 2000).

10.2.2 The Basis of pH Sensing

10.2.2.1 Protonation States of the Chromophore

Four protonation states (Scheme 10.2, **i–iv**) are accessible to the GFP chromophore, owing to the acid–base properties of both the phenol group on the Tyr[66] aromatic ring and the N[66] (drawn in bold in Scheme 10.2) nitrogen on the imidazolinone ring. In the isolated chromophore, N[66] has a pK=1.8–2.4 (Scheme 10.2, **iii** ↔ **i**) whereas the phenol group has a pK=8.1–8.5 (**i** ↔ **ii**) (Bell et al. 2000, Dong et al. 2006); additionally, the phenol group was found to have a pK=6.5 when N[66] is in a quaternary positively charged state similar to the protonated one (Scheme 10.2, **iii** ↔ **iv**) (Dong et al. 2006). These pK values allow the calculation that less than 0.01% of the overall chromophore in the neutral state must be present in the zwitterion form (Scheme 10.2, **iv**). This means that for pH > 3 only the ionization of the phenol group is relevant to determine the ground state of the chromophore (Scheme 10.2, **i** ↔ **ii**), whereas other protonation reactions leading to the fully protonated (**iii**) or zwitterionic (**iv**) states are no more active. The same situation holds also in the folded protein (Elsliger et al. 1999), although the phenol deprotonation takes place at a significantly lower pH value (pK=5.2–7.5) (Llopis et al. 1998, Elsliger et al. 1999, Wachter et al. 2000, Bizzarri et al. 2007). Forms **i** and **ii** are traditionally called states A and B, respectively (Scheme 10.2).

10.2.2.2 pH-Dependent Optical Properties

In wtGFP, state A absorbs at 398 nm and emits at 508 nm, whereas state B absorbs at 475 nm and emits at 503 nm (Tsien 1998). The absorption wavelengths may change in other variants, but the spectral pattern

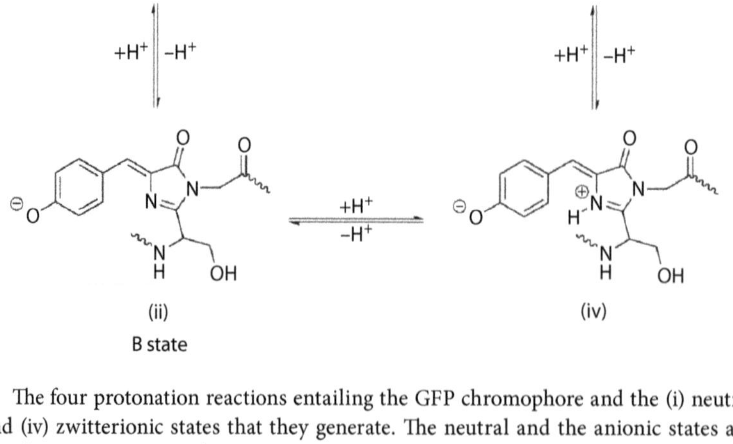

SCHEME 10.2 The four protonation reactions entailing the GFP chromophore and the (i) neutral, (ii) anionic, (iii) cationic, and (iv) zwitterionic states that they generate. The neutral and the anionic states are traditionally referred to as A and B states, respectively.

remains the same: state B absorbs at a much longer wavelength than state A (up to 100–120 nm, Figure 10.2a), whereas their fluorescence emissions are very similar. The absorption characteristics are easily explained in terms of an extended electronic conjugation in the B phenol-anionic state, which lowers the $S_0 \rightarrow S_1$ transition energy (Voityuk et al. 1998). Conversely, the mechanism at the basis of the fluorescence similarity is subtler and was elucidated only by means of fast spectroscopic methods (Chattoraj et al. 1996). The key observation is that A and B states differ not only in the protonation of Tyr[66], but also for the conformation of the residues surrounding the chromophore (Brejc et al. 1997). X-ray structures of wtGFP and other variants show that a peculiar hydrogen-bond network connecting the phenol group to residue Glu[222] is present in A state, whereas it is blocked in B state. When B is excited, it emits directly from the excited state B*. On the contrary, when A is excited, two emission channels are active: (i) direct emission from A* and (ii) deprotonation of A*, owing to the much higher acidity of A* compared to A (Voityuk et al. 1998), with a concomitant proton transfer to Glu[222] (Chattoraj et al. 1996, McAnaney et al. 2002). The latter mechanism is called excited state proton transfer (ESPT) and occurs in a few picoseconds, usually overwhelming the direct A* weak emission that occurs in the 440–460 nm region (Bonsma et al. 2005). ESPT leads to an intermediate excited form, I*, which is characterized by a

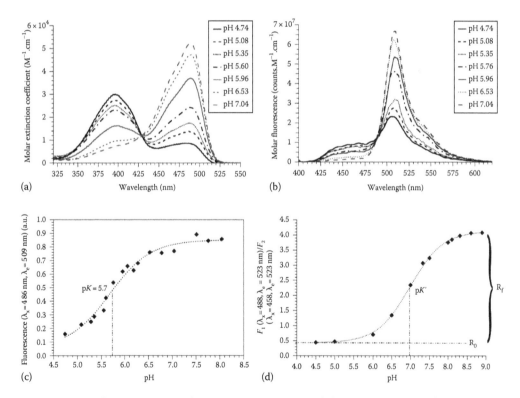

FIGURE 10.2 (a) Absorption spectra of F64L/S65T GFP (EGFP) at different pH. The A state (λ=395 nm) is progressively converted to the B state (λ=486 nm) as the pH is raised. (b) The emission spectra of the EGFP at different pHs by excitation at 395 nm. At a low pH, a detectable fluorescence at 460 nm is observed and stems from the direct emission from A*. Fluorescence at 510 nm derives from the ESPT mechanism. (c) Sigmoidal dependence of fluorescence (λ_x=486 nm, λ_e=509 nm) upon pH for EGFP (black diamonds); the experimental data were fit to Equation 10.3 (dashed line), obtaining pK=5.7. (d) Ratiometric plot for F64L/S65T/T203Y GFP (E^{2}GFP), with set #1: (λ_x=488 nm, λ_e=523 nm) and set #2: (λ_x=458 nm, λ_e=523 nm). Experimental points (black diamonds) were fitted to Equation 10.8 (dashed curve); a graphical legend to highlight the ratiometric parameters is reported. Notably, each point is an average of fluorescence ratios taken at different chloride concentrations (0–200 mM) in order to display ratiometric independence from the quenching activity of the chloride ion.

deprotonated chromophore like B*. Thus, I* resembles B* for the emission energy (Creemers et al. 1999) apart from minor environmental effects [the short timescale of the A* → I* process does not allow the rearrangement of the residues surrounding the chromophore, for instance Thr[203], to the relaxed configuration typical of the anionic state (Chattoraj et al. 1996, Creemers et al. 1999)]. Deactivation of I* leads to I, a metastable ground-state moiety whose anionic chromophore is embedded in an environment typical of the neutral A form. I eventually evolves to A, which is about 7.6 kJ/mol lower in energy (Wiehler et al. 2003), by receiving the proton of Glu[222] back through an internal protonation relay involving Ser[205] and a bound water molecule (Brejc et al. 1997). Recently, a more detailed kinetic analysis revealed that the I species actually comprises two metastable intermediates I_1 and I_2, although I_1 is very short lived (3 ps) compared to I_2 (0.4 ns) (Kennis et al. 2004).

The molecular optimization of wtGFP by a natural selection to play a functional role in *Aequorea victoria* jellyfish has led to an impressive absorption/emission stability in the physiological pH range (pH 5–9). Engineered GFP mutants, however, have no natural role to play and in many cases possess a fast equilibrium (Kneen et al. 1998, Abbruzzetti et al. 2005) between the neutral and anionic forms of the chromophore in the physiological pH interval. Additionally, many GFP mutants (e.g., those devoid of Thr[203]) usually display neither structural relaxation upon chromophore protonation/deprotonation nor one I* state distinguishable from B*. In the following treatment, we shall consider the pH dependence of the simple ground-state A ↔ B equilibrium, implicitly assuming that the structurally/environmentally "non-relaxed" anionic species I (if present), negligibly contribute to the protein absorption spectra on account of their metastable character and short lifetime.

10.2.2.3 Fluorescence Dependence upon pH: Simple One-Site Model

The optical difference of A and B states is at the basis of the use of GFP variants as pH indicators (Kneen et al. 1998). Indeed, a series of distinguishable absorption and fluorescence spectra are obtained as A and B interconvert into each other upon pH changes (Figure 10.2a and b). This behavior is amenable to a simple mathematical description, if: (1) each state maintains its optical characteristics when the pH is varied, and (2) a linear relationship holds between fluorescence and concentration (typical of diluted solutions). Let K be the generic equilibrium constant of the ionization of the phenol group (Scheme 10.2, **i ↔ ii**), the concentrations of the A and B states can be calculated from the mass action law as:

$$[A] = C_0 \cdot \frac{10^{(pK-pH)}}{1+10^{pK-pH}} \tag{10.1}$$

$$[B] = C_0 - [A] \tag{10.2}$$

where C_0 is the overall concentration of the protein. According to our previous hypotheses, the fluorescence excited at λ_x and emitted at λ_e will depend on the pH as

$$F(\lambda_x, \lambda_e) = C_0 \cdot \left(\frac{F_B(\lambda_x, \lambda_e) + F_A(\lambda_x, \lambda_e) \cdot 10^{(pK-pH)}}{1+10^{(pK-pH)}} \right) \tag{10.3}$$

where $F_A(\lambda_x, \lambda_e)$ and $F_B(\lambda_x, \lambda_e)$ are the molar fluorescence emissions of the A and B states, respectively. $F_A(\lambda_x, \lambda_e)$ and $F_B(\lambda_x, \lambda_e)$ account for both the extinction coefficient and quantum yield of the protein as well as the excitation source intensity and collection efficiency of the spectroscopic/imaging apparatus. For most mutants $F_B(\lambda_x, \lambda_e) > F_A(\lambda_x, \lambda_e)$ in a wide interval of wavelengths (this trend being reversed only in proximity to the absorption maximum of A) owing to the larger brightness (the product of extinction coefficient and quantum yield) of B compared to A. Equation 10.3 indicates that the protein fluorescence retains a sigmoidal dependence on pH (Figure 10.2c), in agreement with the typical isotherm for a single-proton dissociation. At pH=pK, the fluorescent signal reaches half of its overall variation [$F_B(\lambda_x$,

λ_e) $- F_A(\lambda_x, \lambda_e)$]. Differentiation of Equation 10.3 also shows that the maximum signal sensitivity to pH changes is found in the pH range [pK – 1; pK+1], where the fluorescence response to pH is nearly linear (maximum deviation: 16%).

10.2.2.4 Fluorescence Dependence upon pH: General Two-Site Model

Some GFP variants display spectral features in contrast with the simple scheme of phenol ionization, although the pH dependence of emission still complies with Equation 10.3. The most striking feature of these variants is the residual presence of the neutral chromophore absorption band at pH » pK (Hanson et al. 2002, Bizzarri et al. 2007). To rationalize the pH-dependent photophysics of all GFPs, we recently proposed a comprehensive protonation model that encompass both the chromophore and a second protonation site X in its proximity (two-site or 2S model, Scheme 10.3) (Bizzarri et al. 2007). The 2S model allows four distinct ground states (A′, A, B, and B′), which correspond to the four possible combinations of the protonation state of the chromophore/X couple. The X site affects the optical characteristics of the protein only if it is thermodynamically coupled to the ionization of the chromophore (the protonation state of X influences the chromophore's pK and vice-versa) (Ullmann 2003). If uncoupled, the X site may play a role in the proton exchange kinetics with the external buffer (Abbruzzetti et al. 2005, Bizzarri et al. 2007), but the optical properties of the protein stems from the simple equilibrium of Scheme 10.2. An inspection of the GFP structure shows that only two residues are proximal enough to the chromophore and ionizable in the stability pH range of the protein: His[148] and Glu[222]. So far, we have not found any mutant for which coupling with the chromophore takes place through X=His[148] (Bizzarri et al. 2007). Indeed, all the "coupled" GFPs we analyzed displayed X=Glu[222] (Bizzarri et al. 2007). Moreover, the latter mutant Glu[222] and the chromophore were found to be so strongly anti-cooperatively coupled (i.e., deprotonation of one forbids the deprotonation of the other) that the fully deprotonated state B′ could not be reached within the stability pH range of the protein. In such a case, it was shown theoretically that the optical response of the protein follows the single ionization of the A′ state (both X and Chro protonated) to an apparent mixed form of A (anionic X, neutral Chro) and B (protonated X, anionic Chro) (Bizzarri et al. 2007). This effect explains why the absorption spectra of these variants at a high

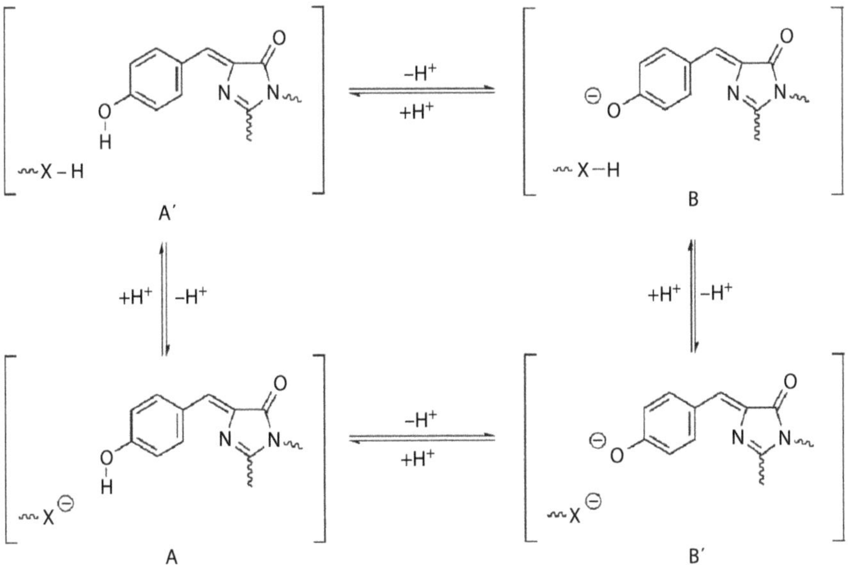

SCHEME 10.3 The 2S model of pH-dependent optically active ground states in GFPs. X represents an ionizable residue nearby the chromophore. If X and the chromophore are uncoupled, only the A′ ↔ B equilibrium determines the optical properties of the protein. Conversely, when the two sites are coupled, mutual inhibition of their ionizations is observed (strong anticooperative coupling). (Modified from Bizzarri, R. et al., *Biochemistry*, 46, 5494, 2007.)

pH state retain features of the neutral chromophore. The strong ionization coupling between Glu[222] and the chromophore is easily understood by considering the presence of two closely spaced negative charges in the B' state, i.e., the anionic Glu[222] and the anionic chromophore. The lack of a significant coupling between His[148] and the chromophore is less intuitive. The shielding of the positive charge on His[148] (A' and B states) from the chromophore can be tentatively ascribed to an H-bonding interaction with the backbone oxygen of the spatially close Asn146, as reported for the S65T mutant at a rather low pH (Elsliger et al. 1999).

So far, the 2S model has been successful in explaining the pH-related optical properties of almost all the GFPs we analyzed experimentally. We noted that thermodynamic coupling arises whenever the Glu[222] residue is not forced into a single protonation state by the H-bonding action of nearby residues, like Thr[65] in S65T GFP (Elsliger et al. 1999). Nonetheless, both "coupled" and "uncoupled" mutants display the same pH-dependent optical behavior pH, described by Equation 10.3. Thus, all GFPs (possessing Tyr[66]) are in principle utilizable as "single-ionization" pH_i indicators. Some "coupled" variants (among which stands wtGFP), however, are characterized by a powerful buffering effect of the Glu[222] residue on chromophore ionization, resulting in a substantial independence of the optical response from the pH in the physiological range. Therefore, these mutants are unsuitable for pH sensing *in vivo*.

10.2.3 Kinetics of GFP Protonation and Resolution of pH_i Imaging Measurements

As described in Section 10.1, pH_i is never constant throughout the cellular body, as different biochemical mechanisms concur to determine the pH values at different locations. Thus, a good pH indicator must be capable of reporting an accurate estimate of pH_i in every subcellular region it comes into, i.e., to provide a realistic pH_i map. So far, we have considered the influence of pH on the optical properties of the indicator (here a GFP) at the *thermodynamic equilibrium* (Equation 10.3). Nothing was said about the characteristic time (τ_c) required to reach the equilibrium state. This value is particularly important as the fluorescent indicator undergoes translational diffusion while monitoring the pH. If the diffusion is much faster than the kinetic relaxation to equilibrium, then the measurement of pH_i could be biased, unless the pH_i is rather homogeneous. In the following we shall treat this effect analytically.

Let us consider the equilibrium between CroH and Cro, indicating the neutral and the anionic chromophore, respectively. This notation is fully compatible with the general 2S-model. Relaxation analysis leads to the following mathematical linkage between τ_c and the kinetic parameters of the CroH/Cro proton exchange:

$$\frac{1}{\tau_c} = k_{on} \cdot \left([Cro]_{eq} + [H^+]_{eq} \right) + k_{off} \tag{10.4}$$

where

$[Cro]_{eq}$ and $[H^+]_{eq}$ are the concentrations of Cro and H$^+$ at equilibrium

k_{on}, k_{off} are the protonation and the deprotonation rate constants, respectively ($K = k_{off}/k_{on}$)

Thus, the larger the rate constants and/or the anionic chromophore concentration, the lower the τ_c. Now, in a time τ_c we must consider that a freely diffusing molecule would travel a linear distance Δw described by (Sonnleitner et al. 1999):

$$\Delta w = \sqrt{4D\tau_c} \tag{10.5}$$

where D is the diffusion constant of the molecule. Therefore, Δw will represent the minimum distance between two space points whose pH_i values can be accurately distinguished, i.e., the resolution of the pH measurement. We shall refer to it as the *kinetic resolution*.

Equation 10.11 shows that increasing the diffusion of the probe (measured by *D*) leads to a larger Δw value. In this perspective, GFP-based pH_i indicators ($D \approx 0.1$–$20\ \mu m^2/s$ (Digman et al. 2005b, Cardarelli et al. 2007, 2008) appear much more advantageous compared to organic probes ($D \approx 100$–$300\ \mu m^2/s$ (Digman et al. 2005b)). For a freely diffusing GFP in the cell cytoplasm (most unfavorable case), $D = 20\ \mu m^2/s$ (Cardarelli et al. 2007) and $\Delta w \approx 9 \cdot (\tau_c)^{0.5}\ \mu m$. Kinetic characterization of a large set of GFP mutants has shown that τ_c is about 1–1.5 ms at the maximum for concentrations similar to those found *in vivo* (Abbruzzetti et al. 2005, Bizzarri et al. 2007). Hence, we have $\Delta w \approx 0.29\ \mu m$. The optical resolution of a microscope system is calculable approximately from Abbe's equation (Hell 2003):

$$\Delta d = \frac{\lambda}{2 \cdot NA} \tag{10.6}$$

where
   λ is the excitation wavelength
   NA is the numeric aperture of the objective

Considering $NA = 1.25$ and $\lambda = 488$ nm (typical imaging conditions of green mutants) we have: $\Delta d \approx 200$ nm, a value quite close to the kinetic resolution. Slowly diffusing GFP constructs ($D \approx 0.1$–$5\ \mu m^2/s$) display a kinetic resolution well below the optical resolution. These examples demonstrate that the fast protonation kinetic and the rather slow diffusion typical of GFP mutants really allow for high-resolution pH_i monitoring *in vivo*. For comparison, pH-dependent organic probes display protonation kinetics faster than diffusion only up to pH 6 (Widengren et al. 1999, Charier et al. 2005).

10.3 GFP-Based pH Indicators

10.3.1 Ratiometric Fluorescent Indicators

Equation 10.3 shows that the emission intensity of a pH indicator depends on its total concentration C_0. Therefore, it is difficult to determine whether observed changes in fluorescence stems from pH changes or indicator concentration. In living specimens, it is nearly impossible to control protein expression (or organic indicator concentration) in several cells at the same time. Additionally, for non-confocal imaging systems, the fluorescence response of the indicator may be affected by the cell thickness, as the probe excitation is proportional to the optical path length. *Ratiometric indicators*, which do not require an independent means of measuring the probe concentration, represent a remarkable way to circumvent all these problems. The theoretical analysis of a general ratiometric pH indicator (here a GFP), in a form equivalent to that described by Grynkiewicz for Ca^{2+}-sensors (Grynkiewicz et al. 1985), is presented in the following.

Let us consider a GFP mutant whose fluorescence obeys Equation 10.3 and two sets of excitation–emission wavelengths or wavelength intervals (λ_{x1}, λ_{e1}) and (λ_{x2}, λ_{e2}). If we take the ratio of fluorescence in these two sets we have:

$$\frac{F(\lambda_{x1}, \lambda_{e1})}{F(\lambda_{x2}, \lambda_{e2})} = \left(\frac{F_\infty(\lambda_{x1}, \lambda_{e1}) + F_0(\lambda_{x1}, \lambda_{e1}) \cdot 10^{(pK-pH)}}{F_\infty(\lambda_{x2}, \lambda_{e2}) + F_0(\lambda_{x2}, \lambda_{e2}) \cdot 10^{(pK-pH)}} \right) \tag{10.7}$$

where the notation F_A and F_B of Equation 10.3 has been replaced by F_0 (a lower molar fluorescence asymptote) and F_∞ (higher molar fluorescence asymptote), in keeping with the more general scheme of the 2S model (Section 10.2.2.4). Equation 10.7 can be recast as:

$$R[1, 2] = R_0[1, 2] \cdot \left(\frac{R_f[1, 2] + 10^{(pK'-pH)}}{1 + 10^{(pK'-pH)}} \right) \tag{10.8}$$

where

$$R[1, 2] = \frac{F\left(\lambda_{x1}, \lambda_{e1}\right)}{F\left(\lambda_{x2}, \lambda_{e2}\right)} \tag{10.9}$$

$$R_0[1, 2] = \frac{F_0\left(\lambda_{x1}, \lambda_{e1}\right)}{F_0\left(\lambda_{x2}, \lambda_{e2}\right)} \tag{10.10}$$

$$R_f[1, 2] = \frac{F_\infty\left(\lambda_{x1}, \lambda_{e1}\right)}{F_0\left(\lambda_{x1}, \lambda_{e1}\right)} \cdot \frac{F_0\left(\lambda_{x2}, \lambda_{e2}\right)}{F_\infty\left(\lambda_{x2}, \lambda_{e2}\right)} \tag{10.11}$$

$$pK' = pK - \log\left[\frac{F_\infty\left(\lambda_{x2}, \lambda_{e2}\right)}{F_0\left(\lambda_{x2}, \lambda_{e2}\right)} \right] \tag{10.12}$$

$R[1, 2]$ is called the *ratiometric fluorescence signal* of sets #1 and #2. Most frequently, ratiometric indicators work either *by excitation* ($\lambda_{x1} \neq \lambda_{x2}$, $\lambda_{e1} = \lambda_{e2}$) or *by emission* ($\lambda_{x1} = \lambda_{x2}$, $\lambda_{e1} \neq \lambda_{e2}$).

Notably, Equations 10.3 and 10.8 have the same functional dependence upon pH. Thus, a plot of $R[1, 2]$ vs pH would show a sigmoidal shape whose lower asymptote is represented by R_0 (the *ratiometric offset*), the amplitude by R_f (the *dynamic range*), and the mid-point by pK' (the *ratiometric pK*) (Figure 10.2d). Differently from Equation 10.3, however, Equation 10.8 does not retain the dependence from the concentration C_0. Furthermore, the ratio $R[1, 2]$ is independent from geometrical features, such as cell or specimen thickness, as well as from general fluorescence variations due to photobleaching effects. Accordingly, it is often said that Equation 10.8 describes a *general or universal calibration curve* under the selected excitation/observation conditions.

R_0 is affected by instrumental characteristics such as the excitation intensity and the detector efficiency, as it represents a fluorescence ratio taken adopting two specific excitation/emission optical sets (Equation 10.10). Instead, Equations 10.11 and 10.12 show that R_f and pK' depend only on the photophysical/thermodynamic properties of the fluorescent protein and on the selection of the excitation/emission sets (λ_{x1}, λ_{e1}) and (λ_{x2}, λ_{e2}). The nature of set (λ_{x2}, λ_{e2}) is extremely important, as it contributes to defining pK', the mid-point of the fluorescence response upon pH (Equation 10.12). Thus, a careful selection of this set is needed to tailor the pH indicator to the desired biological application(s), as the maximum sensitivity of a ratiometric signal to pH occurs in the range [$pK' - 1$; $pK + 1$], where $R[1, 2]$ vs pH is nearly linear.

Only those fluorescent probes characterized by multiple excitation or emission maxima that show opposing changes in fluorescence excitation or emission in response to pH ($R_f \neq 1$) are utilizable as ratiometric pH indicators ($R[1, 2]$ changes with pH). Ideally, GFP mutants are excellent ratiometric indicators by excitation, owing to the large absorption and the excitation difference between the neutral and the anionic chromophore. A restricted number of variants also show significant alteration of the emission spectra upon pH, and can in principle be used as ratiometric indicators by emission. Unfortunately, most GFP mutants display pK values at the margins of the physiological range and/or poor emissivity of the neutral chromophore, and cannot be employed to monitor ratiometrically pH$_i$. We shall discuss the GFP-based ratiometric pH indicators that are really utilizable *in vivo* in Section 10.3.2.2.

10.3.2 GFP-Based pH Indicators Applied *In Vivo*

The first GFP mutant to monitor pH_i was introduced more than 10 years ago (Kneen et al. 1998). Since then, many GFP variants were brought to the attention of the scientific community as genetically encodable pH_i indicators, which in some cases provide new insights into subtle biochemical processes for which the proton concentration has great relevance. The engineered GFP-based pH indicators can be classified into two general families: non-ratiometric and ratiometric. In the following, we shall review the most efficient indicators in each of these two families.

10.3.2.1 Nonratiometric pH Indicators

This family encompasses a large number of GFPs that display good pH-responsiveness in the physiological range and poor emission from the neutral chromophore, and are therefore unsuitable for ratiometric measurements. Notably, these indicators have been used mostly to report *changes* of pH_i rather than absolute pH_i values, owing to the absence of a general calibration curve (Section 10.3.1). Verkman and coworkers introduced in 1998 the F65L/S65T GFP variant (enhanced GFP or EGFP) as a sensitive pH indicator (Kneen et al. 1998). EGFP is characterized by a high emissivity of the anionic state in the green ($\varepsilon_{488}=60,000$ M$^{-1}\cdot$cm^{-1}, $\Phi=0.7$, $\lambda_e=509$ nm, ref. Bizzarri et al. 2007), although its pK (5.8) makes it unsuitable to monitor alkaline subcellular components (e.g., mitochondria). Nonetheless, EGFP has been used to monitor pH_i variations in the cell cytoplasm (Kneen et al. 1998), in the Golgi apparatus (Kneen et al. 1998), and in the synaptic vesicle cycling at nerve terminals (Sankaranarayanan et al. 2000). Recently, EGFP has been proposed as an effective pH_i indicator by using Fluorescence Lifetime Imaging (Nakabayashi et al. 2008).

Several rounds of random mutagenesis yielded two pH-sensitive GFPs, called "Ecliptic" and "Superecliptic" pHlourins (EcGFP and sEcGFP). EcGFP and sEcGFP were demonstrated to have wider applicability as green-emitting pH_i indicators than EGFP, as they display p$K=7.1$–7.2 (Miesenbock et al. 1998, Sankaranarayanan et al. 2000). Additionally, the neutral chromophore form of EcGFP and sEcGFP is almost non-emissive, allowing for a very sensitive detection of biological processes associated with a pH increase. EcGFP and sEcGFP have been proposed as general markers of cell exocytosis (Nakabayashi et al. 2008). Indeed, these mutants were applied to investigate the exocytosis of presynaptic secretory vesicles down to a single event resolution, on account of the fluorescence boost occurring upon a synaptic vesicle fusion to the plasma membrane and the exposure of the lumen, previously acidic, to the neutral extracellular pH (Miesenbock et al. 1998, Sankaranarayanan et al. 2000). In spite of their broad use, no extensive photophysical characterization of these two proteins has been reported, although Fluorescence Correlation Spectroscopy (FCS) experiments supported the existence of a second protonation site near the chromophore in EcGFP, in agreement with the 2S model (Section 10.2.2.4) (Hess et al. 2004).

S65G/S72A/T203Y GFP (EYFP) is a pH responsive variant that has a redshifted absorption and emission of the anionic chromophore with respect to conventional GFPs (Wachter et al. 1998). Likewise, the ecliptic mutants, also the neutral chromophore of EYFP is almost unemissive. Conversely, the anionic chromophore has a high emissivity in the green-yellow region of the spectrum ($\varepsilon_{514}=84,000$ M$^{-1}\cdot$cm^{-1}, $\Phi=0.61$, $\lambda_e=527$ nm, ref. Patterson et al. 2001). On account of its p$K=7.1$, EYFP was shown to be suitable for cytosolic, Golgi, and mitochondrial matrix pH_i measurements by Tsien and coworkers (Llopis et al. 1998). The yellowish fluorescence makes EYFP a good selection for multicolor experiments in tandem with pH-unresponsive cyan or green mutants. Unfortunately, the EYFP emission is severely quenched by chloride (Wachter et al. 2000, Jayaraman et al. 2000), a species rather abundant in some cell types and subcellular organelles. Attempts at reducing the chloride-sensitivity of EYFP by introducing mutations in the chloride-binding pocket afforded mutants with a rather low pK unsuitable for pH monitoring *in vivo* (Griesbeck et al. 2001).

10.3.2.2 Ratiometric pH Indicators

In spite of their relevance, so far only few ratiometric pH indicators based on GFPs have been described. The first, and perhaps most popular one, was firstly described in 1998 by Rothman and coworkers in the same study that introduced EcGFP (Miesenbock et al. 1998). They found that a S202H GFP variant, named Ratiometric pHlourin (RaGFP), displayed a strong increase of the 475 nm excitation band and a decrease of the 395 excitation band upon a pH shift from 7.5 to 5.5. The reasons for which RaGFP is characterized by a pH-dependent optical behavior at odds with all the other GFPs, i.e., the decrease of the anionic chromophore band upon pH rise, have not been elucidated yet.

RaGFP was used as a ratiometric indicator by excitation to measure dynamically the pH of various intracellular compartments, such as the cytoplasm (Karagiannis and Young 2001), peroxisomes (Jankowski et al. 2001), endosomes and the *trans*-Golgi network (Machen et al. 2003), and the presynaptic secretory vesicles (Miesenbock et al. 1998). In a typical experiment, RaGFP is imaged by taking the ratio of the 500–550 nm emission excited at 410 and 470 nm (Karagiannis and Young 2001, Jankowski et al. 2001). Schulte and coworkers reported that RaGFP displays $pK'=6.9$ and $R_f=8.8$ for $\lambda_{x1}=415$ nm, $\lambda_{x2}=475$ nm, and $\lambda_e=508$ nm (Schulte et al. 2006). In the same work, the ratiometric characteristics by the emission of EcGFP were also described. Upon excitation at 400 nm, EcGFP emits at 464 and 511 nm at low and high pH, respectively. Adopting the ratiometric sets (400, 511 nm) and (400, 464 nm), it was found $R_f=28$ and $pK'=7.6$ (Schulte et al. 2006).

In 2002, James Remington and his group introduced four new ratiometric pH indicators, named deGFPs, characterized by the S65T and the H148C(G) and/or the T203C amino acid replacements (Hanson et al. 2002). pK values were found to range from 6.8 to 8.0 (Hanson et al. 2002). The photophysical characterization of deGFPs (Hanson et al. 2002) showed that they retain a significant absorption of the neutral state at a high pH, suggesting a coupling between residue E222 and the chromophore (Bizzarri et al. 2007). Remarkably, at a low pH, excitation of deGFPs at 400 nm resulted in a broad blue fluorescence centered at 460 nm ($\Phi=0.05$–0.1); conversely, at a high pH, the 400 nm-excitation led to green fluorescence ($\lambda_e=516$–518 nm, $\Phi=0.15$–0.35). Ultrafast fluorescence upconversion spectroscopy studies attributed the blue fluorescence to the direct emission of the excited neutral chromophore, whereas the green fluorescence resulted from an efficient ESPT process (McAnaney et al. 2002). Crystal structure analysis allowed the identification of a pH-induced structural rearrangement in the chromophore pocket, presumably at basis of the observed dual emission characteristics (Hanson et al. 2002). In more detail, at a low pH, interruption of the proton relay towards Glu[222] typical of wtGFP forbids ESPT and fluorescence comes only from the excited neutral chromophore. Raising the pH triggers a backbone motion placing Tyr145 and Ser147 in a favorable position to form hydrogen bonds with the chromophore hydroxyl; this configuration established a novel proton relay involving Ser147 and two water molecules and allows ESPT towards the bulk solvent. This result offers a guiding principle for the engineering of GFP-based ratiometric pH indicators. Indeed, the fine modulation of the ESPT rate by a rationale modification of the primary sequence appears a very effective method to change the emission properties of GFP mutants (McAnaney et al. 2002, Shu et al. 2007a,b).

The deGFP4 variant was evaluated as the intracellular ratiometric pH indicator by emission ($\lambda_x=365$ nm, $\lambda_{e1}=475$–525 nm, $\lambda_{e2}=385$–470 nm) in the course of the same study. Although no ratiometric parameters were reported, deGFP4 was shown to possess a much greater dynamic range than the popular SNARF-1 dye (Cody et al. 1993, Dubbin et al. 1993). The emission ratio, however, was rather noisy because of the weak fluorescence signal in the blue wavelength range (Hanson et al. 2002). deGFP4 was also tested as a ratiometric indicator by emission under a two-photon excitation, a topic that will be discussed in Section 10.3.3.2.

In 2006 our group introduced the F64L/S65T/T203Y mutant, named E²GFP, as a very effective and versatile ratiometric pH indicator for intracellular study (Bizzarri et al. 2006). Similarly to deGFPs, E²GFP displayed a proton-dependent optical behavior affected by coupling between the chromophore and the Glu[222] ionization (Bizzarri et al. 2007). Differently from deGFPs, however, the neutral chromophore at a low pH was found to undergo efficient ESPT upon excitation ($\lambda_e=510$ nm, $\Phi=0.22$), although the emission maximum was still significantly blueshifted compared to the emission generated at a higher pH ($\lambda_e=523$ nm, $\Phi=0.91$, Figure 10.3, right). Surprisingly, the crystallographic analysis of the ground-state

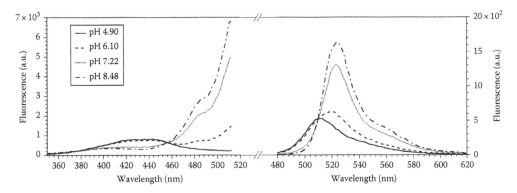

FIGURE 10.3 (*left*) Excitation (λ_e=523 nm) and (*right*) emission (λ_x=473 nm) of E²GFP at four different pH values. The isosbestic points in both series clearly indicate a pH-dependent equilibrium between two forms.

forms and the kinetic pH-jump measurements did not highlight any major structural rearrangement upon chromophore ionization (Bizzarri et al. 2007). These data suggest an ESPT mechanism involving changes in the polarity/electronic distribution of the chromophore cavity upon the A* deprotonation. The emission blueshift could be due to the establishment of a H-bonding interaction between the anionic phenol group in the excited state and a residue nearby, possibly His[148].

We demonstrated that E²GFP is capable of reporting on intracellular pH ratiometrically by emission with a high dynamic range (R_f=5.4 and pK'=7.5 *in vivo* for λ_x=458 nm, λ_{e1}=515–600 nm, λ_{e2}=475–525 nm). Also, E²GFP was shown to be an excellent ratiometric pH indicator by excitation tailored to imaging setups equipped with popular Ar-laser excitation sources.

The presence of an isosbestic point at 460 nm in the pH-plot of the excitation spectra (Figure 10.3, left) made us select the 458 nm Ar-line as λ_{x2}, because at this wavelength $F_\infty \approx F_0$ and Equation 10.9 predict a minimal deviation of pK' from pK (6.9 for E²GFP). Adopting λ_{x1}=488 nm and λ_e=500–600 nm we obtained a large dynamic range of measurement *in vivo* (R_f=10–12) and we were able to observe the cellular alkalinization upon mitosis and the subsequent acidification upon entry in the G1 phase (Bizzarri et al. 2006). E²GFP linked to the transactivator protein of HIV-1, Tat, allowed us to determine the pH in nucleolar and promyelocytic leukemia protein (PML) bodies with a high spatial resolution (Bizzarri et al. 2006). We also observed a significant pH change upon Tat relocation that might be biologically related to the control of HIV-1 transcription (Marcello et al. 2004). Although chloride (and other halides) ions were found to be static quenchers of E²GFP fluorescence (Arosio et al. 2007), the mathematical dependence of fluorescence from the quencher concentration leads to unaffected ratiometric measurements (Figure 10.2d) (Bizzarri et al. 2006).

Recently, our group developed a new ratiometric pH indicator by emission, named E¹GFP, which is spectroscopically similar to E²GFP but more tailored to mildly acidic intracellular compartments (pK'=6.4–6.6). Also, E¹GFP fluorescence was found to be rather unaffected by chloride ions (Arosio 2007). In the same study we were able to demonstrate that E¹GFP is capable of monitoring the vesicle pH in real-time during multi-step endocytosis through to the intracellular endocytic network (Figure 10.4) (Serresi et al. 2009).

So far we have discussed only ratiometric indicators constituted by a single GFP molecule. In 2001, Takeo Awaij and coworkers developed two ratiometric pH sensors (GFpH and YFpH) by fusing in tandem two GFP variants with different pH sensitivities (Awaji et al. 2001). GFpH was constituted by GFPuv, a mutant that displays a low pH-sensitivity and is excitable at 380 nm, and EGFP, excitable at 480 nm. GFpH was observed to emit at 510 nm and displayed pK'=6.2 and $R_f \approx$ 5, suggesting its use as a ratiometric pH indicator by excitation for mildly acidic subcellular components (Awaji et al. 2001). YFpH was constituted by GFPuv and EYFP. Upon a 380 nm excitation, the emission of YFpH was shifted from 509 to 527 nm as the pH was raised. This effect was interpreted as FRET between GFPuv

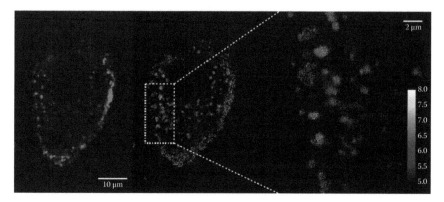

FIGURE 10.4 (See color insert following page 15-30.) Real-time monitoring of endocytosis of Tat- E^IGFP in HeLa cells (Serresi et al. 2009). Images were taken 4 h after the exposure of cells to Tat-E^IGFP. Left image: fluorescence of Tat-E^IGFP on the membrane of a cell and internalized in endocytotic vesicles. Central image: a ratiometric pH_i map of the same cell. Right image: a magnified ratiometric pH_i map of the area enclosed in the white dotted rectangle shown in the central image. The pH_i maps are color-coded according to the lookup table placed on the extreme right of the figure. Note the higher pH experienced by Tat-E^IGFP on the cell membrane (around neutrality) compared to the one associated with internalized vesicles (5.8–6.5). The progressive acidification upon internalization is a hallmark of the dynamic evolution of organelle contents from endosomes and pinosomes to the degradative lysosomal compartments. (Based on Serresi, M. et al., *Anal. Bioanal. Chem.*, 393, 1123, 2009.)

and EYFP, as EYFP is not fluorescent when excited at 380 nm. Conversely, upon 480 nm excitation the emission of YFpH was nearly unchanged, as it derived mostly from EYFP. On account of this behavior, YFpH was proposed as a ratiometric pH indicator by both excitation and emission. The pK' of YFpH was shown to be around 6.5–6.8 in both cases (Awaji et al. 2001). GFpH and YFpH were used to visualize pH changes in the cytosol/nucleus of living cells and during internalization caused by endocytosis upon agonist stimulation (Awaji et al. 2001).

Cyan fluorescent protein and EYFP are two variants widely used for *in vivo* FRET experiments, owing to their large optical complementarity (Chan et al. 2001, Wlodarczyk et al. 2008). Differently to EYFP, CFP is not very sensitive to pH (Llopis et al. 1998), and therefore the CFP-EYFP (or EYFP variants) couple was proposed as a ratiometric pH indicator by excitation and/or emission. CFP-EYFP was shown to possess $pK=6.5$ and was used to detect changes in pH secondary to H^+ efflux into the basolateral space of MDCK cells (Urra et al. 2008). Remarkably, the EYFP sensitivity to Cl^- was exploited to engineer CFP-EYFP ratiometric indicators of both proton and chloride ions (Markova et al. 2008).

Pozzan and coworkers developed an indicator by excitation for the high-pH range (mtAlpHi) by replacing the calmodulin linker of the Ca^{2+}-sensor camgaroo-II (Griesbeck et al. 2001) with a portion of Aequorin. mtAlpHi was used to monitor the pH changes occurring in a variety of physiological and non-physiological situations in mitochondria (Abad et al. 2004). Notably, mtAlpHi (and its parent construct camgaroo-II) comprises an EYFP variant that is poorly sensitive to the quenching effect of chloride ions (Griesbeck et al. 2001).

10.3.3 pH_i Measurement: Instrumentation and Methods

10.3.3.1 Confocal Microimaging Setups

The detection of pH in cultured cells and living organisms has greatly benefited from the recent advancements in microscopy methods. High-resolution pH_i determination requires the high-sensitivity detection of fluorescence, regardless of the microimaging setup. Nonetheless, the major breakthrough in this field was represented by the diffusion of confocal microscopes (Cody et al. 1993, Tsien 2003, Paddock

2008). In a confocal microscope, detection of out-of-focus fluorescence and optical blurring is strongly reduced by the presence of a pinhole in an optically conjugate plane in front of the detector (Diaspro et al. 2006). By means of this configuration, the image quality and the axial resolution are greatly improved compared to conventional (e.g., wide-field) epifluorescence microscopes. The axial resolution (z-axis) can be as low as 0.6–1.5 μm (the *xy* resolution is usually around 0.2–0.5 μm) for measurements carried out with high numerical-aperture objectives. Accordingly, confocal microscopy allows for pH_i detection in focal volumes of less than 1 fL yielding, in principle, high-resolution pH maps in living organisms such as cultured cells.

The properties of the fluorescent probe, in particular its ratiometric capability, are extremely relevant for the selection of the microscopy setup best suited to pH_i measurements. As already stated, non-ratiometric pH indicators are usually applied to monitor pH variations *in vivo* triggered by some biochemical stimuli. These measurements require a stable excitation source as well as the capability to distinguish fluorescence-emission changes in the presence of a strong background noise and/or auto-fluorescence. Thus, any means aimed at maximizing the signal-to-noise ratio (SNR) must be pursued. The best approach involves the use of very sensitive detectors to reveal weak signals, high photon fluxes of excitation (compatible with the bleaching properties of the indicator), and highly selective excitation/dichroic/emission filters.

The measurements by ratiometric indicators need additional considerations. First of all, the micro-imaging system must be flexible in selecting, and fast in switching the excitation sources and/or the collection intervals, in dependence on the ratiometric optical nature of the probe. Nowadays, standard confocal microscopes are usually supplied with multiple excitation laser sources that can be activated sequentially in less than 100 ms, and are therefore well-suited for the excitation ratiometry experiment. It is worth highlighting again that E^2GFP is an optimized indicator for confocal imaging, as it works at its best when excited at 458 and 488 nm, two strong lines of the Ar laser that represent the typical laser source for confocal setups (Bizzarri et al. 2006). Some wide-field microscopes can be supplied with motorized filter wheels that allow the fast switching of excitation wavelengths from a lamp source. In this case, the excitation interval is tunable, although the high axial resolution of confocal setups is lost (and can be recovered only by postprocessing deconvolution).

Confocal microscopes are usually equipped with multiple detectors (photomultipliers, SPADs), permitting the concomitant collection of fluorescence in two or more wavelength intervals, which is the basic requirement of ratiometry by emission. In the most advanced setups, each detector can be tuned to a desired wavelength range of fluorescence collection directly from the software, on account of a complex system of motorized gratings and slits within the microscope (bandwidths can be as low as 5 nm). This instrumental feature is particularly interesting in view of modulating both the dynamic range and the apparent pK' of the ratiometric response (Equations 10.10 and 10.11).

Minimization of the detector noise (image background) is of utmost relevance for ratiometric measurements. Indeed, the ratio between fluorescence values amplifies the electronic error present in each image. The nonlinear (sigmoidal) shape of the calibration curve further amplifies this error when the ratio is converted to pH. This problem is particularly severe for those detectors that have an intrinsically high noise level, such as SPAD, and samples characterized by dim fluorescence. We should point out that the use of photon-counting detectors represents an interesting option to carry out ratiometric measurements, as they are characterized by predictable and low-level Poissonian noise (Digman et al. 2008). Unfortunately, photon-counting detectors are usually not components of standard instruments but rather part of complex microscopy setups devoted to high-sensitivity biophysical measurements.

The easiest, and in many cases the unique way to reduce noise effects on ratiometry is by the use of the maximum excitation intensity compatible with the minimal probe photobleaching. After collection, adequate thresholding of images (by means of some reproducible criteria, e.g., the subtraction of the background plus a fixed multiple of its standard deviation) is essential to obtain a meaningful pH map. The adoption of some pH_i benchmarks such as cells clamped at a given pH by means of ionophores is a further help to validate the measurements of interest.

10.3.3.2 Two-Photon Microimaging Setups

Two-photon microscopy is a laser scanning technique that exploits a nonlinear (quadratic) sample excitation by means of two photons of half energy compared to one photon excitation. On account of its nonlinearity, two-photon excitation (TPE) leads to fluorescence production only within the focal region of the excitation beam (intrinsic confocality). For this property, and for the deep penetration associated to longer wavelength radiation, two-photon microscopy has found a wide use in biological imaging as the best noninvasive means of fluorescence microscopy in tissue explants and living animals (Zipfel et al. 2003, Svoboda and Yasuda 2006). The intrinsic confocality of TPE is also thought to reduce phototoxicity and photobleaching with respect to single-photon excitation. Notably, many commercially available confocal setups can be easily upgraded to multiphoton imaging by interfacing a femto- or pico-second mode-locked infrared laser source (Zipfel et al. 2003). Some of these sources are totally controlled by software and do not require time-consuming manual tuning/mode-locking of the excitation wavelength.

The remarkable characteristics of TPE and the recent fast-growing development of two-photon microscopy methods prompted the development of new fluorescent indicators that respond to nonlinear excitation. In this sense, pH indicators make no exceptions. As anticipated in Section 10.3.2.2, Remington and coworkers validated some deGFPs as two-photon ratiometric pH indicators by emission (Hanson et al. 2002). Significantly, TPE was demonstrated as giving a superior SNR when compared to one-photon excitation in the near UV region (364 nm). This effect was attributed to less autofluorescence under TPE and/or more efficient excitation of the protonated state of deGFP4.

Analogously to deGFP, E^2GFP can act as a two-photon ratiometric pH indicator, by both excitation and emission. The first studies were carried out by Chirico et al., who analyzed the fluorescence of single E^2GFP molecules trapped inside silica gels (Chirico et al. 2004, 2005). Notably, they found $pK \approx 6.5$ from the analysis of single proteins emitting from the neutral and the anionic state as a function of pH. Recently we extended these experiments also to the E^2GFP solution using a tunable IR laser coupled to a confocal microscope. A preliminary excitation (i.e., plot of excitation cross section $\sigma_{\mathrm{GFP}}^{\mathrm{TPE}}$ vs wavelength) and emission spectra are reported in Figure 10.5 for the pH 4.9, 7 and 9.5. Both spectra were normalized by using fluorescein at pH ~12 as the standard. σ_{TPE} has been obtained as:

$$\sigma_{\mathrm{GFP}}^{\mathrm{TPE}} = \frac{C_{\mathrm{st}}\left(F_{\mathrm{GFP}} - F_{\mathrm{back}}\right)}{C_{\mathrm{GFP}}\left(F_{\mathrm{st}} - F_{\mathrm{back}}\right)} \sigma_{\mathrm{st}}^{\mathrm{TPE}} \tag{10.13}$$

where

F_{GFP} and F_{st} are the detected fluorescence intensities of the sample and of the fluorescein

F_{back} is the background signal

C_{GFP} and C_{st} are the protein and fluorescein concentrations and $\sigma_{\mathrm{st}}^{\mathrm{TPE}} = 0.9 \cdot \sigma_{\mathrm{st}}^{\mathrm{TPA}}$ where $\sigma_{\mathrm{st}}^{\mathrm{TPA}}$ is the fluorescein two-photon absorption cross section and 0.9 is its quantum yield (Albota et al. 1998)

This calibration method does not require the knowledge of the second-order temporal coherence of the beam, which depends on the spatial and temporal intensity profile of the excitation pulses (Albota et al. 1998, Heikal et al. 2001).

The TPE spectra of E^2GFP (Figure 10.5) are similar to those measured upon one-photon excitation at half wavelength (Figure 10.3, left). As expected for the chromophore ionization, σ_{TPE} below 880 nm was found to decrease as the pH was raised (and was attributed to the neutral chromophore), while $\sigma_{\mathrm{GFP}}^{\mathrm{TPE}}$ above 930 nm showed the opposite trend (and was attributed to the anionic chromophore). These findings demonstrate the suitability of E^2GFP as a two-photon pH indicator ratiometric by excitation.

Furthermore, the emission band by excitation at 780 nm was found to redshift upon a pH increase. The fluorescence ratio emitted in the 520–550 nm and 475–505 nm ranges (Figure 10.5, shaded areas)

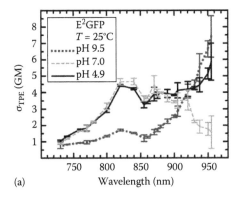
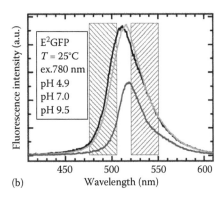

(a)   (b)

FIGURE 10.5 (a) TPE cross-sections σ_{TPE} of E^{2}GFP at pH = 4.9, 7.0, and 9.5 over the 730–960 nm excitation range. Fluorescence emission was collected in the range 450–650 nm and 1 GM = 10^{-50} cm^4 s/photon. Error bars represent standard deviations of 4–10 measurements, taken at different excitation intensities. (b) TPE-fluorescence emission spectra of E^{2}GFP excited at 780 nm and at pH = 4.9, 7.0, and 9.5. The shaded areas show two emission wavelength ranges suitable to use E^{2}GFP as a TPE ratiometric pH indicator by emission.

increased by a factor of 3 when pH was shifted from 4.9 up to 9.5. These data indicate that E^{2}GFP could be suitable also as a two-photon pH indicator ratiometric by emission. Further experiments will be required, however, to validate the use of E^{2}GFP as a TPE pH indicator in living cells.

10.4 Summary and Future Perspectives

Fluorescent proteins, particularly those belonging to the *Aequorea victoria* family (GFPs), are the most relevant fluorescent probes to monitor biochemical processes *in vivo*, owing to the genetic encoding of fluorescence in the protein primary sequence. Accordingly, fusion constructs of GFPs and target proteins are easily engineered by standard methods of molecular biology.

The presence of a phenol ionizable group on the GFP chromophore leads to two protonation ground states (neutral and anionic) characterized by considerably different absorption and fluorescence excitation (and sometime, emission) spectra. This property makes GFP mutants ideally suited to report on intracellular pH. Although in several cases the ionization pK and/or the optical characteristics of the neutral state prevent their practical use as pH$_i$ indicators, some mutants were demonstrated to be effective probes of the proton concentration *in vivo*. Additionally, the photophysics of some GFP mutants allowed the engineering of *ratiometric* pH indicators, either by excitation or by emission. Ratiometric indicators represent the most useful class of fluorescent sensors as they allow a universal calibration within a series of experiments.

In this chapter, we discussed the properties and the reported applications of the principal GFP-based pH indicators developed so far. It is worth noting that none of these indicators proved satisfactory for all *in vivo* uses. Thus, the selection of the indicator best suited for a certain application requires a deep knowledge of the photophysical features, taking into account also the characteristics of the imaging setup.

The development of new pH indicators with improved properties represents a stimulating biophysical task, as it encompasses the physicochemical description of the protonation reactions of the chromophore and the nearby residues. Two aspects are of major importance: the thermodynamics of chromophore ionization (i.e., the pK) and the optical properties of the chromophore in the protonated and the anionic states. We recently showed that the "optical" pK (that is the pK associated to the optical response) is often determined by a thermodynamic balance between the ionization reactions of the chromophore and some nearby residues. The introduction/removal of amino acids that stabilize the neutral/anionic forms via H-bonding (e.g., Thr[203] to stabilize the anionic chromophore) is the best way to tune the optical pK.

Tuning the spectral characteristics of the neutral and the anionic chromophore forms is rather difficult, as no true general photophysical schemes are available. The anionic state is usually very bright, and its excitation/emission can be susceptible to modifications of polarity/electronic density associated with the chromophore cavity (e.g., the yellow emission of Y203 mutants (Wachter et al. 1998)). The neutral chromophore is characterized by two emission pathways: direct emission from its excited state ($\lambda_{em} \approx 460$ nm, Bonsma et al. 2005), and ESPT to yield an excited state resembling that of the anionic chromophore ($\lambda_{em} > 500$ nm). Notably, ESPT relies on an efficient molecular scheme that conveys the proton to a final acceptor; hence, the rate of ESPT can be modulated by modifying the amino acids that constitute this proton relay (Stoner-Ma et al. 2008). Changes in the ESPT rate can affect both the wavelength and the yield of emission.

GFP mutants characterized by pH-dependent detectable emission shifts (possibly on account of different ESPT pathways changing pH) may be promising as TPE pH indicators. Owing to the increasing relevance that multiphoton imaging has in the biological field, we believe that many future scientific efforts will be devoted to the development of these kinds of indicators.

Acknowledgments

We thank Dr. Riccardo Nifosì (Scuola Normale Superiore) for stimulating discussions. Authors also gratefully acknowledge the partial financial support of the Italian Ministry for University and Research (FIRB No. RBLA03ER38).

References

Abad, M. F. C., Di Benedetto, G., Magalhaes, P. J., Filippin, L., and Pozzan, T. 2004. Mitochondrial pH monitored by a new engineered green fluorescent protein mutant. *J. Biol. Chem.*, 279: 11521–11529.

Abbruzzetti, S., Grandi, E., Viappiani, C. et al. 2005. Kinetics of acid-induced spectral changes in the GFPmut2 chromophore. *J. Am. Chem. Soc.*, 127: 626–635.

Albota, M. A., Xu, C., and Webb, W. W. 1998. Two-photon fluorescence excitation cross sections of biomolecular probes from 690 to 960 nm. *Appl. Opt.*, 37: 7352–7356.

Arosio, D., Garau, G., Ricci, F. et al. 2007. Spectroscopic and structural study of proton and halide ion cooperative binding to gfp. *Biophys. J.*, 93: 232–244.

Awaji, T., Hirasawa, A., Shirakawa, H., Tsujimoto, G., and Miyazaki, S. 2001. Novel green fluorescent protein-based ratiometric indicators for monitoring pH in defined intracellular microdomains. *Biochem. Biophys. Res. Commun.*, 289: 457–462.

Bell, A. F., He, X., Wachter, R. M., and Tonge, P. J. 2000. Probing the ground state structure of the green fluorescent protein chromophore using Raman spectroscopy. *Biochemistry*, 39: 4423–4431.

Bizzarri, R., Arcangeli, C., Arosio, D. et al. 2006. Development of a novel GFP-based ratiometric excitation and emission pH indicator for intracellular studies. *Biophys. J.*, 90: 3300–3314.

Bizzarri, R., Nifosi, R., Abbruzzetti, S. et al. 2007. Green fluorescent protein ground states: The influence of a second protonation site near the chromophore. *Biochemistry*, 46: 5494–5504.

Bonsma, S., Purchase, R., Jezowski, S. et al. 2005. Green and red fluorescent proteins: Photo- and thermally induced dynamics probed by site-selective spectroscopy and hole burning. *Chem. Phys. Chem.*, 6: 838–849.

Brejc, K., Sixma, T. K., Kitts, P. A. et al. 1997. Structural basis for dual excitation and photoisomerization of the *Aequorea victoria* green fluorescent protein. *Proc. Natl. Acad. Sci. USA*, 94: 2306–2311.

Cardarelli, F., Serresi, M., Bizzarri, R., Giacca, M., and Beltram, F. 2007. In vivo study of HIV-1 Tat arginine-rich motif unveils its transport properties. *Mol. Ther.*, 15: 1313–1322.

Cardarelli, F., Serresi, M., Bizzarri, R., and Beltram, F. 2008. Tuning the transport properties of HIV-1 Tat arginine-rich motif in living cells. *Traffic*, 9: 528–539.

Carnell, L. and Moore, H. P. 1994. Transport via the regulated secretory pathway in semi-intact PC12 cells: Role of intra-cisternal calcium and pH in the transport and sorting of secretogranin II. *J. Cell Biol.*, 127: 693–705.

Chalfie, M., Tu, Y., Euskirchen, G., Ward, W. W., and Prasher, D. C. 1994. Green fluorescent protein as a marker for gene expression. *Science*, 263: 802–805.

Chan, F. K., Siegel, R. M., Zacharias, D. et al. 2001. Fluorescence resonance energy transfer analysis of cell surface receptor interactions and signaling using spectral variants of the green fluorescent protein. *Cytometry*, 44: 361–368.

Chanat, E. and Huttner, W. B. 1991. Milieu-induced, selective aggregation of regulated secretory proteins in the trans-Golgi network. *J. Cell Biol.*, 115: 1505–1519.

Charier, S., Meglio, A., Alcor, D. et al. 2005. Reactant concentrations from fluorescence correlation spectroscopy with tailored fluorescent probes. An example of local calibration-free pH measurement. *J. Am. Chem. Soc.*, 127: 15491–15505.

Chattoraj, M., King, B. A., Bublitz, G. U., and Boxer, S. G. 1996. Ultra-fast excited state dynamics in green fluorescent protein: Multiple states and proton transfer. *Proc. Natl. Acad. Sci. USA*, 93: 8362–8367.

Chirico, G., Cannone, F., Diaspro, A. et al. 2004. Multiphoton switching dynamics of single green fluorescent proteins. *Phys. Rev. E*, 70.

Chirico, G., Diaspro, A., Cannone, F. et al. 2005. Selective fluorescence recovery after bleaching of single E^2GFP proteins induced by two-photon excitation. *Chem. Phys. Chem.*, 6: 328–335.

Chudakov, D. M., Lukyanov, S., and Lukyanov, K. A. 2005. Fluorescent proteins as a toolkit for in vivo imaging. *Trends Biotechnol.*, 23: 605–613.

Cody, S. H., Dubbin, P. N., Beischer, A. D. et al. 1993. Intracellular pH mapping with Snarf-1 and confocal microscopy.1. A quantitative technique for living tissues and isolated cells. *Micron*, 24: 573–580.

Creemers, T. M., Lock, A. J., Subramaniam, V., Jovin, T. M., and Volker, S. 1999. Three photoconvertible forms of green fluorescent protein identified by spectral hole-burning. *Nat. Struct. Biol.*, 6: 557–560.

Diaspro, A., Bianchini, P., Vicidomini, G. et al. 2006. Multi-photon excitation microscopy. *Biomed. Eng. Online*, 5: 36.

Digman, M. A., Brown, C. M., Sengupta, P. et al. 2005b. Measuring fast dynamics in solutions and cells with a laser scanning microscope. *Biophys. J.*, 89: 1317–1327.

Digman, M. A., Dalal, R., Horwitz, A. F., and Gratton, E. 2008. Mapping the number of molecules and brightness in the laser scanning microscope. *Biophys. J.*, 94: 2320–2332.

Dong, J., Solntsev, K. M., and Tolbert, L. M. 2006. Solvatochromism of the green fluorescence protein chromophore and its derivatives. *J. Am. Chem. Soc.*, 128: 12038–12039.

Dubbin, P. N., Cody, S. H., and Williams, D. A. 1993. Intracellular pH mapping with Snarf-1 and confocal microscopy. 2. pH gradients within single cultured-cells. *Micron*, 24: 581–586.

Elsliger, M. A., Wachter, R. M., Hanson, G. T., Kallio, K., and Remington, S. J. 1999. Structural and spectral response of green fluorescent protein variants to changes in pH. *Biochemistry*, 38: 5296–5301.

Fliegel, L. 2005. The Na^+/H^+ exchanger isoform 1. *Int. J. Biochem. Cell. Biol.*, 37: 33–37.

Griesbeck, O., Baird, G. S., Campbell, R. E., Zacharias, D. A., and Tsien, R. Y. 2001. Reducing the environmental sensitivity of yellow fluorescent protein. Mechanism and applications. *J. Biol. Chem.*, 276: 29188–29194.

Grynkiewicz, G., Poenie, M., and Tsien, R. Y. 1985. A new generation of Ca^{2+} indicators with greatly improved fluorescence properties. *J. Biol. Chem.*, 260: 3440–3450.

Hanson, G. T., McAnaney, T. B., Park, E. S. et al. 2002. Green fluorescent protein variants as ratiometric dual emission pH sensors. 1. Structural characterization and preliminary application. *Biochemistry*, 41: 15477–15488.

Heikal, A., Hess, S., and Webb, W. W. 2001. Multiphoton molecular spectroscopy and excited-state dynamics of enhanced green fluorescent protein (EGFP): Acid–base specificity. *Chem. Phys.*, 274: 37–55.

Hell, S. W. 2003. Toward fluorescence nanoscopy. *Nat. Biotechnol.*, 21: 1347–1355.

Hess, S. T., Heikal, A. A., and Webb, W. W. 2004. Fluorescence photoconversion kinetics in novel green fluorescent protein pH sensors (pHluorins). *J. Phys. Chem. B*, 108: 10138–10148.

Hunte, C., Screpanti, E., Venturi, M. et al. 2005. Structure of a Na^+/H^+ antiporter and insights into mechanism of action and regulation by pH. *Nature*, 435: 1197–1202.

Jahn, R., Lang, T., and Sudhof, T. C. 2003. Membrane fusion. *Cell*, 112: 519–533.

Jankowski, A., Kim, J. H., Collins, R. F. et al. 2001. In situ measurements of the pH of mammalian peroxisomes using the fluorescent protein pHluorin. *J. Biol. Chem.*, 276: 48748–48753.

Jayaraman, S., Haggie, P., Wachter, R. M., Remington, S. J., and Verkman, A. S. 2000. Mechanism and cellular applications of a green fluorescent protein-based halide sensor. *J. Biol. Chem.*, 275: 6047–6050.

Karagiannis, J. and Young, P. G. 2001. Intracellular pH homeostasis during cell-cycle progression and growth state transition in *Schizosaccharomyces pombe*. *J. Cell Sci.*, 114: 2929–2941.

Kellokumpu, S., Sormunen, R., and Kellokumpu, I. 2002. Abnormal glycosylation and altered Golgi structure in colorectal cancer: Dependence on intra Golgi pH. *FEBS Lett.*, 516: 217–224.

Kennis, J. T., Larsen, D. S., van Stokkum, I. H. et al. 2004. Uncovering the hidden ground state of green fluorescent protein. *Proc. Natl. Acad. Sci. USA*, 101: 17988–17993.

Kneen, M., Farinas, J., Li, Y., and Verkman, A. S. 1998. Green fluorescent protein as a noninvasive intracellular pH indicator. *Biophys. J.*, 74: 1591–1599.

Kummer, A. D., Wiehler, J., Rehaber, H. et al. 2000. Effects of threonine 203 replacements on excited-state dynamics and fluorescence properties of the green fluorescent protein (GFP). *J. Phys. Chem. B*, 104: 4791–4798.

Lakowicz, J. R. 1999. *Principles of Fluorescence Spectroscopy*, Chap. 19, pp. 623–672. Oakland: Plenum.

Lin, H. J., Herman, P., and Lakowicz, J. R. 2003. Fluorescence lifetime-resolved pH imaging of living cells. *Cytom. A*, 52A: 77–89.

Llopis, J., McCaffery, J. M., Miyawaki, A., Farquhar, M. G., and Tsien, R. Y. 1998. Measurement of cytosolic, mitochondrial, and Golgi pH in single living cells with green fluorescent proteins. *Proc. Natl. Acad. Sci. USA*, 95: 6803–6808.

Machen, T. E., Leigh, M. J., Taylor, C. et al. 2003. pH of TGN and recycling endosomes of H^+/K^+-ATPase-transfected HEK-293 cells: Implications for pH regulation in the secretory pathway. *Am. J. Physiol.*, 285: C205–C214.

Marcello, A., Lusic, M., Pegoraro, G. et al. 2004. Nuclear organization and the control of HIV-1 transcription. *Gene*, 326: 1–11.

Markova, O., Mukhtarov, M., Real, E., Jacob, Y., and Bregestovski, P. 2008. Genetically encoded chloride indicator with improved sensitivity. *J. Neurosci. Meth.*, 170: 67–76.

McAnaney, T. B., Park, E. S., Hanson, G. T., Remington, S. J., and Boxer, S. G. 2002. Green fluorescent protein variants as ratiometric dual emission pH sensors. 2. Excited-state dynamics. *Biochemistry*, 41: 15489–15494.

Mellman, I. and Warren, G. 2000. The road taken: Past and future foundations of membrane traffic. *Cell*, 100: 99–112.

Miesenbock, G., De Angelis, D. A., and Rothman, J. E. 1998. Visualizing secretion and synaptic transmission with pH-sensitive green fluorescent proteins. *Nature*, 394: 192–195.

Miyawaki, A. 2005. Innovations in the imaging of brain functions using fluorescent proteins. *Neuron*, 48: 189–199.

Nakabayashi, T., Wang, H. P., Kinjo, M., and Ohta, N. 2008. Application of fluorescence lifetime imaging of enhanced green fluorescent protein to intracellular pH measurements. *Photochem. Photobiol. Sci.*, 7: 668–670.

Niwa, H., Inouye, S., Hirano, T. et al. 1996. Chemical nature of the light emitter of the Aequorea green fluorescent protein. *Proc. Natl. Acad. Sci. USA*, 93: 13617–13622.

Ormo, M., Cubitt, A. B., Kallio, K. et al. 1996. Crystal structure of the *Aequorea victoria* green fluorescent protein. *Science*, 273: 1392–1395.

Paddock, S. 2008. Over the rainbow: 25 years of confocal imaging. *BioTechniques*, 44: 643–644, 646, 648.

Patterson, G., Day, R. N., and Piston, D. 2001. Fluorescent protein spectra. *J. Cell Sci.*, 114: 837–838.

Prasher, D. C., Eckenrode, V. K., Ward, W. W., Prendergast, F. G., and Cormier, M. J. 1992. Primary structure of the *Aequorea victoria* green-fluorescent protein. *Gene*, 111: 229–233.

Puri, S., Bachert, C., Fimmel, C. J., and Linstedt, A. D. 2002. Cycling of early Golgi proteins via the cell surface and endosomes upon lumenal pH disruption. *Traffic*, 3: 641–653.

Putney, L. K. and Barber, D. L. 2003. Na-H exchange-dependent increase in intracellular pH times G2/M entry and transition. *J. Biol. Chem.*, 278: 44645–44649.

Rink, T. J., Tsien, R. Y., and Pozzan, T. 1982. Cytoplasmic pH and free Mg^{2+} in lymphocytes. *J. Cell Biol.*, 95: 189–196.

Rybak, S. L., Lanni, F., and Murphy, R. F. 1997. Theoretical considerations on the role of membrane potential in the regulation of endosomal pH. *Biophys. J.*, 73: 674–687.

Sankaranarayanan, S., De Angelis, D., Rothman, J. E., and Ryan, T. A. 2000. The use of pHluorins for optical measurements of presynaptic activity. *Biophys. J.*, 79: 2199–2208.

Schulte, A., Lorenzen, I., Bottcher, M., and Plieth, C. 2006. A novel fluorescent pH probe for expression in plants. *Plant Methods*, 2: 7.

Serresi, M., Bizzarri, R., Cardarelli, F., and Beltram, F. 2009. Real-time measurement of endosomal acidification by a novel genetically encoded biosensor. *Anal. Bioanal. Chem.*, 393: 1123–1133.

Shaner, N. C., Steinbach, P. A., and Tsien, R. Y. 2005. A guide to choosing fluorescent proteins. *Nat. Methods*, 2: 905–909.

Shimomura, O., Johnson, F. H., and Saiga, Y. 1962. Extraction, purification and properties of aequorin, a bioluminescent protein from the luminous hydromedusan, Aequorea. *J. Cell. Comp. Physiol.*, 59: 223–239.

Shu, X., Kallio, K., Shi, X. et al. 2007a. Ultrafast excited-state dynamics in the green fluorescent protein variant S65T/H148D. 1. Mutagenesis and structural studies. *Biochemistry*, 46: 12005–12013.

Shu, X., Leiderman, P., Gepshtein, R. et al. 2007b. An alternative excited-state proton transfer pathway in green fluorescent protein variant S205V. *Protein Sci.*, 16: 2703–2710.

Sonnleitner, A., Schutz, G. J., and Schmidt, T. 1999. Free brownian motion of individual lipid molecules in biomembranes. *Biophys. J.*, 77: 2638–2642.

Srivastava, J., Barber, D. L., and Jacobson, M. P. 2007. Intracellular pH sensors: Design principles and functional significance. *Physiology* (Bethesda, MD), 22: 30–39.

Stoner-Ma, D., Jaye, A. A., Ronayne, K. L. et al. 2008. An alternate proton acceptor for excited-state proton transfer in green fluorescent protein: Rewiring GFP. *J. Am. Chem. Soc.*, 130: 1227–1235.

Sun, H. Y., Wang, N. P., Halkos, M. E. et al. 2004. Involvement of Na^+/H^+ exchanger in hypoxia/re-oxygenation-induced neonatal rat cardiomyocyte apoptosis. *Eur. J. Pharmacol.*, 486: 121–131.

Svoboda, K. and Yasuda, R. 2006. Principles of two-photon excitation microscopy and its applications to neuroscience. *Neuron*, 50: 823–839.

Tsien, R. Y. 1989. Fluorescent indicators of ion concentrations. *Meth. Cell Biol.*, 30: 127–156.

Tsien, R. Y. 1998. The green fluorescent protein. *Annu. Rev. Biochem.*, 67: 509–544.

Tsien, R. Y. 2003. Imagining imaging's future. *Nat. Rev. Mol. Cell Biol.*, Suppl: SS16–SS21.

Ullmann, G. M. 2003. Relations between protonation constants and titration curves in polyprotic acids: A critical view. *J. Phys. Chem. B*, 107: 1263–1271.

Urra, J., Sandoval, M., Cornejo, I. et al. 2008. A genetically encoded ratiometric sensor to measure extracellular pH in microdomains bounded by basolateral membranes of epithelial cells. *Pflugers Arch.*, 457 (1): 233–242.

Voityuk, A. A., Michel-Beyerle, M. E., and Rosch, N. 1998. Quantum chemical modeling of structure and absorption spectra of the chromophore in green fluorescent proteins. *Chem. Phys.*, 231: 13–25.

Wachter, R. M. 2007. Chromogenic cross-link formation in green fluorescent protein. *Acc. Chem. Res.,* 40: 120–127.

Wachter, R. M., Elsliger, M. A., Kallio, K., Hanson, G. T., and Remington, S. J. 1998. Structural basis of spectral shifts in the yellow-emission variants of green fluorescent protein. *Structure,* 6: 1267–1277.

Wachter, R. M., Yarbrough, D., Kallio, K., and Remington, S. J. 2000. Crystallographic and energetic analysis of binding of selected anions to the yellow variants of green fluorescent protein. *J. Mol. Biol.,* 301: 157–171.

Wakabayashi, I., Poteser, M., and Groschner, K. 2006. Intracellular pH as a determinant of vascular smooth muscle function. *J. Vasc. Res.,* 43: 238–250.

Wang, H., Singh, D., and Fliegel, L. 1997. The Na^+/H^+ antiporter potentiates growth and retinoic acid-induced differentiation of P19 embryonal carcinoma cells. *J. Biol. Chem.,* 272: 26545–26549.

Ward, W. W. and Bokman, S. H. 1982. Reversible denaturation of Aequorea green-fluorescent protein: Physical separation and characterization of the renatured protein. *Biochemistry,* 21: 4535–4540.

Webber, N. M., Litvinenko, K. L., and Meech, S. R. 2001. Radiationless relaxation in a synthetic analogue of the green fluorescent protein chromophore. *J. Phys. Chem. B,* 105: 8036–8039.

Widengren, J., Terry, B., and Rigler, R. 1999. Protonation kinetics of GFP and FITC investigated by FCS – aspects of the use of fluorescent indicators for measuring pH. *Chem. Phys.,* 249: 259–271.

Wiehler, J., Jung, G., Seebacher, C., Zumbusch, A., and Steipe, B. 2003. Mutagenic stabilization of the photocycle intermediate of green fluorescent protein (GFP). *Chem. Bio. Chem.,* 4: 1164–1171.

Wlodarczyk, J., Woehler, A., Kobe, F. et al. 2008. Analysis of FRET signals in the presence of free donors and acceptors. *Biophys. J.,* 94: 986–1000.

Zhang, J., Campbell, R. E., Ting, A. Y., and Tsien, R. Y. 2002. Creating new fluorescent probes for cell biology. *Nat. Rev. Mol. Cell. Biol.,* 3: 906–918.

Zhang, L., Patel, H. N., Lappe, J. W., and Wachter, R. M. 2006. Reaction progress of chromophore biogenesis in green fluorescent protein. *J. Am. Chem. Soc.,* 128: 4766–4772.

Zipfel, W. R., Williams, R. M., and Webb, W. W. 2003. Nonlinear magic: Multiphoton microscopy in the biosciences. *Nat. Biotechnol.,* 21: 1369–1377.

11

Fluorescence Photoactivation Localization Microscopy

11.1 Introduction 11-1
11.2 Background 11-1
  Definition of Technical Terms • History
11.3 Presentation of the Method 11-5
  Theory • Materials • Methods
11.4 Critical Discussion 11-19
  Information That Can Be Obtained with FPALM • Microscope
  Position Stability and Drift
11.5 Summary 11-21
11.6 Future Perspective 11-21
Acknowledgments..11-21
References 11-22

Manasa V. Gudheti
Travis J. Gould
Samuel T. Hess

University of Maine

11.1 Introduction

Diffraction limits the resolution that can be obtained with visible light microscopy to ~200 nm, and has limited biological applications of light microscopy for more than a century. Recently, novel super-resolution techniques have succeeded in breaking the diffraction barrier. These techniques utilize the principles of observation volume confinement, modulated illumination, and single molecule localization. Fluorescence photoactivation localization microscopy (FPALM) is an example of such a super-resolution light microscopy technique, which can image fixed or living biological samples with a demonstrated lateral resolution of ~20–40 nm (Hess et al. 2006, 2007, Gould et al. 2008). FPALM involves repeated cycles of photoactivation, localization, and photobleaching of many sparse subsets of photoactivatable molecules. FPALM is based on the principle that the positions of single molecules can be localized with a precision better than the diffraction-limited resolution. Final images are rendered by plotting the positions and intensities of localized molecules, yielding a map of the distribution of those molecules with an optimal resolution given by the localization precision. Here, we present a detailed description of FPALM along with examples of results obtained from imaging biological samples.

11.2 Background

11.2.1 Definition of Technical Terms

Diffraction-limited spot: The finite-sized image of a single point-like object as generated by a lens-based imaging system due to diffraction.

Resolution (R_0): The smallest distance of separation at which two objects are recognized as being separate, usually defined by the Rayleigh criterion. This minimum distance depends on the wavelength of light and the numerical aperture (NA) of the objective lens.

Point spread function (PSF): The two- or three-dimensional spatial intensity distribution obtained from imaging a single point-like object. The size of the PSF is approximately equal to the smallest recognizable feature in a normal light microscope. The standard deviation of the PSF (symbol s) normalized to unity at its maximum is approximately equal to $s \sim 0.37R_0$.

Full-width at half maximum (FWHM): The width of the PSF at half its maximal value.

$1/e^2$ radius (r_0): The distance from the center of the PSF at which the intensity drops to $1/e^2$ (13.5%) of the maximal value. $FWHM = 1.17r_0$.

Super-resolution or ultra-high-resolution microscopy: Imaging technique that can resolve features in a sample that are smaller than the diffraction-limited resolution.

Localization: Finding the position of an object, i.e., the center of the PSF. The localization can be obtained with a higher precision than $\pm R_0$ (the resolution).

Fluorescence quantum yield (Φ): The ratio of the number of fluorescence photons emitted to the number of photons absorbed (typically by a fluorophore).

Photoactivation quantum yield (Φ_{PA}): The probability that a photoactivatable molecule will be converted to its active state from its pre-activated (inactive) state after absorbing a photon.

Photobleaching quantum yield (Φ_B): The probability per excitation that the fluorophore is permanently converted into a nonfluorescent form.

Photoactivatable fluorophores: A fluorophore, which is initially nonfluorescent within a given spectral window, even under illumination at the standard excitation wavelength, until it is activated (usually using UV or blue light). It then behaves as a normal (conventional) fluorophore with a given excitation and emission spectrum. Photoactivation can be either a reversible or irreversible process.

Photoswitchable fluorophores: Similar to a photoactivatable fluorophore, except that the photoactivation process alters the spectral properties of the fluorophore by changing the emission from one set of wavelengths to another (rather than from a dark state to a fluorescent state). This process can be either reversible or irreversible.

Readout activation: Conversion of a photoactivatable probe into its activated state when illuminated by the readout beam in the absence of activation light.

Spontaneous activation: Conversion of a photoactivatable fluorophore into an activated state in the absence of illumination (readout or activation light).

11.2.2 History

The wave nature of light and the NA of the objective lens impose a limitation on the resolution that can be obtained by traditional light microscopy techniques. The resolution in widefield microscopy can be quantified using the Rayleigh criterion (Born and Wolf 1997)

$$R_0 = 0.61\lambda/NA \tag{11.1}$$

where
  λ is the wavelength of the detected photons
  NA is the numerical aperture of the lens system

At best, a resolution of ~200 nm can be obtained using far-field (i.e., lens based) visible light microscopy techniques. High-resolution images can be obtained by employing alternative techniques with lower effective wavelengths (e.g., electron and x-ray microscopy and tomography) or near-field optics (e.g., near-field scanning optical microscopy).

Despite being limited in resolution, traditional light microscopy has been extensively used for imaging biological samples, which can include live biological specimens noninvasively in two and three dimensions as a function of time with single molecule sensitivity (Lakowicz 1983, Pawley 1995). Confocal laser-scanning fluorescence microscopy enhances resolution by a factor of $\sqrt{2}$ over widefield by using diffraction-limited laser illumination and a detector aperture (Pawley 1995) while providing three-dimensional and temporal imaging capabilities. However, the resolution in confocal microscopy is still limited to some fraction of the wavelength.

Two-photon microscopy offers several advantages for biological imaging, such as reduced out-of-focus background, photobleaching, and photodamage by confining the excitation to a small volume in the focal plane (Denk et al. 1990, Xu et al. 1996). In principle, the inherent three-dimensional resolution of two-photon laser scanning microscopy can be further enhanced by coupling two-photon excitation with a confocal pinhole (Denk et al. 1990). However, in practice, the resolution is similar to that obtained with an optimized confocal microscope setup. Therefore, the resolution obtained from far-field microscopy techniques can be approximately quantified by $\lambda/2n$, where n is the refractive index of the medium (Hell 2007).

Electron microscopy (EM) is a powerful way to obtain high-resolution images of biological samples. Unfortunately, although EM affords atomic scale resolution, it has not yet been made compatible with live specimen imaging. For example, samples typically need to be subjected to elaborate preparation procedures, such as chemical fixation, embedding, and sectioning, prior to EM imaging, and are imaged in vacuum to facilitate electron propagation.

Near-field techniques like near-field optical scanning microscopy (NSOM) provide enhanced resolution by using the "near-field" electromagnetic waves found at distances less than λ from the object. Under visible light illumination, a resolution of at least 12 nm (better than $\lambda/40$) has been obtained (Betzig et al. 1991, Betzig and Trautman 1992). Even higher resolutions can be obtained by using the near-field interactions between a sharp fiber-optic probe and a sample (Betzig and Trautman 1992). NSOM has found some use in imaging biological specimens (Hwang et al. 1998), but is limited by the requirement of close proximity between the probe and the sample (Hell 2007).

11.2.2.1 Super-Resolution Microscopy Techniques

Super-resolution or ultra-high-resolution light microscopy techniques break the diffraction barrier and enable imaging of features that are well below the classical light resolution limit of ~200 nm. A brief summary of super-resolution techniques is given in the following text; for further reading, several reviews are available (Walter et al. 2008, Hell 2007, Gustafsson 2008, Fernandez-Suarez and Ting 2008).

Stimulated emission depletion (STED) fluorescence microscopy first demonstrated that the diffraction limit can be broken (Hell and Wichmann 1994). STED combines a doughnut-shaped laser beam with the excitation beam to drive the molecules near the periphery of the excitation volume to the ground state before they can fluoresce. Only molecules at the null (center) of the doughnut-shaped beam remain in the excited state long enough to fluoresce, resulting in emission from a highly confined volume. Spatial resolution close to 16 nm has been demonstrated (Westphal and Hell 2005) in a laser-scanning STED microscope using visible wavelengths. Using STED with nonlinear deconvolution, a focal plane resolution of 15–20 nm has been achieved in fixed biological samples (Donnert et al. 2006). Recently, presynaptic vesicles in live neurons were tracked at video-rate with a resolution of 62 nm (Westphal et al. 2008) using STED, thus demonstrating its applicability to imaging live biological specimens. By exploiting the optically driven transitions between states with drastically different emission properties, such as photoswitching of fluorescent proteins (Hofmann et al. 2005), the concept of STED has been generalized to other reversible saturable optical fluorescence transition (RESOLFT) (Hell et al. 2003) techniques such as ground state depletion (GSD) microscopy (Hell and Kroug 1995, Folling et al. 2008) to achieve enhanced resolution.

4Pi microscopy (Schrader and Hell 1996) and I^{5}M (Gustafsson et al. 1999) provide a 3–7 fold improved resolution in the axial resolution by increasing the aperture solid angle of the microscope; this is achieved by coherently combining the wavefronts of two opposing objectives. The lateral resolution in 4Pi is still diffraction limited although an axial resolution of 100 nm has been obtained (Egner et al. 2002). Structured illumination microscopy (SIM) (Gustafsson et al. 1999, 2008) utilizes a spatially structured excitation profile to encode high resolution information into the images through spatial frequency mixing. Mathematical deconvolution is then used to extract the information, thus resulting in resolutions better than 50 nm (Gustafsson 2005). A three-dimensional resolution of 100 nm has been achieved by combining I^{5}M with structured illumination in a technique referred to as I^{5}S microscopy (Shao et al. 2008).

11.2.2.2 Single Molecule Localization and Reconstruction

Single molecule localization and reconstruction microscopy, also referred to as pointillist microscopy (Scherer 2006), encompasses an array of techniques that provide enhanced resolution by localizing single molecules that are separated by more than the resolution limit. Localizing involves finding the center (i.e., position) of the molecule from an image of its PSF; a localization precision as high as 1 nm has been demonstrated for single fluorescent molecules (Kural et al. 2005, Yildiz et al. 2003). Single molecule techniques rely on the manipulation of the optical properties of the labeled molecules within the specimen to achieve a higher localization precision and consequently a higher resolution. In order for molecules to be a localized with high precision, they need to be distinctly (spatially) separated. This can be achieved in two ways: (1) by labeling the specimen with a very low density of fluorophores wherein the molecules to be localized are distinctly separated; this concept is utilized in fluorescent speckle microscopy (FSM) (Waterman-Storer et al. 1998) and other techniques that exploit fluorescence intermittency (Lidke et al. 2005, Lagerholm et al. 2006). Other methods have exploited stepwise photobleaching to localize single molecules (Gordon et al. 2004, Qu et al. 2004), although these methods have required small numbers of probe molecules to be located within a diffraction-limited area. The localization of diffusing probe molecules that become fluorescent as they bind to a target molecule and subsequently photobleach has also been demonstrated (Sharonov and Hochstrasser 2006), although control over the density of fluorescent molecules in this approach still requires the adjustment of the concentration of probe molecules; (2) by using photoactivatable and/or photoswitchable fluorophores that can be switched between the bright and dark states thereby providing an optical control over the fluorescent properties of the sample.

FPALM is an example of a technique that utilizes photoactivatable/photoswitchable fluorophores to achieve resolution below the diffraction limit (Hess et al. 2006, Gould et al. 2008) in live biological samples (Hess et al. 2007). A sparse subset of molecules are photoactivated to a bright state, their positions localized, and subsequently photobleached irreversibly or photoswitched reversibly to a dark state. This cycle is repeated many times to gather information about a large number (10^3–10^6) of molecules. Finally, the tabulated positions and intensities of all the molecules are used to reconstruct an image that has a higher resolution than the classical diffraction limit. Photoactivated localization microscopy (PALM) (Betzig et al. 2006, Shroff et al. 2007, 2008), Stochastic optical reconstruction microscopy (STORM) (Rust et al. 2006, Bates et al. 2007, Huang et al. 2008), and PALM with independently running acquisition (PALMIRA) (Egner et al. 2007) are other techniques that operate on similar principles as FPALM.

FPALM can be used in a standard widefield-fluorescence (nonscanning) geometry, does not require ultrafast-pulsed lasers or image deconvolution, and does not rely on nonlinear excitation when compared to the confined and modulated illumination methods. The use of genetically encoded fluorescent markers such as photoactivatable green fluorescent protein (PA-GFP (Patterson and Lippincott-Schwartz 2002)) provides flexibility by allowing existing fluorescent protein (FP) constructs to be converted into photoactivatable versions using standard molecular biology procedures.

11.3 Presentation of the Method

11.3.1 Theory

In a typical widefield fluorescence microscopy setup, all the fluorescent molecules are visible at the same time, which leads to diffraction-limited blurring of the sample as a result of the overlap of PSFs of the different molecules. When only a few molecules are visible such that the PSFs of the molecules do not overlap, the molecules will be well separated and distinct from one another. In such a scenario, the positions (centers) of the molecules can be identified (this is also referred to as localizing a molecule) with a very high precision. An object can be localized with greater precision (smaller error) than the diffracted-limited resolution. The two-dimensional localization precision (σ_{xy}) of a fluorescent object is given by (Thompson et al. 2002)

$$\sigma_{xy}^2 = \frac{s^2 + q^2/12}{N} + \frac{8\pi s^4 b^2}{q^2 N^2} \qquad (11.2)$$

where
  s is the standard deviation of the PSF (which is proportional to the resolution)
  N is the total number of photons collected (not photons per pixel)
  q is the effective size of an image pixel within the sample focal plane
  b is the background noise per pixel (not background intensity)

Equation 11.2 demonstrates that single fluorescent molecules can be localized with a significantly better precision than $\pm R_0$. The localization precision can be improved by increasing the number of detected photons and decreasing the background.

Figure 11.1 illustrates the concept and operating principle of FPALM. In FPALM, it is essential to have control over the number of active (fluorescent) molecules such that only a sparse subset of the total number of fluorescent molecules is visible at any given time. This control is achieved by using photoactivatable probes, which can be optically switched from an inactive (nonfluorescent state) to an active (potentially fluorescent) state. Photoswitchable probes that change their emission from one color to another upon activation can also be utilized. Typically, molecules are switched to an optically active state using an activation laser, which has a different (higher) frequency than the readout laser. The readout laser is used to image and photobleach the activated molecules. It should be noted that the activation process is stochastic, i.e., molecules are randomly photoactivated. The number of molecules that are active (on average) can be controlled by adjusting the intensity of the activation laser. This number must be kept small enough to minimize the probability of having molecules with overlapping PSFs. Quantitatively, this requires that L_0, the average separation between molecules should be larger than R_0, where R_0 is the diffraction-limited resolution given by Equation 11.1. Because of the difficulty in analyzing images of multiple emitters in close spatial proximity, molecules that are closer than this minimum distance can either be excluded from analysis or analyzed separately with more sophisticated methods. The cycle of activation, imaging, and photobleaching is repeated until the sample is depleted of photoactivatable fluorescent molecules, or until a sufficient number of molecules have been imaged.

In practice, the sample is placed on the stage of a widefield fluorescence microscope and is illuminated continuously with the readout laser. The activation laser can be used to illuminate the sample in either a continuous or pulsed mode to switch on subsets of fluorescent molecules. The images of the molecules are captured using a high-sensitivity charge-coupled device (CCD) camera, which is sensitive enough to detect single fluorescent molecules. Typically a sequential recording of images (a movie) of a certain number of frames is taken. The frames are then analyzed using custom or commercial image

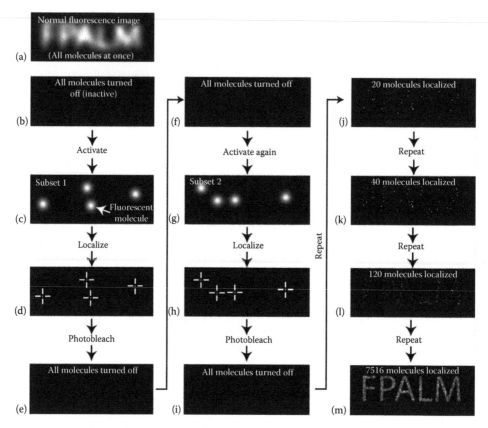

FIGURE 11.1 Concept of fluorescence photoactivation localization microscopy (FPALM). (a) In normal fluorescence microscopy, large numbers of fluorescent molecules are visible at once, and diffraction blurs objects smaller than 200–250 nm, obscuring fine details. In FPALM, light is used to limit the number of visible (fluorescent) molecules. In contrast to normal fluorescent molecules, FPALM uses photoactivatable or photoswitchable fluorescent probes, which are initially nonfluorescent (inactive) within a particular range of wavelengths. (b) Even under normal illumination, inactive molecules are invisible. (c) The low-intensity 405 nm activation laser converts a small subset of inactive molecules into active ones (large spots). (d) Active molecules are imaged and localized (crosses) to precisely determine their positions (small spots). (e) Photobleaching turns active molecules permanently off. (f–i) Starting with the remaining inactive molecules, the process of activation, imaging, localization, and photobleaching is repeated, this time yielding the coordinates of a new subset of molecules. The process is repeated until enough molecules have been localized to reveal the structure of the sample. The plotted positions of the localized molecules constitute the FPALM image. (j–m) Simulated FPALM images of a sample with increasing numbers of molecules illustrate how the data are built up iteratively. FPALM can be used to image a variety of samples in two or three dimensions.

processing algorithms to determine the positions and intensities of the molecules. This information is then subsequently used to produce a map (image) of the distribution of molecules with a localization-based resolution given by Equation 11.3.

11.3.1.1 Factors That Determine Localization Precision

Equation 11.2 quantifies the factors that determine the localization precision. The first term in the equation corresponds to the contribution to σ_{xy} from shot noise and pixelization noise in the absence of background. The second term corresponds to uncertainties introduced due to background noise. Maximizing the number of detected photons (N) will reduce the contribution to the localization uncertainty from both terms. Background noise may arise from scattered light, autofluorescence from the sample, weak fluorescence from inactive molecules, camera electronics, and other sources. Therefore, it is crucial to minimize background noise and maximize the number of detected photons to improve the localization precision obtained by FPALM. The pixel size q has a small influence with large values of q decreasing the localization precision (Thompson et al. 2002). When $q \gg r_0$, the pixel size limits the localization precision since the location of a molecule entirely within a single pixel cannot be clearly identified.

A term referred to as the effective localization-based resolution (r_L) is used (Gould et al. 2008, Hess et al. 2009) to describe the smallest structure that can be imaged by FPALM, taking into account both the localization precision (σ_{xy}) and the molecular density (or sparseness):

$$r_L^2 = \sigma_{xy}^2 + r_{NN}^2 \qquad (11.3)$$

where r_{NN} is the nearest neighbor distance between molecules, which is dependent on the molecular density. If $\sigma_{xy} \ll r_{NN}$, then the resolution is limited by sparseness and if $\sigma_{xy} \gg r_{NN}$, the resolution is limited by localization precision. An alternative definition of the effective resolution has been given in terms of the Nyquist criterion (Shroff et al. 2008).

11.3.2 Materials

11.3.2.1 Choice of Photoactivatable Fluorophores

The choice of fluorescent probe depends on its photophysical properties, its biochemical properties, and the intended application. It is ideal to have probes with high photoactivation quantum yields and low rates of spontaneous activation (Hess et al. 2006). Reviews of photoactivatable probes, photoswitchable fluorophores (Lukyanov et al. 2005), and other probes used in super-resolution imaging offer guidelines for biological fluorescence applications (Fernandez-Suarez and Ting 2008, Walter et al. 2008). Probes should have a large contrast ratio; that is, the fluorescence from the active state should be significantly higher than the contribution from the inactive state. To prevent the accumulation of a large number of fluorescent molecules in a field of view, the probe should have a small but finite photobleaching quantum yield. Under imaging conditions, the number of photobleaching molecules per frame (including deactivation in the case of reversible photoactivation) must be greater than or equal to the number of activated molecules per frame (Hess et al. 2006). In case multiple probes are used, their emission spectra should be well separated to prevent spectral overlap.

11.3.2.2 Sample Preparation

11.3.2.2.1 Protein Expression vs. Antibody Labeling

Using genetically encoded fluorescent proteins enables relatively noninvasive labeling and imaging of biological structures of interest. The fluorescent proteins are bound to the target of interest and hence the localization accuracy is better than with antibody labeling where non-specific binding could occur and the antibody itself is much larger than a GFP. Some of the currently available photoactivatable fluorescent proteins (PA-FPs) are PA-GFP (Patterson and Lippincott-Schwartz 2002), PAmRFP1-1(Verkhusha and Sorkin 2005), and Dronpa (Ando et al. 2004) and photoswitcable fluorescent proteins (PS-FPs) are PS-CFP2 (Chudakov et al. 2004), EosFP (Wiedenmann et al. 2004), Dendra2 (Chudakov et al. 2007), Kaede (Ando et al. 2002), and KikGR (Tsutsui et al. 2005). It was recently demonstrated (Biteen et al. 2008) that EYP, a conventional fluorescent protein, can also be used to perform single molecule–based super-resolution imaging.

Organic fluorophores comprising of photo-caged dyes such as caged-FITC (Juette et al. 2008) and caged Q-rhodamine (Gee et al. 2001), pairs of probes that constitute a molecular switch (referred to as STORM probes) such as cyanine dye pair (e.g., Cy3–Cy5) combinations (Bates et al. 2005, 2007) and photoswitchable traditional dyes such as rhodamine B (Folling et al. 2007), Cy5 and Alexa 647

(Heilemann et al. 2008, Bates et al. 2007) have been used in single molecule localization microscopy. A new class of photoactivatable azido-DCDHF fluorogenic single molecules (Lord et al. 2008) and covalently bound heterodimers of Cy3–Cy5 (Conley et al. 2008) were recently developed. A modified method introduced to synthesize STORM probes (Huang et al. 2008) allows them to be attached as a single, functionalized entity to a protein (antibody). These probes generally have a better quantum yield and emit a greater number of photons per molecule when compared to the fluorescent proteins.

The desirable characteristics of photoactivatable fluorophores (general term encompassing both the photoactivatable and photoswitchable varieties) are (1) large absorption coefficient at the activation and readout excitation wavelengths, (2) large fluorescence quantum yield in the active state, (3) large quantum yield for activation, (4) small, but finite quantum yield for photobleaching to ensure that the active molecules eventually photobleach thus allowing new molecules to be activated and localized, (5) high contrast ratio such that the fluorescence from the inactive state is negligible in comparison to the active state, (6) high photon emission rate that leads to a better localization precision due to the shorter acquisition times required for detecting a larger number of photons, (7) low readout activation, (8) negligible spontaneous activation compared to light-induced activation, (9) efficient expression in cell line of interest and minimum target perturbation, and (10) efficient folding into a potentially fluorescent form (i.e., chromophore maturation) at incubation temperature. Cells are either transfected using genetically encoded photoactivatable/photoswitchable proteins or are subjected to antibody labeling using organic fluorophores.

11.3.2.2.2 Protocols for Minimizing Background

During sample preparation, care is taken to mitigate the effect of reagents that would contribute to the background. For example, cells are grown in phenol red free growth media. UV-bleached buffer (e.g., PBS) and reagents made with UV-bleached water are used as media for imaging. Cell chambers are wrapped in an opaque but not airtight covering (i.e., aluminum foil) to prevent inadvertent photoactivation in the incubator due to stray UV or ambient light, especially at wavelengths that could cause inadvertent activation of the probe molecules.

Since the localization precision is affected by the background, sources that contribute to the background have to be carefully monitored. The main sources of background are (1) cellular autofluorescence, (2) media used to rinse or immerse the sample such as buffers, (3) immersion fluid used to bead the objective, and (4) coverslips and slides.

To minimize background from cellular autofluorescence, cells are grown in phenol red free media. When imaging fixed cells, use of UV-bleached buffers can reduce the background. A typical procedure to make UV-bleached reagent is as follows: 50 mL of the reagent contained in a thoroughly cleaned glass beaker is exposed to a 500 W UV lamp for ~15–20 min. The same protocol is used to make UV-bleached HPLC (high purity liquid chromatography) water, which can be used to make reagents for live cell applications and also for beading the objective. The UV-bleached water can be used to clean the coverslips and slides, or a more elaborate procedure (Ha 2001) can be employed. The integrity of reagents that are stored in plastic containers will usually get increasingly compromised over time (i.e., a few days) due to the leaching of fluorescent contaminants. Also, repeatedly exposing samples and reagents to the atmosphere can cause them to become contaminated. This can be mitigated by minimizing sample exposure to air and by creating smaller aliquots of the reagents.

11.3.2.3 Choice of Filters

Appropriate filter set combinations that successfully accommodate the photophysical properties of the fluorophore(s) being used will contribute to optimal FPALM imaging. A dichroic mirror that reflects both the activation and readout lasers and maximally transmits the desired fluorescence from the sample must be used. Emission filters that selectively transmit light of a particular wavelength range and reduce background and scattered laser light are used to maximize the desired signal from the sample. Dichroics, laser cleanup and excitation filters, and emission filters can be purchased from Chroma Technology (Rockingham, VT), Omega Optical (Brattleboro, VT), Semrock (Rochester, NY), or other manufacturers. As in standard fluorescence imaging, if multiple fluorophores are used, additional dichroics and emission filters with the appropriate properties will be needed to separate the fluorescence signals, attenuate the laser(s) by many orders of magnitude, and prevent cross talk.

11.3.2.4 Typical FPALM Setup

The typical FPALM setup (shown in Figure 11.2) consists of a fluorescence microscope with two lasers (activation and readout), high-numerical aperture (high-NA) objective lens, and a sensitive CCD camera capable of imaging single fluorescent molecules. A high-NA objective lens (typically ≥ 1.2 NA water immersion or ≥ 1.4 NA oil immersion) is used to improve collection efficiency, which helps to maximize the number of detected photons (N_{det}). The sample is placed on the stage of the microscope. Molecules in the focal plane (approx. 1 μm thick, depending on the depth of field) are imaged and localized.

11.3.2.4.1 Detection Devices

For single molecule imaging, a high-sensitivity imaging detector is required. High quantum efficiency CCD cameras are often used for this purpose. Some of the factors that need to be considered when

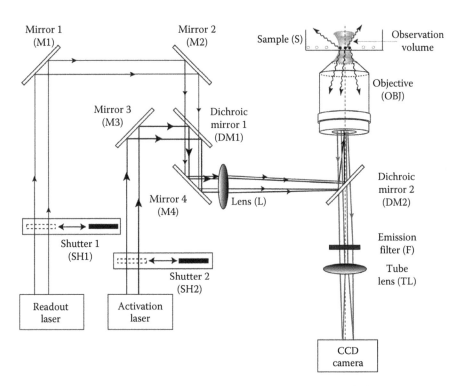

FIGURE 11.2 Experimental setup for FPALM. FPALM typically uses a widefield fluorescence microscope equipped with a high numerical-aperture objective lens (OBJ) and a high-sensitivity charge-coupled device (CCD) camera for single-molecule imaging. The geometry also includes readout and activation lasers with electronically controlled shutters (SH1 and SH2), dichroic mirrors (DM1 and DM2), steering mirrors (M1, M2, M3, and M4), an emission filter (F) and a lens (L). Additional components that are also part of the setup such as neutral density filters for attenuating the lasers, the microscope stage, and the condenser are not shown. The activation laser beam is reflected by a dichroic mirror (DM1) and becomes collinear with the readout laser beam (passed by DM1). Both beams are focused by a lens (L) and reflected by a second dichroic mirror (DM2) to form a focus in the back aperture of the objective (OBJ), which causes a large area of the sample (S) to be illuminated. Some emitted fluorescence photons (black wavy arrows) are collected by the same objective, pass through DM2 and an emission filter (F), and are focused by the microscope tube lens (TL) to form an image on the camera. Drawing is not to scale.

selecting a camera are the maximum frame rate, frame and pixel size, detection efficiency, and noise (i.e., dark current noise and readout noise). Back-thinned CCD cameras exhibit high quantum efficiencies, up to about 90% across much of the visible spectrum. High frame rates (>30 frames/s) are useful when imaging live specimens. In fixed specimens, a wider range of frame rates is possible (>5 frames/s) as long as microscope drift is minimized or corrected. The two most significant sources of noise with CCD cameras are dark current noise and readout noise. Cooling the CCD chip is an effective way to reduce dark current noise since dark current is the result of a thermal process. Electron-multiplying CCDs (EMCCDs) offer high sensitivity, high frame rates (>30 frames/s), and extremely low readout noise. EMCCDs, also referred to as on-chip multiplication cameras, apply gain (amplification) during the line-shifting process prior to analog-to-digital conversion. The amplification of the signal on the chip before readout results in a great reduction of the readout noise when compared to conventional CCDs. The frame size of the camera refers to the number of pixels available to capture data. A larger frame size (256×256 or 512×512 pixels vs. 128×128 pixels) will be useful for capturing features over larger length scales. Since a trade-off exists between camera speed and frame size, an optimal combination based on the sample and camera properties is selected. For example, when imaging live cells, a smaller frame size and faster speed are selected to ensure that the dynamics of the diffusing species are captured. Since the localization precision decreases as the pixel size increases, a camera with a small pixel size is preferred, but too small a pixel size is not recommended because readout noise and background noise per pixel will begin to interfere with the identification and localization of single molecules when pixel sizes are made too small. An effective pixel size (this is computed by dividing the physical pixel size on the camera by the magnification achieved by the microscope setup) of ~80–120 nm works well for many FPALM applications.

Prices of equipment: Inverted fluorescence microscope: $15–25k, high-NA objective lens: $7k, electron-multiplying charge-coupled device (EMCCD) camera: $25–40k; activation laser: $5–10k, readout laser: $5–20k, miscellaneous optics, mounts, excitation and emission filters, dichroic mirrors: $15k, computer and analysis software: $2–3k, computer controlled shutters: $1k each; autofocus: $15k; additional laser lines: $3–40k; small computer cluster for analysis: $5–20k.

11.3.3 Methods

11.3.3.1 Setup and Data Acquisition

11.3.3.1.1 Alignment

The activation and readout laser beams are aligned to be collinear and centered in the back aperture of the microscope objective lens. A lens (L) is then placed at one focal length from the objective back focal plane, such that the beams are focused to a spot at the center of the objective back aperture. The introduction of this lens helps to spread the laser beams over a larger area of the sample. Step 1: The alignment procedure starts by obtaining Köhler illumination of the sample with the transmitted light lamp and condenser and imaging a reticule (scale) positioned at the center of the field with a low-magnification objective (e.g., 10×). For this step, lens L should be absent from the beam path. Step 2: The image of the reticule is projected onto mirror M2 and mirror M1 is adjusted to center the beam on the center of the reticule. Step 3: Now the image of the reticule is focused onto the periscope (mirror M4), which is usually located behind the microscope and is used to raise the height of the beam from table level to the height of the center of DM2 (inside the microscope). The readout laser alignment mirrors (mirrors M1 and M2) are adjusted to center the beam on the center of the reticule on mirror M4. Step 4: The process is repeated with the activation laser using the alignment mirror M3 and DM1. Note that this procedure can be continued if more than two lasers are used. The beams should be fairly close to the center of the reticule image on the periscope. Fine adjustments can be made using the laser dichroics, which can be adjusted independently. If coarse adjustments are needed, repeat Steps 2 and 3. The periscope mirror and dichroics are adjusted to ensure that the laser beams are collinear coming out of the objective lens that is used for FPALM imaging (typically a high-NA and high magnification objective; e.g., 60×/1.2 NA water immersion). Step 5: Now, the lens (L) is placed at one focal length from the objective back focal plane and centered in the beam path. Using the CCD camera with minimal gain and *low intensity illumination*, the now expanded laser beam profile at the sample focal plane is viewed by focusing into a fluorophore (dye) solution with spectral properties that are compatible with the filter sets being used. Care should be taken to ensure that the beam profile does not saturate the camera. The laser beam profiles can be aligned with one another using the camera view. Ideally, the readout laser and activation lasers should illuminate the same area at the sample, with the centers of the beams aligned with one another. The laser beam profiles are typically Gaussian in the sample focal plane as shown in Figure 11.3.

Electronic shutters are installed in front of the lasers for time-dependent computer control. Some fluorophores are spontaneously activated by the readout laser itself thereby facilitating FPALM data acquisition with just one laser (Egner et al. 2007, Hess et al. 2006, 2007, Geisler et al. 2007), but reducing user control over the density of active molecules. Normally, the activation is turned on intermittently when necessary (e.g., when not enough activated molecules are visible) or can be continuously turned on at lower intensity. UV-treated HPLC grade water is used to bead the objective to minimize the contribution to background from the immersion liquid.

11.3.3.1.2 Choice of Sample Region

Transmitted light and epifluorescence are used to find a region of interest (ROI) in the sample. Any light striking the sample should be filtered beforehand to eliminate wavelengths that could unintentionally photoactivate the sample. The sample is typically positioned at the center of the readout and activation lasers. Manually marking the boundaries of the beam profiles on the display monitor can be helpful in setting up for data acquisition. After selecting an ROI using the lamp, the sample is *briefly* viewed under readout laser illumination to ensure that there are indeed single molecules that can be imaged. It is useful to be able to identify a region where the cells have been effectively transfected/labeled when selecting an ROI for FPALM imaging. Comparing the transfected sample of interest with an untransfected sample enables one to effectively identify transfected cells. This can also be confirmed by pulsing the activation laser briefly and looking for an increase in fluorescence, which continues after the activation laser is turned off. The coverslip-proximal portion of the sample is identified by focusing completely out of the sample and then gradually focusing in until the first molecules become visible. Single molecules can be identified by their discrete on–off behavior, and by their apparent size and shape, which are given by the

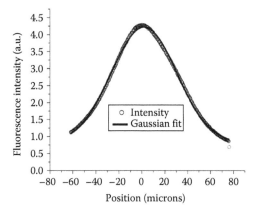

FIGURE 11.3 Gaussian fit of a 496 nm readout laser beam intensity profile. In a typical FPALM setup, the activation and readout lasers have beam profiles that are approximately Gaussian in the sample focal plane. The fluorescent intensity obtained by illuminating a dye sample with the readout laser (496 nm Argon ion laser line) was fitted using a Gaussian function, obtaining a $1/e^2$ radius of ~53 μm corresponding to a full-width at half maximum (FWHM) of ~62 μm. In many cases, a smaller beam profile width is appropriate to achieve a higher illumination intensity. Use of a thin sample helps one to measure the profile of the laser in the focal plane.

diffraction-limited PSF. Small numbers of fluorescent molecules are often visible on a coverslip even in unlabeled sample regions, which is one way to find the surface of the coverslip; then, one can scan laterally to find a transfected cell. Such a technique can be useful when imaging features that are located adjacent to the coverslip such as plasma membrane–bound proteins. When imaging features such as proteins that are intracellular (not plasma membrane associated), prior knowledge of the cellular architecture and distribution of the molecules of interest are immensely helpful in selecting and identifying the correct structures. Another way of keeping track of the focal plane being imaged is to find the coverslip-proximal region first and measure the axial distance (using a z stage or the focus knob) to reach the ROI.

11.3.3.1.3 Recording Data

An optimal density of molecules needs to be active in order to keep molecules separated enough to allow localization, but to maintain high rate of data acquisition. Typically, a density of active molecules of roughly $0.1–1.0/\mu m^2$ is an effective compromise (Gould et al. 2009).

Frame acquisition times and overall acquisition length (total number of frames) vary based on the photophysical properties of the fluorophore used and the resolution desired. The exposure times of 100–150 ms are generally sufficient for fixed samples and yield a demonstrated resolution of ~40 nm for PA-GFP cells illuminated at ~600–1000 W/cm² (Hess et al. 2007). Exposure times of 20–50 ms are sufficient for samples labeled with Dendra2 and illuminated at ~5000–10000 W/cm² to yield a resolution of 20–30 nm (Gould et al. 2008). When imaging live cells, shorter exposure times per frame (2–100 ms) are needed to prevent the blurring of the image due to diffusion of the single molecules and allow reconstruction of full images in 2–10 s. After acquiring FPALM data, a transmitted light (preferably DIC or phase contrast) image of the ROI is recorded to enable the identification of the cellular features imaged within the context of the whole cell.

11.3.3.2 Data Analysis

11.3.3.2.1 Determining Optical Parameters

Some of the optical parameters that need to be determined before proceeding with the analysis are (a) image pixel size (b) pixel value/photon conversion factor, and (c) $1/e^2$ radius of the PSF.

The image pixel size (q) is the effective size of a camera pixel within the sample. The value of q is determined by the physical size of pixels on the camera chip (q_{chip}) and the total magnification (M) according to the relation $q = q_{chip}/M$. Since q is in the numerator of Equation 11.2, a small value of q helps to improve the localization precision, but too small a pixel size makes molecules harder to distinguish from background and readout noise. An EMCCD and microscope configured to have an effective pixel size of 80–120 nm works well for many FPALM applications.

The pixel value/photon conversion factor represents the background-subtracted pixel value that is equivalent to one photon in any given image pixel. To determine this value, the average intensity and variance in intensity are measured using the camera as a function of the source intensity. A stable source such as a battery-powered light or LED is preferred, although the microscope lamp can be used as long as it has steady output. For photons, measurement of the variance versus the mean intensity yields a slope of unity. Thus, the measured slope of variance vs. mean for a given EM gain setting is equal to the pixel/photon conversion factor (Thompson et al. 2002). A more sophisticated method for camera calibration is described elsewhere (Friedman et al. 2006).

$1/e^2$ radius of PSF: The experimental PSF of a single molecule is determined by imaging fluorescent beads of a known size (much smaller than the PSF, i.e., 100 nm or smaller) using the camera. This step also serves as a measure of the diffraction-limited resolution that can be compared against theoretical calculations.

11.3.3.2.2 Background Subtraction

Before acquiring data, consideration should be made to minimize background light that could reach the CCD camera. Fluorescence from inactive molecules, out-of-focus active molecules, the immersion liquid, a dirty or dye-contaminated objective lens, the coverslip, and scattered light are some common internal sources that contribute to the background. When imaging cells, some additional sources of background include autofluorescence, and fluorescence from the media due to ingredients such as phenol red, serum, and residual transfection reagents. Choosing the appropriate filters will also reduce background.

A background subtraction is typically performed before the image of a molecule is analyzed to determine the position of the molecule. First, the zero-level (the pixel value measured with the camera when no light is striking its surface) is subtracted. The background can be either uniform (spatially invariant) or nonuniform. The simplest method is uniform baseline subtraction in which a single value (potentially time-dependent, but spatially independent) is subtracted from every pixel within the given image. This value is typically chosen as the average pixel value from a region in the image where there are no visible probe molecules. Two examples of nonuniform background subtraction methods are time-averaged widefield image (Hess et al. 2007) and rolling ball (Sternberg 1983). In the time-averaged widefield image method, the average widefield image (which provides the position dependence of the background profile) is computed by averaging all frames from the acquisition that will be analyzed, and then dividing this spatial profile by its own average intensity (so that the profile has an average pixel value of one). Then from each frame analyzed, the average widefield profile is subtracted, multiplied by (typically 95% of) the average intensity of that given frame (this provides the time dependence of the background profile).

Background subtraction using the rolling ball algorithm is performed by "rolling" a sphere of given radius (larger than the radius of the image of a single molecule) underneath the three-dimensional surface generated from the intensity of the image mapped into a height (z) as a function of the image coordinates (x and y) (Sternberg 1983). The advantage of this technique is that each frame is subjected to an independent background subtraction.

11.3.3.2.3 Setting Thresholds and Localizing Molecules

Thresholds need to be set to establish the criteria for what constitutes a single molecule. Thresholds are typically set as pixel values or numbers of photons per pixel. Each frame in the time-lapse movie is analyzed sequentially to find single molecules and tabulate their positions and intensities. The first step is the selection of a ROI from a background-subtracted image. Each pixel of the ROI is scanned to identify objects above an initial threshold. Each identified object is enclosed by and centered within a square (typically 5×5 pixel) box that must satisfy additional criteria to be ultimately recognized as a single molecule. A minimum of (typically 3–5) pixels must exceed a second threshold (typically 50%–80% the initial threshold, depending on the size of the image of the single molecule relative to one pixel) and no more than a certain maximum number of pixel values (typically 8–15) may exceed a third threshold (typically between 50%–100% of the initial threshold) within this box or this object will be rejected as being too dim or too large, respectively, to be a single molecule. Regions that satisfy the above criteria are then analyzed to determine the coordinates and intensity of the molecule within the ROI. The center of mass coordinates are used as the initial guess for a least-squares Gaussian fit of the image of the PSF of the single molecule. The x and y coordinates, the amplitude, and the width of the Gaussian, as well as a constant offset are used as the least-squares fitting parameters (Gould et al. 2009, Hess et al. 2006). The amplitude and width of the least-squares Gaussian fit are used to calculate the number of photons detected from that molecule.

An alternate algorithm (Egner et al. 2007) that requires only one threshold can also be used. The first step in this algorithm is identifying and enclosing the brightest pixel in the image within a square box with the pixel in the center (typically 5×5 pixel box size). The next brightest pixel is then identified (excluding the previously boxed region) and this process is continued until all the pixels above the threshold have been identified. This is done for all the frames in a sequential fashion. Each boxed area is subsequently fitted using a Gaussian, which yields the intensity and the position of the molecule, using the chi-squared, the fitted width of the Gaussian, and uncertainties in the fit parameters as a measure of the quality of the fit. In this case, fits with too large a chi-squared, too large or small a Gaussian width, or large uncertainties in the fitting parameters are discarded as unsuccessful attempts at localization.

One of the crucial questions facing the FPALM experimenter is how to confirm that the analysis algorithm is correctly localizing single molecules using a given current set of threshold values and background subtraction. How does one go about identifying a single molecule and what can one expect? The image of each molecule will be spread over a certain area, which is dependent on the magnification. The PSF of a single molecule will consist of a few bright pixels surrounded by dimmer pixels. If there are too many bright pixels adjacent to each other, the probability of that particular object being a single molecule is relatively small. On the other hand, the background could potentially contribute to a contiguous area of bright or moderately fluorescent pixels. Figure 11.4 shows a raw frame and a background-subtracted frame obtained during a typical FPALM acquisition. Also shown are examples of localized single molecules.

When determining the initial threshold values, it is best to become familiar with how a single molecule looks by observing a sample with a known sparse distribution of fluorescent molecules, or the image of known single molecules in an image-processing program such as ImageJ or Adobe Photoshop. Based on the histogram of intensity values, one can estimate the optimal threshold values to use for analysis.

11.3.3.2.4 Image Rendering

Once all the positions and intensities of the molecules have been localized and tabulated, the information is then used to render an image of the molecules. FPALM images can be rendered using three methods: (1) unweighted, point-like plots of the positions of localized molecules, where each localized molecule is plotted with the same intensity, or (2) weighted plots of localized molecules, plotted as a spot with a Gaussian profile, with amplitude proportional to the number of photons detected from each

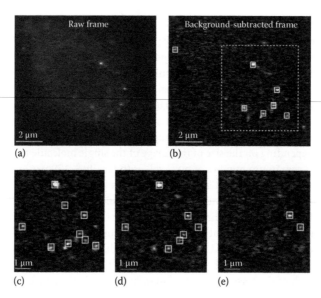

FIGURE 11.4 Single-molecule images of EosFP-Actin. HAb2 cells were transfected with EosFP-actin, fixed, and imaged using a 532 nm readout laser. In this case, the readout laser spontaneously activated the molecules thus facilitating data acquisition even in the absence of an activation laser. Panel (a) shows a raw frame from a time-series acquisition and panel (b) shows the background-subtracted frame using the rolling ball background subtraction technique. The acquisition time per frame was 15.57 ms. The white square boxes enclose the image of a single molecule localized by the analysis software. (c), (d), and (e) are the images of localized molecules in the subsequent frames in the region highlighted by the dashed rectangle in (b). Note that most molecules are visible for a few frames before getting photobleached.

molecule, and radius equal to the calculated or experimentally determined localization-based resolution. Localization precision can be experimentally determined by repeatedly measuring the position of the same (immobile) molecule using multiple observations. The standard deviation of the measured positions would determine the experimental localization precision of the molecule, or (3) density plot generated by binning all molecules that have been localized within a grid of selected size (e.g., 50 nm) of square pixels. The number of molecules localized within each pixel contributes proportionally to the intensity of each pixel. The comparison of images rendered by using the above methods is shown in the supplementary material of Hess et al. (2007, 2009).

11.3.3.3 Flowchart and a Typical Experimental and Analysis Protocol

Figure 11.5 shows the sequence of steps involved in FPALM. Each of the individual steps is described in detail.

1. *Sample preparation*: (1a) Choose the photoactivatable fluorophore that would be suitable for the system. For cells, either transfection and/or antibody labeling can be used. (1b) Transfect or label the sample with the appropriate fluorophore. (1c) If an immersion medium is used, ensure that it has a minimal contribution to fluorescence background at the wavelengths being detected. For example, while imaging cells, phenol red free media and UV-bleached buffers are used to minimize the contribution to the background. Oxidized products that can sometimes form in reagents that are illuminated with high intensities of UV light might not be compatible with live cells. In such a case, using a buffer that is made with UV-bleached water might be more suitable.

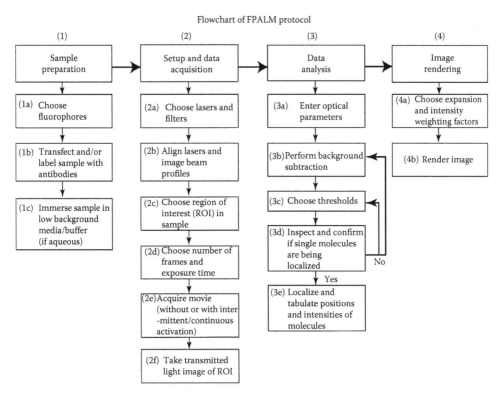

FIGURE 11.5 Flowchart of the FPALM protocol. The FPALM protocol can be divided into four main stages: (1) sample preparation, (2) setup and data acquisition, (3) data analysis, and (4) image rendering. Each stage in turn consists of additional steps to be completed before moving on to the next stage. Each step is discussed in detail within the text.

2. *Setup and data acquisition*: (2a) Choose the activation and readout lasers that would be appropriate for the fluorophores. As for other fluorescence methods, choose excitation, emission, and dichroic filters that provide optimal imaging conditions and prevent spectral overlap when multiple fluorophores are used. (2b) The readout and activation lasers are aligned to the same spot on the periscope, which is generally located behind the microscope, by adjusting the steering mirrors and dichroics. The lasers are focused on the center of the objective back aperture by adjusting the lens L (see Figure 11.2). Minor adjustments can be made to ensure that the beams are concentric by making minor adjustments to the dichroics while observing the image of a fluorescent dye projected onto the CCD camera. The readout beam can be spread out over an area larger than the desired ROI to yield a nearly uniform illumination intensity within the ROI. (2c) Choose a ROI that has been labeled with the photoactivatable fluorophore. (2d) Choose the number of frames and exposure time for the movie acquisition. When imaging fixed samples, an exposure time of 100–150 ms is sufficient (lower values are ideal when sample drift is a concern) but for live cells shorter exposure times (2–100 ms) are sometimes needed to effectively capture the dynamics of moving single molecules. The length of the movie acquired would be approximately equal to the product of the number of frames and exposure time per frame. A typical acquisition for fixed cells would consist of 5000–10,000 frames at 30–50 ms exposure time and for live cells would be 500–10,000 frames at 5–50 ms. (2e) Allow the readout laser to illuminate the sample (i.e., open the shutter) and immediately acquire a movie. The activation laser can illuminate the sample continuously for the duration of the experiment (once the acquisition has begun), or can be intermittently shuttered, illuminating the sample briefly when the density of molecules becomes low (fewer than *roughly* $0.1/\mu m^2$). In some cases, the readout laser can spontaneously activate the sample thereby obviating the need for an activation laser. It is useful to have electronically controlled shutters in the laser beam paths for automated control. (2f) After the FPALM movie has been recorded, take a transmitted light (preferably DIC or phase contrast) image of the same ROI to enable an accurate identification of the imaged features.

3. *Data analysis*: The recorded movie is now sequentially analyzed frame by frame using a single-molecule identification and localization algorithm (usually custom software). (3a) Optical parameters to be used in the analysis: image pixel size, pixel to photon conversion factor, and the $1/e^2$ radius of the PSF must be determined before analysis. One can decide if the entire field of view or only a subregion is to be analyzed. This will determine how long the analysis takes. (3b) Choose a background subtraction technique. In most cases, the widefield sum or rolling ball subtraction method will be sufficient. (3c) Choose threshold(s) to efficiently identify single molecules. (3d) It is very important to investigate if the threshold(s) used are optimal and that the algorithm is indeed successfully localizing single molecules. False positive localizations, where the algorithm identifies objects that are not single molecules, are to be avoided especially; false negative identifications discard actual single molecules and decrease the overall density of the final image. It is crucial to adjust the thresholds accordingly to minimize both false positives and negatives, but it is generally better to miss a small percentage of molecules rather than falsely identify objects that are not actually single molecules. (3e) Localize and tabulate positions and intensities of molecules from all the frames.

4. *Image rendering*: The final FPALM image is rendered using the tabulated positions and intensities of the molecules. (4a) Since the FPALM image has a better resolution than its widefield counterpart, it is common to render the FPALM image with a greater number of pixels (magnification factor, also called expansion factor). For example, if an image of size 100×100 pixels^2 is rendered with an expansion factor of 4, the expanded image would have a size of 400×400 pixels^2. Rendering an image in this manner is necessary to enable the smallest features of the data to be shown. A linear intensity weighting factor (analogous to the brightness) is used to determine the overall intensities of the plotted molecules. (4b) FPALM images can be rendered using one of three methods (see Section 11.3.3.2.4).

Typical FPALM protocol: An example of a typical FPALM protocol is as follows: Cells are grown in chambers with a #1.5 coverslip bottom (e.g., Nunc Lab-Tek II growth chambers #12-565-8 from Fisher Healthcare, Houston, Texas) in phenol red free growth media, transfected with PA-FP-tagged protein of interest, and fixed if desired. Before starting data acquisition, the cells are washed with UV-irradiated phosphate buffered saline (PBS) and the chamber is filled with the same to minimize background fluorescence. The cells are illuminated with 5–25 mW of power from a continuous-wave readout laser (typically a diode laser at 556 nm for excitation of Dendra2 and an Argon ion laser at 488 nm for PA-GFP), spread over an area of ~300–400 μm^2 to yield ~1200–8500 W/cm². For activation, 0.1 μW–1.5 mW of power at 405 nm is spread over 200 μm^2 to yield 0.05–750 W/cm². The sample is continuously illuminated with the readout laser during the acquisition, and the activation laser is turned on either intermittently (pulses of ~1–10 s) or continuously when the density of molecules is fewer than ~0.1/μm^2. The photoactivated molecules are visualized using a CCD camera. For example, to generate the rendered image in Figure 11.6A, 2000 frames at 20 ms acquisition time were recorded using a Cascade 512B CCD camera (an Andor iXon+ DU897DCS-BV EMCCD is typically used). The image pixel size was 80 nm/pixel and the conversion factor from pixel value to photons was 20.6. The images were rendered with an expansion factor of 12. Figure 11.7 compares the images of the protein hemagglutinin tagged with PA-GFP in a live fibroblast cell obtained by FPALM and widefield microscopy.

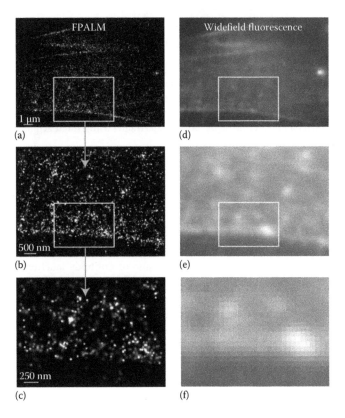

FIGURE 11.6 Fixed cell imaging of Dendra2-Actin. FPALM (a, b, c) and widefield (d, e, f) images of actin tagged with Dendra2 in a fixed HAb2 fibroblast. (a) A large number of molecules were localized ($n = 10,540$). (b) Zoom of boxed region in (a) ($n = 2861$). (c) Zoom of boxed region in (b) ($n = 615$). (d–f) Widefield fluorescence images of the corresponding FPALM images on the left. Due to the large number of molecules localized, the images span a range of length scales and illustrate the enhanced resolution obtained with an FPALM when compared to their respective widefield images. A time-lapse movie of 2000 frames with a frame acquisition time of 20 ms was recorded to obtain the images.

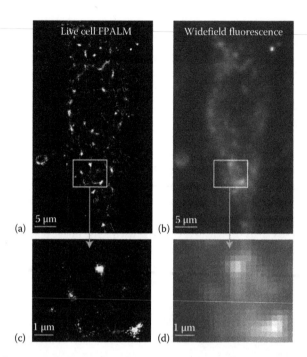

FIGURE 11.7 Live cell imaging of PA-GFP-HA. FPALM (a) and widefield fluorescence (b) images of the influenza protein hemagglutinin (HA) tagged with PA-GFP in a live HAb2 fibroblast cell acquired at room temperature. Panels (c) and (d) show zoomed images of the boxed regions. Note the significantly enhanced resolution in the FPALM images (a, c) when compared to their respective widefield images (b, d). The total number of molecules rendered in (a) = 9742 and (c) = 1005. A time-lapse movie of 500 frames with a frame acquisition time of 150 ms was recorded to obtain the images.

11.3.3.4 Troubleshooting

This section focuses on the possible troubleshooting steps that might need to be performed to ensure that the FPALM setup is functioning optimally.

Step 1a. When using multiple fluorophores, ensure that that there is minimal spectral overlap by choosing the appropriate filter combinations.

Step 1b. It is useful to ascertain that the transfection and/or antibody labeling is indeed successful. Comparing the transfected sample with an untransfected sample will be helpful for positively identifying labeled cells. Looking for blinking or photobleaching molecules while illuminating the sample with the activation laser is also useful as a way to confirm the presence of single molecules. The color of the fluorescence emission can also be a useful assay for success.

Step 1c. Ensure that the sample is immersed in media or buffer that contributes minimally to the background fluorescence. Cells should be plated in phenol red free media and UV-bleached PBS should be used whenever possible. To ensure that fluorescent contaminants from the objective do not contribute to the background, clean it thoroughly with UV-bleached water and use UV-bleached water for beading it. Reagents can become contaminated by exposure to the atmosphere or by the leaching of fluorescent molecules over time from plastic storage bottles. Using freshly prepared reagents (e.g., UV-bleaching the buffer on the day of use) will greatly minimize their contribution to the background.

Step 2a. Choose the appropriate readout and activation laser (often, in practice, the PA-FP is chosen based on the available lasers). If the readout laser being used is not optimized for the excitation maximum of the fluorophore, the intensity of the molecules might be lower than expected due to a reduced excitation rate. In this case, increase the intensity of the readout laser to more effectively excite the sample.

Step 2b. Ensure that the readout and activation lasers overlap at the sample and that their centers are aligned. Use a beam expander, if necessary, to match the beam diameters.

Step 2c. When choosing an ROI for acquiring data, ensure that there are sufficient photoactivatable molecules by checking for discrete ("on–off") transitions between the bright and dark states (also see step 1b). The coverslip can be located by completely focusing out of the sample and then gradually focusing in (from below in an inverted microscope) until molecules are first observed. Cleaning coverslips (Ha 2001) prior to use will minimize their contribution to background. Measuring the spectrum of emission from the sample excited by the readout laser can be useful to confirm that the fluorescence emitted is coming from the probe-labeled molecules of interest.

Step 2e. Before acquiring data, check for the following scenarios: (1) too high a density of active molecules due to inadvertent prior photoactivation or due to high protein expression levels in that area. In this case, delay acquisition until the molecules have sufficiently photobleached to the optimal density (2) too low a density of active molecules (fewer than $\sim 0.1/\mu m^2$); in this case use either intermittent pulses of (~ 1 s) or continuous illumination with the activation laser, adjusting the intensity (power) to yield a balance between activation and photobleaching that gives ~ 0.1–1.0 molecule/μm^2 of illuminated area.

Step 3c. The thresholds chosen should minimize the number of false positives and false negatives obtained with the localization algorithm. If too many false positives (which can lead to artifacts-see critical discussion) are obtained, increase the threshold value and/or increase background subtraction. Reducing the threshold and/or background subtraction values will help when too many false negatives are obtained.

11.4 Critical Discussion

Calibration of FPALM setup: Fluorescent beads of a known size can be used to calibrate the FPALM setup and determine an experimental value of the PSF. It is also useful to image a system of known configuration to test that the setup is functioning optimally. A surface with known nanostructure imaged by an independent method (Hess et al. 2006, Gould et al. 2008) can be a useful calibration sample.

Artifacts that can arise due to suboptimal conditions: (1) Pixelization artifacts due to failed localization and (2) low density of localized molecules. FPALM can be used to image specimens with a very high resolution, but care has to be taken to prevent or minimize artifacts. There is a possibility of obtaining artifacts in the rendered image due to improperly selected thresholds and/or nonoptimal background subtraction. Figure 11.8 shows an example of a fixed fibroblast tagged with PA-GFP-HA. Panels A and B have the same thresholds but different background subtraction methods. Under optimal conditions, the rendered image would resemble that seen in panel B. In panel A, the widefield sum subtraction technique was employed, which in this case led to pixelization artifacts (incorrect localizations of molecules along the edges of the image pixels). Pixelization could also occur due to low thresholds wherein some molecules identified cannot be successfully fitted, and the coordinates obtained are erroneous. The pixelization artifact can be identified by checking if the initial guess at the coordinates for the molecule has a strong effect on final molecular coordinates, and often can be eliminated by increasing thresholds to include only the brighter molecules, or in some cases by using a different (nonuniform) background subtraction method. In panels B and C, the rolling ball background subtraction method was used, but a higher threshold was used in panel C, leading to elimination of the artifact, and also a smaller number of localized molecules. The goal of FPALM is to localize a sufficient number of molecules to reliably image structures at a resolution below the diffraction limit. Large uncertainties result when the density of molecules localized is too low to reliably identify the structures. Even if the molecules are localized with great precision, the resulting image will have an uncertainty in the density, which is quite large if the number of molecules within a given area is small. Thus, there will be a trade-off between generating an image with a large number of molecules and an image that includes molecules with the most precisely defined positions.

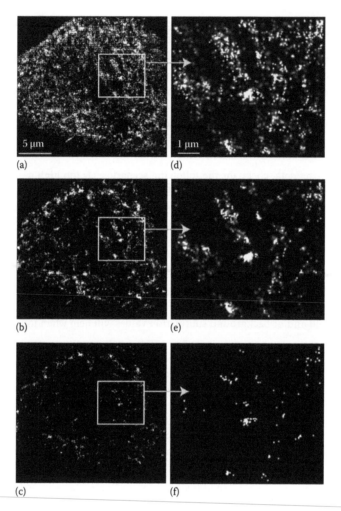

FIGURE 11.8 Adjustment of FPALM analysis parameters to avoid artifacts and optimize rendered image. The final rendered imaged that is obtained with FPALM is dependent on various parameters. The threshold values selected determine whether a localized molecule is included in the final image as indeed being a single molecule or excluded because it fails to meet the requirements. The two most common artifacts obtained are panel (a) pixelization due to low thresholds set during analysis or nonuniform background and panel (c) low number of molecules (more false negatives) localized due to a very high threshold. Using an optimal set of parameters the final FPALM image obtained in this particular case would resemble panel (b). (d–f) zoomed-in images of boxed region of the corresponding images on their left. Rolling ball subtraction technique was used for obtaining (b) and (c) and wide-field sum background subtraction was used for (a). The number of molecules and corresponding thresholds in pixel values localized were (a) 22,328 molecules, 2,000 pixel value (equivalent to ~97 photons) for threshold; (b) 10,092 molecules, 2,000 pixel value for threshold; (c) 2,073 molecules, 3,500 pixel value (~170 photons) for threshold. The pixel to photon conversion factor was 20.6. Note that pixelization is more striking in panel (a) when compared to its zoomed-in image (d). The scale bar in (a) applies to (b) and (c); scale bar in (d) applies to (e) and (f).

Light exposure: An exposure of living biological specimens to laser radiation, especially UV radiation, can cause photodamage and alter biological function. In FPALM experiments, readout laser intensities are of the order ~10^3–10^4 W/cm^2, compared to intensities of ~5×10^5 W/cm^2 in confocal microscopy. As when using any light-based microscopy or spectroscopy method, proper control experiments are required to confirm that the illumination of the sample has not affected the biological processes of interest.

11.4.1 Information That Can Be Obtained with FPALM

In addition to generating a map of the positions of labeled molecules, FPALM also measures the number of molecules, and the number of photons emitted by each molecule. The distribution of single-molecule fluorescence intensities can be used to assess probe performance and reveal population heterogeneities that are not easily accessible by typical light microscopy methods. For example, using FPALM it would be possible to determine the number of labeled receptor molecules binding to a ligand in a diffraction-limited region. Knowing that there are multiple populations (heterogeneity) can be relevant to biological processes where binding stoichiometry or aggregation is important. The effect of environmental variables such as pH, ion concentration, and probe dipole moment orientation on fluorophore photophysical properties can also be used to monitor the configuration of a biological system.

11.4.2 Microscope Position Stability and Drift

The extent of lateral and axial drifts of the microscope needs to be considered, especially during longer acquisitions. Placing the microscope on a vibration isolation table provides reasonable stability over timescales of minutes. A lateral drift of ~7 nm for a time period of 20 min is acceptable for live cell imaging (Hess et al. 2006). For two-dimensional imaging, axial drift is also important and can be adjusted manually or using an automatic focus to be <1 µm. Time-lapse transmitted light imaging of immobilized fluorescent beads (such as polystyrene spheres) can be used to characterize the extent of drift.

11.5 Summary

FPALM is a super-resolution light microscopy technique that can be used to obtain images of living and biological specimens at a resolution that surpasses the classical diffraction limit; resolutions in the range of 20–40 nm have been demonstrated. When compared to other super-resolution techniques, FPALM has the advantages that it can be used with a standard widefield microscope setup, is compatible with live cell imaging, provides single molecule information in addition to an image, does not require ultrafast pulsed lasers, and does not rely on nonlinear excitation. FPALM is based on the principles of stochastic photoactivation, high sensitivity fluorescence detection, and localization of single photoactivatable fluorophores. FPALM uses repeated cycles of activation, localization, and photobleaching to identify and localize large numbers of molecules within a sample. The localization information obtained is then used to build an image with a resolution determined by the localization precision and molecular density, rather than by diffraction.

11.6 Future Perspective

The concept of FPALM has been extended to perform super-resolution imaging of multiple species (Bates et al. 2007, Shroff et al. 2007, Huang et al. 2008, Bock et al. 2007, Gould and Hess, unpublished) in biological samples, three-dimensional imaging (Huang et al. 2008, Juette et al. 2008), and simultaneous super-resolution imaging of molecular positions and anisotropies (Gould et al. 2008). The capabilities of FPALM can now be used to address a variety of biological problems. The technique is also applicable to imaging nonbiological systems. With continuing advances in technologies for single-molecule detection, and an increasing variety of photoactivatable fluorophores, the potential applications of FPALM are rapidly expanding.

Acknowledgments

The authors thank Joshua Zimmerberg and Paul Blank for the argon laser and CCD camera; Vladislav Verkhusha for the Dendra2 construct; Joerg Wiedenmann and Uli Nienhaus, University of Ulm, for

EosFP constructs and purified protein; George Patterson for the PA-GFP construct; Patricia Byard and Thomas Tripp for professional services; and Siyath Gunewardene, Joerg Bewersdorf, Erik Dasilva, and Seth Bolduc for useful discussions. This work was supported by National Institutes of Health grant K25AI65459, National Science Foundation grant CHE-0722759, and funds from the University of Maine Office of the vice president for Research. T.J.G. benefited from an MEIF Doctoral Dissertation fellowship.

References

Ando, R., Hama, H., Yamamoto-Hino, M., Mizuno, H., and Miyawaki, A. 2002. An optical marker based on the UV-induced green-to-red photoconversion of a fluorescent protein. *Proceedings of the National Academy of Sciences of the United States of America* 99: 12651–12656.

Ando, R., Mizuno, H., and Miyawaki, A. 2004. Regulated fast nucleocytoplasmic shuttling observed by reversible protein highlighting. *Science* 306: 1370–1373.

Bates, M., Blosser, T. R., and Zhuang, X. W. 2005. Short-range spectroscopic ruler based on a single-molecule optical switch. *Physical Review Letters* 94: 108101.

Bates, M., Huang, B., Dempsey, G. T., and Zhuang, X. W. 2007. Multicolor super-resolution imaging with photo-switchable fluorescent probes. *Science* 317: 1749–1753.

Betzig, E., Trautman, J. K., Harris, T. D., Weiner, J. S., and Kostelak, R. L. 1991. Breaking the diffraction barrier—Optical microscopy on a nanometric scale. *Science* 251: 1468–1470.

Betzig, E. and Trautman, J. K. 1992. Near-field optics: Microscopy, spectroscopy, and surface modification beyond the diffraction limit. *Science* 257: 189–195.

Betzig, E., Patterson, G. H., Sougrat, R. et al. 2006. Imaging intracellular fluorescent proteins at nanometer resolution. *Science* 313: 1642–1645.

Biteen, J. S., Thompson, M. A., Tselentis, N. K. et al. 2008. Super-resolution imaging in live *Caulobacter crescentus* cells using photoswitchable EYFP. *Nature Methods* 5: 947–949.

Bock, H., Geisler, C., Wurm, C. A. et al. 2007. Two-color far-field fluorescence nanoscopy based on photoswitchable emitters. *Applied Physics B—Lasers and Optics* 88: 161–165.

Born, M. and Wolf, E. 1997. *Principles of Optics: Electromagnetic Theory of Propagation, Interference and Diffraction of Light.* Cambridge, UK: Cambridge University Press.

Chudakov, D. M., Verkhusha, V. V., Staroverov, D. B. et al. 2004. Photoswitchable cyan fluorescent protein for protein tracking. *Nature Biotechnology* 22: 1435–1439.

Chudakov, D. M., Lukyanov, S., and Lukyanov, K. A. 2007. Tracking intracellular protein movements using photoswitchable fluorescent proteins PS-CFP2 and Dendra2. *Nature Protocols* 2: 2024–2032.

Conley, N. R., Biteen, J. S., and Moerner, W. E. 2008. Cy3–Cy5 covalent heterodimers for single-molecule photoswitching. *Journal of Physical Chemistry B* 112: 11878–11880.

Denk, W., Strickler, J. H., and Webb, W. W. 1990. 2-photon laser scanning fluorescence microscopy. *Science* 248: 73–76.

Donnert, G., Keller, J., Medda, R. et al. 2006. Macromolecular-scale resolution in biological fluorescence microscopy. *Proceedings of the National Academy of Sciences of the United States of America* 103: 11440–11445.

Egner, A., Jakobs, S., and Hell, S. W. 2002. Fast 100-nm resolution three-dimensional microscope reveals structural plasticity of mitochondria in live yeast. *Proceedings of the National Academy of Sciences of the United States of America* 99: 3370–3375.

Egner, A., Geisler, C., Von Middendorff, C. et al. 2007. Fluorescence nanoscopy in whole cells by asynchronous localization of photoswitching emitters. *Biophysical Journal* 93: 3285–3290.

Fernandez-Suarez, M. and Ting, A. Y. 2008. Fluorescent probes for super-resolution imaging in living cells. *Nature Reviews Molecular Cell Biology* 9: 929–943.

Folling, J., Belov, V., Kunetsky, R. et al. 2007. Photochromic rhodamines provide nanoscopy with optical sectioning. *Angewandte Chemie—International Edition* 46: 6266–6270.

Folling, J., Bossi, M., Bock, H. et al. 2008. Fluorescence nanoscopy by ground-state depletion and single-molecule return. *Nature Methods* 5: 943–945.

Friedman, L. J., Chung, J., and Gelles, J. 2006. Viewing dynamic assembly of molecular complexes by multi-wavelength single-molecule fluorescence. *Biophysical Journal* 91: 1023–1031.

Gee, K. R., Weinberg, E. S., and Kozlowski, D. J. 2001. Caged Q-rhodamine dextran: A new photoactivated fluorescent tracer. *Bioorganic & Medicinal Chemistry Letters* 11: 2181–2183.

Geisler, C., Schonle, A., von Middendorff, C. et al. 2007. Resolution of lambda/10 in fluorescence microscopy using fast single molecule photo-switching. *Applied Physics A—Materials Science & Processing* 88: 223–226.

Gordon, M. P., Ha, T., and Selvin, P. R. 2004. Single-molecule high-resolution imaging with photobleaching. *Proceedings of the National Academy of Sciences of the United States of America* 101: 6462–6465.

Gould, T. J. and Hess, S. T. 2008. Nanoscale biological fluorescence imaging: Breaking the diffraction barrier. *Methods in Cell Biology* 89: 329–358.

Gould, T. J., Gunewardene, M. S., Gudheti, M. V. et al. 2008. Nanoscale imaging of molecular positions and anisotropies. *Nature Methods* 5: 1027–1030.

Gould, T. J., Verkhusha, V. V., and Hess, S. T. 2009. Imaging biological structures with fluorescence photo-activation localization microscopy. *Nature Protocols* 4: 291–308.

Gustafsson, M. G. L. 2005. Nonlinear structured-illumination microscopy: Wide-field fluorescence imaging with theoretically unlimited resolution. *Proceedings of the National Academy of Sciences of the United States of America* 102: 13081–13086.

Gustafsson, M. G. L. 2008. Super-resolution light microscopy goes live. *Nature Methods* 5: 385–387.

Gustafsson, M. G. L., Agard, D. A., and Sedat, J. W. 1999. (IM)-M-5: 3D widefield light microscopy with better than 100 nm axial resolution. *Journal of Microscopy—Oxford* 195: 10–16.

Gustafsson, M. G. L., Shao, L., Carlton, P. M. et al. 2008. Three-dimensional resolution doubling in wide-field fluorescence microscopy by structured illumination. *Biophysical Journal* 94: 4957–4970.

Ha, T. 2001. Single-molecule fluorescence resonance energy transfer. *Methods* 25: 78–86.

Heilemann, M., van de Linde, S., Schüttpelz, M. et al. 2008. Subdiffraction-resolution fluorescence imaging with conventional fluorescent probes. *Angewandte Chemie—International Edition* 47: 6172–6176.

Hell, S. W. 2007. Far-field optical nanoscopy. *Science* 316: 1153–1158.

Hell, S. W. and Kroug, M. 1995. Ground-state-depletion fluorescence microscopy—A concept for breaking the diffraction resolution limit. *Applied Physics B—Lasers and Optics* 60: 495–497.

Hell, S. W. and Wichmann, J. 1994. Breaking the diffraction resolution limit by stimulated-emission—stimulated-emission-depletion fluorescence microscopy. *Optics Letters* 19: 780–782.

Hell, S. W., Jakobs, S., and Kastrup, L. 2003. Imaging and writing at the nanoscale with focused visible light through saturable optical transitions. *Applied Physics A—Materials Science & Processing* 77: 859–860.

Hess, S. T., Girirajan, T. P. K., and Mason, M. D. 2006. Ultra-high resolution imaging by fluorescence photo-toactivation localization microscopy. *Biophysical Journal* 91: 4258–4272.

Hess, S. T., Gould, T. J., Gudheti, M. V. et al. 2007. Dynamic clustered distribution of hemagglutinin resolved at 40 nm in living cell membranes discriminates between raft theories. *Proceedings of the National Academy of Sciences of the United States of America* 104: 17370–17375.

Hess, S. T., Gould, T. J., Gunewardene, M. S. et al. 2009. Ultra-high resolution of biomolecules by fluorescence photoactivation localization microscopy (FPALM). *Methods in Molecular Biology*, 544:483–522.

Hofmann, M., Eggeling, C., Jakobs, S., and Hell, S. W. 2005. Breaking the diffraction barrier in fluorescence microscopy at low light intensities by using reversibly photoswitchable proteins. *Proceedings of the National Academy of Sciences of the United States of America* 102: 17565–17569.

Huang, B., Jones, S. A., Brandenburg, B., and Zhuang, X. 2008a. Whole-cell 3D STORM reveals interactions between cellular structures with nanometer-scale resolution. *Nature Methods* 5: 1047–1052.

Huang, B., Wang, W., Bates, M., and Zhuang, X. 2008b. Three-dimensional super-resolution imaging by stochastic optical reconstruction microscopy. *Science* 319: 810–813.

Hwang, J., Gheber, L. A., Margolis, L., and Edidin, M. 1998. Domains in cell plasma membranes investigated by near-field scanning optical microscopy. *Biophysical Journal* 74: 2184–2190.

Juette, M. F., Gould, T. J., Lessard, M. D. et al. 2008. Three-dimensional sub-100 nm resolution fluorescence microscopy of thick samples. *Nature Methods* 5: 527–529.

Kural, C., Kim, H., Syed, S. et al. 2005. Kinesin and dynein move a peroxisome in vivo: A tug-of-war or coordinated movement? *Science* 308: 1469–1472.

Lagerholm, B. C., Averett, L., Weinreb, G. E., Jacobson, K., and Thompson, N. L. 2006. Analysis method for measuring submicroscopic distances with blinking quantum dots. *Biophysical Journal* 91: 3050–3060.

Lakowicz, J. R. 1983. *Principles of Fluorescence Spectroscopy*. New York: Plenum Press.

Lidke, K. A., Rieger, B., Jovin, T. M., and Heintzmann, R. 2005. Superresolution by localization of quantum dots using blinking statistics. *Optics Express* 13: 7052–7062.

Lord, S. J., Conley, N. R., Lee, H. L. D. et al. 2008. A photoactivatable push–pull fluorophore for single-molecule imaging in live cells. *Journal of the American Chemical Society* 130: 9204–9205.

Lukyanov, K. A., Chudakov, D. M., Lukyanov, S., and Verkhusha, V. V. 2005. Photoactivatable fluorescent proteins. *Nature Reviews Molecular Cell Biology* 6: 885–891.

Patterson, G. H. and Lippincott-Schwartz, J. 2002. A photoactivatable GFP for selective photolabeling of proteins and cells. *Science* 297: 1873–1877.

Pawley, J. B. 1995. *Handbook of Biological Confocal Microscopy*. New York: Plenum Press.

Qu, X. H., Wu, D., Mets, L., and Scherer, N. F. 2004. Nanometer-localized multiple single-molecule fluorescence microscopy. *Proceedings of the National Academy of Sciences of the United States of America* 101: 11298–11303.

Rust, M. J., Bates, M., and Zhuang, X. W. 2006. Sub-diffraction-limit imaging by stochastic optical reconstruction microscopy (STORM). *Nature Methods* 3: 793–795.

Scherer, N. F. 2006. Imaging—Pointillist microscopy. *Nature Nanotechnology* 1: 19–20.

Schrader, M. and Hell, S. W. 1996. 4PI-confocal images with axial superresolution. *J Microsc* 183: 189–193.

Shao, L., Isaac, B., Uzawa, S. et al. 2008. (IS)-S-5: Wide-field light microscopy with 100-nm-scale resolution in three dimensions. *Biophysical Journal* 94: 4971–4983.

Sharonov, A. and Hochstrasser, R. M. 2006. Wide-field subdiffraction imaging by accumulated binding of diffusing probes. *Proceedings of the National Academy of Sciences of the United States of America* 103: 18911–18916.

Shroff, H., Galbraith, C. G., Galbraith, J. A. et al. 2007. Dual-color superresolution imaging of genetically expressed probes within individual adhesion complexes. *Proceedings of the National Academy of Sciences of the United States of America* 104: 20308–20313.

Shroff, H., Galbraith, C. G., Galbraith, J. A., and Betzig, E. 2008. Live-cell photoactivated localization microscopy of nanoscale adhesion dynamics. *Nature Methods* 5: 417–423.

Sternberg, S. R. 1983. Biomedical image processing. *IEEE Computer* 16(1): 22–34.

Thompson, R. E., Larson, D. R., and Webb, W. W. 2002. Precise nanometer localization analysis for individual fluorescent probes. *Biophysical Journal* 82: 2775–2783.

Tsutsui, H., Karasawa, S., Shimizu, H., Nukina, N., and Miyawaki, A. 2005. Semi-rational engineering of a coral fluorescent protein into an efficient highlighter. *Embo Reports* 6: 233–238.

Verkhusha, V. V. and Sorkin, A. 2005. Conversion of the monomeric red fluorescent protein into a photoactivatable probe. *Chemistry & Biology* 12: 279–285.

Walter, N. G., Huang, C. Y., Manzo, A. J., and Sobhy, M. A. 2008. Do-it-yourself guide: How to use the modern single-molecule toolkit. *Nature Methods* 5: 475–489.

Waterman-Storer, C. M., Desai, A., Bulinski, J. C., and Salmon, E. D. 1998. Fluorescent speckle microscopy, a method to visualize the dynamics of protein assemblies in living cells. *Current Biology* 8: 1227–1230.

Westphal, V. and Hell, S. W. 2005. Nanoscale resolution in the focal plane of an optical microscope. *Physical Review Letters* 94: 143903.

Westphal, V., Rizzoli, S. O., Lauterbach, M. A. et al. 2008. Video-rate far-field optical nanoscopy dissects synaptic vesicle movement. *Science* 320: 246–249.

Wiedenmann, J., Ivanchenko, S., Oswald, F. et al. 2004. EosFP, a fluorescent marker protein with UV-inducible green-to-red fluorescence conversion. *Proceedings of the National Academy of Sciences of the United States of America* 101: 15905–15910.

Xu, C., Zipfel, W., Shear, J. B., Williams, R. M., and Webb, W. W. 1996. Multiphoton fluorescence excitation: New spectral windows for biological nonlinear microscopy. *Proceedings of the National Academy of Sciences of the United States of America* 93: 10763–10768.

Yildiz, A., Forkey, J. N., McKinney, S. A. et al. 2003. Myosin V walks hand-over-hand: Single fluorophore imaging with 1.5-nm localization. *Science* 300: 2061–2065.

12

Molecular Resolution of Cellular Biochemistry and Physiology by FRET/FLIM

12.1	Introduction	12-1
12.2	Background	12-2
	FRET	
12.3	Presentation of State of the Art	12-6
	Intensity-Based Measurements: Intramolecular FRET • Intensity-Based Measurements: Intermolecular FRET • FRET Measurements from Fluorescence Kinetics	
12.4	Critical Discussion	12-19
	Shortcomings and Possible Pitfalls of FRET	
12.5	Summary	12-21
12.6	Future Perspective	12-21
	Instrumentation • Labels and Sensing Paradigms	
	Acknowledgment	12-23
	References	12-23

Fred S. Wouters
Gertrude Bunt

University of Göttingen

12.1 Introduction

Modern cell biology aims at understanding the inner workings of cells. This is an ambitious goal given the fact that the molecular machines that drive cellular reactions are unlike any counterpart in the macroscopic world. Molecular machines represent solutions to problems that are not optimized for their efficiency or for the minimum number of components that industrial design guides necessarily would impose. Rather, they were assembled ad hoc from components that were already available and their defining properties are robustness and adaptability, giving rise to properties like feed-back loops, cross talk, and redundancy (Coveney and Fowler, 2005; Ma'ayan and Iyengar, 2006). Furthermore, cellular machines can add, remove, and change their protein components, or reassign their function during full operation. These specifications complicate the identification of individual components and the relevance of their behaviors, i.e., cellular biochemistry, and the elucidation of the long-range causality of their actions down the chain of events to the desired effect, i.e., cellular physiology.

Fluorescence microscopy has played a pivotal role in the investigation of proteins in cells, particularly during the last ~15 years when the availability of the green fluorescent protein (GFP) (Tsien, 1998; Matz et al., 1999) and its ever-growing collection of spectral variants revolutionized the field. These genetically encoded fluorophores, which require no cofactors for the generation of fluorescence, can be fused by simple molecular biology techniques to virtually any protein of interest, and can then be expressed to follow the localization and dynamics of the tagged protein after expression in a wide variety of cell types, tissues, and organisms. The fluorescent proteins have enabled and revolutionized cell biology and have dramatically increased our knowledge of cells. For this reason, their discovery and pioneering use in biology were honored with this year's Nobel Prize in chemistry.

Due to the multimodal nature of fluorescence (Lakowicz, 1999), the fluorescent proteins carry more information than only their brightness, which give location and concentration information. Spectral information allows the simultaneous detection of multiple labeled proteins, and fluorescence anisotropy carries information on rotational velocity. One other modality of fluorescence, the fluorescence lifetime, i.e., the average duration of the excited state, or the delay time between absorption and emission of a photon, carries information on the direct molecular environment of the fluorescent label. These different modalities can be used and combined to extract detailed information on the "lifestyle" of the labeled protein, and from that, on its role in cellular physiology. Importantly, these measurements are performed in the context of intact living cells; high crowding conditions, compartmentalization, transport processes, and enzymatic processes set the stage and define the conditions for the activity and behavior of the investigated proteins. In a recent survey, 75% of cell biological researchers favored microscopy as a tool to study cells. Fluorescence microscopes were the most popular choice (85%), followed by confocal microscopes (57%) (Netterwald, 2008).

An understanding of molecular mechanisms requires the elucidation of changes in *connected* events upon exposure of living cells to stimuli. This demands the rigorous quantification of events. Therefore, where microscopy's primary role was once to provide the contrast between different regions, this contrast now has to be based on a solid physical basis that can be quantified.

With regard to obtaining quantitative information on the biochemical status, activity, and functions of proteins in cells, perhaps one of the most powerful fluorescence microscopy techniques is based on the photophysical phenomenon of Förster resonance energy transfer (FRET) (Clegg, 1996; Förster, 1948). In FRET, the energy stored in the excited state of the fluorescent molecule can be transferred by dipole–dipole interactions to populate the excited state of an extremely nearby located second fluorophore, provided this second fluorophore possesses suitable spectral properties that allow it to accept the energy content that is donated by the first fluorophore. Its exquisite dependence on the separation distance between the fluorophores confers its remarkable sub-nanometer resolution for their proximity, which makes this technique a unique tool for the investigation of interactions, conformational change, and posttranslational modifications of proteins. FRET, thus, provides the tools needed to reduce the action of complicated molecular machines to simple, measurable, biochemical events.

The occurrence of FRET has a number of characteristic consequences for the fluorescence parameters of the coupled donor–acceptor pair that can be used as diagnostic signs and that can be quantified in a microscope. One particularly useful implementation of FRET imaging uses the fluorescence lifetime as quantification metric. fluorescence lifetime imaging microscopy (FLIM) offers decisive advantages for the quantification of FRET (Clegg and Schneider, 1996; Esposito and Wouters, 2004b).

This chapter will explore the process of FRET, its detection and quantification in microscopy—especially using FLIM, and its use in the generation of integrative bioassays.

12.2 Background

12.2.1 FRET

12.2.1.1 Description

The process of FRET was first discovered from an observed loss of polarization in the emission of a fluorophore solution—upon polarized excitation—at high fluorophore concentrations (Förster, 1948). It was concluded that this was caused by the transfer of the excited state energy from the photoselected pool to fluorophores that did not share this strict photoselection. These measurements were the experimental basis for the formulation of the transfer process and its properties by Theodor Förster. Today, polarization-based or the related anisotropy-based FRET measurements are often used in spectrofluorometry.

It is also very common in the drug screening industry—in the form of a fluorometric-imaging plate reader (FLIPR) instrument (Sullivan et al., 1999)—as the measurement is very simple and robust, and because it allows assays to be performed in a "homogeneous format," i.e., requiring no washing steps to remove unbound components. The measurement requires the measurement of fluorescence emission in the polarization plane perpendicular and parallel to the direction of polarization of the excitation light. The same detection principle applies to microscopy, although here the distortions in the polarization of the light paths from the use of high numerical aperture objectives need to be considered.

The measurement of FRET by the loss of polarization represents a special case as the excited state energy is transferred between spectrally identical molecules. This is called homo-FRET, and is also often referred to as (intermolecular) energy migration, a term that was also originally used by Theodor Förster: "zwischenmolekulare Energiewanderung."

However, FRET pairs are usually composed of spectrally different fluorophores, i.e., hetero-FRET. There are strict requirements for the spectral properties of fluorescent species that can undergo FRET and, thereby, for the choice of suitable fluorophore pairs. The accepting species should be capable of accommodating the energy content that is donated. For fluorescent molecules, where energy levels are expressed by the excitation and emission spectra, this requirement is matched when there is a sufficiently large overlap between the donor emission spectrum and the acceptor excitation spectrum. However, this optimization is directly counteracted by the experimental requirement for the selective spectral acquisition of both species. These opposing demands thus create a compromise that has to be considered for all new FRET pairs.

For efficient FRET, the donor molecule has to possess a high quantum yield (Q, the probability of photon emission per absorbed photon), so that most of the excited state energy is available for transfer and is not lost to the environment in a nonradiative manner (e.g., by quenching). The acceptor molecule has to be highly absorbing in the energy/spectral range where the donor molecule has its maximal emission, i.e., has to possess a large molar extinction coefficient (ε_λ) at those wavelengths.

The excited state energy of the donor is transferred by the coupling of the electromagnetic oscillations of the transition dipoles in the FRET pair. However, the description of the properties of FRET dyes in radiative terms can lead to the common misconception that the donor would emit a photon that is then absorbed by an acceptor. This process, called trivial reabsorption, only occurs for extremely high concentrations, where in fact FRET dominates the fate of fluorescence and prevents the emission of photons.

Although FRET is often referred to as "fluorescence resonance energy transfer," it is not a fluorescent process, and in honor of his work on this topic, the "F" in FRET nowadays refers to Förster himself.

12.2.1.2 Distance and Orientation

The process of FRET (E) shows a sixth order dependence on the separation distance between the donor and acceptor fluorophores.

$$E = \frac{R_0^6}{R_0^6 + R^6} \tag{12.1}$$

where
 R is the separation distance
 R_0 is the Förster or critical distance: the distance where 50% of the molecules undergo FRET, i.e., where the FRET efficiency is 50%

This distance can be calculated from the known spectral qualities of the fluorophores and serves as a quality parameter for efficient FRET coupling.

$$R_0^6 = \frac{\kappa^2}{n^4} Q_D J; \quad J = \int f_D(\lambda)\varepsilon_A(\lambda)\lambda^4 \, d\lambda \tag{12.2}$$

where

    κ^2 is the orientation factor

    n is the refractive index of the medium

    Q_D is the donor quantum yield

    J is the overlap integral in which f_D is the normalized donor fluorescence

    ε_A is the molar extinction coefficient of the acceptor

The sixth order dependence creates exquisite sensitivity around the Förster distance. This gives rise to two important properties of the FRET process: (1) FRET constitutes an optical switch between a high efficiency to an undetectably low efficiency, and (2) This R_0 switching point lies in the distance range of average protein size, i.e., between 2 and 6 nm for most FRET pairs in use in biomedical research, and provides Ångstrom resolution in this distance window (Figure 12.1). Consequentially, FRET is ideally suited for probing protein–protein interactions and small structural rearrangements within proteins or multi-protein structures. In fact, a high FRET efficiency represents the ultimate co-localization.

In terms of resolution, FRET microscopy is unique in that FRET becomes easier to detect at smaller separation distances, approaching 100% for a distance of zero nanometer. One could therefore envisage FRET microscopy as a form of super-resolution microscopy method among those that broke the diffraction limit in far-field optical imaging in recent years (Betzig et al., 2006; Hell and Wichmann, 1994; Klar et al., 2000). However, one should realize that this process does not provide an image of high spatial resolution, but rather provides an additional layer of information on top of the normal diffraction-limited image.

Importantly, as FRET is a resonant process between dipoles it is highly sensitive to the relative orientation between the fluorophores. The dipole orientation is described in the spatial parameter κ^2 that varies between 0 for perpendicularly oriented dipoles and 4 for a collinear orientation. Parallel-oriented dipoles possess a κ^2 value of 1. This orientation factor scales linearly in the description of the Förster distance (see Equation 12.1), and thus in the efficiency of FRET. In most discussions on the distance dependence of FRET, it is assumed that the fluorophores can sample all spatial orientations during the time in which they can transfer energy, i.e., during the time when the donor is in the excited state, giving rise to an average value for κ^2 equal to $^2/_3$ (Dale et al., 1979). However, this assumption cannot always be respected, especially in the case of the fluorescent proteins where the fluorophore is immobilized inside a β-strand barrel. Furthermore, a fluorescent protein moiety folds together with the protein to which it is fused and might interact with its surface. Unfortunately, statistical analysis shows that under conditions where free movement cannot be assured, it is very much more likely to obtain a low

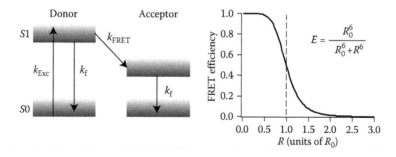

FIGURE 12.1 Fluorescence transitions and distance dependence of FRET. Left: Simplified Jablonski diagram of energy levels in a coupled donor and acceptor fluorophore system. Rates shown are k_{Exc}, rate of excitation, k_f, rate of fluorescence emission, and k_{FRET}, the rate of energy transfer. Right: Efficiency of FRET as a function of fluorophore separation distance expressed in units of the Förster distance (R_0) under the assumption of complete spatial freedom during the duration of the lifetime ($\kappa^2 = ^2/_3$).

than a high κ^2 value (Van der Meer, 1999, 2002), making it more difficult to measure efficient FRET. In generating FRET constructs, it is therefore strongly advised to fuse the fluorescent protein to the protein of interest using a linker of a few amino acids, preferably using a small inert sequence such as to generate a flexible connection. The conformational restriction of the transition dipole is also the reason why intramolecular FRET, i.e., where the donor and acceptor fluorophores are contained in the same polypeptide, is generally more susceptible to chemical fixation than intermolecular FRET (Anikovsky et al., 2008), especially in the case of small fusion peptides.

12.2.1.3 Consequences

The occurrence of FRET in a coupled pair of fluorophores carries a number of photophysical consequences that can be used for its detection and quantification. As discussed earlier, the loss of polarization in the emission of a fluorophore can be observed for homo-FRET.

In homo-FRET, the fluorophores are spectrally identical. For hetero-FRET, however, spectral changes occur. As the donor excited state can, in addition to fluorescence emission, also depopulate via FRET, less energy is dissipated via its radiative pathway. As a consequence, the emission yield of the donor will be reduced compared to the situation where there is no FRET. This effect is called "quenched donor emission." In other words, the brightness per unit concentration fluorophore, i.e., its quantum yield, will be reduced. The quantum yield is given by the radiative decay rate divided by the sum of all decay rates. From this, it can be easily appreciated that the addition of an additional depopulation rate in the form of FRET coupling will reduce the quantum yield:

$$Q_D = k_f/(k_n + k_f) \text{ for the situation without FRET} \tag{12.3}$$

$$Q_D = k_f/(k_n + k_f + k_{FRET}) \text{ when FRET occurs} \tag{12.4}$$

where
  k_f is the rate of fluorescence emission
  k_n is the rate of nonradiative decay
  k_{FRET} is the rate of FRET

Secondly, upon excitation of the donor, energy will be transferred to the acceptor to populate its excited state. The acceptor thus becomes "excited" by FRET and will fluoresce without having been directly excited. This process is called "sensitized acceptor emission" (Figure 12.1).

Other observables of FRET are related to changes in the duration of the excited state of the donor fluorophore. As the FRET-coupled donor experiences an additional depopulation channel in the form of the FRET process by which it can return to the ground state, the average time it spends in the excited state will reduce. This average time is called the fluorescence lifetime τ, and is typically in the order of nanoseconds for most biologically relevant fluorophores. The fluorescence lifetime is a constant and robust fluorescence metric that is characteristic for a given fluorophore and is independent of the concentration of the fluorophore and rather insensitive to trivial factors. The lifetime does depend on temperature, but this can be easily controlled in a microscope setting, and quenching processes will reduce its value. The lifetime relates to the efficiency of the fluorophore to generate a fluorescent photon from the energy absorbed by excitation. That is, the observed donor lifetime (τ_{obs}) is directly related to the quantum yield (Q_D) by its ratio with the radiative lifetime (τ_{rad}), i.e., the lifetime that the fluorophore would exhibit in the absence of nonradiative decay. As the latter is a (hypothetical) constant, it merely scales the actual lifetime with the quantum yield:

$$Q_D = \tau_{obs}/\tau_{rad}; \quad \tau_{obs} = 1 \Big/ \sum k_{depopulation} \tag{12.5}$$

so that from Equations 12.3 and 12.4 and

$$E = k_{\text{FRET}} \Big/ \sum k_{\text{depopulation.}} \tag{12.6}$$

It follows that

$$E = 1 - \frac{Q_{\text{FRET}}}{Q_{\text{non-FRET}}} = 1 - \frac{F_{\text{DA}}}{F_{\text{D}}} = 1 - \frac{\tau_{\text{DA}}}{\tau_{\text{D}}} \tag{12.7}$$

wherein the equivalence between the changes in Q and the fluorescence F_{D} is due to the fact that the molar extinction coefficient is not affected by FRET. The subscripts DA and D refer to the FRET (donor and acceptor) situation and non-FRET (only donor) control situation, respectively. The changes in lifetime are obvious from Equation 12.5.

Fluorophores exhibit a normal distribution of lifetimes. This compares favorably to emission spectra that are typically skewed for most fluorophores, and that are typically broader than the lifetime distribution. This means that changes in lifetimes can be detected more easily than shifts in emission wavelength. Also, in contrast to the spectral changes in the emission, the FRET-induced shortening of the fluorescence lifetime is proportional to the efficiency of FRET, and the lifetime is independent of the concentration of the fluorophore and the light path, both of which are variable in cells. Furthermore, a FRET determination from the donor lifetime requires a measurement in only one spectral window. Most importantly, in contrast to the sensitized emission, lifetimes are intrinsically calibrated in that both extremes of the (linear) scale are known and/or experimentally achievable: with no FRET, the lifetime is equal to the characteristic lifetime of the fluorophore, known from literature and from control experiments and with 100% FRET, the lifetime is equal to zero. Therefore, the measurement of the donor fluorescence lifetime provides a robust and sensitive quantitative determination of the efficiency of FRET. As the linear relation of the lifetime and the quantum yield indicates, lifetime determinations are actually equivalent to the measurements of donor quenching by FRET (see Equations 12.5 through 12.7). However, the measurement of the brightness of the donor per unit concentration by lifetimes does not require a concentration reference, i.e., is not a ratio measurement.

One final consequence of the shortening of the duration of the excited state of the donor is that the donor-excited state becomes, per unit time, less vulnerable for excited state reactions that would render it nonfluorescent. These reactions are known as "bleaching." However, as the bleaching rate can be assumed to be constant at constant illumination power, this irreversible depopulation route contains information on the entire photophysical cycle. In fact, because the probability of photobleaching is relatively low, yielding a rate that is several orders of magnitudes lower than the other transition rates, it transfers information on fast processes like FRET to a conveniently long timescale. From the bleaching decay at constant illumination, the bleaching time constant(/s) (τ_{bleach}) can be obtained on a pixel-by-pixel basis that is (/are) inversely proportional to the fluorescence lifetime(/s).

$$I(t) = I_{\text{background}} + I(0) \bullet e^{-t/\tau_{\text{bleach}}} \tag{12.8}$$

12.3 Presentation of State of the Art

12.3.1 Intensity-Based Measurements: Intramolecular FRET

Sensitized emission is related to the concentration of acceptors that participate in FRET and receive donor energy. The opposite fluorescence gains of both spectral sides allow the simple and highly sensitive, concentration-independent measurement of dividing the sensitized acceptor emission by the quenched donor emission.

However, these measurements suffer from spectral contamination. Emission spectra tail at the red side, so that donor photons that "bleed-through" into the acceptor channel can be confused with the sensitized emission photons. Excitation spectra tail at the blue edge. Donor excitation will therefore generate acceptor photons by direct excitation, which contaminate the acceptor photons that were indeed generated by FRET (Figure 12.2). As long as the donor and acceptor concentrations have a fixed stoichiometry, as for instance given in the case for the popular FRET probes where the donor and acceptor fluorophores are contained in the same polypeptide chain, this spectral contamination is at least constant, and changes in the ratio or sensitized emission can be attributed to FRET. In other words, contrast is generated by FRET, but its quantitative value is limited.

For this reason, these types of measurements are often performed in regions of interest in the cell for the investigation of kinetic parameters of responses.

Intramolecular FRET measurements thus present a special case that is favorable for simple ratio measurements (Bunt and Wouters, 2004). This has made this type of FRET assay very popular; one could simply sandwich a conformationally responsive sensing element between a suitable pair of fluorescent proteins, typically a cyan fluorescent protein (CFP) donor and a yellow fluorescent protein (YFP) acceptor, and measure the spectral intensity changes as a result of FRET.

The "plug-and-play" character of these types of assays has provided a versatile toolbox for the cell biologist. As protein activation is often accompanied by relatively large conformational changes in the protein, and these conformationally active sites are often known, they can be immediately used for the construction of a biosensor. Often, these sensors contain two domains that are grafted from the same or different signaling/effector proteins, one that responds to a biochemical modification, e.g., a post-translational modification like a phosphorylation, and another that specifically recognizes and binds to the first domain in its modified form. Another popular example is the generation of proteolytic sensors, e.g., for the activity of apoptotic cell death–associated executioner caspases. Here, a small recognition peptide is cleaved, generating a large FRET contrast between the intact sensor (FRET) and the cleaved sensor (no FRET). The latter type of sensors typically have a larger dynamic range than the conformationally sensitive sensors where a no-FRET condition is not normally achieved; the donor and acceptor moieties remain linked throughout the trajectory from an inactive to an active state of the sensing domain. Rather, the change in dipole orientation dominates the response of these sensors, and this

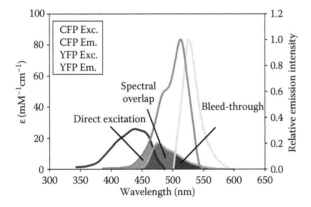

FIGURE 12.2 Excitation and emission spectra of CFP and YFP fluorescent proteins. Excitation spectra are scaled to the molar extinction coefficient, emission spectra to the product of the molar extinction coefficient and the quantum yield, i.e., to the relative brightness of the fluorophores. Shown is a 1:1 stoichiometric complex. The spectral overlap required for FRET, and the spectral contaminants: donor bleed-through and acceptor direct excitation are shown.

cannot be easily and rationally optimized. For this reason, intramolecular sensors take a lot of effort to optimize their dynamic range by the mutation of the linkers, truncation of the sensing domain, and other manipulations.

There are three detection schemes that can be followed for intramolecular sensors (Figure 12.3):

1. The most common scheme is the ratio of donor emission and acceptor emission upon excitation of the donor. This ratio is highly sensitive to FRET because the changes in the donor (decrease with increasing FRET) and acceptor (increase with increasing FRET) emission yields are opposite. In a wide-field microscope, this requires switching the emission filter. Real-time imaging of the ratio is possible with an imaging beam splitter, which projects both spectral signals side by side on the same CCD chip, or by the use of a dual camera arrangement. In a confocal microscope, the dual emission information is generated simultaneously on two spectrally dedicated detectors and can be treated separately to form a real-time ratio image.

2. A disadvantage of the previous scheme is that it is very sensitive toward noise. As the sensitized acceptor fluorescence can be very low, noise will be amplified. An alternative, more robust, scheme therefore involves referencing the quenched donor emission to the (larger) directly excited acceptor emission, i.e., dividing the donor emission upon excitation of the donor and

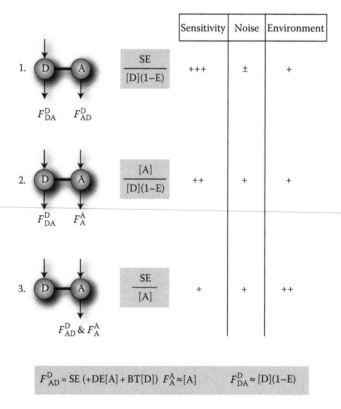

$$F_{AD}^{D} = SE\ (+DE[A] + BT[D])\quad F_{A}^{A} \approx [A]\qquad F_{DA}^{D} \approx [D](1-E)$$

FIGURE 12.3 Different ratiometric detection schemes for intramolecular FRET constructs. Arrows show excitation and emission combinations. Fluorescence results are denoted with excitation wavelength (superscript D or A to indicate optimal wavelengths for donor or acceptor, respectively), emission wavelength (First subscript position, D or A), and the presence of the acceptor, i.e., the FRET case, is indicated for the donor fluorescence by an A in the second subscript position. Ratios of fluorescence products are shown in the order in which an increase in FRET results in an increase in ratio. Relative detection sensitivity, noise sensitivity, and sensitivity for environmental factors that differentially affect the fluorophores (e.g., pH for CFP-YFP, see text) are indicated. + signs indicate favorable conditions; E, FRET efficiency; SE, sensitized emission; DE, direct excitation.

the acceptor emission upon excitation of the acceptor (as measure for the concentration of the sensor). Whether the improved signal-to-noise level outweighs the loss in sensitivity by this method has to be established on a case-by-case basis. This alternative ratio measurement is similar to the FqRET approach that was first introduced by Allen to measure actin capping (Allen, 2003). FqRET stands for fluorescence quenching resonance energy transfer. In this approach, FRET was measured from the quenching of a donor fluorophore undergoing FRET to a red-shifted nonfluorescent acceptor. The extent of FRET-induced quenching could thus no longer be obtained by comparison with the sensitized emission or direct acceptor excitation/emission, as this molecule did not fluoresce. Allen solved this by introducing a third dye, blueshifted with respect to the donor dye that was attached to the same molecule carrying the donor fluorophore. The division of the donor fluorescence by the fluorescence of the reference dye here provided the normalization of donor emission to the concentration of the donor-labeled molecule. In a later implementation, where we generated a nonfluorescent mutant of the YFP for efficient FRET with GFP, we decided to normalize the FRET-quenched donor (GFP) fluorescence by a red-shifted dye (Cy5 on an antibody that recognized the GFP-labeled protein) (Ganesan et al., 2006). This has the advantage that we exclude the possibility of FRET between the reference dye and the donor dye by introducing a large spectral distance between them. This alternative emission ratio measurement for "classical" intramolecular FRET probes can thus be seen as a special case of FqRET, where the optically inert concentration reference dye is the acceptor fluorophore.

3. One potential problem with the detection of FRET from the ratio of the donor- and acceptor-derived signals is that their emission might be affected differently inside the cell. For instance, YFP is sensitive toward pH and halide ions, whereas CFP is not, and both fluorescent proteins might experience different folding kinetics that is also dependent on their molecular environment. When a cell enters a cell death path, its cytoplasm will acidify. FRET measurements on CFP–YFP sensors will report on a reduction of FRET as the FRET-generated sensitized emission of the YFP acceptor will appeared to have diminished. Differential effects might also be created by the addition of pharmacological compounds. Furthermore, these compounds are often fluorescent themselves, and contribute differentially to both emission channels from which FRET is determined. In this third detection scheme, these effects are minimized by sampling only one emission channel and varying the excitation between donor and acceptor. The two components generated now are (1) the emission of YFP upon excitation of CFP (containing FRET and some direct excitation) and (2) the emission of YFP upon excitation of YFP (a measure of the concentration of YFP and therefore, of the sensor). Since FRET and sensor concentration are read from the same emission channel, the impact of factors that affect the donor and acceptor fluorescence differentially is reduced. Figure 12.4(a) shows the application of the pH-insensitive ratio method for the proteolytic cleavage of a FRET biosensor during apoptotic cell death. The sensitivity of this scheme is again lower than the classical first method, and might also be lower than the second method: The information about FRET is contained in the reduction in donor emission and in the extent of sensitized emission. However, in this method, the donor fluorescence is not sampled and the sensitized emission is typically contaminated with some direct acceptor excitation, reducing the sensitivity toward FRET. Nevertheless, there might be situations where this loss of sensitivity is compensated by the insensitivity toward differential environmental factors. Furthermore, the measurement is easier to perform in most wide-field microscopes, as there is no need to change the emission filters between both measurements (which would carry the risk of mechanical disturbances and would take some time), or to project two spectral images on the same or different CCD chips (with pixel-shift issues to solve). Instead, the excitation wavelength in the coupling of the microscope is switched. Switching laser or modern light-emitting-diode sources, or changing the wavelength of a monochromator can be done very rapidly. Even changing the exciter filter for a mercury lamp can be done rapidly by the use of a motorized filter wheel that does not have to be physically incorporated into the microscope, as is the case for emission filters.

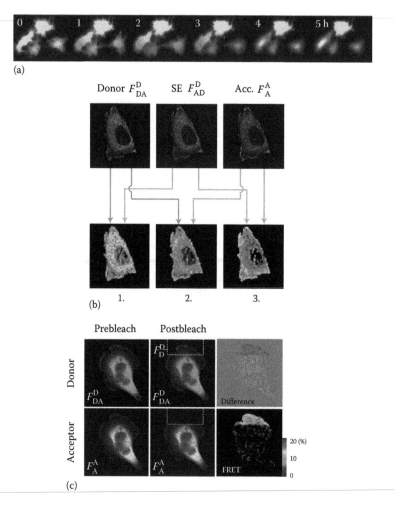

FIGURE 12.4 (See color insert following page 15-30.) FRET measurement examples in cell biology. (a) MCF7 mammary carcinoma cells expressing a CFP–YFP fusion sensor for caspase 3 proteolytic activity. Cleavage of the DEVD recognition peptide releases both fluorophores, at the cost of FRET. Caspase 3 was activated by an induction of apoptosis by incubation with Fas ligand for the indicated time periods. FRET was measured from the ratio of acceptor direct excitation over sensitized emission (Method 3 in Figure 12.3) so that the ratio increases with decreasing FRET, i.e., increasing caspase activity. (b) Conformational activation of focal adhesion kinase (FAK). An intramolecular biosensor for activation-correlated conformational change in the FERM domain of FAK ("FERM sensor") shows higher FRET in the focal adhesions as in the cytoplasm. The three ratio methods presented in the text and in Figure 12.3 are shown. The high intracellular signal in method 1 is most likely caused by substantial autofluorescence in the CFP image. (c) Confirmation of FRET in the FERM sensor by acceptor photobleaching (see Section 12.3).

Figure 12.4(b) illustrates the application of all three methods to a recently published biosensor for a conformational change in the FERM domain of focal adhesion kinase (FAK) upon integrin activation (Papuscheva et al., 2009). The second method was used in the publication.

Besides the relative simplicity in their generation and detection, these intramolecular sensors also have the advantage of being "innocent bystanders." Their borrowed signaling/effector protein domains typically only perform a passive, responsive, role and do not participate in the cellular machinery. For this reason, they can be expressed in high concentrations to report with high sensitivity to a given biochemical activity in the cell. Other than buffering the activity of enzymes or the concentration of recognized intermediate products, they are not expected to grossly perturb the physiology of the cell. This is a central argument for all FRET—and indeed all other fluorescently or otherwise labeled protein that is introduced exogenously into the intracellular environment. Perturbation should always be minimized and verified by the experimenter, and this has driven the development of brighter fluorescent probes and more sensitive detection equipment (like electron multiplication CCD cameras). If possible, the expression levels of exogenously added proteins should ideally not exceed the endogenous level, which can be conveniently measured by immunofluorescence labeling of the host protein upon transfection. The relative immunofluorescence levels between the transfected (also expressing the fluorescence from the fluorescent protein fusion) and untransfected cells give the relative overexpression level.

12.3.2 Intensity-Based Measurements: Intermolecular FRET

When two fluorescently labeled proteins meet in the cell, their concentration relationship is not given. When acceptors outnumber donors, direct acceptor excitation contaminates the FRET signal, when donors outnumber acceptors, donor bleed-through dominates.

The solution to this problem is to measure the concentrations of the donor and acceptor species by their fluorescence emission upon direct excitation at their optimum wavelengths, and calculating the extent of their spectral contamination to the FRET signal (Wouters et al., 2001). This requires one sample that only expresses the donor fluorophore that is excited at its excitation optimum and whose emission in the acceptor window is measured. The amount of signal relative to the emission in the proper donor window represents the fraction donor bleed-through. A sample that expresses only the acceptor fluorophore is used for the quantification of acceptor direct excitation: the emission in the acceptor emission window upon excitation at the donor excitation maximum wavelength is related to the emission upon excitation at the optimal wavelength for acceptor excitation. These two constant, microscope- and filter-specific factors can be used to correct the sensitized emission, i.e., the "FRET" measurement. For the estimation of these two effects, two additional acquisitions in the donor–acceptor FRET sample are required: the concentration of the donor is estimated by direct excitation/emission detection of the donor fluorophore, which is multiplied by the bleed-through correction factor to obtain its contribution to the FRET measurement. Note that this is quenched by an amount that is proportional to FRET, but the bleed-through contamination is also caused by the remaining donor fluorescence. The concentration of the acceptor is estimated by direct excitation/emission detection of the acceptor fluorophore, which is multiplied by the direct excitation correction factor to obtain its contribution to the FRET measurement. A simple subtraction of these two contributions from the FRET signal returns the spectrally pure sensitized emission.

The same information can be obtained from spectrally resolved measurements. Rather than centering the detection windows on the maxima of the donor and acceptor wavelengths, one can define additional windows to span the entire visual spectrum occupied by both emissive species, or scan the entire wavelength tract as in cuvette-based spectrofluorometry by the use of spectrographs, tunable filters, or an interferometer-based spectral detector. By linear unmixing of the emission, the contributions of the donor and the acceptor can be deconvolved and used for the estimation of FRET (Wlodarczyk et al., 2008; Zimmermann et al., 2002).

After successful correction of the sensitized emission signal, the remaining challenge is to express the result in a concentration-independent manner. Numerous algorithms have been put forward that aim at deriving a value that relates to the FRET efficiency and/or the concentration of the fluorescent species that undergo FRET. However, these might be too specialist or elaborate for the purpose of obtaining reproducible FRET contrast. The normalization of sensitized emission to concentration is an important step as it can change the meaning of the data. There are three commonly used approaches for the normalization: (1) to acceptor concentration, (2) to donor concentration, and (3) to the concentration of the complex. Some knowledge on the relative concentrations of the fluorescent proteins is required to decide on the proper approach. Acceptor normalization produces a signal that expresses the fraction of the acceptors that fluoresce by sensitized emission. Donor normalization relates sensitized emission to the remaining, quenched, donor fluorescence, i.e., to the donor molecules that successfully couple via FRET, albeit in a less intuitive manner. Apart from interpretation differences between these normalization ratios, the most important guiding factor that should be considered is the relative abundance of both fluorescently labeled FRET partners. For the efficient detection of sensitized emission, all acceptors should have the chance to interact with a donor, i.e., the amount of donor-labeled protein should exceed that of the acceptor (by a factor of about three). This requirement can be set experimentally, and it is therefore important to know the relative expression of both proteins in the biological setting, and to label the most abundant protein with the donor fluorophore. The other reason for this preferred relation is that large amounts of endogenous unlabeled proteins would compete with the donor-labeled counterparts to degrade the FRET efficiency. Normalization to the donor fluorescence now loses relevance, especially if its expression level is very much higher than that of the acceptor. On the other hand, if the acceptor is in excess, not only does it outcompete the donors that couple for FRET, reducing the amount of FRET that could have been achieved in the sample, but the normalization to the acceptor concentration also becomes less informative for the same reason. In this case, normalization to the donor concentration would be more suited. It is therefore important to consider the issue of relative abundance when designing and analyzing FRET experiments. Theoretically, the best normalization parameter is the concentration of the complex. However, this concentration is not known as the FRET experiment is performed with the purpose of estimating the concentration of interacting proteins in the first place. Nevertheless, the probability of its formation depends on the concentration of both donor and acceptor species and therefore scales with the product of their concentrations. For convenience, the square root of the product of both fluorescent signals is typically used as normalization parameter.

Since a sensitized emission measurement already contains information on the donor and acceptor concentrations in the calculation of the spectral contaminations from these species, it might be informative to show all three normalization signals, as they contain the biologically relevant information on the local prevalence of FRET partners and the formation of complexes.

In this respect, it should be noted that it is generally good practice to investigate FRET data for correlations with donor/acceptor ratios. With increasing donor/acceptor ratio, FRET by sensitized emission should initially increase and then reach a plateau value that indicates a saturation of the acceptors, i.e., the maximum concentration of FRET-coupled pairs. The shape of this curve can be used to discriminate between dimers and higher-order oligomers/aggregates of interacting proteins: dimers are less sensitive to increases in donor concentrations as they only couple to one partner, oligomers can couple with more partners, and consequently FRET is correlated to the concentration of donors. It should be noted that failure to reach a saturation level is indicative of a nonspecific interaction. The donor/acceptor ratio can be further be experimentally diversified in the sample(s) by the selective photobleaching of increasing amounts of acceptor fluorophore. It is also important to realize that the same arguments as above hold for FRET measurements on the donor side, i.e., donor lifetime, donor photobleaching, and donor quenching (/acceptor photobleaching, see Section 12.3.3) measurements, but with inverse donor–acceptor dependence: acceptor excess for efficient FRET participation, and saturation with increasing acceptor/donor ratios.

The corrections for spectral contamination that are required for a normalized sensitized emission experiment can introduce substantial calculation noise, as many arithmetic steps are needed. Especially if (one of) the signals are (/is) dim, this can create a result that is no longer meaningful. Unfortunately, there is no straightforward way to estimate and set an acceptable threshold level of noise in sensitized emission experiments, also because the distribution of a zero FRET reference cannot be evaluated, e.g., by leaving out the acceptor. In contrast, control measurements for donor quenching, bleaching, and lifetime measurements produce value distributions that center symmetrically, ideally Gaussian, around the zero average. The spread of these distributions relates to the experimental error and allows a proper statistical interpretation of value changes connected with FRET events.

Intensity-based FRET experiments on large numbers of cells allow high-resolution correlations, precise measurements, and statistics in flow cytometry (Nagy et al., 2006). The conditions of reduced complexity and large sample sizes permit relatively unproblematic and straightforward FRET measurements from emission intensities.

The most quantitative treatment of sensitized emission involves a reference transformation of sensitized emission to the corresponding amount of donor quenching (Van Rheenen et al., 2004). This is achieved by calibration using an intramolecular FRET construct where the FRET efficiency can be changed experimentally, i.e., a sensor for an inducible biochemical reaction, or by an acceptor photobleaching step. A convenient calibration sensor is the calcium-sensitive cameleon probe (Palmer and Tsien, 2006), but proteolytic sensors for inducible proteases, such as apoptotic caspases, can also be used. It is important to note that sensitized emission and donor quenching are related by their difference in molar extinction coefficient. It is therefore essential to do the transformation measurement using a sensor that contains the same fluorophores as in the FRET experiment. By this transformation, the problem of unknown scaling of sensitized emission by the absence of a reference point at 100% FRET is elegantly circumvented. This way, true FRET efficiencies can be determined from sensitized emission measurements.

There are basic differences between intensity-based FRET measurements performed in a wide-field versus a confocal microscope. One advantage of the confocal microscope is that, due to the simultaneous acquisition of donor and acceptor emissions during scanning, the measurement is less sensitive toward motion artifacts in the cell. The solution for the wide-field detection is to use an imaging beam splitter "dual view" arrangement to split the donor and acceptor spectral information and project these side by side on the same CCD chip (or on two spectrally dedicated cameras). Furthermore, the FRET information is obtained from a discrete optical plane in the confocal microscope, and not, as is the case for wide-field systems, integrated in the axial direction. For spatially restricted signals, e.g., a membrane-based response, the information can be obtained without contamination from out-of-focus signals. Disadvantages of the confocal measurement are the restriction to the available laser lines, e.g., an optimal laser line for CFP excitation is not widely available in most confocal microscopes, and the dependence of spectral cross talk on detector gain settings between experiments. The latter point creates the inconvenient demand for bleed-through and acceptor direct excitation measurements in each experiment. With some preparation, the cells under investigation can be co-seeded with the donor-only and acceptor-only expressing cells for in-experiment control measurements, that also provide an immediate visual feedback for the successful subtraction of the spectral contamination contributions to the sensitized emission.

A very elegant way to isolate the directly excited acceptor emission, which is typically the largest contamination (e.g., 50%–70% of the total signal in the "FRET channel" for a CFP–YFP couple) is to include polarization information (Rizzo and Piston, 2005). Directly excited acceptor emission maintains the polarization from the excitation light, whereas sensitized emission is depolarized because it is generated by FRET.

The combination of polarization and spectral information is extremely powerful in the special case of FRET measurements on single molecules. Here, stoichiometry issues that confound sensitized emission measurements are nonexistent as information on the number of fluorescent molecules in a sparsely labeled sample is immediately accessible: two molecules fluoresce twice as bright. The anti-correlation of donor quenching and acceptor-sensitized emission is diagnostic for FRET in this simplified non-ensemble situation. By splitting the donor and acceptor information also in the two perpendicular polarization directions, information on the relative orientation of the transition dipoles in the fluorophores of the interacting single molecules is obtained that can be used to set limits to the κ^2 orientation factor, and thus on the separation distance range (Dale et al., 1979). Here, FRET can thus truly be used as spectroscopic ruler. This form of multimodal microscopy (Webb et al., 2006; Schaffer et al., 1999) has become a powerful tool for the investigation of structural aspects and dynamics of (supra)molecular complexes at a resolution that is comparable to crystallographic data.

12.3.3 FRET Measurements from Fluorescence Kinetics

12.3.3.1 Photobleaching

One of first FRET measurement from fluorescence kinetics was the quantification of donor photobleaching rates (Jovin and Arndt-Jovin, 1989). As explained in the background section, FRET effectively protects the donor fluorophore from photobleaching by reducing the time it spends in the excited state, where photobleaching takes place. The measurement is simple, requiring only the repetitive timed acquisition of fluorescence images during continuous illumination with constant power. The images are then fitted on a pixel-by-pixel basis to an exponential decay function (Equation 12.8) in order to derive its pre-exponential time constant, the bleaching lifetime. This value is inversely proportional to the fluorescence lifetime, i.e., the duration of the excited state. The advantage of donor photobleaching, the transfer of information on the extremely rapid fluorescence transitions by a slow, cumulative readout is also its biggest disadvantage. The measurement typically takes minutes, in which time the object is not allowed to move.

This time cost, of course, implies that the measurement is not suited for living cells, and its practical implementation requires exceptionally stable microscope mechanical properties and illumination power stability. Still, the measurement is immediately concentration independent and quantitative. One improvement on the method is to scan/bleach the sample in a single line in a confocal microscope in xt-modus, revisiting the same line repeatedly in time, and bleaching the line very rapidly. The resulting image then shows in one dimension the spatial information through a cell along the scanned line, and in the other dimension the bleaching kinetics.

Unlike the direct acceptor excitation problem in FRET, where the exclusive excitation of a blue-shifted fluorophore is not possible without the simultaneous cross-excitation of a red-shifted fluorophore, a red-shifted fluorophore can be excited without the excitation of a blue-shifted fluorophore. This provides a convenient way to create a region of interest where the donor emission can be measured in the presence (F_{DA}, the donor fluorescence before acceptor bleaching) and absence (F_D, the donor fluorescence after photobleaching) of the acceptor. FRET quenches the donor fluorescence F_D by a factor equal to the FRET efficiency (E, ranging from 0 to 1), so that the remaining donor fluorescence F_{DA} is reduced by a factor $1 - E$. This reduction is recovered by the bleaching of the acceptor; the FRET efficiency is equal to the increase in donor fluorescence normalized to the unquenched donor fluorescence F_D.

$$F_{DA} = F_D - E \bullet F_D = (1 - E)F_D;$$

$$E = 1 - \frac{F_{DA}}{F_D} = \frac{F_D - F_{DA}}{F_D} \tag{12.9}$$

This acceptor photobleaching method (Bastiaens et al., 1996, 1997; Miyawaki and Tsien, 2000; Wouters et al., 1998) has now become a standard FRET microscopy method, as this simple measurement, conveniently compatible with the scanning operation of a confocal microscope, is the easiest method to derive the FRET efficiency quantitatively. For an example, see Figure 12.4(c). This measurement is typically performed as end-point measurement, where the majority of the acceptor is bleached in one step, and the resulting donor unquenching is acquired in a single measurement. The donor measurement could, however, also be performed during the photobleaching of the acceptor (Van Munster et al., 2005), e.g., to obtain information on the difference between the dimeric (rapid unquenching) and oligomeric interactions (gradual unquenching) as was explained for sensitized emission FRET microscopy above.

When generating a region where the acceptor is photobleached, the remainder of the cell serves as a negative control. Here, the difference is zero and the resulting FRET efficiency should be distributed sharply around zero. If the distribution is shifted to nonzero values, the cell has moved between the two measurements. Translations in the plane of focus can be corrected for by image correlation routines, but drifts in the axial direction are difficult to accommodate. Opening the confocal pinhole or taking a small number of z-images could alleviate this problem, but some effort should also be invested in finding sources for the mechanical instability of the microscope, mostly related to temperature differences.

Acceptor photobleaching has been combined with fluorescence recovery after photobleaching (FRAP) measurements of protein motility to obtain information on the diffusion/transport kinetics of a FRET complex (Van Royen et al., 2007).

12.3.3.2 Fluorescence Lifetime

The relation of the fluorescence lifetime with the quantum yield (Equation 12.5) shows that it represents a measurement of the molecular brightness of fluorophores. This is a highly useful additional metric of fluorescence in itself, and it can be used to separate fluorescent species with similar spectral properties (Pepperkok et al., 1999). The fact that different lifetime species combine linearly in the observed lifetime of a sample can be used to estimate the absolute concentration of a fluorophore if mixed with a known concentration of a fluorophore with a different lifetime. This approach can be used for normalization purposed in advanced optical cellular biosensors.

However, lifetime imaging has its major application in the quantification of quantum yield changes, as occurs during FRET, as the fluorescence lifetime directly provides information on the presence of additional nonradiative transition rates. A typical example of FRET–FLIM is shown in Figure 12.5. Many fluorescent probes for small analytes and other environmental conditions change their quantum yields, which translates into changes in emission intensity. However, these changes are difficult to quantify in cells where the light path is variable and an isotropic distribution of the probe cannot be guaranteed. This has prompted the development of ratiometric probes that correct for concentration differences. When reporter dyes change emission by changes in quantum yield, lifetime imaging can thus be used directly for their quantification. Of course, when these probes change their fluorescence brightness by changes in extinction coefficients, the number of emitters change, but without affecting their lifetimes.

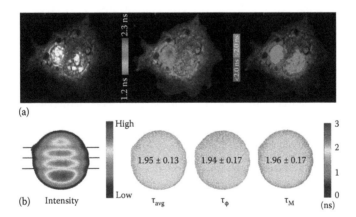

FIGURE 12.5 (See color insert following page 15-30.)   Lifetime detection of FRET. (a) Hela cell co-expressing CFP in the nucleus (especially in the nucleoli) and a CFP–YFP fusion protein construct in the cytoplasm. Left: Fluorescence intensity distribution of the CFP fluorescence. Middle: 2-photon TCSPC FLIM image of the CFP fluorescence lifetime. Note the high, unperturbed lifetimes in the nucleoli, the slightly lowered lifetimes in the remainder of the nucleus caused by mixing in with the lower lifetime of the fusion construct as it passively enters the nucleus from the cytosol. Right: simple thresholding of the lifetime at 2.0 ns suffices to segment the nucleus from the cytosol. (b) Demonstration that lifetimes are independent of fluorophore concentration. GFP, covalently coupled to a sepharose bead, is bleached in a confocal microscope in three lines to locally reduce the concentration of fluorescent species. The lifetime calculated from the phase shift (τ_f) and demodulation (τ_M) in wide-field frequency-domain FLIM is identical over the entire bead. Note that both lifetimes are very similar; this is an indication for the single lifetime nature of GFP, i.e., of the absence of lifetime heterogeneity.

This can be seen for instance for the protonation of the chromophore of fluorescent proteins at lower pH, most dramatically seen for the YFP variant, where brightness is lost at low pH due to a loss in molar extinction coefficient. One can still use FLIM for the quantification of absorption-based changes by transferring the information of a reduction in extinction coefficient to a second fluorophore by FRET. In this case, the loss of extinction coefficient in an acceptor fluorophore will dequench the donor fluorescence by a linear loss of FRET coupling, raising the donor lifetime accordingly. We recently generated a pH sensor, dubbed pHameleon that uses this principle in a high-FRET intramolecular CFP–YFP construct (Esposito et al., 2009). Simple exchange of the YFP moiety from the less pH-sensitive Venus to EYFP and the highly pH-sensitive H148G mutant of YFP allowed the construction of highly sensitive and intrinsically calibrated sensors with optimal sensitivity around three pK_as for use in different pH regimes. More generally, this principle of "spectral overlap engineering" is also useful for situations where the acceptor undergoes a spectral shift that, by itself, cannot necessarily be detected by lifetime changes directly, but can be gauged through FRET coupling.

The reduction in donor lifetime is connected to an increase in the fluorescence lifetime of the sensitized acceptor emission. The reason is that even though the excited state depopulation rates of the acceptor fluorophore are not changed by FRET, its excited state is only populated after the energy was transferred from the donor. Its fluorescence therefore experiences a delay that is highly diagnostic for FRET (Harpur et al., 2001). It should be noted, however, that the same limitations described for to sensitized emission intensity-based imaging now apply as directly excited acceptors will exhibit their characteristic fluorescence lifetimes, i.e., they do not experience the FRET-induced delays. However, in this case, the error will bias the measurement to reduced FRET (false negative), unlike the intensity measurement that exaggerates FRET (false positive). The former situation is clearly preferable for experimentation. For this reason, the application of this acceptor ingrowth signature is typically restricted to intramolecular FRET sensors. For optimal sensitivity, the FRET coupling in the pHlameleon sensor was in the range of 80%–90% due to a highly favorable dipole orientation between the fluorophores, leaving very little donor fluorescence for FLIM. In this case, we used the time delay in energy transfer to read out FRET from the acceptor lifetime. Spectrally resolved lifetime detection of the entire emission bandwidth of the donor and acceptor fluorescence, called a decay-associated emission spectrum, can be used to observe the negative pre-exponential factor corresponding to this ingrowth process in the wavelength region where the acceptor emission dominates (Jose et al., 2007).

Interestingly, this delayed acceptor–sensitized emission effect, although not used very often in academic research, forms the basis of the most widely adopted FRET measurement in the screening industry: homogeneous time-resolved fluorescence (HTRF) (Mathis, 1999). In this elegant application, a very long lifetime phosphorescent dye, typically a lanthanide chelate, is used as the donor for a short lifetime fluorescent organic acceptor dye. As the donor exhibits a lifetime in the millisecond range, compared to the nanosecond lifetime of the acceptor, a simple time-gated fluorescence measurement is capable of completely separating sensitized emission from direct excitation. After excitation of the donor, a timed delay allows for the complete decay of directly excited acceptor, and the sensitized emission, which decays with the characteristics of the long lifetime donor, is acquired.

This is, however, the only example of direct decay imaging as most fluorophores that are in use for biological imaging exhibit nanosecond lifetimes.

There are several ways to image the fluorescence lifetime that can be broadly divided into time- and frequency-domain techniques. Traditionally, time-domain (TD) techniques are mainly used in scanning, e.g., confocal and two-photon microscopes, and frequency-domain (FD) techniques in widefield microscopes. However, both techniques can also be implemented in the other type of microscope (Esposito and Wouters, 2004a; Esposito et al., 2007b).

Both approaches are functionally fundamentally equivalent and their data can be inter-converted and analyzed with tools that were initially developed for the other technique. Both techniques probe the distortion of time-encoded intensity patterns in the excitation by comparison with the resulting emission pattern.

TD systems probe the impulse response; the excited state is populated by a (train of) short (dirac-type, practically femtosecond) pulse(s), and the emitted photons are fitted according to their arrival times in the detector. A fluorophore with a single lifetime will, due to the stochastic nature of fluorescence emission, emit photons that experienced varying delays. Their emission probability distribution builds a single-exponential decay with a pre-exponential factor tau (τ) that is equal to the fluorescence lifetime, i.e., the time at which the excited state population (the intensity $I(t)$) has decayed to $1/e$ of its original size I_0. Multiple lifetime species increase the number of exponential factors accordingly.

$$I(t) = \int \left\{ I_0 \left(\alpha_1 e^{-t/\tau_1} \right) + \left(\alpha_2 e^{-t/\tau_2} \right) + \cdots + \left(\alpha_n e^{-t/\tau_n} \right) \right\} dt \tag{12.10}$$

Depending on a priori knowledge of the system, and/or statistical considerations, the data are fitted to models that account for these multiple species to obtain their lifetimes and fractional sizes (α) on a pixel-by-pixel basis. For example, FRET from a single exponential donor creates two lifetime species, one with an unperturbed lifetime, and one with a FRET-reduced lifetime. In principle, the FRET efficiency in the complex is not always the parameter of interest, because it depends on the, mostly unknown, approach of the fluorophores that were attached to the interacting proteins. Only when changes in the approach, i.e., distance and angle, are the objects of investigation does the magnitude of FRET efficiency change carry direct information. Mostly, however, the relative fraction of molecules that engage in an interaction *is* the biologically relevant parameter. Multiple exponential fitting can provide access to that information. It should be noted that the number of photons emitted that are required for the correct fitting of two species is very much higher than that required for the correct fitting of a single species and, correspondingly, takes a longer acquisition time. Also, specific models are available for the fitting of distributions of species, rather than discrete mixes of invariant lifetimes, that might better represent the biological problem. It is conceivable that lifetimes are heterogenous, as differences in microenvironment and folding are to be expected for fluorescent proteins and there are likely to be many ways in which two proteins can approach and dock onto each other, generating small differences in their FRET efficiencies. These solutions are described somewhere else (Esposito et al., 2005a).

There are principally two ways in which the timing information of emission photons can be obtained: in time-correlated single photon counting (TCSPC), an electronic timer circuit is started at the time of the pulse, which is stopped at the arrival of the first emitted photon. The difference time is logged and used for the construction of the decay profile. In time gating, different timing circuits are started one after the other at the time of the pulse and all photons that fall inside a defined time window are counted. This way, the decay profile is built as a histogram where the number of bins corresponds to the number of timed counting windows. The latter method is faster, but contains less precise information, that however suffices for the determination of an average lifetime.

These methods require pulsed lasers, e.g., mode-locked gas lasers, laser diodes, or a two-photon laser. The instrument response, i.e., the deviation from the ideal case of an infinitely short laser pulse and infinitely fast detection, has to be well characterized and sets the limit to the affordable excitation switching time. Also, the affordable emission rate is limited as higher photon fluxes increase the uncertainty in the assignment of photons to pulses, i.e., it becomes more difficult to discriminate between a fast photon from the last, or a slow photon from the pulse before last. This "pulse pile-up" effect and the scanning nature of TD systems make them relatively slow.

In FD techniques, the harmonic response of the fluorescence is probed. A high frequency (ω is circular frequency) intensity-modulated excitation signal (I) serves as time-encoded pattern (φ_0 is the phase bias, m is the modulation depth).

$$I = I_0 \left\{ 1 + m_0 \cos \left(\omega t + \varphi_0 \right) \right\} \tag{12.11}$$

The harmonic content of the signal (of every periodic form) undergoes a typical distortion where the emission signal, of the same frequency, shifts in phase and demodulates, i.e., loses amplitude in its excursions from the average (I_0) level.

$$\varphi = \arctan(\omega\tau); \quad m = \sqrt{\frac{1}{1+\omega^2\tau^2}} \tag{12.12}$$

These differences can be detected by cross-correlation techniques. Here, changes in the emission are detected by mixing, i.e., multiplication, of the emission signal by the excitation signal. An image intensifier performs this mixing in each pixel of the image simultaneously. In the image intensifier, emission photons impinging on the photocathode will generate photoelectrons that are attracted by the anode. By varying the charge on the photocathode, the probability of the escape of the electrons can be modulated: a positive charge will "close" the gate as all electrons are reabsorbed. By modulating the voltage on the cathode, the emission signal is thus multiplied by the excitation. The electrons then travel down the slanted micrometer-sized holes in a glass substrate, the multi-channel plate (MCP), where each channel amplifies the amount of electrons at each collision, similar to the mode of operation of photomultiplier tubes. The electrons are driven through the MCP by a potential difference that sets the level of amplification, which is the original purpose of the intensifier. The electrons are converted back into photons when they collide with the anode phosphor screen, and a CCD camera captures the phosphorescence image. The phosphor screen acts as a slow integrating device, i.e., as a low-pass filter that integrates all high-frequency information signal; the intensity in each pixel of the CCD camera thus depends on the product of both signals as well as the fluorescence distribution in the sample. In addition to the modulation mode, the intensifier can also be used for gating, effectively turning the device into a very fast shutter, allowing time-gated wide-field imaging, as explained earlier for a scanning microscope.

The major two cross-correlation techniques are homodyning and heterodyning. In heterodyning, the intensifier is modulated at a frequency that is slightly different from the excitation frequency. This results in a beating frequency equal to the difference of the excitation and detector frequencies that is slow enough to pass through the phosphor screen. This difference frequency contains the same phase shift and demodulation information as the original high frequency emission signal, but can be sampled in time by the CCD camera. In the homodyning approach, the modulation frequency of the detector equals that of the excitation signal. The CCD camera now records an integrated steady-state image. By shifting the phase of the MCP gain, the changes in the emission signal can be determined in a manner similar to the operation of a lock-in amplifier, i.e., the product of both signals is optimal when they are in phase, and minimal when they are in opposite phase. Effectively, the emission signal is now transferred from time- to phase-space and can be sampled with arbitrary resolution by the choice of the number of phase steps.

These methods each have their advantages and disadvantages that are discussed in more detail elsewhere (Esposito et al., 2007b), but are summed up best by a quote from Paul French (Imperial College London): "The time domain gives the highest lifetime accuracy per photon and the frequency domain the highest accuracy per second."

As the fluorescence lifetime is an additional metric of fluorescence, its combination with other parameters holds the promise of increasing the parallelization and/or information content of fluorescence measurements. Its narrow normal distribution can be used to multiply the number of measurements at a given number of spectrally different windows by the number of separable lifetimes; easily by a factor of three to four. The combination of hyperspectral information and lifetime detection (De Beule et al., 2007), e.g., by the use of a line-scanning spectrometer, dramatically increases the information content of the collected emission light. Recent developments in white light laser technology, i.e., the supercontinuum generation in holey waveguides, have provided an affordable pulsed tunable excitation source for lifetime measurements as a function of excitation wavelength. This allows the generation of an excitation–emission lifetime matrix that represents an incredibly detailed and unique spectral fingerprint of fluorescence signals (Owen et al., 2007). One last example of the combinatorial nature of lifetime measurements is the generation of a microscope capable of FLIM-STED (Auksorius et al., 2008) that combines the sub-nanometer event resolution offered by the lifetime with true spatial super-resolution imaging on the scale of tens of nanometers.

12.4 Critical Discussion

12.4.1 Shortcomings and Possible Pitfalls of FRET

One of the major difficulties of FRET is the combination of conceptual simplicity with experimental intricacy. The different issues with the κ^2 orientation factor, spectral contamination, and a lack of generalized and commonly adopted analysis approaches, i.e., a lack of standardization, complicate the interpretation of FRET data.

FRET, by its extreme dependence on separation distance and transition dipole orientation of the fluorophores, suffers from false negative detection. That is, the absence of proof for an interaction or conformational change can never be taken as proof of its absence. This might be considered a small price to pay considering its inherent lack of false positives, but given the size of proteins and the size of the fluorescent protein tags, the investigation of, for instance, the nanomechanical behavior of supramolecular complexes can become very difficult. As FRET physically derives from the weak electromagnetic coupling of two dipoles, it can be appreciated that the distance dependence of FRET can be reduced by the introduction of additional dipoles. In the plane of a membrane, the dipoles form a 2-dimensional array, and the distance dependence of FRET is reduced to the fourth power:

$$k_{\text{FRET}} = \left(\frac{\pi}{2}\right)n_{\text{A}}x^{-4} \tag{12.13}$$

where

 n_{A} is the dimensionless planar acceptor concentration defined as $n_{\text{A}} = N_{\text{A}}R_0^2$ (N_{A} is the area concentration of acceptors in nanometer^{-2})

 x is the distance of a single donor fluorophore (in units of R_0) to an infinite acceptor plane (Bastiaens et al., 1990)

Furthermore, the probability for nonspecific interactions due to random encounters, also called "concentration FRET" is increased as the number of molecules per unit volume in a membrane increases dramatically with the total number of molecules. This crowding effect should be considered in membrane-based FRET studies.

A specific problem for acceptor photobleaching is the formation of a photoproduct with spectral properties that overlap with the donor. This unfortunate condition has been observed for the popular combination of CFP and YFP (Valentin et al., 2005). The latter became apparent with the more widespread availability of deep-blue (405 nm) laser diodes in scanning confocal microscopes in recent years. This wavelength is also suboptimal for CFP excitation, but CFP here still has a higher molar extinction coefficient than at the 458 nm Argon laser line that is routinely used for its excitation (its optimum lies at 435 nm). However, the photoproduct generated by YFP bleaching at 514 nm is optimally excited at 405 nm and unfortunately emits in the same spectral range as CFP, generating a large erroneous contribution to the unquenched donor image from which the FRET efficiency is then overestimated. This effect is very small when the sample is excited at the more common 458 nm Argon laser line (Figure 12.6).

Lifetime analysis suffers less from ambiguity in the interpretation and comparison of results as it directly delivers a physical entity and not mere contrast as is the case for intensity measurements. Nevertheless, especially noise contributions can seriously upset the fitting procedure and results of

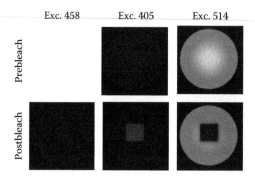

FIGURE 12.6 (See color insert following page 15-30.) Photobleaching YFP creates a photoproduct with CFP emission properties. Shown is a sepharose bead covalently conjugated with EYFP. After photobleaching YFP at 514 nm, a photoproduct is generated that emits in the CFP channel when excited at 405 nm. The excitation of this photoproduct is considerably less efficient at 548 nm, which allows the separation of dequenched CFP emission from this artifact.

lifetime maps. Care should therefore be taken with lifetime values associated with low fluorescence intensities, usually present at the cell periphery.

Recent developments in the statistical interpretation of lifetimes have attempted to extract information on the relative concentration of the different or interacting species and their FRET efficiencies; the reader is referred to the references in Wouters and Esposito (2008). However, these approaches have relied on assumptions that may not always be realistic. These problems were solved by an elegant representation of lifetime information that can be immediately visually inspected and interpreted with respect to the fluorescence composition of the sample. In this "phasor" approach, the lifetime information of each pixel is mapped in a polar coordinate space (Digman et al., 2008). This space has the form of a half circle, the universal circle, as no point can exist outside the circle (in the absence of photochemical reactions). All possible single lifetimes lie on, and describe, the half circle. With increasing heterogeneity, the points move toward the center. The distribution of the data cloud of single lifetimes is symmetrical, allowing the accurate determination of its average value at its center of mass. In addition to this information, the phasors adhere to a convenient addition rule: all linear combinations of two lifetimes lie on the straight line connecting both lifetimes, with a position proportional to their mixing ratio. This can be geometrically extended to more components. FRET being a nonlinear process follows a distinct nonlinear trajectory: it would slide from a given position on the half circle (0% FRET, one donor component) to the rightmost corner (100% FRET, zero lifetime). However, since the fluorescence close to 100% FRET is dominated by autofluorescence, it never arrives here, but curves toward a (deep, as autofluorescence is highly multi-component) position inside the circle. This latter point can be experimentally determined, and the FRET efficiency and fractional contribution can be simply read off from the position on the trajectory. The starting point will also be offset by autofluorescence, but since this represents a linear mixing condition, it can be seen and quantified by drawing a line between the pure components. This system is therefore fully defined, and immediately interpretable without any fitting routine or assumption. Its power becomes really obvious when the lifetime information of multiple cells for a given condition is accumulated in one phasor plot to reveal different species with separate lifetime footprints in the plot. These populations can be mapped in the original images and vice versa, allowing a unique kind of truly global, guided discovery process where the user navigates between the image space (biology-based contrast) and the phasor space (physics-based contrast) to gain insight from the synthesis of this information that is not accessible by the comparison of lifetime distribution histograms alone (Wouters and Esposito, 2008).

12.5 Summary

FRET is a highly valuable imaging tool for the modern life sciences. It allows the generation of sophisticated bioassays for use in living cells and can provide a (sub) nanometer view of the architecture of multi-protein complexes. Especially in combination with the steadily increasing knowledge on the components of pathways and complexes, FRET can contribute significantly to the elucidation of the working of cellular molecular machines. The different photophysical consequences of FRET can be used for its detection and quantification. These techniques can be broadly divided into those that use the spectral intensity changes in the coupled fluorophore pair and those that are based on time-resolved fluorescence measurements, i.e., those that determine the kinetics of the emission process. The determination of the duration of the donor excited state, i.e., the donor lifetime, by FLIM represents one of the most sensitive and robust methods that allows the full quantification of FRET, including the coupling efficiency and the fractional contribution of the components in the FRET process. This information is necessary to derive at a quantitative understanding of the molecular mechanistic basis of a process, its connection to—and causality with—other processes. Recent developments in FLIM have made the technique more accessible to the life sciences and we expect this technique to become a standard tool in the life sciences.

12.6 Future Perspective

12.6.1 Instrumentation

Speed and simplicity of operation are important issues for the major growth areas of FLIM–FRET detection, drug discovery, and molecular diagnostics, where the demand for multiplexed high-content assays could already be met by the current state of the art. Especially the emerging concept of theranostics, i.e., the coupled application of diagnostics and therapeutics in the form of "personalized medicine" that allows the identification of the proper treatment for a precisely defined disease condition—and the monitoring and adjustment during the treatment—requires mechanism-based insights into the disease etiology and progression that can only be obtained by highly sophisticated integrative and multiparameter assays as, ultimately, diseases are biochemical pathway and physiological network disorders.

A major effort is directed toward increasing its speed and throughput. Throughput can be increased by FLIM parallelization by the use of multiple counting units, i.e., channel multiplexing (16 channels are currently commercially available) in the time domain, or multibeam scanning (Niesner et al., 2007). We have automated our FLIM microscope to ultra-high-throughput levels allowing the investigation of all fluorescent objects in a sample regardless of its dimensions (Esposito et al., 2007a).

The rapid lifetime algorithm (Moore et al., 2004) results in fast acquisition times. In the time domain, this is achieved by time gating with only two gates. In the frequency domain, this is equivalent to the acquisition of two images taken at opposite phase. This principle was realized in an all-solid state frequency-domain hardware solution that was recently developed in our laboratory (Esposito et al., 2005b, 2006). This detector consists of an array of lock-in pixels that operate on the basis of phase-dependent charge separation. Two collection bins are integrated with each pixel that accumulate the charge generated at its photosensitive area. The charge of these bins is modulated in synchrony with the excitation signal and at opposite phase. They collect the photoelectrons as they are shifted between the bins. At the end of the exposure, the two charges are read out over all pixels, generating two images from which the fluorescence lifetime is calculated. Directly modulated cameras (see also Mitchell et al., 2002a,b) are bound to become the next generation of FLIM cameras that replace the image intensifier. As their operation is intrinsically parallel, in contrast to the serial mode of currently adopted FLIM implementations (multiple phase images, time-gated images, or repeated scans over pixels in TCSPC), movement artifacts can no longer occur.

12.6.2 Labels and Sensing Paradigms

As discussed earlier, by the environmental sensitivity of the fluorescence lifetime, the imaging contrast in FLIM provides information on the nano-environment of the fluorophore. Besides FRET, where the contrast is designed to report on the molecular proximity of two paired fluorophores, FLIM can be used to investigate the fluorescence properties of histological staining in tissues (Eliceiri et al., 2003) and of vital labeling with fluorescent compounds. An example is the staining of cancer cells with the fluorescent metabolite 5-ALA whose uptake and specific metabolism generates diagnostic spectral footprints (Kress et al., 2003). A major effort is also directed toward label-free measurements in tissues, using the many endogenous fluorophores in cells and extracellular matrix to identify, for instance, cancerous tissue. This application has important implications for tumor resection surgery as it can visualize the borders of the tumor with healthy tissue (Galletly et al., 2008; Munro et al., 2005). Especially the availability of affordable 355 nm frequency–tripled ytterbium fiber lasers opens up the possibility for routine testing in the field of oncology.

Extraneous fluorophores increase the optical contrast in fluorescence detection. One of the major challenges in designing FRET sensors is the choice of suitably matched fluorophore pairs. Here, the simultaneous and mutually exclusive demands for high spectral overlap between donor emission and acceptor excitation and high spectral separation between the donor and acceptor emissions asks for a carefully balanced compromise. However, for FLIM, all relevant information on FRET is already contained in the donor fluorescence. The acceptor therefore does not necessarily have to be a fluorophore. For this reason, we recently constructed a nonfluorescent acceptor YFP-based chromoprotein that still forms the chromophore, but loses its excited state energy by rapid internal conversion to the environment faster than its lifetime, i.e., it does not have a chance to emit photons, thereby acting as energy transfer "sink" (Ganesan et al., 2006). The use of nonfluorescent chromophore acceptors carries a number of advantages: (1) the entire emission spectrum of the donor can be collected causing a higher signal-to-noise level in the measurement, (2) the red part of the spectrum, normally occupied by the acceptor fluorescence is liberated for another fluorophore or FRET pair, (3) the spectral overlap between donor and emission can be optimized without consideration for emission separation, (4) the acceptor excited state is populated by FRET where fluorescent acceptors can undergo photobleaching. Unfortunately most popular FRET pairs suffer from an unfavorable photostability relationship where the acceptor is more susceptible than the donor (CFP–YFP, Cy3–Cy5, GFP–RFP). However, the extremely short lifetime of the acceptor chromoprotein protects it from FRET-induced photobleaching, and (5) because of its very short lifetime, the process of FRET is not "frustrated" at high donor excitation probability (especially in confocal scanning microscopy) where the acceptor has to depopulate its excited state before a new quantum of energy can be accepted from the same donor.

Recent advances in in vivo bioconjugation strategies for the site-directed labeling of proteins offer an alternative to the use of fluorescent proteins (Bunt and Wouters, 2004; Chen and Ting, 2005; Lin and Wang, 2008). Chemical labeling offers a wide choice of fluorophores with superior spectral qualities and/or specific sensing properties. These methods allow the probing of the molecular environment of specific proteins by the use of sensing fluorophores, also uniquely in a pulse-chase design experiment. Through the use of orthogonal labeling chemistries, dyes can be multiplexed within one cell. Additionally, site-directed labeling allows the construction of stoichiometrically defined FRET pairs (Meyer et al., 2006) inasmuch as the ratio of the donor and acceptor dyes can be conveniently and precisely set by their relative concentration in the labeling reaction. This way, equal FRET participation of the labeled components can be assured between different cells and FRET titration curves for the determination of, e.g., the oligomeric status of protein complexes can be investigated. Importantly, labeling chemistries that rely on the addition of a labeling enzyme to the cell culture or that involve the use of membrane-impermeable dyes can selectively label the surface-exposed fraction of a given protein, for instance the membrane-resident pool of a membrane receptor, without labeling the intracellular pool (Gralle et al., 2009). We expect to see an increase in these types of assays, with dyes that possess unique properties. We also expect to see a revival of immunofluorescence-based investigations on live cells and with antibodies carrying novel probes and FRET dyes. The versatility of synthetic fluorophores will hopefully stimulate the generation of sophisticated and multiplexed integrative bioassays that can uncover the causalities and quantitative relationships between individual biochemical events and components that make up the molecular physiological machinery of the living cell.

Acknowledgment

We would like to thank E. Papusheva, A.M. Dani, A. Esposito, and J.T. Gonçalves for their help in generating the images.

References

Allen, P. G. (2003) Actin filament uncapping localizes to ruffling lamellae and rocketing vesicles. *Nat Cell Biol*, 5, 972–979.

Anikovsky, M., Dale, L., Ferguson, S., and Petersen, N. (2008) Resonance energy transfer in cells: A new look at fixation effect and receptor aggregation on cell membrane. *Biophys J*, 95, 1349–1359.

Auksorius, E., Boruah, B. R., Dunsby, C., Lanigan, P. M. P., Kennedy, G., Neil, M. A. A., and French, P. M. W. (2008) Stimulated emission depletion microscopy with a supercontinuum source and fluorescence lifetime imaging. *Opt. Lett.*, 33, 113–115.

Bastiaens, P. I. H., De Beus, A., Lacker, M., Somerharju, P., Vaukhonen, M., and Eisinger, J. (1990) Resonance energy transfer from a cylindrical distribution of donors to a plane of acceptors. Location of apo-B100 protein on the human low-density lipoprotein particle. *Biophys J*, 58, 665–675.

Bastiaens, P. I. H., Majoul, I. V., Verveer, P. J., Soling, H. D., and Jovin, T. M. (1996) Imaging the intracellular trafficking and state of the AB5 quaternary structure of cholera toxin. *Embo J*, 15, 4246–4253.

Bastiaens, P. I. H., Wouters, F. S., and Jovin, T. M. (1997) Imaging the molecular state of proteins in cells by fluorescence resonance energy transfer. Sequential photobleaching of Förster donor–acceptor pairs. *Proceedings of the 2nd Hamamatsu International Symposium on Biomolecular Mechanisms and Photonics: Cell–Cell Communications* pp. 77–82 (Research Foundation for Opto-Science and Technology, Hamamatsu, Japan, Feb. 20–22, 1995).

Betzig, E., Patterson, G. H., Sougrat, R., Lindwasser, O. W., Olenych, S., Bonifacino et al. (2006) Imaging intracellular fluorescent proteins at nanometer resolution. *Science*, 313, 1642–1645.

Bunt, G. and Wouters, F. S. (2004) Visualization of molecular activities inside living cells with fluorescent labels. *Int Rev Cytol*, 237, 205–277.

Chen, I. and Ting, A. Y. (2005) Site-specific labeling of proteins with small molecules in live cells. *Curr Opin Biotechnol*, 16, 35–40.

Clegg, R. M. (1996) Fluorescence resonance energy transfer. In *Fluorescense Imaging Spectroscopy and Microscopy*, Wang, X. F. and Herman, B. (Eds.), London, UK, John Wiley & Sons.

Clegg, R. M. and Schneider, P. C. (1996) Fluorescence lifetime-resolved imaging microscopy: A general description of lifetime-resolved imaging measurements. In Slavik, J. (Ed.) *Fluorescence Microscopy and Fluorescence Probes*, 1st edn., New York, Plenum Press.

Coveney, P. V. and Fowler, P. W. (2005) Modelling biological complexity: A physical scientist's perspective. *J R Soc Interface*, 2, 267–280.

Dale, R. E., Eisinger, J., and Blumberg, W. E. (1979) The orientational freedom of molecular probes. The orientation factor in intramolecular energy transfer. *Biophys J*, 26, 161–193.

De Beule, P., Owen, D. M., Manning, H. B., Talbot, C. B., Requejo-isidro, J., Dunsby et al. (2007) Rapid hyperspectral fluorescence lifetime imaging. *Microsc Res Tech*, 70, 481–484.

Digman, M. A., Caiolfa, V. R., Zamai, M., and Gratton, E. (2008) The phasor approach to fluorescence lifetime imaging analysis. *Biophys J*, 94, L14–L16.

Eliceiri, K. W., Fan, C. H., Lyons, G. E., and White, J. G. (2003) Analysis of histology specimens using lifetime multiphoton microscopy. *J Biomed Opt*, 8, 376–380.

Esposito, A. and Wouters, F. S. (2004a) Fluorescence lifetime imaging (unit 4.14). In *Current Protocols in Cell Biology*, New York, John Wiley & Sons.

Esposito, A. and Wouters, F. S. (2004b) Fluorescence lifetime imaging microscopy. In Bonifacino, J. S., Dasso, M., Harford, J. B., Lippincott-Schwartz, J., and Yamada, K. M. (Eds.), *Current Protocols in Cell Biology*, New York, John Wiley & Sons.

Esposito, A., Gerritsen, H. C., and Wouters, F. S. (2005a) Fluorescence lifetime heterogeneity resolution in the frequency domain by lifetime moments analysis. *Biophys J*, 89, 4286–4299.

Esposito, A., Oggier, T., Gerritsen, H. C., Lustenberger, F., and Wouters, F. S. (2005b) All-solid-state lock-in imaging for wide-field fluorescence lifetime sensing. *Opt. Express*, 13, 9812–9821.

Esposito, A., Gerritsen, H. C., Oggier, T., Lustenberger, F., and Wouters, F. S. (2006) Innovating lifetime microscopy: A compact and simple tool for life sciences, screening, and diagnostics. *J Biomed Opt*, 11, 34016.

Esposito, A., Dohm, C. P., Bähr, M., and Wouters, F. S. (2007a) Unsupervised fluorescence lifetime imaging microscopy for high content and high throughput screening. *Mol Cell Proteomics*, 6, 1446–1454.

Esposito, A., Gerritsen, H. C., and Wouters, F. S. (2007b) Optimizing frequency-domain fluorescence lifetime sensing for high-throughput applications: Photon economy and acquisition speed. *J Opt Soc Am A: Opt Image Sci Vis*, 24, 3261–3273.

Esposito, A., Gralle, M., Dani, M. A., Lange, D., and Wouters, F. S. (2009) pHlameleons: A family of FRET-based protein sensors for quantitative pH imaging. *Biochemistry*, 47, 13115–13126.

Förster, T. (1948) Zwischenmolekulare Energiewanderung und Fluoreszenz. *Ann Phys*, 2, 55–75.

Galletly, N. P., McGinty, J., Dunsby, C., Teixeira, F., Requejo-Isidro, J., Munro et al. (2008) Fluorescence lifetime imaging distinguishes basal cell carcinoma from surrounding uninvolved skin. *Br J Dermatol*, 159, 152–161.

Ganesan, S., Ameer-Beg, S. M., Ng, T. T., Vojnovic, B., and Wouters, F. S. (2006) A dark yellow fluorescent protein (YFP)-based resonance energy-accepting chromoprotein (REACh) for Forster resonance energy transfer with GFP. *Proc Natl Acad Sci USA*, 103, 4089–4094.

Gralle, M., Botelho, M. G., and Wouters, F. S. (2009) Neuroprotective secreted amyloid precursor protein acts by disrupting amyloid precursor protein dimers. *J. Biol. Chem.*, 284, 15016–25025.

Harpur, A. G., Wouters, F. S., and Bastiaens, P. I. (2001) Imaging FRET between spectrally similar GFP molecules in single cells. *Nat Biotechnol*, 19, 167–169.

Hell, S. W. and Wichmann, N. J. (1994) Breaking the diffraction resolution limit by stimulated-emission—stimulated-emission-depletion fluorescence microscopy. *Opt Lett* 19, 780–782.

Jose, M., Nair, D. K., Reissner, C., Hartig, R., and Zuschratter, W. (2007) Photophysics of Clomeleon by FLIM: Discriminating excited state reactions along neuronal development. *Biophys J*, 92, 2237–2254.

Jovin, T. M. and Arndt-Jovin, D. J. (1989) Luminescence digital imaging microscopy. *Annu Rev Biophys Biophys Chem*, 18, 271–308.

Klar, T. A., Jakobs, S., Dyba, M., Egner, A., and Hell, S. W. (2000) Fluorescence microscopy with diffraction resolution barrier broken by stimulated emission. *Proc Natl Acad Sci USA*, 97, 8206–8210.

Kress, M., Meier, T., Steiner, R., Dolp, F., Erdmann, R., Ortmann, U., and Rück, A. (2003) Time-resolved microspectrofluorometry and fluorescence lifetime imaging of photosensitizers using picosecond pulsed diode lasers in laser scanning microscopes. *J Biomed Opt*, 8, 26–32.

Lakowicz, J. R. (1999) *Principles of Fluorescence Spectroscopy*, New York, Kluwer Academic/Plenum Publishers.

Lin, M. Z. and Wang, L. (2008) Selective labeling of proteins with chemical probes in living cells. *Physiology (Bethesda)*, 23, 131–141.

Ma'ayan, A. and Iyengar, R. (2006) From components to regulatory motifs in signalling networks. *Brief Funct Genomic Proteomic*, 5, 57–61.

Mathis, G. (1999) HTRF(R) technology. *J Biomol Screen*, 4, 309–314.

Matz, M. V., Fradkov, A. F., Labas, Y. A., Savitsky, A. P., Zaraisky, A. G., Markelov, M. L., and Lukyanov, S. A. (1999) Fluorescent proteins from nonbioluminescent Anthozoa species. *Nat Biotechnol*, 17, 969–973.

Meyer, B. H., Martinez, K. L., Segura, J. M., Pascoal, L, P., Hovius, R., George, N. et. al. (2006) Covalent labeling of cell-surface proteins for in-vivo FRET studies. *FEBS Lett*, 580, 1654–1658.

Mitchell, A. C., Wall, J. E., Murray, J. G., and Morgan, C. G. (2002a) Direct modulation of the effective sensitivity of a CCD detector: A new approach to time-resolved fluorescence imaging. *J Microsc*, 206, 225–232.

Mitchell, A. C., Wall, J. E., Murray, J. G., and Morgan, C. G. (2002b) Measurement of nanosecond time-resolved fluorescence with a directly gated interline CCD camera. *J Microsc*, 206, 233–238.

Miyawaki, A. and Tsien, R. Y. (2000) Monitoring protein conformations and interactions by fluorescence resonance energy transfer between mutants of green fluorescent protein. *Methods Enzymol*, 327, 472–500.

Moore, C., Chan, S. P., Demas, J. N., and Degraff, B. A. (2004) Comparison of methods for rapid evaluation of lifetimes of exponential decays. *Appl Spectrosc*, 58, 603–607.

Munro, I., McGinty, J., Galletly, N., Requejo-Isidro, J., Lanigan, P., Elson, M. et al. (2005) Toward the clinical application of time-domain fluorescence lifetime imaging. *J Biomed Opt*, 10, 051403.

Nagy, P., Vereb, G., Damjanovich, S., Matyus, L., and Szollosi, J. (2006) Measuring FRET in flow cytometry and microscopy. *Curr Protoc Cytom,* Chapter 12, Unit 12 8.

Netterwald, J. (2008) Emerging trends in cell biological research. *Drug Discov Dev*, 10, 39.

Niesner, R., Andresen, V., Neumann, J., Spiecker, H., and Gunzer, M. (2007) The power of single and multibeam two-photon microscopy for high-resolution and high-speed deep tissue and intravital imaging. *Biophys J*, 93, 2519–2529.

Owen, D. M., Auksorius, E., Manning, H. B., Talbot, C. B., De Beule, P. A., Dunsby, C. et. al. (2007) Excitation-resolved hyperspectral fluorescence lifetime imaging using a UV-extended supercontinuum source. *Opt Lett*, 32, 3408–3410.

Palmer, A. E. and Tsien, R. Y. (2006) Measuring calcium signaling using genetically targetable fluorescent indicators. *Nat Protoc*, 1, 1057–1065.

Papuscheva, E., mello de Queiroz, F., Dalous, J., Han, Y.-Y., Esposito, A., Jares-Erijman, E. A., Jovin, T. M., and Bunt, G. (2009) Dynamic conformational changes in the FERM domain of FAK are involved in focal adhesion behavior during cell spreading and motility. *J Cell Biol*, 122, 656–666.

Pepperkok, R., Squire, A., Geley, S., and Bastiaens, P. I. H. (1999) Simultaneous detection of multiple green fluorescent proteins in live cells by fluorescence lifetime imaging microscopy. *Curr Biol*, 9, 269–272.

Rizzo, M. A. and Piston, D. W. (2005) High-contrast imaging of fluorescent protein FRET by fluorescence polarization microscopy. *Biophys J*, 88, L14–L16.

Schaffer, J., Volkmer, A., Eeggeling, C., Subramaniam, V., Striker, G., and Seidel, C. A. M. (1999) Identification of single molecules in aqueous solutions by time-resolved fluorescence anisotropy. *J Phys Chem A*, 3, 331–336.

Sullivan, E., Tucker, E. M., and Dale, I. L. (1999) Measurement of [Ca^{2+}] using the fluorometric imaging plate reader (FLIPR). *Methods Mol Biol*, 114, 125–133.

Tsien, R. Y. (1998) The green fluorescent protein. *Annu Rev Biochem*, 67, 509–544.

Valentin, G., Verheggen, C., Piolot, T., Neel, H., Coppey-Moisan, M., and Bertrand, E. (2005) Photoconversion of YFP into a CFP-like species during acceptor photobleaching FRET experiments. *Nat Methods*, 2, 801.

Van der Meer, B. W. (1999) Orientational aspects in pair energy transfer. In Andrews, D. L. and Demidov, A. A. (Eds.), *Resonance Energy Transfer*, London, UK, Wiley & Sons.

Van der Meer, B. W. (2002) Kappa-squared: From nuisance to new sense. *Rev Mol Biotechnol*, 82, 181–196.

Van Munster, E. B., Kremers, G. J., Adjobo-Hermans, M. J., and Gadella, T. W. J., Jr. (2005) Fluorescence resonance energy transfer (FRET) measurement by gradual acceptor photobleaching. *J Microsc*, 218, 253–262.

Van Rheenen, J., Langeslag, M., and Jalink, K. (2004) Correcting confocal acquisition to optimize imaging of fluorescence resonance energy transfer by sensitized emission. *Biophys J*, 86, 2517–2529.

Van Royen, M. E., Cunha, S. M., Brink, M. C., Mattern, K. A., Nigg, A. L., Dubbink, H. J. et al. (2007) Compartmentalization of androgen receptor protein–protein interactions in living cells. *J Cell Biol*, 177, 63–72.

Webb, S. E., Needham, S. R., Roberts, S. K., and Martin-Fernandez, M. L. (2006) Multidimensional single-molecule imaging in live cells using total-internal-reflection fluorescence microscopy. *Opt Lett*, 31, 2157–2159.

Wlodarczyk, J., Woehler, A., Kobe, F., Ponimaskin, E., Zeug, A., and Neher, E. (2008) Analysis of FRET signals in the presence of free donors and acceptors. *Biophys J*, 94, 986–1000.

Wouters, F. S., Bastiaens, P. I., Wirtz, K. W., and Jovin, T. M. (1998) FRET microscopy demonstrates molecular association of non-specific lipid transfer protein (nsL-TP) with fatty acid oxidation enzymes in peroxisomes. *Embo J*, 17, 7179–7189.

Wouters, F. S. and Esposito, A. (2008) Quantitative analysis of fluorescence lifetime imaging made easy. *HFSP J*, 2, 7–11.

Wouters, F. S., Verveer, P. J., and Bastiaens, P. I. (2001) Imaging biochemistry inside cells. *Trends Cell Biol*, 11, 203–211.

Zimmermann, T., Rietdorf, J., Girod, A., Georget, V., and Pepperkok, R. (2002) Spectral imaging and linear un-mixing enables improved FRET efficiency with a novel GFP2-YFP FRET pair. *FEBS Lett*, 531, 245–249.

13

FRET-Based Determination of Protein Complex Structure at Nanometer Length Scale in Living Cells

13.1 Introduction 13-1

13.2 Nomenclature and Definitions of the Physical Quantities...... 13-2
Fluorescence and FRET in Dimeric Complexes • RET Efficiency in Oligomeric Complexes of Arbitrary Size and Geometry

13.3 Determination of FRET Efficiency from Fluorescence Intensity Measurements 13-6
Fluorescence Intensities at Fixed Emission Wavelengths • Spectral Decomposition of Intensities Measured at Multiple Emission Wavelengths • Instrumentation for FRET Imaging with Single-Molecule or Molecular Complex Sensitivity

13.4 Determination of Protein Structure In Vivo 13-11
Experimental Investigations • Determination of the Protein Complex Structure from Apparent FRET Efficiency • Comparison of Pixel-Level Spectrally Resolved FRET with Other Methods

Acknowledgments..13-16

References 13-16

Valericǎ Raicu
University of Wisconsin–Milwaukee

13.1 Introduction

One of the most widely used methods for studying protein distributions and interactions in living cells is based on the detection of light from fluorescent molecules. If a molecule (called "acceptor") capable of absorbing light at wavelengths at which another one (called "donor") emits, lies within a distance of less than 10 nm (or a hundred millionth of a meter) of the excited donor, the donor energy can be transferred to the acceptor through a non-radiative process called resonance energy transfer (RET) (Clegg 1996, Selvin 2000, Lakowicz 2006, Raicu and Popescu 2008). If the energy transfer occurs via long-range interaction between the transition dipoles of the donor and acceptor, the process is called Förster resonance energy transfer (FRET), after Theodore Förster, who was the first to successfully model the RET using quantum mechanics calculations (Lakowicz 2006).

RET reduces the donor emission and excites the acceptor. If the acceptor is fluorescent, it will eventually lose its excitation either through non-radiative internal conversion or through the emission of light (which will be red-shifted when compared to the donor emission). Here again, the energy transfer process is called FRET, where the letter "F" stands this time for "fluorescence." An experimenter staring at a sample that contains donor and acceptor molecules excited by light will detect a strong emission of fluorescence from the donors if all the donor molecules are far from any potential acceptors, but less emission from the donors if there are acceptors nearby, as these would drain energy away from the donor.

By highlighting the proteins of interest with fluorescent tags, which can act as acceptors (A) and donors (D) of energy through FRET, it is possible to determine distances between parts of a molecule, to probe whether two or more macromolecules form molecular associations (or complexes), and to determine the distribution of such complexes in living cells (Eggeling et al. 2005, Wallrabe and Periasamy 2005, Merzlyakov et al. 2006, Caorsi et al. 2007, Lee et al. 2007, Maurel et al. 2008). Recent progress in the FRET theory and technology has led to the possibility to determine the size (i.e., the number of monomers) and structure (i.e., the relative positions of the monomers) of protein complexes in living cells with molecular resolution (Raicu et al. 2009). This chapter integrates the theoretical and experimental aspects into a unitary description of FRET in an attempt to provide an overall picture of the advances that led to the development of a FRET-based method capable of performing the structural determinations of protein complexes in living cells.

13.2 Nomenclature and Definitions of the Physical Quantities

13.2.1 Fluorescence and FRET in Dimeric Complexes

The aim of any FRET investigation is to relate the apparent efficiency of energy transfer in a mixture of free monomers and oligomers to the spectroscopic properties of D and A (e.g., quantum yields in the presence and absence of FRET), on the one hand, and to experimentally measurable parameters (e.g., fluorescence intensities or lifetimes), on the other hand. The first topic is dealt with in this section, while the second will be treated subsequently. If not stated otherwise, the discussion in this section will refer to RET, and not to FRET, in order to emphasize the general character of the theory. In the next section, which deals with fluorescent acceptors and donors, we will return to the use of the acronym "FRET."

In order to relate RET to physical properties of the molecules under investigation, let us first consider the simple case of one donor and one acceptor of energy. The quantum yields of acceptors and donors, or the rate of emission of photons by the excited molecules are expressed as (Raicu 2007, Raicu and Popescu 2008)

$$Q^D = \frac{\Gamma^{r,D}}{\Gamma^{r,D} + \Gamma^{nr,D}},$$ (13.1a)

and

$$Q^A = \frac{\Gamma^{r,A}}{\Gamma^{r,A} + \Gamma^{nr,A}},$$ (13.1b)

where

$\Gamma^{r,X}$ and $\Gamma^{nr,X}$ (X=D, A) are the rate constants for de-excitation through radiative (i.e., photon emission) and non-radiative (e.g., internal conversion) processes

$\left(\Gamma^{r,X} + \Gamma^{nr,X}\right)^{-1} = \tau_X$ is the lifetime of the excited state of species X (or the fluorescence lifetime)

If excitation energy may be transferred from D to A, the additional pathway for de-excitation of the donor changes its quantum yield, which is given by

$$Q^{DA} = \frac{\Gamma^{r,D}}{\Gamma^{r,D} + \Gamma^{nr,D} + \Gamma^{RET}}, \tag{13.2}$$

where $\left(\Gamma^{r,D} + \Gamma^{nr,D} + \Gamma^{RET}\right)^{-1} = \tau_{DA}$ is the fluorescence lifetime of D in the presence of A (or FRET). If the mechanism of energy transfer is of a dipolar (or Förster) type, the rate constant of non-radiative transfer from D to A is given by

$$\Gamma^{FRET} = (\Gamma^{r,D} + \Gamma^{nr,D})(R_0/r)^6, \tag{13.3}$$

where
    r is the D–A separation
    R_0 is the well-known Förster distance (Clegg 1996, Selvin 2000, Lakowicz 2006)

Since RET only introduces a new pathway for de-excitation of the donor, it only affects the excitation of the acceptor (see the following text) and not its quantum yield.

The proportion of excitations dissipated through RET by the donor, called RET efficiency, is

$$E = \frac{\Gamma^{RET}}{\Gamma^{r,D} + \Gamma^{nr,D} + \Gamma^{RET}}. \tag{13.4}$$

The extra term in the sum of de-excitation rate constants in the denominator modifies the lifetime of the donor such that $\tau_{DA} < \tau_D$, while τ_A remains unchanged.

Using the definitions of the lifetimes introduced earlier, Equation 13.4 becomes

$$E = 1 - \frac{\tau_{DA}}{\tau_D}. \tag{13.5}$$

This equation provides a sensitive means for detecting RET from *fluorescence lifetime measurements* by measuring the donor lifetimes in the presence and the absence of the acceptor.

Furthermore, with the help of Equation 13.3, Equation 13.4 can also be written as a function of the distance between interacting molecules, namely,

$$E = \frac{R_0^6}{R_0^6 + r^6}. \tag{13.6}$$

This equation relates the RET efficiency to the distance between D and A, and it is sometimes used to determine the distance between donors and acceptors (Stryer 1967). Care has to be exercised, however, when the relative orientation of the donor and acceptor transition dipoles is not well known; this may lead to errors in the determination of the distance (r) since this orientation enters the expression for R_0 (Clegg 1996, Lakowicz 2006).

By combining Equation 13.1a with Equations 13.3 and 13.4, we obtain a relation,

$$Q^{DA} = Q^D(1 - E), \tag{13.7}$$

which indicates that the donor emission is reduced through RET. This reduction, known as *donor quenching*, can be used to quantify the interaction between D and A using the measurements of donor fluorescence intensity in the presence and absence of acceptor. To be precise, in practice, it is actually the donor de-quenching that provides information on the energy transfer. This is realized usually by detecting the signal from the donors before and after photobleaching the acceptors by repeated excitation with light of an appropriate wavelength.

The excitation rate constants of A and D in the absence of RET are, respectively (Song et al. 1995),

$$\Gamma^{ex,A} = I_0\left(\lambda_{ex}\right)/(hcN_A)\varepsilon^A(\lambda_{ex}),$$ (13.8a)

and

$$\Gamma^{ex,D} = I_0\left(\lambda_{ex}\right)/(hcN_A)\varepsilon^D(\lambda_{ex}),$$ (13.8b)

where

$I_0(\lambda_{ex})$, $\varepsilon^A(\lambda_{ex})$, and $\varepsilon^D(\lambda_{ex})$ are the intensity of the incident radiation and the absorption cross sections at excitation wavelength λ_{ex}

h is Planck's constant

c is the speed of light

N_A is Avogadro's number

In the presence of RET, the excitation rate constant of the donor remains unchanged, while the that of the acceptor increases according to

$$\Gamma^{ex,AD} = \Gamma^{ex,A} + \Gamma^{ex,D}E.$$ (13.9)

This increased acceptor excitation rate, called *acceptor-sensitized emission*, can be used to detect RET from acceptor emission intensity measurements.

13.2.2 RET Efficiency in Oligomeric Complexes of Arbitrary Size and Geometry

For a protein complex having a particular configuration (or geometry), q (such as the one in Figure 13.1), and containing n monomers, of which k are identical donors and $n-k$ are identical acceptors, the quantum yield of the ith donor is given by the expression (Raicu 2007):

$$Q^{DA}_{i,k,n,q} = \frac{\Gamma^{r,D}}{\Gamma^{r,D} + \Gamma^{nr,D} + \sum_{j=1}^{n-k}\Gamma^{RET}_{i,j,q}} = \frac{Q^D}{1 + \sum_{j=1}^{n-k}\Gamma^{RET}_{i,j,q}/(\Gamma^{r,D} + \Gamma^{nr,D})},$$ (13.10)

where

j is a summation index for acceptors

$\Gamma^{RET}_{i,j,q} = (\Gamma^{r,D} + \Gamma^{nr,D})(R^0_{i,j,q}/r_{i,j,q})^6$ is the rate constant for RET between the single pairs of D and A

Note that, in this more general case, there may be a different Förster radius for each D–A pair, since the orientation factor may be different for each pair (Clegg 1996, Lakowicz 2006).

Assuming that the donors do not compete with one another for transferring energy to acceptors, viz., the donor excitation rate is so low that only one donor in a complex is in an excited state at any period of time, the RET efficiency for each donor may be written as (Raicu 2007)

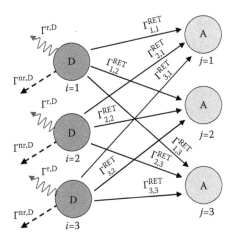

FIGURE 13.1 Illustration of the various ways in which excited donors may lose their excitations. Significance of the symbols: i and j are counting indices for donors and acceptors, respectively; Γ with various superscripts denotes rate constants of de-excitation; superscript "r" denotes a radiative process; superscript "nr" denotes a non-radiative process other than transfer to the acceptor (such as internal conversion); superscript "RET" denotes a non-radiative process of transfer from D to A. (Adapted from Raicu, V., *J. Biol. Phys.*, 33, 109, 2007. With permission.)

$$E_{i,k,n,q} = \frac{\sum_{j=1}^{n-k} \Gamma_{i,j,q}^{RET} \big/ (\Gamma^{r,D} + \Gamma^{nr,D})}{1 + \sum_{j=1}^{n-k} \Gamma_{i,j,q}^{RET} \big/ (\Gamma^{r,D} + \Gamma^{nr,D})}, \tag{13.11}$$

where the $\Gamma_{i,j,q}^{RET} \big/ (\Gamma^{r,D} + \Gamma^{nr,D}) = (R_{i,j,q}^{0}/r_{i,j,q})^{6}$. Equation 13.11 is a generalization of Equation 13.4 for the case of oligomers containing more than two monomers, in which several pathways exist for de-excitation through RET.

Substituting for $\sum_{j=1}^{n-k} \Gamma_{i,j,q}^{RET} \big/ (\Gamma^{r,D} + \Gamma^{nr,D})$ between Equations 13.10 and 13.11, an equation for the quantum yield of D is obtained,

$$Q_{i,k,n,q}^{DA} = Q^{D} \left(1 - E_{i,k,n,q}\right), \tag{13.12}$$

which is similar to Equation 13.7.

Finally, the quantum yield of each acceptor in the complex remains the same as in the absence of RET, while the excitation rate constant changes, due to excitation through RET, into

$$\Gamma_{j,k,q}^{ex,AD} = \Gamma^{ex,A} + \Gamma^{ex,D} \sum_{i=1}^{k} \frac{\Gamma_{i,j,q}^{RET} \big/ (\Gamma^{r,D} + \Gamma^{nr,D})}{1 + \sum_{j=1}^{n-k} \Gamma_{i,j,q}^{RET} \big/ (\Gamma^{r,D} + \Gamma^{nr,D})}, \tag{13.13}$$

where the sum with respect to i includes the pair-wise RET efficiency between each donor and the acceptor j, for the case of successive excitations of donors (i.e., no competition).

13.3 Determination of FRET Efficiency from Fluorescence Intensity Measurements

13.3.1 Fluorescence Intensities at Fixed Emission Wavelengths

To illustrate the practical way in which FRET efficiency determinations are done, let us consider two populations of fluorescent molecules (D and A) mixed together in a solution or a living cell, and excited by light with an arbitrary wavelength, λ_{ex} (Figure 13.2a). Initially, all molecules of the two populations are located at large distances from one another (compared to R_0), and no energy transfer takes place between them (as illustrated in Figure 13.2b). When observed at an arbitrary emission wavelength, λ_{em}, the fluorescence intensities of the two populations add up to give the measured fluorescence (see Figure 13.2a),

$$I^m(\lambda_{em}, \lambda_{ex}) \equiv I^D(\lambda_{em}, \lambda_{ex}) + I^A(\lambda_{em}, \lambda_{ex}). \tag{13.14}$$

Let us now assume that a D molecule moves into close proximity of an A molecule, as illustrated in Figure 13.2c, so that FRET may occur between those molecules. In that case, the acceptor emission is

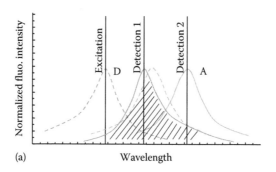

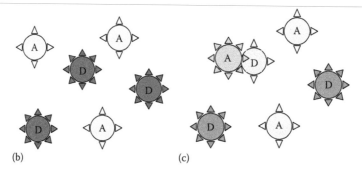

FIGURE 13.2 (a) Excitation (dashed lines) and emission (solid lines) spectra of donor (D) and acceptor (A) molecules overlap over a broad range of wavelengths. An extensive overlap between D emission and A excitation is required, which increases the probability of energy transfer. (b) Excitation spectra in (a) indicate that, if D and A are separated by large distances (relative to R_0), the donors are more often excited than the acceptors and emit more as a result. This differential emission level is represented by the different number of "light rays" (i.e., triangles) leaving the molecules (circles). (c) When a donor is brought into the close proximity of an acceptor molecule, FRET may occur. The FRET efficiency depends, among other things, on the degree of spectral overlap between D emission and A excitation (Lakowicz 2006), represented by the hatched area within the spectra in (a). FRET renders the acceptor brighter and the donor dimmer, as represented by the number of "rays" (triangles) emanating from each molecule (circle).

enhanced, while the donor emission is diminished. As a result of this effect, Equation 13.14 has to be replaced by

$$I^{DA}(\lambda_{em},\lambda_{ex})+I^{AD}(\lambda_{em},\lambda_{ex})=I^{D}(\lambda_{em},\lambda_{ex})-I^{D}(\lambda_{em},FRET)+I^{A}(\lambda_{em},\lambda_{ex})+I^{A}(\lambda_{em},FRET) \qquad (13.15)$$

where

I^{DA} is the intensity of donor fluorescence when the acceptor is present
I^{AD} is the intensity of acceptor fluorescence when the donor is present
$I^{D}(\lambda_{em}, FRET)$ is the FRET-induced reduction in donor intensity
$I^{A}(\lambda_{em}, FRET)$ is the gain in acceptor emission

Equation 13.15 indicates that the measured fluorescence intensity at any emission wavelength depends on the excitation wavelength. In addition, both types of fluorescent molecules are excited, although to different degrees (as illustrated in Figure 13.2a). This effect is called "spectral cross talk," and it should be avoided in experiments. This depends on the Stokes shift of the donors and acceptors (i.e., the difference between the "center of mass" of excitation and emission spectra), which varies from molecule to molecule. Also note that the A signal at the emission wavelength selected in Figure 13.2 is stronger than the D signal, but there is practically no single wavelength at which only D or only A can be detected. This is the so-called spectral bleed-through. The effect of the spectral cross talk and bleed-through on the experimental determination of FRET efficiency has been one of the most important driving forces in the evolution of FRET techniques, as it will become apparent in the following text.

FRET efficiency is one of the most important parameters in FRET studies because it depends on the distance between the interacting fluorophores, and thereby can provide information on the configuration of a protein complex. In steady-state (or intensity-based) fluorescence studies, several methods have been introduced that rely on different observation strategies to calculate some indices that directly relate to the FRET efficiency. Published methods have been summarized and comparatively analyzed in numerous publications (Gordon et al. 2001, Berney and Danuser 2003, Zal and Gascoine 2004).

Many of those methods relied on combinations of emission and detection filters, as well as on internal and/or external calibrations of the detection systems (employing standard fluorescence samples), in order to permit compensation for the spectral crosstalk and bleed-through. Following such filter-based methods, one may use two separate emission filters and detect the fluorescence intensity with each of them, first at $\lambda_{em,1}$ corresponding to "Detection 1" in Figure 13.2a, and then at $\lambda_{em,2}$ corresponding to "Detection 2." At $\lambda_{em,1}$, donor emission is relatively low, and Equation 13.15 becomes approximately

$$I^{AD}(\lambda_{em,2},\lambda_{ex}) \cong I^{A}(\lambda_{em,2},\lambda_{ex})+I^{A}(\lambda_{em,2},FRET), \qquad (13.16)$$

while for $\lambda_{em,2}$, acceptor emission is lower and Equation 13.15 becomes

$$I^{DA}(\lambda_{em,1},\lambda_{ex}) \cong I^{D}(\lambda_{em,1},\lambda_{ex})-I^{D}(\lambda_{em,1},FRET). \qquad (13.17)$$

Equations 13.16 and 13.17 permit calculation of two similarly defined but quantitatively different *apparent FRET efficiencies* (E_{app}) as the fractional transfer of energy from D to A molecules (Lakowicz 2006, Raicu 2007),

$$E_{app}^{Dq} \equiv \frac{I^{D}(\lambda_{em},FRET)}{I^{D}(\lambda_{em},\lambda_{ex})}=\left[1-\frac{I^{DA}(\lambda_{em},\lambda_{ex})}{I^{D}(\lambda_{em},\lambda_{ex})}\right], \qquad (13.18)$$

and as the fractional increase in A emission in the presence of D,

$$E_{\text{app}}^{\text{Ase}} \equiv \frac{I^{\text{A}}(\lambda_{\text{em}}, \text{FRET})}{I^{\text{A}}(\lambda_{\text{em}}, \lambda_{\text{ex}})} = \left[\frac{I^{\text{AD}}(\lambda_{\text{em}}, \lambda_{\text{ex}})}{I^{\text{A}}(\lambda_{\text{em}}, \lambda_{\text{ex}})} - 1 \right],$$
(13.19)

where superscripts Dq and Ase stand for donor quenching and acceptor-sensitized emission. In this context, the term "apparent" (subscript app) signifies that the efficiency determined from Equations 13.18 and 13.19 is a global measure of the efficiency of the energy transfer in the populations of proteins containing *large numbers* of donors and acceptors, only a fraction of which are interacting.

The message to be taken from Equations 13.18 and 13.19 is that an efficiency of the energy transfer can be determined from measurements on one fluorescent species (e.g., D) in the presence (I^{DA}) and absence (I^{D}) of the other species (e.g., A). The determination of I^{D} and I^{A} has been done by employing acceptor photobleaching (Ha et al. 1996, Wouters et al. 1998, Zal and Gascoine 2004, Raicu et al. 2005) and external calibration (Gordon et al. 1998, Hoppe et al. 2002, Edelman et al. 2003) of acceptor fluorescence, respectively. The first method may also incorporate corrections for inadvertent donor photobleaching (Zal and Gascoine 2004, Raicu et al. 2005).

Several issues can be raised in connection with FRET index–based methods, one of which is that, by their nature, FRET indices are dependent upon the characteristics of the instruments being used and on the local concentrations of the interacting species. In using such methods, one has to employ complex calibration procedures to accommodate the limited validity of Equations 13.18 and 13.19. However, this has not precluded the use of FRET indices methods in some imaging applications.

The determination of FRET efficiency from acceptor photobleaching requires cumbersome corrections for inadvertent donor photobleaching (Raicu et al. 2005), and it is inapplicable to kinetic studies, because the bleached acceptor is not available for further use. In its turn, the determination of FRET efficiency from sensitized emission necessitates the separate measurements of acceptor concentration using a second excitation wavelength. In both methods, therefore, one successively scans the sample at two different excitation wavelengths. Because the time between such scans is long compared to the timescale of molecular diffusion in cells, the local composition of the sample may change from scan to scan, and this has so far confined the applicability of FRET-based stoichiometry to the determination of cellular averages. We will describe next how recent improvements in instrumentation and data analysis have overcome this problem to allow the determination of protein complex structure *in vivo*.

13.3.2 Spectral Decomposition of Intensities Measured at Multiple Emission Wavelengths

Recent publications have demonstrated the feasibility of spectrally resolved fluorescence microscopy (Lansford et al. 2001, Haraguchi et al. 2002) and demonstrated its use in quantitative FRET studies (Zimmermann et al. 2002, Neher and Neher 2004, Raicu et al. 2005, Thaler et al. 2005, Chen et al. 2007, Merzlyakov et al. 2007). The practical power of the spectral method comes from the following two unique features: (1) its ability to exactly extract the donor and acceptor signals from composite spectra and (2) the possibility to relate the experimentally measured intensities to other spectral and molecular parameters, which can then be used to determine the FRET efficiency. We will tackle these aspects one by one in the following text.

   1. In this method, the measurements are performed in two steps. First, in experiments on samples containing *only A or only D molecules*, the fluorescence intensity of the samples is measured at several wavelengths, λ_{em}, to obtain emission spectra (i.e., a set of emission intensities at several wavelengths spread uniformly over the emission range). The intensities in these spectra are divided by their respective maximum intensity values to obtain individual fluorescence spectra,

$$s^{A} = \begin{bmatrix} i_1^{A} & \cdots & i_n^{A} \\ \lambda_{\mathrm{em},1} & \cdots & \lambda_{\mathrm{em},n} \end{bmatrix}, \tag{13.20a}$$

$$s^{D} = \begin{bmatrix} i_1^{D} & \cdots & i_n^{D} \\ \lambda_{\mathrm{em},1} & \cdots & \lambda_{\mathrm{em},n} \end{bmatrix}, \tag{13.20b}$$

where i with various subscripts and superscripts are normalized intensities. Then, the fluorescence intensity of the sample containing *both A and D* is measured at the same wavelengths and the composite fluorescence spectrum,

$$S^{m} = \begin{bmatrix} I_1^{m} & \cdots & I_n^{m} \\ \lambda_{\mathrm{em},1} & \cdots & \lambda_{\mathrm{em},n} \end{bmatrix}, \tag{13.20c}$$

is recorded. The measured composite spectrum S^{m} is related to individual s^{A} and s^{D} spectra through the following relation

$$S^{m} = k^{DA}(\lambda_{\mathrm{ex}})s^{D} + k^{AD}(\lambda_{\mathrm{ex}})s^{A}, \tag{13.21}$$

where $k^{DA}(\lambda_{\mathrm{ex}})$ and $k^{AD}(\lambda_{\mathrm{ex}})$ provide the measures of the fluorescence emission from donors in the presence of acceptors and from acceptors in the presence of donors, respectively.

By using a least-squares minimization procedure, it is possible to determine $k^{DA}(\lambda_{\mathrm{ex}})$ and $k^{AD}(\lambda_{\mathrm{ex}})$ (Epe et al. 1983, Raicu et al. 2005). We will show next that these quantities together with the integrated elementary spectra and the quantum yields of the D and A molecules are all that one needs to determine the FRET efficiency in spectrally resolved FRET.

2. By integrating Equation 13.15 over the range of emission wavelengths that encompasses both fluorescent species (A and D), we obtain the following expression for the total number of photons emitted from both fluorescence species in the presence of FRET:

$$F^{DA}(\lambda_{\mathrm{ex}}) + F^{AD}(\lambda_{\mathrm{ex}}) = F^{D}(\lambda_{\mathrm{ex}}) - F^{D}(\mathrm{FRET}) + F^{A}(\lambda_{\mathrm{ex}}) + F^{A}(\mathrm{FRET}). \tag{13.22a}$$

In this equation,

$$F^{DA}(\lambda_{\mathrm{ex}}) = k^{DA}(\lambda_{\mathrm{ex}}) \int_{\lambda_{\mathrm{em}}} i^{D}(\lambda_{\mathrm{em}})\mathrm{d}\lambda_{\mathrm{em}} = k^{DA}(\lambda_{\mathrm{ex}})w^{D} \tag{13.22b}$$

and

$$F^{AD}(\lambda_{\mathrm{ex}}) = k^{AD}(\lambda_{\mathrm{ex}}) \int_{\lambda_{\mathrm{em}}} i^{A}(\lambda_{\mathrm{em}})\mathrm{d}\lambda_{\mathrm{em}} = k^{AD}(\lambda_{\mathrm{ex}})w^{A} \tag{13.22c}$$

were derived with the help of Equation 13.21. In Equations 13.22b and 13.22c, $i^{A}(\lambda_{\mathrm{ex}})$ and $i^{D}(\lambda_{\mathrm{ex}})$ are the normalized (to their maxima) emission intensities of A and D upon excitation at wavelength λ_{ex}, and w^{A} and w^{D} are the integrals of the elementary spectra of A and D, respectively. Equation 13.22a may be separated into the following equations:

$$F^{DA}(\lambda_{ex}) = F^D(\lambda_{ex}) - F^D(\text{FRET}), \tag{13.23a}$$

$$F^{AD}(\lambda_{ex}) = F^A(\lambda_{ex}) + F^A(\text{FRET}) \tag{13.23b}$$

Noticing that a fraction $N^{FRET}Q^D = F^D(\text{FRET})$ (where Q^D is the quantum yield of D) of the total number of excitations, N^{FRET}, would have been emitted as photons by the donor in the absence of FRET, while a fraction $N^{FRET}Q^A = F^A(\text{FRET})$ (with Q^A the quantum yield of A) of N^{FRET} is emitted as photons by the acceptor if FRET occurs, we obtain a third necessary equation (Raicu et al. 2009),

$$F^D(\text{FRET}) = F^A(\text{FRET}) Q^D / Q^A. \tag{13.23c}$$

Further, if the acceptor is only excited through FRET (i.e., $F^A(\lambda_{ex}) = 0$), Equations 13.23a through c may be used to solve for $F^D(\lambda_{ex})$ and $F^D(\text{FRET})$. Under this condition and with the use of Equations 13.22b and c, one obtains immediately the following equation:

$$E_{app} \equiv \frac{F^D(\text{FRET})}{F^D(\lambda_{ex})} = \frac{1}{1 + \left(Q^A k^{DA}(\lambda_{ex}) w^D\right) / \left(Q^D k^{AD}(\lambda_{ex}) w^A\right)}. \tag{13.24}$$

Equation 13.24 together with the parameters $k^{DA}(\lambda_{ex})$ and $k^{AD}(\lambda_{ex})$ determined from experiment as described under item (1) above and the spectral integrals of the individual fluorescent species determined separately may be used to determine the apparent FRET efficiency, E_{app}, from the fluorescence spectra of each pixel in an image. The value of E_{app}, therefore, is obtained without recourse to acceptor photobleaching or the external calibration of acceptor emission. This general equation applies to random as well as oligomeric interactions, and to both homo- and hetero-oligomers of arbitrary size. Note that Equation 13.24 may be used only if the acceptor and donor fluorescence spectra are acquired in a parallel fashion for each image pixel; otherwise, the molecular makeup of a cellular region would change while scanning between emission wavelengths.

It is very important to remark here that, in the spectrally resolved FRET method described in this section, the separation of Equation 13.22a into Equations 13.23a and b is only done for mathematical reasons in order to derive Equation 13.24. Unlike the case of measurements at two fixed wavelengths described in the previous section, here we do not need to perform that separation experimentally, since the measured intensities do not enter expression (13.24). Instead, Equation 13.24 relies on $k^{DA}(\lambda_{ex})$ and $k^{AD}(\lambda_{ex})$, which are determined through a spectral decomposition method from the experimental data. Therefore, while methods based on filters provide only an approximation of the FRET efficiency, the spectral method is exact (within the range of experimental uncertainties).

13.3.3 Instrumentation for FRET Imaging with Single-Molecule or Molecular Complex Sensitivity

A breakthrough in the development of imaging techniques has been the introduction of the two-photon laser scanning microscope (Denk et al. 1990, Zipfel et al. 2003, Masters and So 2004, Diaspro et al. 2005). This microscope exploits the ability of fluorescent molecules to absorb at once two photons (Masters and So 2004) having together the necessary energy to induce an electronic excitation of the molecule (equivalent to the excitation of a single, more energetic photon). The two photons do not have to present equal energies (or wavelengths); their wavelengths only have to obey the following relationship: $1/\lambda_{abs} = 1/\lambda_1 + 1/\lambda_2$. In addition, this effect is possible in principle for an arbitrary number of photons. The conditions needed for a multiphoton process to occur are created experimentally by tightly focusing the beam of a pulsed (subpicosecond) laser through a high numerical aperture objective and thus confining the photons spatially and temporally. Because the intensity of the light is only high enough for the two-photon excitation process to occur in a small volume around the focal point, image sectioning without a confocal pinhole is achieved.

Some of the advantages of multiphoton imaging over, e.g., confocal microscopy are (1) the number of photons arriving at the detector is larger than when using detection pinhole (as in the case of confocal microscopes) because image sectioning occurs without a need to use confocal pinholes that attenuate the signal; (2) any photodestruction (i.e., photobleaching) of the sample by the exciting beam is confined to the sample layer being imaged, leaving intact the rest of the sample for further study; (3) biological materials are more transparent at longer wavelengths and thus images can be obtained from the deeper layers of biological tissues; (4) with two-photon microscopes, the signal does not need to be descanned and it is thus possible to use two-dimensional detectors and attain higher image acquisition speeds than when using point detectors such as photomultiplier tubes.

Using a two-photon excitation scheme, the research group led by this author has built a microscope with spectral resolution (two-photon microscope [TPM]-SR) (Raicu et al. 2007–2009), which satisfies all the requirements and approximations invoked in the theory presented in the previous section, particularly the spectral resolution and the selective excitation of the donor molecules. The instrument uses femtosecond light pulses with wavelengths ranging from about 700 to 860 nm for excitation. When necessary, we can also use a pulse compressor for pre-compensation of the group velocity dispersion (Fork and Martinzez 1984) and a pulse shaper (see Fig. 20 of Weiner 2000) for selective excitation of fluorescent molecules (Cruzblanca 2006, Ogilvie et al. 2006). The detection is performed in a nondescanned mode, in which the fluorescence is projected through a transmission grating onto a cooled electron-multiplying CCD camera with single-photon sensitivity. This instrument currently may be used to acquire up to 310 images (over the visible range of 400–650 nm) with 300 pixel × 500 pixel resolution within ~3 min, and smaller, single-cell images (with ~100 pixel × 100 pixel area) in less than 30 s for the whole set of 310 images. The speed may be further increased by reducing the spectral resolution, or by using faster scanning protocols (Kim et al. 1999). Increased speed may become necessary for the dynamic investigations of protein complexes.

13.4 Determination of Protein Structure In Vivo

13.4.1 Experimental Investigations

Employing a TPM-SR of the type described earlier, it has been possible to obtain fluorescent images of yeast cells expressing the fusion proteins Ste2p-GFP$_2$ and Ste2p-YFP (first row of Figure 13.3). GFP$_2$, which was used as a donor, has a single-photon excitation maximum at ~400 nm (Zimmermann et al. 2002) and a two-photon excitation at ~800 nm (Raicu 2007), while YFP (Tsien 1998, Raicu et al. 2005) was used as an acceptor, since its excitation spectrum almost perfectly overlaps with the donor emission.

Using the separately measured spectral properties of the two GFP variants, and the method for the determination of E_{app} described in Section 13.3.2 together with the known ratio of the quantum yields (Zimmermann et al. 2002, Raicu et al. 2005) and the spectral integrals of the two fluorescent tags (determined from measurements on singly expressed GFP$_2$- and YFP-tagged Ste2p proteins), an apparent FRET efficiency was obtained for each image pixel (second row of Figure 13.3). As seen in Figure 13.3, FRET was present in the plasma membrane as well as in internal membranes (e.g., in the ER and/or transport vesicles) in agreement with previous observations on this same membrane receptor (Overton et al. 2005). However, it is not clear at this time the extent to which molecular crowding within internal membranes leads to stochastic FRET.

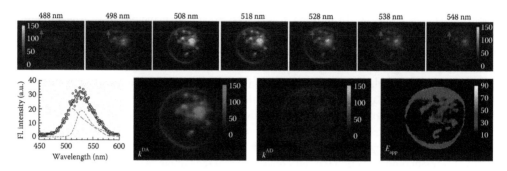

FIGURE 13.3 (See color insert following page 15-30.) First row: Microphotographs of yeast cells coexpressing the fusion proteins Ste2p-GFP$_2$ and Ste2p-YFP, obtained at seven selected emission wavelengths with a TPM-SR upon excitation at 800 nm (±15 nm). Second row: Analysis of the images in the top row: First panel from left: Fluorescence spectrum from a single pixel located on the cell membrane and indicated by the red arrows in the first row. Circles, measured intensities; long-dashed lines, decomposed donor-only (GFP$_2$ or Ste2p-GFP$_2$) spectrum; short-dashed lines, acceptor-only (YFP or Ste2p-YFP) spectrum; solid lines, sums of individual spectra. Second panel: donor-only image. Third panel: acceptor-only image. Fourth panel: spatial distribution of the apparent FRET efficiency. (Adapted from Raicu, V. et al., *Nat. Photon.*, 3, 107, 2009. With permission.)

When plotting the number of pixels showing the same E_{app} values against E_{app}, histograms such as that shown in Figure 13.4 were obtained, which could be fitted to a sum of five Gaussian functions. We will show in the following text that, while the resolution of all images obtained with the TPM-SR is limited by diffraction, as is the case with any other classical light-based microscope, a further analysis of the data using a pertinent theory allows one to determine structural information at nanometer length scale. This gain in resolution is due to the very nonlinear dependence of FRET on distance. This allows one to interpret the E_{app} in a manner that provides information about the size (i.e., the number of monomers) and the structure (i.e., the relative disposition of the monomers) of protein complexes.

In the next section, we will (1) derive mathematical expressions for the apparent FRET efficiencies expected for different types of protein complex structures, (2) obtain an inventory of the different expected values of E_{app} for each type of structure, and (3) determine possible correlations among those expected values. Then, we will show that this method of analysis might allow one to determine the most likely structure of a protein complex that is compatible with the E_{app} histogram presented in Figure 13.4.

13.4.2 Determination of the Protein Complex Structure from Apparent FRET Efficiency

Let us begin by making the assumption that the concentration of protein complexes is so low that one voxel in the sample space contains a single protein complex. This means that the photons originating from a voxel are all collected in a single pixel in the image plane. In practice, the photons originating from a single voxel are collected in several pixels in the image plane, due to the point spread function of the optical setup. This introduces image blur and constrains the use of the theory to extremely low concentrations of proteins, which in turn demands the use of very sensitive detectors, such as electron-multiplying CCD cameras (Raicu et al. 2009). For higher concentrations of proteins and protein complexes, the image blur has to be corrected for, and this can be done by employing numerical methods

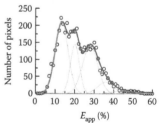

FIGURE 13.4 Experimental distribution of E_{app} for the cell shown in Figure 13.3 (points), and theoretical best-fit to a sum of five Gaussian functions shown individually (thin lines) or as a sum (thick lines). (Adapted from Raicu, V. et al., *Nat. Photonics*, 3, 107, 2009. With permission.)

(Neher and Neher 2004, Hoppe et al. 2008). In the image analysis presented in this chapter, the approximation applies due to very low concentration of proteins and the high sensitivity of the camera.

With the above approximation, one may derive expressions for the apparent FRET efficiency in protein complexes of various sizes (i.e., dimers, trimers, tetramers, etc.), and configurations (e.g., linear versus square-shaped tetramers), and then test such models against the experimental data in order to infer the structure that best fits the data. In such cases, the apparent FRET efficiency has the meaning of an average efficiency per donor in a complex. Let us begin by assuming a tetrameric complex of the type presented in Figure 13.5, in which each monomer (acceptor or donor) occupies a vertex of a square. Assuming that at most one donor in the complex is in an excited state for any time period and that the energy transfer takes place between the chromophores located at the centers of the disks symbolizing the acceptors and the donors, the apparent efficiency per donor is obtained from Equation 13.11 as

$$E_{\mathrm{app}} \equiv \frac{1}{3}\sum_{i} E_{i,k,n,q} = \frac{1}{3}\left[2\frac{\Gamma_{\mathrm{s}}^{\mathrm{FRET}}/(\Gamma^{\mathrm{r,D}}+\Gamma^{\mathrm{nr,D}})}{1+\Gamma_{\mathrm{s}}^{\mathrm{FRET}}/(\Gamma^{\mathrm{r,D}}+\Gamma^{\mathrm{nr,D}})} + \frac{\Gamma_{\mathrm{d}}^{\mathrm{FRET}}/(\Gamma^{\mathrm{r,D}}+\Gamma^{\mathrm{nr,D}})}{1+\Gamma_{\mathrm{d}}^{\mathrm{FRET}}/(\Gamma^{\mathrm{r,D}}+\Gamma^{\mathrm{nr,D}})}\right]. \tag{13.25}$$

The factor two in front of the first fraction accounts for the fact that the FRET efficiencies for donors two and three with acceptor one are identical, assuming circular symmetry of the problem, i.e., that either static or dynamic averaging of the orientation factor applies (Clegg 1996). Since the sidewise rate of transfer between donors and the acceptors, $\Gamma_{\mathrm{s}}^{\mathrm{FRET}}$, is eight times larger that the one across the diagonal, $\Gamma_{\mathrm{d}}^{\mathrm{FRET}}$, the second term in Equation 13.25 can be neglected to a first approximation. Equation 13.25 therefore becomes

$$E_{\mathrm{app}} = \frac{2}{3}\frac{\Gamma_{\mathrm{s}}^{\mathrm{FRET}}/(\Gamma^{\mathrm{r,D}}+\Gamma^{\mathrm{nr,D}})}{1+\Gamma_{\mathrm{s}}^{\mathrm{FRET}}/(\Gamma^{\mathrm{r,D}}+\Gamma^{\mathrm{nr,D}})} = \frac{2}{3}E_{\mathrm{p}}, \tag{13.26}$$

wherein we have introduced the notation $E_{\mathrm{p}} = [\Gamma_{\mathrm{s}}^{\mathrm{FRET}}/(\Gamma^{\mathrm{r,D}}+\Gamma^{\mathrm{nr,D}})]/[1+\Gamma_{\mathrm{s}}^{\mathrm{FRET}}/(\Gamma^{\mathrm{r,D}}+\Gamma^{\mathrm{nr,D}})]$. The symbol E_{p} denotes the pair-wise FRET efficiency, which is equal to the FRET efficiency between a single donor and a single acceptor in a dimeric complex.

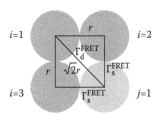

FIGURE 13.5 Illustration of the relative distances and rate constants of de-excitation through FRET within a square-shaped tetramer ($n=4$) containing three donors ($k=3$) and one acceptor ($n-k=1$). Note the differing rate constants of de-excitation along the sides of the square, $\Gamma_{\mathrm{s}}^{\mathrm{FRET}} = (\Gamma^{\mathrm{r,D}}+\Gamma^{\mathrm{nr,D}})(R_0/r)^6$, and across the diagonal, $\Gamma_{\mathrm{d}}^{\mathrm{FRET}} = (1/8)(\Gamma^{\mathrm{r,D}}+\Gamma^{\mathrm{nr,D}})(R_0/r)^6$.

Depending on the number of acceptors and donors as well as on their relative positions within a complex, square tetramers could assume several configurations, the configuration shown in Figure 13.5 being only one possibility. Following the above line of reasoning, one may easily obtain E_{app} expressions for practically any combination of the donor and acceptor numbers within any oligomeric configuration. Figure 13.6 shows all possible FRET-productive configurations for the linear and triangular trimers as well as for square, parallelogram, and linear tetramers, together with the expressions for the apparent (or average) FRET efficiencies associated with each configuration. To obtain those expressions, the following approximations were made, depending on the particular case investigated: (1) For linear configurations, only the nearest neighbors are involved in FRET. In other words, if an acceptor and a donor are separated by another acceptor or donor, the energy transfer that may take place between them is neglected, because Γ^{FRET} is $2^6=64$ times lower in that case compared to Γ^{FRET} for nearest neighbors. (2) For parallelograms, only the energy transfer taking place between donors and acceptors located across the long diagonal is neglected.

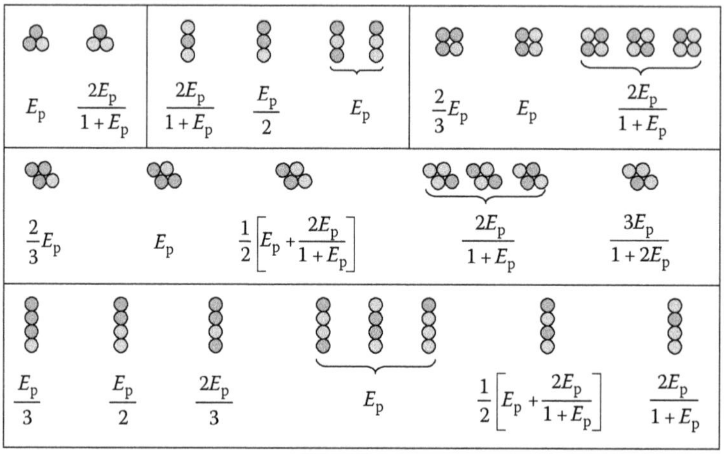

E_p	$\dfrac{2E_p}{1+E_p}$	$\dfrac{2E_p}{1+E_p}$	$\dfrac{E_p}{2}$	E_p	$\dfrac{2}{3}E_p$	E_p	$\dfrac{2E_p}{1+E_p}$

$\dfrac{2}{3}E_p$	E_p	$\dfrac{1}{2}\left[E_p+\dfrac{2E_p}{1+E_p}\right]$	$\dfrac{2E_p}{1+E_p}$	$\dfrac{3E_p}{1+2E_p}$

$\dfrac{E_p}{3}$	$\dfrac{E_p}{2}$	$\dfrac{2E_p}{3}$	E_p	$\dfrac{1}{2}\left[E_p+\dfrac{2E_p}{1+E_p}\right]$	$\dfrac{2E_p}{1+E_p}$

FIGURE 13.6 Examples of oligomeric complexes and the relationships between the FRET efficiencies associated with individual configurations and that between a single donor and a single acceptor within a dimer (E_p). Note that only FRET-productive configurations (i.e., complexes containing at least one donor and at least one acceptor) are shown.

To summarize the results presented in Figure 13.6, we note that each oligomeric size and structure is characterized by a different number of mathematical expressions for FRET efficiencies. These features allow one to test various oligomeric models against the experimental data, and thereby to find the one that best explains the data. In the particular case of the Ste2p receptor discussed earlier (see Figures 13.3 and 13.4), it was found necessary to fit the data with a sum of five Gaussian functions; that in itself indicates that the parallelogram-shaped tetramer is the best candidate for describing the structure of the Ste2p complex. The case for this structure was strengthened further by the fact that the values of E_{app} corresponding to the peaks of the first, third, fourth, and fifth Gaussians in Figure 13.4 were very close to those calculated from the E_{app} value for the second peak (which was identified as the pair-wise efficiency, E_p) according to the expressions for the FRET efficiencies in Figure 13.6. The data in Figure 13.4 were inconsistent with the other structural models presented in Figure 13.6, either in terms of the predicted number of different peaks, or in terms of the predicted relationships among the different values of E_{app} (Raicu et al. 2009).

13.4.3 Comparison of Pixel-Level Spectrally Resolved FRET with Other Methods

FRET efficiency may also be determined using fluorescence lifetime imaging (FLIM) (Elangovan et al. 2002, Bacskai et al. 2003, Duncan et al. 2004, Wallrabe and Periasamy 2005, Spriet et al. 2008). This method relies on the fact that excited donor molecules spend less time in that state if they can transfer energy to an unexcited nearby acceptor, since FRET is a more rapid pathway for losing energy than the spontaneous de-excitation of the donor. If the lifetime of the donor in the absence of acceptors is known, it is possible to determine the FRET efficiency using Equation 13.5. In general, it seems that FLIM faces significant difficulties in detecting interactions between more than two partners, since signal-to-noise considerations require binning of data from multiple pixels. That in turn requires fitting with multiple exponentials, a well-known ill-posed mathematical problem. For instance, in the case of the simplest tetrameric complexes (i.e., the square and linear shapes) one expects no less than four different lifetimes: one for donor alone, and three for the three different FRET efficiencies listed in Figure 13.6. With a parallelogram-shaped structure, the problem becomes even more complicated. This situation may improve in the future by developing FLIM techniques with higher photon-collection efficiency (so that binning will not be necessary), as well as more robust data-fitting procedures that are less sensitive to noise.

Other methods based on RET include rapidly alternating between two or three excitation wavelengths to discriminate between D and A excitation for molecules in solution (Eggeling et al. 2005, Lee et al. 2007) and bioluminescence resonance energy transfer (BRET) (Xu et al. 1999, Angers et al. 2000, Pfleger and Eidne 2006), which has been used to avoid direct acceptor excitation altogether. To date, neither of these methods has been extended to permit determination of protein complex structure. BRET recently has achieved enough sensitivity to permit single-cell or even subcellular investigations (Hoshino et al. 2007, Xu et al. 2007, Coulon et al. 2008). Unfortunately, single-cell BRET techniques currently provide no quantitative information (in the sense described in this chapter); rather, they indicate only the regions of a cell wherein the proteins of interest interact with one another. The interactions between proteins are detected through the so-called BRET ratio, which is defined as the ratio of fluorescence intensities at two different wavelengths. The physical significance of the BRET ratio is unclear, however, as it does not relate directly to the RET efficiency. The lack of full spectral resolution currently prevents BRET from being a quantitative technique. Moreover, the original advantage of BRET over FRET—namely, that BRET avoids direct excitation of the acceptors—is no longer relevant, since there are at least two different methods by which this problem can be circumvented in FRET (see previous text). The situation may change if the sensitivity of BRET, i.e., the amount of detectable signal increases by at least an order of magnitude compared to that obtained with current BRET techniques; such a gain in sensitivity could then be expended on the required spectral, spatial, and temporal resolutions.

There are also limitations that affect all types of known RET-based methods (including FRET, BRET, and FLIM), when these are compared to other techniques that are not based on fluorescence.

One problem that is sometimes mentioned is that the presence of fluorescence tags may affect the function of the proteins under investigation. Although there is no much experimental evidence to substantiate this concern (due to the lack of alternative methods to compare to), the problem may exist in principle.

Another potential concern is that the precise shape of measured fluorescence spectra may vary from cell to cell due to the fluorescent molecules' sensitivity to pH and other physical and chemical parameters (Tsien 1998, Patterson et al. 2001) that may vary between individual cells. These variations can be minimized by (1) experimentally controlling the properties of the cellular medium (Raicu et al. 2005), such as the content of glucose that ultimately determines the cell's internal pH, by (2) placing the fluorescent tag on the external side of the membrane, where the pH can be controlled better, or by (3) creating fluorescent molecules that are less sensitive to the pH. However, more detailed investigations should be carried out in the future in order for this effect to be better quantified.

Finally, accurate FRET, BRET, or FLIM measurements often require the high expression levels of proteins, which may inadvertently enhance FRET through random encounters of donors and acceptors with no particular biological significance; this is often called "stochastic FRET" (Wolber and Hudson 1979, Dewey and Hammes 1980, Kenworthy and Edidin 1998, Meyer et al. 2006), and it poses significant difficulties to measurements based on cellular or even multiple cellular averages. Our preliminary investigations of the distributions of FRET efficiencies in stochastic FRET (D.R. Singh and V. Raicu, unpublished) indicate that pixel-level FRET might offer a possibility to discriminate between functional and stochastic interactions by looking at the differences in the distributions of FRET efficiencies of stochastic versus functional interactions. By contrast, using cellular averages, i.e., averages over entire E_{app} distributions, it is usually difficult to discriminate between stochastic and functional interactions (Kenworthy and Edidin 1998), because the amount of information carried by the distribution of E_{app} values is reduced in the process of averaging.

In conclusion, notwithstanding the potential limitations, pixel-level FRET with spectral resolution provides an excellent means for determining, in living cells and on a nanometer scale, both the size of a protein complex and the structural arrangement of its constituent subunits. Although information about the size of a complex may also be obtained with methods that rely on cellular averages, the structural information is lost owing to the inability of such approaches to deliver distributions of FRET efficiencies.

Acknowledgments

The author thanks James W. Wells for critical reading of the manuscript, Deo Singh and Sasmita Rath for double-checking the derivations of the equations presented in this chapter, and Michael Stoneman for help with formatting figures and references and for proofreading the manuscript. This work was supported in part by a grant from the Wisconsin Institute for Biomedical and Health Technologies (Grant No. W620), and seed funds from the UWM Research Growth Initiative (Grant No. 101X014).

References

Angers, S., Salahpour, A., Joly, E. et al. 2000. Detection of beta 2-adrenergic receptor dimerization in living cells using bioluminescence resonance energy transfer (BRET). *Proceedings of the National Academy of Sciences of the United States of America* 97: 3684–3689.

Bacskai, B. J., Skoch, J., Hickey, G. A., Allen, R., and Hyman, B. T. 2003. Fluorescence resonance energy transfer determinations using multiphoton fluorescence lifetime imaging microscopy to characterize amyloid-beta plaques. *Journal of Biomedical Optics* 8: 368–375.

Berney, C. and Danuser, G. 2003. FRET or no FRET: A quantitative comparison. *Biophysical Journal* 84: 3992–4010.

Caorsi, V., Ronzitti, E., Vicidomini, G., Krol, S., McConnell, G., and Diaspro, A. 2007. FRET measurements on fuzzy fluorescent nanostructures. *Microscopy Research and Technique* 70: 452–458.

Chen, Y., Mauldin, J. P., Day, R. N., and Periasamy, A. 2007. Characterization of spectral FRET imaging microscopy for monitoring nuclear protein interactions. *Journal of Microscopy* 228: 139–152.

Clegg, R. M. 1996. Fluorescence resonance energy transfer. In *Fluorescence Imaging Spectroscopy and Microscopy*, X. F. Wang and B. Herman, eds. New York: Wiley-Interscience.

Coulon, V., Audet, M., Homburger, V. et al. 2008. Subcellular imaging of dynamic protein interactions by bioluminescence resonance energy transfer. *Biophysical Journal* 94: 1001–1009.

Cruzblanca, H. 2006. An M2-like muscarinic receptor enhances a delayed rectifier K+ current in rat sympathetic neurones. *British Journal of Pharmacology* 149: 441–449.

Denk, W., Strickler, J. H., and Webb, W. W. 1990. Two-photon laser scanning fluorescence microscopy. *Science* 248: 73–76.

Dewey, T. G. and Hammes, G. G. 1980. Calculation on fluorescence resonance energy transfer on surfaces. *Biophysical Journal* 32: 1023–1035.

Diaspro, A., Chirico, G., and Collini, M. 2005. Two-photon fluorescence excitation and related techniques in biological microscopy. *Quarterly Reviews of Biophysics* 38: 97–166.

Duncan, R. R., Bergmann, A., Cousin, M. A., Apps, D. K., and Shipston, M. J. 2004. Multi-dimensional time-correlated single photon counting (TCSPC) fluorescence lifetime imaging microscopy (FLIM) to detect FRET in cells. *Journal of Microscopy* 215: 1–12.

Edelman, L. M., Cheong, R., and Kahn, J. D. 2003. Fluorescence resonance energy transfer over approximately 130 basepairs in hyperstable Lac repressor-DNA loops. *Biophysical Journal* 84: 1131–1145.

Eggeling, C., Kask, P., Winkler, D., and Jager, S. 2005. Rapid analysis of Forster resonance energy transfer by two-color global fluorescence correlation spectroscopy: Trypsin proteinase reaction. *Biophysical Journal* 89: 605–618.

Elangovan, M., Day, R. N., and Periasamy, A. 2002. Nanosecond fluorescence resonance energy transfer-fluorescence lifetime imaging microscopy to localize the protein interactions in a single living cell. *Journal of Microscopy* 205: 3–14.

Epe, B., Steinhauser, K. G., and Woolley, P. 1983. Theory of measurement of Foster-type energy transfer in macromolecules. *Proceedings of the National Academy of Sciences of the United States of America* 80: 2579–2783.

Fork, R. L. and Martinzez, O. E. 1984. Negative dispersion using pairs of prisms. *Optics Letters* 9: 150–152.

Gordon, G. W., Berry, G., Liang, X. H., Levine, B., and Herman, B. 1998. Quantitative fluorescence resonance energy transfer measurements using fluorescence microscopy. *Biophysical Journal* 74: 2702–2713.

Gordon, A. S., Yao, L., Jiang, Z., Fishburn, C. S., Fuchs, S., and Diamond, I. 2001. Ethanol acts synergistically with a D2 dopamine agonist to cause translocation of protein kinase C. *Molecular Pharmacology* 59: 153–160.

Ha, T., Enderle, T., Ogletree, D. F., Chemla, D. S., Selvin, P. R., and Weiss, S. 1996. Probing the interaction between two single molecules: Fluorescence resonance energy transfer between a single donor and a single acceptor. *Proceedings of the National Academy of Sciences of the United States of America* 93: 6264–6268.

Haraguchi, T., Shimi, T., Koujin, T., Hashiguchi, N., and Hiraoka, Y. 2002. Spectral imaging fluorescence microscopy. *Genes Cells* 7: 881–887.

Hoppe, A., Christensen, K., and Swanson, J. A. 2002. Fluorescence resonance energy transfer-based stoichiometry in living cells. *Biophysical Journal* 83: 3652–3664.

Hoppe, A. D., Shorte, S. L., Swanson, J. A., and Heintzmann, R. 2008. Three-dimensional FRET reconstruction microscopy for analysis of dynamic molecular interactions in live cells. *Biophysical Journal* 95: 400–418.

Hoshino, H., Nakajima, Y., and Ohmiya, Y. 2007. Luciferase-YFP fusion tag with enhanced emission for single-cell luminescence imaging. *Nature Methods* 4: 637–639.

Kenworthy, A. K. and Edidin, M. 1998. Distribution of a glycosylphosphatidylinositol-anchored protein at the apical surface of MDCK cells examined at a resolution of <100 A using imaging fluorescence resonance energy transfer. *Journal of Cell Biology* 142: 69–84.

Kim, K. H., Buehler, C., and So, P. T. C. 1999. High-speed, two-photon scanning microscope. *Applied Optics* 38: 6004–6009.

Lakowicz, J. R. 2006. *Principles of Fluorescence Spectroscopy*, pp. 443–506. New York: Springer.

Lansford, R., Bearman, G., and Fraser, S. E. 2001. Resolution of multiple green fluorescent protein color variants and dyes using two-photon microscopy and imaging spectroscopy. *Journal of Biomedical Optics* 6: 311–318.

Lee, N. K., Kapanidis, A. N., Koh, H. R. et al. 2007. Three-color alternating-laser excitation of single molecules: Monitoring multiple interactions and distances. *Biophysical Journal* 92: 303–312.

Masters, B. R. and So, P. T. C. 2004. Antecedents of two-photon excitation laser scanning microscopy. *Microscopy Research and Technique* 63: 3–11.

Maurel, D., Comps-Agrar, L., Brock, C. et al. 2008. Cell-surface protein–protein interaction analysis with time-resolved FRET and snap-tag technologies: Application to GPCR oligomerization. *Nature Methods* 5: 561–567.

Merzlyakov, M., Li, E., Casas, R., and Hristova, K. 2006. Spectral Forster resonance energy transfer detection of protein interactions in surface-supported bilayers. *Langmuir* 22: 6986–6992.

Merzlyakov, M., Chen, L., and Hristova, K. 2007. Studies of receptor tyrosine kinase transmembrane domain interactions: The EmEx-FRET method. *Journal of Membrane Biology* 215: 93–103.

Meyer, B. H., Segura, J.-M., Martinez, K. L. et al. 2006. FRET imaging reveals that functional neurokinin-1 receptors are monomeric and reside in membrane microdomains of live cells. *Proceedings of the National Academy of Sciences of the United States of America* 103: 2138–2143.

Neher, R. and Neher, E. 2004. Optimizing imaging parameters for the separation of multiple labels in a fluorescence image. *Journal of Microscopy* 213: 46–62.

Ogilvie, J. P., Debarre, D., Solinas, X., Martin, J. L., Beaurepaire, E., and Joffre, M. 2006. Use of coherent control for selective two-photon fluorescence microscopy in live organisms. *Optics Express* 14: 759–766.

Overton, M. C., Chinault, S. L., and Blumer, K. J. 2005. Oligomerization of G-protein-coupled receptors: Lessons from the yeast *Saccharomyces cerevisiae*. *Eukaryotic Cell* 4: 1963–1970.

Patterson, G., Day, R. N., and Piston, D. 2001. Fluorescent protein spectra. *Journal of Cell Science* 114: 837–838.

Pfleger, K. D. and Eidne, K. A. 2006. Illuminating insights into protein–protein interactions using bioluminescence resonance energy transfer (BRET). *Nature Methods* 3: 165–174.

Raicu, V. 2007. Efficiency of resonance energy transfer in homo-oloigomeric complexes of proteins. *Journal of Biological Physics* 33: 109–127.

Raicu, V. and Popescu, A. 2008. *Integrated Molecular and Cellular Biophysics*, pp. 195–208. London, UK: Springer, pp. 195–208.

Raicu, V., Jansma, D. B., Miller, R. J., and Friesen, J. D. 2005. Protein interaction quantified in vivo by spectrally resolved fluorescence resonance energy transfer. *Biochemical Journal* 385: 265–277.

Raicu, V., Fung, R., Melnichuk, M., Chaturvedi, A., and Gillman, D. 2007. Combined spectrally-resolved multiphoton microscopy and transmission microscopy employing a high-sensitivity electron-multiplying CCD camera. *Multiphoton Microscopy in the Biomedical Sciences VII, Proc. of SPIE*, San Jose, CA, 6442: 64420M-1–64220M-5.

Raicu, V., Chaturvedi, A., Stoneman, M. R. et al. 2008. Determination of two-photon excitation and emission spectra of fluorescent molecules in single living cells. *Proceedings of SPIE, Multiphoton Microscopy in the Biomedical Sciences VIII*, San Jose, CA, 6860: 686018-1–686018-8.

Raicu, V., Stoneman, M. R., Fung, F. et al. 2009. Determination of supramolecular structure and spatial distribution of protein complexes in living cells. *Nature Photonics* 3: 107–113.

Selvin, P. R. 2000. The renaissance of fluorescence resonance energy transfer. *Nature Structural Biology* 7: 730–734.

Song, L., Hennink, E. J., Young, I. T., and Tanke, H. J. 1995. Photobleaching kinetics of fluorescein in quantitative fluorescence microscopy. *Biophysical Journal* 68: 2588–2600.

Spriet, C., Trinel, D., Riquet, F., Vandenbunder, B., Usson, Y., and Heliot, L. 2008. Enhanced FRET contrast in lifetime imaging. *Cytometry A* 73: 745–753.

Stryer, L. 1967. Fluorescence spectroscopy of proteins. *Science* 162: 526–533.

Thaler, C., Koushik, S. V., Blank, P. S., and Vogel, S. S. 2005. Quantitative multiphoton spectral imaging and its use for measuring resonance energy transfer. *Biophysical Journal* 89: 2736–2749.

Tsien, R. Y. 1998. The green fluorescent protein. *Annual Review of Biochemistry* 67: 509–544.

Wallrabe, H. and Periasamy, A. 2005. Imaging protein molecules using FRET and FLIM microscopy. *Current Opinion in Biotechnology* 16: 19–27.

Weiner, A. M. 2000. Femtosecond pulse shaping using spatial light modulators. *Review of Scientific Instruments* 71: 1929–1960.

Wolber, P. K. and Hudson, B. S. 1979. An analytic solution to the Förster energy transfer problem in two dimensions. *Biophysical Journal* 28: 197–210.

Wouters, F. S., Bastiaens, P. I., Wirtz, K. W., and Jovin, T. M. 1998. FRET microscopy demonstrates molecular association of non-specific lipid transfer protein (nsL-TP) with fatty acid oxidation enzymes in peroxisomes. *The EMBO Journal* 17: 7179–7189.

Xu, Y., Piston, D. W., and Johnson, C. H. 1999. A bioluminescence resonance energy transfer (BRET) system: Application to interacting circadian clock proteins. *Proceedings of the National Academy of Sciences of the United States of America* 96: 151–156.

Xu, X., Soutto, M., Xie, Q. et al. 2007. Imaging protein interactions with bioluminescence resonance energy transfer (BRET) in plant and mammalian cells and tissues. *Proceedings of the National Academy of Sciences of the United States of America* 104: 10264–10269.

Zal, T. and Gascoine, N. R. 2004. Photobleaching-corrected FRET efficiency imaging of live cells. *Biophysical Journal* 86: 3923–3939.

Zimmermann, T., Rietdorf, J., Girod, A., Georget, V., and Pepperkok, R. 2002. Spectral imaging and linear un-mixing enables improved FRET efficiency with a novel GFP2-YFP FRET pair. *FEBS Letters* 531: 245–249.

Zipfel, W. R., Williams, R. M., and Webb, W. W. 2003. Nonlinear magic: Multiphoton microscopy in the biosciences. *Nature Biotechnology* 21: 1369–1377.

14

Automation in Multidimensional Fluorescence Microscopy: Novel Instrumentation and Applications in Biomedical Research

14.1 Introduction 14-1
14.2 Instrumentation for High-Throughput/Content
Microscopy 14-3
Image Screening Microscopy • Confocal High-Content Screening
Microscopy • Flow Cytometry and Laser-Scanning Cytometry
14.3 Automated Image Analysis in High Content Microscopy......14-13
14.4 Applications in Biomedical and Oncological Research 14-15
14.5 Future Perspectives 14-17
Acknowledgment 14-17
References 14-18

Mario Faretta
*European Institute
of Oncology*

14.1 Introduction

Starting from the birth of the light microscope, optical microscopy always accompanied research in cell biology providing an invaluable tool for the observation of life. The increased complexity of the requests stemming from the analysis of living matter was paralleled by a continuous evolution in optical technology, which determined a dramatic rise in the number of observable dimensions in the experimental variables universe. The two-dimensional view of the widefield microscope evolved into the optical sectioning ability of the confocal microscope opening the way to *in vivo* intracellular tomography. In parallel, the synergy between fluorescence microscopy and cell biology was maximized by the revolutionary discovery of green fluorescent protein (GFP). The possibility to introduce a fluorescent tag for protein localization in living cells dramatically increased the impact of fluorescence microscopy in life sciences. At the same time, computer-controlled acquisitions became a reality providing the possibility to program specific observation conditions at selected time points. The ability to monitor events in living cells with minimal perturbations thus gave origin to the spatiotemporal universe in the biological world.

Finally, the evolution of the optical microscope introduced the opportunity to browse through additional dimensions in the cell universe. Fluorescence manipulation by fluorescence recovery after photoleaching and photoactivation protocols, fluorescence resonance energy transfer, spectral imaging, and fluorescence lifetime introduced an innovative perspective in the approach to the structure/function problem. Besides providing a characterization of the cellular and intracellular structures at good resolution, fluorescence microscopy acquired the ability to perform functional analysis directly looking at molecular mobility, biomolecular interactions in living cells and organisms, and evaluating metabolic parameters through the analysis of photophysical properties of the living environment.

The technological evolution thus made the observables universe in physics and microscopy applied to life sciences very similar, at least in terms of their number of dimensions. Similarities do not stop here. Quantum mechanics foundation posed precise conditions on the precision of the measurement of physical variables. Similarly fluorescence microscopy faced intrinsic limitations in its analysis capability. An uncertainty principle is also applicable to fluorescence microscopy making it impossible to maximize the resolution of a measured parameter (space, time, intensity, etc.) without affecting the precision of the others. Thus, even the latest developments in "far-field nanoscopy," namely, stimulated emission depletion (STED) microscopy, 4Pi microscopy, photoactivated localization microscopy (PALM), and stochastic optical reconstruction microscopy (STORM) went over the diffraction barrier (Hell, 2007, 2009) partially impinging on temporal resolution, field of view, and signal-to-noise ratio.

The optimization of the analysis conditions and the determination of the required resolution have to consider additional requirements from life sciences. The concept of "statistical resolution" acquired more and more relevance after the completion of the human genome sequencing. This project revolutionized the approach to the comprehension of biological mechanisms, providing an enormous amount of information to be decrypted by associating a function to the sequenced genomic material. Throughput, that is the number of analyzed samples, and content, the number of simultaneously observed parameters, are now inevitably coupled to sensitivity and spatial and temporal resolutions in defining the optimal performances of fluorescence-based observation technologies. Automation consequently invaded the field of technological development in fluorescence microscopy. The robotization of the elementary steps normally executed by microscope users evolved to create fully automated acquisition procedures, starting from sample positioning, target identification, data collection, storage, and annotation up to image analysis and active machine-learning procedures for phenotype recognition.

In addition to the obvious need to process a relevant amount of samples, the increase in throughput is intrinsically wrapped in many of the most recent hot spots in biomedical and cancer researches. The view of cancer as an anomalous tissue generated from the differentiation of committed progenitor cells and maintaining a self-renewal capability, thanks to a reservoir of multipotent "cancer stem cells" (Kakarala and Wicha, 2007; Vermeulen et al., 2008), is now largely accepted. The physiological and cancer stem cells compartment is poorly represented from the statistical point of view, as low as a fraction of percentile point. The characterization of its properties *in vitro* and *in vivo* requires the collection of large amount of data to acquire statistical relevance. The same requirement for an extremely large field of view and, consequently, a high number of acquired images originates traditionally from histopathological tissue analysis. The ability to apply diffraction-limited observation to areas whose dimension remains informative at histological level calls for a high degree of automation in mosaic acquisition (Figure 14.1) and relevant computational efforts for a correct processing of the acquired data.

A full exploitation of the potential of a fluorescence microscopy–based analysis simultaneously requires the maximization of the number of information retrievable from the single sample observation. Spectral imaging enormously increased the available combination of fluorescence tags to be simultaneously employed for protein marking, including tissue autofluorescence as an informative parameter in the global analysis.

Similarly to the optimization of spatial resolution in *in vivo* microscopy, the increase in content and throughput has to face the paradigm of the living cell as a Schroedinger's Cat. Fluorescence observation inevitably causes cell poisoning due to free radicals production. Consequently, spatial resolution, temporal resolution, signal-to-noise ratio, and a number of samplings have to be set maintaining the principle of minimal perturbation of the observed system.

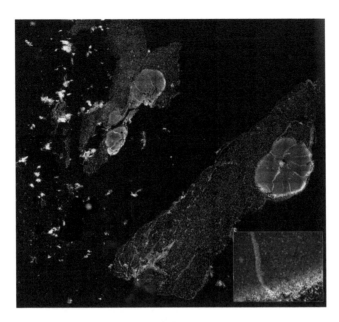

FIGURE 14.1 (See color insert following page 15-30.) Mosaic reconstruction of a tissue slice for high content analysis. A mouse ovary tissue slice was fixed and stained for three color parameters analysis (DNA:DAPI. BrdU: FITC anti BrdU. CK5: Cy5 anti CK5). A Scan∧R acquisition station assembled on an upright motorized microscope was employed for acquisition. Magnification for each field of view is 10× and the entire scanned area is 2 cm × 2 cm obtained by a mosaic reconstruction of 25 × 30 images. The inset shows the detail of a single acquired field. (Sample kindly provided by Dr. Alessandra Insinga, IEO, Milan, Italy.)

All these considerations led to the birth of a specialized branch of optical microscopy devoted to high-content image analysis. In this chapter, an overview of the most recent developments in the field of high-throughput and content microscopy are presented. Initially, the general hardware requirements for the realization of a robotized microscopy station are discussed. Particular attention is devoted to some crucial points in the definition of an acquisition protocol, namely, speed optimization, autofocusing, and optical arrangements (widefield versus confocal).

A general discussion of the problems related to image analysis is presented according to the type of analyzed sample (cell populations, tissues, etc.).

The chapter concludes with a brief summary of the most diffused high-content applications in the basic and clinical biomedical researches.

14.2 Instrumentation for High-Throughput/ Content Microscopy

14.2.1 Image Screening Microscopy

The need to manage thousands of samples pushed for the automation of the entire workflow of image acquisition and analysis making available a large number of commercial and open source solutions (Tables 14.1 and 14.2).

However, the request for robotization starts from sample preparation. Liquid-handling systems designed to manipulate cells for culture and stimulation, and to eventually execute staining procedures were consequently coupled to the microscope. The robotized acquisition system for high-throughput screening then included automatic sample loaders to complete the integration procedure (Carpenter, 2007). The preparation of samples for imaging and, in particular, expression of DNA or RNA libraries

TABLE 14.1　Commercially Available High-Throughput/Content Imaging Screening Systems

Manufacturer	System	Type	Web Site	Analysis Software
Molecular Devices	ImageXpress[MICRO TM] ImageXpress ULTRA	Widefield and point-scanning confocal Imaging screening system	http://www.moleculardevices. com/pages/instruments/ imagexpress_micro.html http://www.moleculardevices. com/pages/instruments/ imagexpress_ultra.html	MetaXpress Application modules + Acuity Xpress
Molecular Devices	Isocyte	Laser scanning cytometer	http://www.blueshiftbiotech. com/products/isocyte	Isocyte control software Integrated data analysis package
Perkin Elmer	Opera series	Widefield and spinning disk confocal Imaging screening system	http://las.perkinelmer.com/ imaging/Products/HCS/ default.htm	Acappella analysis software
GE Healthcare	IN Cell Analyzer Series	Widefield and line-scanning confocal Imaging screening system	http://www.blacore.com/ high-content-analysis/ product-range/Overview/	IN Cell Investigator software + IN Cell Miner HCM
Thermo Scientific	ArrayScanHCSReader	Widefield and structured light illumination Imaging screening system	http://www.cellomics.com/ menu/instrumentation	Cellomics bioapplications
BD Biosciences	BD Pathway	Widefield and white light scanning disk Imaging screening system	http://www.bdbiosciences.com/ bioimaging/cell_biology/	BD Pathway software
Olympus	Scan^R	Widefield imaging screening system	http://www.microscopy. olympus.eu/microscopes/ Life_Science_Microscopes_ scan_R.htm	Scan^R analysis
Amnis	ImageStream	Image streaming cytometer	http://www.amnis.com	IDEAS
Compucyte	iCyte series	Laser scanning cytometer	http://www.compucyte.com/	iCyte cell analysis software
TTP Labtech	Acumen eX3	Microplate cytometer	http://www.ttplabtech.com/ products/acumen/index.html	Integrated data analysis package
Leica Microsystems	SP5 Matrix Screener	Point-scanning spectral confocal microscope	http://www.leica-microsystems. com	

of fluorescence constructs in living cells, stimulated the development of new transfection procedures. The immobilization of retroviral constructs and new techniques to deliver target DNA for exogenous expression will not be the subject of this chapter, and we invite the reader to consult the large amount of literature produced in the field (Stewart et al., 2003; Wheeler et al., 2004, 2005; Moffat and Sabatini, 2006; Moffat et al., 2006; Erfle et al., 2007, 2008). However, it is worthwhile to mention them herein to underline that their development contributed to the miniaturization of the support for screenings. High-throughput screening did not solely rely on multiwell plates: a single microscope slide can now host

TABLE 14.2 Academic Resources for HTS/HCA

Web Site	Description
http://sen.sourceforge.net/state.html	Open Source project for development of computational tools for high content screening
http://www.cellprofiler.org/	Free image software for image-based phenotype screening
http://www.micro-manager.org/	Open Source software for control of automated microscopes
http://rsb.info.nih.gov/ij/	Open Source software for image analysis
http://www.biohts.com/	Free resource website for Drug Discovery
http://liebel-lab.fzk.de/liebelwiki/index.php/Main_Page	Web site of Urban Liebel Lab at the Karlsruhe Institute of Technology
http://www.pasteur.or.kr/group/technology_index.jsp?c_seq=3	Web site of the screening-related program at the Institute Pasterur in Korea
http://web.wi.mit.edu/sabatini/pub/research_arrays.html	Web site of Cell micro-array Technology at the David Sabatini lab

from hundreds to thousands of cell populations expressing different biological constructs. Automatic sample loading had to face the problem of the wide variety of supports optimized according to the goal of the analysis, e.g., making slide spotting preferable for high resolution and optically sectioned examination. The coupling of robotized loaders to the microscope involves a particular consideration due to the impact on the design of the incubation system for live cell imaging. According to the throughput, temperature, humidity, and oxygenation control can demand for enclosure chambers for the microscope up to the realization of dedicated sterile rooms with controlled environmental parameters. The design of the microscope itself can undergo dramatic rearrangements due to the housing of mechanical parts devoted to sample management.

The acquisition software for image screenings possesses routines to optimize sampling according to the employed support and to the task of the analysis. The most common procedures are generally based on the acquisition of few images for each type of construct by random or targeted acquisition in a single well to be repeated over an array covering the entire plate. To maximize throughput, intelligent acquisition protocols can be implemented: to avoid the accumulation of useless data, image storage can be driven by the recognition of specific events such as the presence of cells in the field of view.

A correct design of the fluorescence components is fundamental for the establishment of the conditions for a quantitative analysis. Automated microscopy stations largely benefited from the new light sources that replaced the standard mercury arc lamps as metal halide light sources. In addition to increased lifetime and stability, they provided a redistributed intensity output over the emitted spectrum. Favoring the blue wavelengths range these lamps demonstrated a better efficiency in the excitation of GFP when compared to traditional illumination sources at parity of total power output. As an alternative choice, xenon and mercury xenon high-power arc lamps were generally employed to increase the available illumination power, in order to optimize excitation when using long working distance low numerical aperture objectives.

The minimization of intensity fluctuations became a goal of primary importance considering the long duration of screening experiments and the need for relative quantification, both in kinetics experiments and in the evaluation of steady-state expression levels in different biological backgrounds. The fiber coupling of the light source to the microscope body contributed to the stabilization of the illuminating field. Simultaneously a direct measurement of the emitted power by photodiodes can be introduced to directly monitor this critical parameter.

In some cases, the maximization of the excitation efficiency called for a redesign of the illumination geometry. In a microscope, the field of view is generally optimized for the ocular observation leading to the maximization of the objective field number. The illumination apparatus alignment is consequently devoted to the homogenous coverage of its entire extension. However, image screening completely excluded direct human observation, relying exclusively on CCD-acquired data. Some microscope manufacturers modified the illumination setup in order to refocus illumination to a restricted region corresponding to the CCD sensor chip–covered area, thus gaining energy density and preserving the sample from deleterious side effects of not informative exposure.

The minimization of the phototoxic side effects stemming from the use of increased energy densities called in parallel for the introduction of an active modulation of the excitation intensity. Large sets of neutral density filters mounted on high-speed filter wheels or grating devices provided an almost continuous selection of the illumination power.

The spectral selection of the excitation and emission wavelengths plays obviously a pivotal role in determining the system efficiency in terms of collected signal intensity, and temporal and statistical resolutions. Confocal microscopy experienced a revolution with the introduction of dispersive elements in the scanheads leading to the possibility of a photophysical characterization of the emitted light. Two-photon laser sources and more recently white light lasers (Betz et al., 2005) then turned the modern microscope into a sort of high-spatial-resolution spectrophotometer with imaging capability. Even if high content analysis heavily relies on an increased spectral resolution and maximal flexibility in terms of optimal detection of the widest range of fluorophores, spectral analysis in widefield screening microscopy is commonly based on standard optical filters probably due to the difficulties in coupling between parallelized acquisition of spatial and spectral information. Consequently, high-speed excitation and emission filter wheels (in the range of tens of milliseconds exchange time) generally ensure the required flexibility in the spectrum coverage.

The design of the optimal configuration of an image screening microscopy station requires a careful consideration of the filter repertoire. The evolution of the technology for the production of optical filters provided a wide variety of multiband dichroic and emission filters.

Multiband filters call for at least two main advantages, namely, increased time resolution and the minimization of mechanical parts movement. However, the acquisition of different spectral bands through the same optical element exposes to the risk of spectral bleedthrough. In considering optimized spectral separation, it is necessary to take into account the relevance of tails in the excitation spectrum. Many red fluorescent proteins, including the popular mCherry, present for example a relevant absorption cross section in the blue range where GFP is normally excited, increasing the risk of spectral contamination of signals when multiband filters are employed in the signal detection. The presence of an unbalanced expression of the tagged proteins, together with the extended spectral bandwidth collected, amplify the contribution of excitation tails creating dangerous artifacts in the evaluation of colocalization experiments.

The use of multiband dichroic filters in conjunction with emission filter wheels constitutes a possible solution to overcome the aforementioned limitations. Emission filter wheels allow the selection of optimal collection ranges with minimal time wasting without involving movements of the microscope turret. In fact, if filter switching resides in the range of tens of milliseconds, dichroic cube replacement requires in the best cases hundreds of milliseconds. Multiband dichroic filters present possible negative side effects in the collection efficiency due to the multiple-reflected spectral regions. Besides time-consuming operations, the use of filter configurations designed for the maximization of single fluorochromes collection efficiency implies a tremendous workload for microscope components linked to the thousands of operations executed during a high-throughput acquisition procedure. To conclude, it is worthwhile mentioning the progress achieved in optical filter–coating technology to optimize the collection and reflection efficiencies, with a simultaneous minimization of the chromatic shifts that potentially affect multicolor acquisition.

Spatial resolution is obviously one of the key parameters to be correctly fixed before the acquisition of millions of images. The choice of the right objective must consider all the features determining its overall performances, namely, working distance, magnification, aberration correction, type of transmitted light detection method (e.g., phase contrast, differential interference contrast...), and numerical aperture.

Although optically speaking plastic is one of the worst substrates, particular biological requirements, like the use of cell types with problematic adhesive properties on glass, make its employment sometimes highly convenient. In that case, the use of high working-distance objectives becomes mandatory. At the same time, correction collars are required to compensate for differences in thickness and refractive index. Uncorrected observation leads in fact to chromatic and spherical aberrations and decreased collection efficiency, thus causing image degradation. To compensate for the presence of uncorrected focal shifts, the acquisition procedure can take advantage of focus motorization. The objective can thus be repositioned on the optimal axial coordinate before image collection takes place. Such repositioning is a relevant operation in the combined acquisition of transmitted light and fluorescence signals.

Maintaining a fixed magnification, high working distance is almost inevitably associated with a decreased numerical aperture, i.e., diminished spatial resolution and brightness. The choice of the correct support material must consequently take into consideration all the "biological" and "technical" parameters to optimize sample preparation, preservation, and observation.

The maximization of the optical resolution cannot be considered an *a priori* target in the definition of the correct acquisition procedure in an image-screening protocol. Magnification and numerical aperture, once fixed, determine the depth of field according to the formula

$$d = \frac{\lambda \cdot n}{NA^2} + \frac{n}{M \cdot NA} \cdot e,$$

where

d is the depth of field, i.e., the thickness of the sample slice in focus
λ is the light wavelength
n is the refractive index of the immersion medium ($n = 1$ for air)
NA is the numerical aperture
M is the magnification
e is the minimal distance spatially resolvable by the detector (i.e., the physical pixel size)

The choice of an extended depth of field instead of an increased resolution can be justified when the required spatial resolution is far from the diffraction limit, and the need of precise signal quantification prevails. The simultaneous evaluation of intracellular localization and protein expression provides a common and useful requirement of a correct balance between the two aforementioned parameters. The use of a long working-distance objective with moderate numerical aperture, intermediate magnification, and increased depth of field, is in this case fundamental for collecting the fluorescence generated from the whole cell volume, maintaining the ability to spatially resolve the protein localization across the main cell compartments.

The depth of field of the employed objective obviously has a deep impact on the search of the optimal focal plane. Autofocusing constitutes indeed one of the main challenges for automatic image acquisition. The existing approaches are based on software or hardware solutions or a mixture of them.

Laser-based autofocus systems (Liron et al., 2006) possess a mode of operation similar to CD players. The objective focus is defined by maximizing the reflected light signal impinging on a detector behind a pinhole. A red laser is employed in order to avoid interference in the spectral range of the usually employed fluorophores. Plastic substrates are not compatible with this kind of tools due to their refringence and absorption properties. The detected interface varies according to the variation in the refractive indexes. Dry objectives focus on the air–glass interface at the bottom of the plate, whereas the use of oil immersion objectives poses the maximal reflection in correspondence of the cell–glass attachment surface where the drop in refractive index takes place. The motorization of the z-axis allows for a correct repositioning by introducing an offset (corresponding in some cases to the substrate thickness) before the collection of the image. Two different strategies can be adopted in laser-based autofocusing. A continuous feedback can be established through electronic circuits in order to maintain the focal position constantly fixed. This way a higher temporal resolution can be obtained thanks to the small axial shift required for focus maintenance. Continuous autofocusing can be very helpful in acquisition protocols where time-lapse data collection has to be performed over different spatial positions in sequence. Also, mosaic acquisition spanning over contiguous fields of view can retrieve great benefits from such a procedure. Some laser-based autofocus systems do not establish a continuous feedback and repeat the search of the major reflection surface at every sampling point. In this case, the required time interval grows from tens to some hundreds of milliseconds.

The limitations of hardware autofocusing stem from the targeting of a reference point, applying eventually a fixed shift, independent of the biological target. When the system is challenged in live cell–imaging acquisitions, the biological significance of the collected image can be dramatically affected by cell detachment during mitosis. In that case, the collection of multiple planes of interest can contribute to the problem solution at the cost of a sometimes relevant increase in data storage of useless information.

Computational autofocusing can provide a higher level of flexibility that is able to be adapted to the different biological end points of a targeted screening. The term autofocus can indeed be quite limiting when used for microscopy applied to the observation of life. Without entering the enormous universe of image autofocus algorithms, it is worthwhile mentioning herein two large classes, namely, contrast- and object-based recognition of the plane of interest. Image contrast analysis provides excellent performances both in transmitted light and epifluorescence microscopy. Focus search in phase contrast images presents the obvious advantage of a reduced phototoxicity avoiding repeated exposure of cells to fluorescence. Image contrast maximization can also provide excellent performances when the focus target is, like in hardware-based autofocusing, the attachment surface. When the biological conditions make the contrast gradient insufficient to determine the right plane of interest, e.g., in the presence of a low number of cells in large fields of view, the detection of the reflected fluorescence can provide an efficient tool simulating laser-based autofocusing engines. An increase in contrast on the targeted interface, for example by physical scratching of the surface, can contribute to establish a highly efficient and fast autofocusing procedure.

Fluorescence objects recognition can be used as a first approach to the realization of a robotized "intelligent" acquisition. The presence of biological targets in the analyzed field of view can be employed to optimize data collection avoiding the storage of useless images. Object analysis is generally based on mean fluorescence intensity and dimension of pixel clusters. The focal plane is in this case determined taking into account not only intensity gradients between axial positions, but also the average intensity level of some targeted biological entities together with their physical extensions. The term "biological entity" has been used in place of "cell" because object recognition can select different phenotypes, not necessarily coinciding with the criteria "presence of cells in the field view." The mean intensity and size of a DNA associated signal, as DAPI or histone H2B, can be set to specifically identify condensed chromatin for the recognition of the cell mitosis stage. Considering that the fraction of mitotic cells reaches few percentile points in an exponentially growing cell populations, the use of image analysis–driven autofocus allows to select only the field of view of interest. This provides the additional advantage to focus on the correct image z-plane that is normally shifted up due to the detachment prior to cell division. The interplay between acquisition and analysis is one of the most promising solutions in the field of automated fluorescence microscopy thanks to the high level of flexibility. Such flexibility is granted by the possibility to completely separate the autofocus procedure from the image collection, e.g., changing the spatial resolution when passing from the first (autofocus) to the second (image-acquisition) step of the protocol. Image acquisition parameters completely differ in the two phases. Imaging for focus determination requires the minimization of exposure, both for time saving and sample preservation, simultaneously pushing for sensitivity and contrast. Spatial resolution and sampling are not usually an issue, being the autofocus procedure limited to cell recognition. The extensive usage of CCD camera binning is thus not only allowed but strongly suggested. On the other hand, an image collection for data analysis maintains the requirements for sensitivity and contrast, interchanging the role of temporal and spatial resolutions.

The maximization of the field of view is an almost mandatory requirement in high content screening, and consequently the use of a moderate magnification with high numerical aperture is frequently adopted as a solution. This choice poses restrictions on the adopted spatial sampling frequency that must obey the Nyquist theorem statement to get data really limited only by the optical resolution. A pixel dimension that is at least half of the optical resolution is the minimal required sampling frequency. Binning and time exposure must be set for the optimization of contrast, and inevitably impinge on the choice of an optimized camera for a high content–screening platform.

Pushed sensitivity for very fast imaging is not generally a "must" for a CCD camera dedicated to automated widefield microscopy. The aforementioned requirements for field of view, spatial sampling, and sensitivity make front-illuminated detection preferable to back-thinned CCD thanks to the reduced physical pixel size and wider extension of the digital format. In general, back-illuminated cameras possess smaller (in terms of number of pixels) sensor chips with larger pixels, someway sacrificing spatial sampling in favor of sensitivity and readout speed, making them more suitable for high speed confocal imaging (see Section 14.2.2). However, living cell applications pose restrictions on the minimal sensitivity adoptable in *in vivo* screenings. CCD cooling can provide great improvements in the acquired signal-to-noise ratio, reducing illumination intensity and exposure time, and thus phototoxicity effects. This feature is generally available at good cost–performance ratio in commercial high digital-resolution cameras.

14.2.2 Confocal High-Content Screening Microscopy

Image-based phenotype screenings are generally characterized by a low-resolution approach employing long working-distance objectives with extended depth of field, thus conducting the spatial analysis to a pure two-dimensional view in favor of signal quantification. However, as happened for conventional fluorescence microscopy, high content screening (HCS) microscopy is experiencing a growing pressure to acquire optical-sectioning capability to perform a real three-dimensional analysis of the analyzed samples. To better understand why confocal microscopy initially encountered difficulties in coupling to high content screening, it is necessary to focus the attention on some of its intrinsic limitations. Laser scanning confocal microscopy achieves enhanced resolution by the sequential detection of spatially filtered signals from illuminated diffraction-limited regions. The reconstruction of the image implies scanning over the field of view of interest at the desired spatial sampling (number and reciprocal distance of the illuminated regions). Volumetric reconstruction is then obtained by the repetition of the operations, refocusing to different axial coordinates through the sample. Generally, the scanning procedure constrains the residence time of the laser beam and consequently the detection interval to 2–10 μs (considering a standard digital resolution of 512 pixels and a line-scanning frequency of 200–1000 Hz). The deriving temporal resolution is extremely low, in the order of a couple of frames per second. This number *per se* could not be considered as a limiting factor if not extended to a full 3D stack acquisition. The throughput required for a statistically significant analysis of an image-based screening will imply enormous acquisition times, making such an analysis impracticable. The time variable is not the only limitation to be considered in a comparison between widefield and confocal point-scanning approach in the image-screening context. Point-by-point illumination and detection based on photomultipliers impinge on the system sensitivity considering the limited quantum efficiency of the detector. The intrinsic requirement of signal quantification in HCS applications points against a conventional optical-sectioning technique and makes it totally unfeasible if one considers the dramatic phototoxicity related to the use of highly focused laser sources. The spatial parallelization of the scanning beam paved the road to introduce optical-sectioning devices in high content imaging. Nipkow disk–based confocal scanheads are presently one of the most adopted solutions for collection of large amounts of 3D information (Graf et al., 2005; Toomre and Pawley, 2006). Illumination is achieved by projecting a homogenous photon flux onto a spinning disk with a series of circular holes positioned in the image plane conjugated to the objective focus. The emitted fluorescence is then collected through the same circular pinhole pattern allowing for optical sectioning. Drawbacks in the realization of such an optical setup stem from the heavy loss of light due to the spatial filtering exerted by the disk, cutting out a relevant percentage of the illumination energy flux. In addition to improvements linked to the introduction of microlenses focusing the beam through the pinholes and to the advent of compact high power laser diodes, the real explosion in the popularity of the technique has been obtained thanks to the introduction of cooled and fast sensible cameras. While forced sensitivity is not a mandatory requirement for conventional high content screening, enhanced photon detection is absolutely necessary in confocal applications. The aforementioned limitations in spatial sampling related to the employment of back-illuminated sensor chips are still valid and even reinforced in this context where the gain in spatial resolution demands for smaller pixel size to fulfil the Nyquist theorem. Consequently, the choice of back-illuminated cameras must be carefully considered for high content applications, particularly for high-resolution automated microscopy in the histology field. Their relevant impact in sensitivity and temporal resolution must be appropriately evaluated in counterbalancing a potential loss in spatial resolution. Use of electron multiplying (EM) CCD cameras can provide unpaired performances in fast (live) cell imaging in back-illuminated detection schemes while granting, in front-illuminated sensor chips, a reduction in pixel size and thus a better spatial sampling without compromising speed and sensitivity.

CCD-based detection is maintained in the structured illumination approach to optical sectioning (Juskaitis et al., 1996) employed in the past in robotized microscopy solutions. The adopted scheme is based on the introduction of a grid in the optical pathway producing a striped illumination in the focal plane. The easy computation performed on at least three images recorded after phase shifting the grating of equally spaced angles provides an efficient removal of out-of-focus light. The need of repeated imaging of the same field of view for sectioned images reconstruction obviously reduces the temporal performances of the system making it not optimized for observation of living cells. However, its simplicity and low implementation costs however make it worth of consideration for intermediate throughput applications. The possibility to rely on standard light sources could be a valuable addition, e.g., when considering the design of high content applications based on the simultaneous detection of a large number of fluorophores.

Spectral resolution and enhanced and flexible ranges of detectable wavelengths revitalized the employment of point scanning laser confocal microscopes equipped with fast (resonance) scanning devices. This class of instruments reaches acquisition rates of up to 30 frames/s in spectrally resolving setups based on the presence of dispersive elements such as grids or prisms. When considering the acquisition speed of such systems, it is necessary to bear in mind that spectral acquisition at full resolution in CCD-based parallelized scanning can only be exploited by sequential image acquisitions. This leads to a reduced temporal resolution when increasing the spectral complexity. The implementation of a spectral analysis capability in a CCD framework implies the use of spectroscopic cameras with dedicated formats of the camera chip, onto which a spectrally decomposed signal is projected thanks to the action of dispersive elements. This way spatial resolution is sacrificed to acquire the possibility to browse through the emission spectrum of fluorescence. The use of liquid crystal optical filters (Levenson and Mansfield, 2006) in combination with EMCCD detectors seems to constitute a valid opportunity for fast spectral imaging but it is now limited by the high loss of light in its polarization-based action.

In addition to increased spectral capability, point scanning devices offer the advantage of incomparable flexibility in active remodeling of the field of view and of the adopted spatial sampling frequency laying the bases for an image analysis–driven acquisition. When moving from *in vitro*-based screening to high content *in vivo* microscopy performed on tissue sections or other thick specimens, point scanning schemes present advantages in compensating the photon scattering of emitted light, particularly if coupled to multiphoton fluorescence excitation. Independently from the trajectories of the collected photons, point detection allows associating a precise spatial coordinate to the detected light, corresponding to the position of the volume targeted by the focused infrared laser beam. The parallelization of the excitation beam became a "must" also in nonlinear microscopy techniques when the content scales up passing from a tissue slice–based analysis to the observation of entire organs (Kim et al., 2007; Ragan et al., 2007). In this context, it is interesting to note that automation enables in some cases to bypass limited penetration depth of light in comparison to histological dimensions. Coordinated coupling of the imaging process with automated thick slices cutting allowed for the reconstruction of fluorescence signal from a mouse whole heart.

A recent elegant work in live cell imaging of zebra fish embryo development paved the way for the contribution of the new "alternative" microscopies to the field of high content analysis (Keller et al., 2008). Selective plane illumination microscopy (SPIM) employs a cylindrical lens to create a planar illuminate region inside the sample from the side. Signal detection occurs through an objective positioned as usual perpendicularly to the illuminated plane. The confinement in excitation gives to the technique an intrinsic optical sectioning ability, which demonstrated its usefulness in the analysis of very thick but low scattering samples, such as Zebra fish, Medaka killer fish, *Xenopus*, *Drosophila*, and *C. elegans*. In the last version, the implementation of a digital-scanned laser light-sheet microscope allowed the observation of the first 24 h of development of a zebra fish embryo at excellent spatial and temporal resolutions, high signal-to-noise ratio, and in absence of relevant phototoxicity. Although at a superficial glance the impact on high content microscopy could be underestimated, the number of collected data makes it astonishing. More than 400,000 images per embryo were collected with a continuous imaging process performed at a speed of 10 million voxels/s to track the growth of 16,000 cells during the observation period.

14.2.3 Flow Cytometry and Laser-Scanning Cytometry

For several years, a fluorescence-based analysis of large population cohorts has been associated to flow cytometry. A flow cytometer employs laser-induced fluorescence to analyze the content of targeted biomolecules, namely, DNA, RNA and/or proteins of interest. Cells in flow are hydrodynamically focused in a stream to pass through a light beam one by one. In a very short time interval, from minutes to tens of seconds, thousands of cells can be analyzed. The optical setup of a flow cytometer presents several similarities when compared to a laser-scanning confocal microscope. The main feature is the signal point detection that forces to use high power light sources, namely, lasers, and leads simultaneously to the employment of photomultipliers as detectors. The flowing of the cells through the interrogation region, i.e., the focus of the objective collecting emitted fluorescence or side-scattered light, reverts the concept of beam scanning with sample-driven scanning, as occurred in the first confocal microscopes. The reading time is consequently limited to some milliseconds thus explaining the choice of a high photon density and sensible detectors. Flow cytometry realizes the maximization of throughput and signal collection despite spatial resolution. In some way, a flow cytometer can be thought as a zero spatially resolving confocal microscope with extremely high light collection capability and "statistical" resolution. The maximization of the number of collected photons is obtained through the choice of extended depth of focus objectives. Flow cytometry also provides a tremendous content by spectral analysis, which provided up to 18 different color parameters analysis (Perfetto et al., 2004). The concept of spectral unmixing, now largely applied in spectral confocal microscopy, was first developed for the flow cytometric signal compensation to correct spectral bleedthrough, thanks to a linear spectral deconvolution matrix.

Although the robotization of flow cytometers to adapt to screening requests has been in some cases considered as a possible development, the strong analytical limitation imposed by the absence of spatial resolution represented for years a dramatic obstacle for the entrance of flow cytometry in the world of image-screening techniques. Different approaches have been adopted to create a fusion between the flow cytometry and the microscopy universe giving rise to laser scanning cytometers (LSC), microplate cytometer, and image streaming cytometers. All these instruments enable to associate the statistical significance of a flow cytometric analysis to the image collection of a large cell population.

Laser scanning and microplate cytometers (Kamentsky et al., 1997; Darzynkiewicz et al., 1999a; Kamentsky, 2001; Auld et al., 2006; Bowen and Wylie, 2006; Pozarowski et al., 2006) works on immobilized cells performing a scan of an extended surface, a microscope slide, or a multiwell plate respectively, in order to collect low-resolution images of thousands of cells. The complete procedure is generally achieved by the combination of an optical (mirror-based laser scanning along the x-axis) action and mechanical (motorized stage moving along y direction) action. In general, the laser beam dimension is expanded to cover areas of some microns down to half a micron in order to create a good resolving power. Microplate cytometers tend to realize a flow cytometry–based approach to high-throughput drug screening limiting their optical configuration flexibility in favor of the ability to perform a single cell–based analysis over an entire well in a limited time interval. Again, the specific design matched to multiwells substrates reflects the choice of minimizing the loss in statistical resolution and throughput to acquire imaging capability. A fixed magnification is generally adopted, coupled to a specialized lens that grants flat-field-extended area illumination and a high depth of focus in order to avoid autofocusing during intrawell scans. LSCs employ similar scanning modalities, but they generally use a microscope optical setup for sample illumination gaining the possibility to vary spatial sampling and resolution in spite of a diminished scanning speed. Laser-based autofocusing is thus frequently part of the instrument setup in order to maintain the optimal observation plane at different depths of field. The image-forming capability and the cell population analysis capability for the two types of instruments rely on the clustering of the point-detected signal to reconstruct a single cell, i.e., a single statistical event. LSCs had a strong impact on the dissemination of high content screening techniques by extending cytometry analysis to the field of histopathology (Gerstner et al., 2004; Peterson et al., 2008). Its initial slide-based approach (recent versions have been adapted to different substrates such as multiwell plates) was naturally associated to tissue biopsies observation employing an approach (flow cytometry) known for years in the clinical studies.

Laser scanning and microplate cytometers essentially conserved the optical detection scheme of a flow cytometer replacing the hydrodynamic circuit required for in-flow analysis with the optomechanical scanning of immobilized samples. Image streaming cytometers (Zuba-Surma et al., 2007) employs the innovation introduced by time-delayed integration (TDI) of specialized CCD camera to maintain the possibility to work in flow. While the hydrodynamic focusing is consequently conserved, optical detection scheme is completely redesigned. TDI CCDs have been created for fast motion-tracking applications. In a streaming cytometer laser, side illumination induces fluorescence in the interception volume, and the signal collected by a fixed objective is projected onto the CCD pixel array. Synchronization between CCD charge transfer rate and object speed (flow rate) allows signal integration to preserve sensitivity and image quality achieving a continuous acquisition with a detection time interval of 10 ms per object and a resulting rate in the order of hundreds of cells per second. An extended depth of focus techniques can be used to perform 3D imaging and quantification.

To preserve the flow cytometer ability to perform multispectral analysis, a diffraction grating is employed to decompose the emitted light into different spectral bands and to project them onto separated CCDs. The use of high sensitivity detectors and increased integration time coupled to in-flow analysis pushes for the minimization of the total acquisition time. Even if at a superficial glance an image-streaming approach seems to grant for an increased statistical throughput per unit of time, it has to be considered that single shot analysis due to wasting of the cells can be someway limiting. The content of LSC-based analysis can be dramatically enhanced by repeated analysis of the sample applying the ability to relocate objects immobilized on a slide with good spatial precision. Polychromatic staining, iterative restaining, differential photobleaching, photoactivation, and photodeconstruction are all useful protocols to achieve hyperchromatic analysis (Mittag et al., 2005, 2006a,b; Laffers et al., 2006).

Finally, it is worthwhile mentioning a novel confocal axial tomography technique in suspension, which hopefully can contribute to pave the road to an alternative and efficient 3D high resolution imaging of living cells in suspension (Renaud et al., 2008). Even if the throughput and content are still far from the above-mentioned techniques the great potential of this new imaging methodology is to couple increased 3D resolution to the possibility to recover and purify the phenotype of interest for further analysis. Cell sorting is a routine technique in flow cytometry allowing for cell population discrimination and physical isolation based on the total amount per cell of a targeted antigen. The aforementioned cell tomography approach could open the possibility of a viable selection based on highly resolved spatial observation of intracellular signals opening the way for an automated 3D "active" imaging cytometry.

14.3 Automated Image Analysis in High Content Microscopy

The enormous amount of images collected by automated microscopy systems requires a strong effort in the creation of software solutions able to process them and to provide a comprehensive analysis of the main parameters of interest for the identification of the targeted biological end points. The present section will be dedicated to a general overview of the main general points that an automated analysis of fluorescence microscopy data has to tackle, without entering details regarding the single computational tool.

Data annotation and browsing is certainly the first topic to be addressed in high content analysis (HCA). In addition to the data regarding the spatial distribution of a fluorescence signal, every single image has appended to itself a collection of news regarding its identity that has to be taken into account for the correct processing. Most software packages solve the problem of the correct identification and location of data just by a self-explaining name choice. Spectral identity (channel or fluorophore name), axial position (sequential index inside Z stacks), temporal location (sequential index of a time sequence), spatial location (index of row and column of a mosaic acquisition, or intra- and inter-well position in a multiwell screening) can all be efficiently encoded in a single filename allowing for the correct and fast retrieval and reconstruction of the data flow. The need to manage huge amount of images and to save their acquisition history reinforced the request for a standardized image format validating the efforts of the Open Microscopy Environment Project, to create a worldwide-shared way to encode imaging data (Goldberg et al., 2005; Schiffmann et al., 2006).

Signal quantification being one of the major targets of image-screening analysis, possible artifacts introduced in the image acquisition step have to be amended before undertaking every other computational effort. Although the hardware solutions pointed at ameliorating the homogeneity of illumination of the field of view, the correction of the uneven spatial distribution of the exciting photons has to be frequently introduced, also considering the wide range of substrates employed in the different screenings. The use of plastic wells is sometimes inescapable, with the consequent loss of a real "flat-field" illumination due to the increased light scattering. The correction of the excitation intensity distribution is also useful to increase the sensitivity in the measurement of a specific biological parameter. DNA content evaluation, that is measurement of the intensity distribution of a nuclear dye in a population of objects (cells), is one of the most adopted tools for cell cycle progression analysis in cytometry (Figure 14.2). DNA index, i.e., the amount of DNA in tumor versus normal cells, is another parameter measured by evaluating the intensity of a DNA binding fluorochrome staining that is largely employed in clinical applications. The extended field of view required by a statistically significant cell population analysis frequently calls for a low spatial sampling and thus a reduced object dimension in pixels. A disparity in the illumination homogeneity can introduce a wider standard deviation with respect to the average of the population if not properly corrected, thus reducing the sensitivity in the discrimination of different cell ensembles.

The same argument applies to the problem of background correction, i.e., the estimation of an informative level of intensity to be employed in object recognition and signal quantification.

Intensity thresholding is the most common and efficient approach adopted in cell recognition particularly when applied to nuclear structures. A correct decomposition of cell clustering is however a big challenge in image-screening analysis. The number of new algorithms is now growing to achieve efficient image segmentation for object identification (Lamprecht et al., 2007; Li et al., 2007a; Gudla et al., 2008). The task requires increasing efforts when applied to the reconstruction of 3D objects (Belien et al., 2002; Ploeger et al., 2004; Li et al., 2007c; Lin et al., 2007; McCullough et al., 2008). Moving out

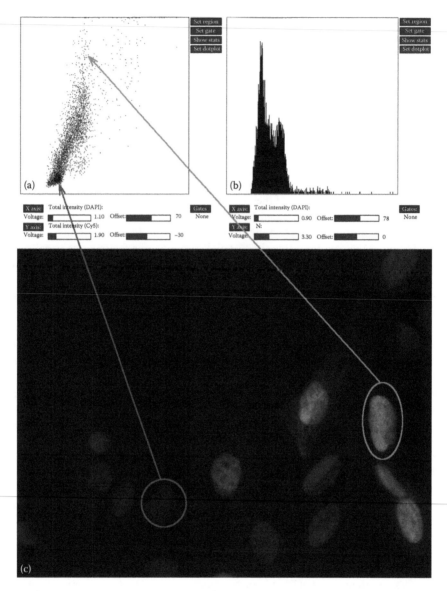

FIGURE 14.2 (See color insert following page 15-30.) The image cytometry analysis of cell cycle distribution. Bivariate DNA (DAPI)/cyclin A (anti-Cyclin A rabbit polyclonal antibody + cy3 conjugated goat anti-rabbit secondary antibody) content in exponentially growing U2OS osteosarcoma cell line. Analysis was performed using an ImageJ macro written by the author. In the figure, an example of the link between the shown cytogram in (a) (DNA versus Cyclin A) and a sampled image in (c) reflecting the position of the outlined cells in the graph based on evaluation of their fluorescence intensity in the two channels is shown. The histogram on the right panel (b) shows the DNA distribution of the analyzed cell population. (Cells kindly provided by Dr. Sara Barozzi, IEO, Milan, Italy.)

the field of image screening applied to drug or molecular screening in cell lines, the difficulties in the creation of image segmentation tools grow dramatically when entering histopathological analysis. The high density of biological material, the diminished signal-to-noise ratio due to the increased light scattering linked to a frequent demand for a 3D analysis calls for specific computational procedures for object identification in tissues (Kriete and Boyce, 2005).

A subsequent step naturally following cell identification is the location and characterization of intracellular structures for the characterization of a biological phenotype. The need of an automated interpretation of subcellular location patterns in two and three dimensions rises for example to address the protein structure–function problem in the screening of large DNA or RNA libraries (Jiang et al., 2007a; Simpson et al., 2007; Jones et al., 2009). Over the past years, new machine-learning algorithms were created to instruct computers to memorize signal distribution characterizing intracellular compartmentalization patterns (DeBernardi et al., 2006; Jiang et al., 2007a). Subcellular analysis requires observation close to the diffraction limit to achieve a better capability to distinguish biological features. Besides the aforementioned illumination and background corrections, additional computational procedures can be applied to ameliorate the classification performances. Spatial deconvolution, normally employed to acquire an optical-sectioning ability in widefield microscopy, can consequently be very useful to achieve an increased signal-to-noise ratio and real spatial resolution even in data originated by confocal systems. Moreover, high temporal sampling and reduced excitation energy typical of *in vivo* microscopy can produce extremely low contrast images, which can largely benefit from the application of deconvolution algorithms before undertaking further analysis.

Living cells analysis introduced a further level of complexity requiring object identification and trajectory reconstruction in the spatiotemporal universe (Harder et al., 2006; Neumann et al., 2006; Jiang et al., 2007b; Padfield et al., 2009). The main biological end points to be targeted in a four-dimensional analysis relate to cell ability to migrate across the space and to generate daughters through cell divisions. The next frontier for high content analysis will be to definitively acquire the ability to perform such an analysis in a complex three-dimensional context such as living organism tissues. *In vivo* microscopy poses the request to extend automated analysis to the spectral dimension. Hyperspectral analysis, tissue autofluorescence identification, and classification and evaluation of resonance fluorescence energy transfer events to identify molecular interactors are all assays where the introduction of spectral deconvolution can further enhance the content of retrievable information.

14.4 Applications in Biomedical and Oncological Research

High-throughput screening in biomedicine and cancer research has been associated for many years with the selection of new potential therapeutic molecules in the big pharma industry, sometimes employing fluorescence-based assays without imaging capability such as the detection of emitted photon from single wells in multiwell plates. The development of automation in fluorescence microscopy and image cytometry described in the previous sections paved the way for image screening also in the field of drug selection (Carpenter, 2007). In parallel the discovery that dsRNA can direct gene-specific silencing through RNA interference provided a new tool to dissect the function of genes. Large RNAi library stimulated indeed the development of the first noncommercial platforms for image-based phenotype screening (Wheeler et al., 2004; Neumann et al., 2006). Genetic function discovery also employed an innovative platform obtained by the immobilization of cDNAs on substrates to perform the reverse transfection of cell arrays (Ziauddin and Sabatini, 2001).

The biological hits for both drug and genetic screens are related to the alteration of cell growth ability and toxicity by monitoring mitotic phase defects, cell cycle progression, and detecting apoptotic cell death. Thanks to the continuous dissemination of image-screening approaches in the research area, the number of *ad hoc* developed assays for the characterization of specific molecular pathways is increasing enormously. In addition to G-protein-coupled receptors (Heilker et al., 2009) that interested the drug discovery field, high content screening was used to investigate specific molecular networks mediating inflammatory response (NFkB, p38) (Bertelsen and Sanfridson, 2005; Ross et al., 2006; Zanella et al., 2007), such as mitogen-activated signaling pathway (MAP kinases/p38, estrogen growth factor receptor) (Ross et al., 2006; Wang and Xie, 2007), cell cycle control, and apoptosis (FOXO) (Zanella et al., 2008).

In the frame of assays targeted to the comprehension of the role of single molecules in cellular circuitries based on model cell lines, the panel of investigated biological phenotypes is growing rapidly.

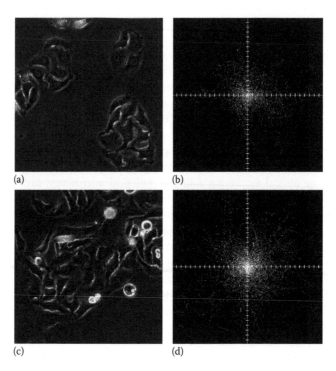

FIGURE 14.3 (See color insert following page 15-30.) Automated tracking of cell movements. Migratory phenotype upon overexpression of RAB5 [panels (c), (d)] protein in comparison to wild type HeLa cells [panels (a), (b)] was evaluated by automated GFP fluorescence tracking. Image segmentation for cell identification and reconstruction of trajectories was applied to 49 fields/well, transfected respectively with GFP or GFP + RAB5 expression vectors. "Spider Graphs" in panels (b) (ctrl) and (d) (RAB5) report the tracked movements of the two cell populations. (Cells kindly provided by Dr. Andrea Palamidessi, IFOM, Milan, Italy.)

The analysis of neurites outgrowth and the evaluation of other morphological features related to neural differentiation, as synaptogenesis, or to the pathogenesis of neurogenerative diseases, as formation of inclusion bodies, entered the list of high content assays in neuroscience (Mosch et al., 2006; Li et al., 2007b, 2008; Zhang et al., 2007a,b; Dragunow, 2008; Byrne et al., 2009).

Living cells–automated observation then allowed developing a systematic approach to the analysis of genes and molecular factors involved in the regulation of cell migration (Figure 14.3) (Fotos et al., 2006; Palamidessi et al., 2008; Simpson et al., 2008; Chen et al., 2009).

The study of cell cycle and apoptosis, particularly in laser-scanning cytometry, also inherited the large ensemble of assays originally developed for flow cytometry, namely, DNA content evaluation, apoptosis marker detection, phosphoH2AX measurement for DNA damage estimation to cite some of the most diffused applications (Bedner et al., 1999; Darzynkiewicz et al., 1999b, 2006; Abdel-Moneim et al., 2000; Smolewski et al., 2001).

Laser-scanning cytometry also contributed to elevate the level of biological complexity by introducing automation in the field of histopathological analysis. Immunological research consequently transduced the classical flow cytometric immunophenotyping for the identification of cell compartments of the immune system on slide assays (Tarnok and Gerstner, 2003; Mittag et al., 2005; Gerstner et al., 2006) allowing for the dissection of the response *in situ* on tissue biopsies (Harnett, 2007). The natural extension to a real *in vivo* analysis is one of the main stimuli in employing high content–automated approaches to a three-dimensional analysis of developing model organisms by also adapting it to a real screening environment as in the study of the angiogenesis process in zebra fish (Vogt et al., 2009). The aforementioned whole-organ analysis of organs like the adult heart (Ragan et al., 2007) together with the tracking of immune system populations in the thymic cortex (Chen et al., 2009) provided examples as automated analysis based on nonlinear microscopy could provide a dramatic breakthrough in the field of *in vivo* biological imaging. To conclude this brief overview of the emerging perspectives for HCA in biomedical research, it is worthwhile mentioning the great impact it can play in stem cell biology. The extremely rare frequency of stem cells, the wide number of parameters required for their identification and the complexity of the interaction established with the environment hosting them, makes high content–automated approaches an invaluable tool for their study both *in vitro* and *in vivo* (Oswald et al., 2004; Lenz et al., 2005; Shen et al., 2008; Ratajczak et al., 2009).

14.5 Future Perspectives

The complexity dictated by the mechanisms at the base of life continuously demands for the generation and analysis of an always increasing data flow. The result is the massive development of automated solutions able to dramatically extend the throughput and the type of approaches to the dissection of the structure–function problem. Imaging, and in particular fluorescence microscopy, was largely involved in the development of automated solutions, making it now an invaluable tool in large-scale analysis. Image screenings became an ideal solution to target the analysis of biological phenotypes, contributing to clarify that the generation of incredibly large amounts of data does not stem exclusively from a high number of samples. The observation of biologically complex phenomena, implying three-dimensional analysis over time targeted to the measurement of multiple parameters, can thus require collection of thousands of images and their subsequent elaboration. The development of new hardware solutions able to increase the spatial, temporal, and "statistical" resolutions is strongly desired to acquire the collection capability of more and more informative data. In addition to the potential improvements that far-field nanoscopy can provide for the *in vitro* characterization of biological phenotypes at molecular levels, new insights are strongly desired to further expand high content fluorescence analysis to the living matter at the level of tissues, organs, and organisms. The throughput of the analyzed events must take into account the ability to target poorly represented subsets, e.g., the stem cell compartment, requiring the simultaneous measurement of multiple biological end points to dissect the complex network of interactions occurring *in vivo*. For the success of the operation sophisticated analysis protocols are required to develop *automata* able to instruct themselves to the recognition of the enormous heterogeneity of biological phenotypes.

The extremely difficult retrieval of statistically significant samples, both for the amount of biological material required and for the throughput of the subsequent analysis, is now pushing forward a further step in the evolution of the existing solutions. The acquisition and analysis phase, till now considered as separated compartments, are converging toward a direct connection. A continuous feedback can select the best modalities of data collection for the optimization of the performances of the computational tools required for data interpretation. The challenge for the future is the development of a new generation of robotized microscopes not simply devoted to the iteration of a sequence of elementary steps but able to develop an automated decision-making process. The use of machine learning procedures starting during data collection points toward a continuous adaptation of the employed experimental strategy to target *in vivo*-selected cell phenotypes as occurring in the newborn computer aided microscopy recently commercialized by one of the world-leader manufacturers in the fluorescence microscopy field.

Acknowledgment

The author is grateful to Dr. Francesca Ballarini for her precious contribution in the manuscript revision.

References

Abdel-Moneim, I., Melamed, M. R., Darzynkiewicz, Z. et al. (2000) Proliferation and apoptosis in solid tumors. Analysis by laser scanning cytometry. *Anal Quant Cytol Histol*, 22, 393–397.

Auld, D. S., Johnson, R. L., Zhang, Y. Q. et al. (2006) Fluorescent protein-based cellular assays analyzed by laser-scanning microplate cytometry in 1536-well plate format. *Methods Enzymol*, 414, 566–589.

Bedner, E., Li, X., Gorczyca, W. et al. (1999) Analysis of apoptosis by laser scanning cytometry. *Cytometry*, 35, 181–195.

Belien, J. A., Van Ginkel, H. A., Tekola, P. et al. (2002) Confocal DNA cytometry: A contour-based segmentation algorithm for automated three-dimensional image segmentation. *Cytometry*, 49, 12–21.

Bertelsen, M. and Sanfridson, A. (2005) Inflammatory pathway analysis using a high content screening platform. *Assay Drug Dev Technol*, 3, 261–271.

Betz, T., Teipel, J., Koch, D. et al. (2005) Excitation beyond the monochromatic laser limit: Simultaneous 3-D confocal and multiphoton microscopy with a tapered fiber as white-light laser source. *J Biomed Opt*, 10, 054009.

Bowen, W. P. and Wylie, P. G. (2006) Application of laser-scanning fluorescence microplate cytometry in high content screening. *Assay Drug Dev Technol*, 4, 209–221.

Byrne, U. T., Ross, J. M., Faull, R. L. et al. (2009) High-throughput quantification of Alzheimer's disease pathological markers in the post-mortem human brain. *J Neurosci Methods*, 176, 298–309.

Carpenter, A. E. (2007) Image-based chemical screening. *Nat Chem Biol*, 3, 461–465.

Chen, Y., Ladi, E., Herzmark, P. et al. (2009) Automated 5-D analysis of cell migration and interaction in the thymic cortex from time-lapse sequences of 3-D multi-channel multi-photon images. *J Immunol Methods*, 340, 65–80.

Darzynkiewicz, Z., Bedner, E., Li, X. et al. (1999a) Laser-scanning cytometry: A new instrumentation with many applications. *Exp Cell Res*, 249, 1–12.

Darzynkiewicz, Z., Li, X., and Bedner, E. (1999b) Detection of DNA strand breakage in the analysis of apoptosis and cell proliferation by flow and laser scanning cytometry. *Methods Mol Biol*, 113, 607–619.

Darzynkiewicz, Z., Huang, X., and Okafuji, M. (2006) Detection of DNA strand breaks by flow and laser scanning cytometry in studies of apoptosis and cell proliferation (DNA replication). *Methods Mol Biol*, 314, 81–93.

Debernardi, M. A., Hewitt, S. M., and Kriete, A. (2006) Automated confocal imaging and high-content screening for cytomics. In Pawley, J. B. (Ed.) *Handbook of Biological Confocal Microscopy*, 3rd edn., pp. 809–818, New York, Springer Science + Business Media.

Dragunow, M. (2008) High-content analysis in neuroscience. *Nat Rev Neurosci*, 9, 779–788.

Erfle, H., Neumann, B., Liebel, U. et al. (2007) Reverse transfection on cell arrays for high content screening microscopy. *Nat Protoc*, 2, 392–399.

Erfle, H., Neumann, B., Rogers, P. et al. (2008) Work flow for multiplexing siRNA assays by solid-phase reverse transfection in multiwell plates. *J Biomol Screen*, 13, 575–580.

Fotos, J. S., Patel, V. P., Karin, N. J. et al. (2006) Automated time-lapse microscopy and high-resolution tracking of cell migration. *Cytotechnology*, 51, 7–19.

Gerstner, A. O., Trumpfheller, C., Racz, P. et al. (2004) Quantitative histology by multicolor slide-based cytometry. *Cytometry A*, 59, 210–219.

Gerstner, A. O., Mittag, A., Laffers, W. et al. (2006) Comparison of immunophenotyping by slide-based cytometry and by flow cytometry. *J Immunol Methods*, 311, 130–138.

Goldberg, I. G., Allan, C., Burel, J. M. et al. (2005) The open microscopy environment (OME) data model and XML file: Open tools for informatics and quantitative analysis in biological imaging. *Genome Biol*, 6, R47.

Graf, R., Rietdorf, J., and Zimmermann, T. (2005) Live cell spinning disk microscopy. *Adv Biochem Eng Biotechnol*, 95, 57–75.

Gudla, P. R., Nandy, K., Collins, J. et al. (2008) A high-throughput system for segmenting nuclei using multiscale techniques. *Cytometry A*, 73, 451–466.

Harder, N., Mora-Bermudez, F., Godinez, W. J. et al. (2006) Automated analysis of the mitotic phases of human cells in 3D fluorescence microscopy image sequences. *Med Image Comput Comput Assist Interv Int Conf Med Image Comput Comput Assist Interv*, 9, 840–848.

Harnett, M. M. (2007) Laser scanning cytometry: Understanding the immune system in situ. *Nat Rev Immunol*, 7, 897–904.

Heilker, R., Wolff, M., Tautermann, C. S. et al. (2009) G-protein-coupled receptor-focused drug discovery using a target class platform approach. *Drug Discov Today*, 14, 4–5.

Hell, S. W. (2007) Far-field optical nanoscopy. *Science*, 316, 1153–1158.

Hell, S. W. (2009) Microscopy and its focal switch. *Nat Methods*, 6, 24–32.

Jiang, S., Zhou, X., Kirchhausen, T. et al. (2007a) Detection of molecular particles in live cells via machine learning. *Cytometry A*, 71, 563–575.

Jiang, S., Zhou, X., Kirchhausen, T. et al. (2007b) Tracking molecular particles in live cells using fuzzy rule-based system. *Cytometry A*, 71, 576–584.

Jones, T. R., Carpenter, A. E., Lamprecht, M. R. et al. (2009) Scoring diverse cellular morphologies in image-based screens with iterative feedback and machine learning. *Proc Natl Acad Sci USA*, 106, 1826–1831.

Juskaitis, R., Wilson, T., Neil, M. A. et al. (1996) Efficient real-time confocal microscopy with white light sources. *Nature*, 383, 804–806.

Kakarala, M. and Wicha, M. S. (2007) Cancer stem cells: Implications for cancer treatment and prevention. *Cancer J*, 13, 271–275.

Kamentsky, L. A. (2001) Laser scanning cytometry. *Methods Cell Biol*, 63, 51–87.

Kamentsky, L. A., Burger, D. E., Gershman, R. J. et al. (1997) Slide-based laser scanning cytometry. *Acta Cytol*, 41, 123–143.

Keller, P. J., Schmidt, A. D., Wittbrodt, J. et al. (2008) Reconstruction of zebrafish early embryonic development by scanned light sheet microscopy. *Science*, 322, 1065–1069.

Kim, K. H., Ragan, T., Previte, M. J. et al. (2007) Three-dimensional tissue cytometer based on high-speed multiphoton microscopy. *Cytometry A*, 71, 991–1002.

Kriete, A. and Boyce, K. (2005) Automated tissue analysis—A bioinformatics perspective. *Methods Inf Med*, 44, 32–37.

Laffers, W., Mittag, A., Lenz, D. et al. (2006) Iterative restaining as a pivotal tool for n-color immunophenotyping by slide-based cytometry. *Cytometry A*, 69, 127–130.

Lamprecht, M. R., Sabatini, D. M., and Carpenter, A. E. (2007) CellProfiler: Free, versatile software for automated biological image analysis. *Biotechniques*, 42, 71–75.

Lenz, D., Lenk, K., Mittag, A. et al. (2005) Detection and quantification of endothelial progenitor cells by flow and laser scanning cytometry. *J Biol Regul Homeost Agents*, 19, 180–187.

Levenson, R. M. and Mansfield, J. R. (2006) Multispectral imaging in biology and medicine: Slices of life. *Cytometry A*, 69, 748–758.

Li, F., Zhou, X., Ma, J. et al. (2007a) An automated feedback system with the hybrid model of scoring and classification for solving over-segmentation problems in RNAi high content screening. *J Microsc*, 226, 121–132.

Li, F., Zhou, X., Zhu, J. et al. (2007b) High content image analysis for human H4 neuroglioma cells exposed to CuO nanoparticles. *BMC Biotechnol*, 7, 66.

Li, G., Liu, T., Tarokh, A. et al. (2007c) 3D cell nuclei segmentation based on gradient flow tracking. *BMC Cell Biol*, 8, 40.

Li, F., Zhou, X., Zhu, J. et al. (2008) Workflow and methods of high-content time-lapse analysis for quantifying intracellular calcium signals. *Neuroinformatics*, 6, 97–108.

Lin, G., Chawla, M. K., Olson, K. et al. (2007) A multi-model approach to simultaneous segmentation and classification of heterogeneous populations of cell nuclei in 3D confocal microscope images. *Cytometry A*, 71, 724–736.

Liron, Y., Paran, Y., Zatorsky, N. G. et al. (2006) Laser autofocusing system for high-resolution cell biological imaging. *J Microsc*, 221, 145–151.

McCullough, D. P., Gudla, P. R., Harris, B. S. et al. (2008) Segmentation of whole cells and cell nuclei from 3-D optical microscope images using dynamic programming. *IEEE Trans Med Imaging*, 27, 723–734.

Mittag, A., Lenz, D., Gerstner, A. O. et al. (2005) Polychromatic (eight-color) slide-based cytometry for the phenotyping of leukocyte, NK, and NKT subsets. *Cytometry A*, 65, 103–115.

Mittag, A., Lenz, D., Bocsi, J. et al. (2006a) Sequential photobleaching of fluorochromes for polychromatic slide-based cytometry. *Cytometry A*, 69, 139–141.

Mittag, A., Lenz, D., Gerstner, A. O. et al. (2006b) Hyperchromatic cytometry principles for cytomics using slide based cytometry. *Cytometry A*, 69, 691–703.

Moffat, J., Grueneberg, D. A., Yang, X. et al. (2006) A lentiviral RNAi library for human and mouse genes applied to an arrayed viral high-content screen. *Cell*, 124, 1283–1298.

Moffat, J. and Sabatini, D. M. (2006) Building mammalian signalling pathways with RNAi screens. *Nat Rev Mol Cell Biol*, 7, 177–187.

Mosch, B., Mittag, A., Lenz, D. et al. (2006) Laser scanning cytometry in human brain slices. *Cytometry A*, 69, 135–138.

Neumann, B., Held, M., Liebel, U. et al. (2006) High-throughput RNAi screening by time-lapse imaging of live human cells. *Nat Methods*, 3, 385–390.

Oswald, J., Jorgensen, B., Pompe, T. et al. (2004) Comparison of flow cytometry and laser scanning cytometry for the analysis of CD34+ hematopoietic stem cells. *Cytometry A*, 57, 100–107.

Padfield, D., Rittscher, J., Thomas, N. et al. (2009) Spatio-temporal cell cycle phase analysis using level sets and fast marching methods. *Med Image Anal*, 13, 143–155.

Palamidessi, A., Frittoli, E., Garre, M. et al. (2008) Endocytic trafficking of Rac is required for the spatial restriction of signaling in cell migration. *Cell*, 134, 135–147.

Perfetto, S. P., Chattopadhyay, P. K., and Roederer, M. (2004) Seventeen-colour flow cytometry: Unravelling the immune system. *Nat Rev Immunol*, 4, 648–655.

Peterson, R. A., Krull, D. L., and Butler, L. (2008) Applications of laser scanning cytometry in immunohistochemistry and routine histopathology. *Toxicol Pathol*, 36, 117–132.

Ploeger, L. S., Belien, J. A., Poulin, N. M. et al. (2004) Confocal 3D DNA cytometry: Assessment of required coefficient of variation by computer simulation. *Cell Oncol*, 26, 93–99.

Pozarowski, P., Holden, E., and Darzynkiewicz, Z. (2006) Laser scanning cytometry: Principles and applications. *Methods Mol Biol*, 319, 165–192.

Ragan, T., Sylvan, J. D., Kim, K. H. et al. (2007) High-resolution whole organ imaging using two-photon tissue cytometry. *J Biomed Opt*, 12, 014015.

Ratajczak, M. Z., Kucia, M., Ratajczak, J. et al. (2009) A multi-instrumental approach to identify and purify very small embryonic like stem cells (VSELs) from adult tissues. *Micron*, 40, 386–393.

Renaud, O., Vina, J., Yu, Y. et al. (2008) High-resolution 3-D imaging of living cells in suspension using confocal axial tomography. *Biotechnol J*, 3, 53–62.

Ross, S., Chen, T., Yu, V. et al. (2006) High-content screening analysis of the p38 pathway: Profiling of structurally related p38alpha kinase inhibitors using cell-based assays. *Assay Drug Dev Technol*, 4, 397–409.

Schiffmann, D. A., Dikovskaya, D., Appleton, P. L. et al. (2006) Open microscopy environment and findspots: Integrating image informatics with quantitative multidimensional image analysis. *Biotechniques*, 41, 199–208.

Shen, Q., Wang, Y., Kokovay, E. et al. (2008) Adult SVZ stem cells lie in a vascular niche: A quantitative analysis of niche cell-cell interactions. *Cell Stem Cell*, 3, 289–300.

Simpson, J. C., Cetin, C., Erfle, H. et al. (2007) An RNAi screening platform to identify secretion machinery in mammalian cells. *J Biotechnol*, 129, 352–365.

Simpson, K. J., Selfors, L. M., Bui, J. et al. (2008) Identification of genes that regulate epithelial cell migration using an siRNA screening approach. *Nat Cell Biol*, 10, 1027–1038.

Smolewski, P., Grabarek, J., Kamentsky, L. A. et al. (2001) Bivariate analysis of cellular DNA versus RNA content by laser scanning cytometry using the product of signal subtraction (differential fluorescence) as a separate parameter. *Cytometry*, 45, 73–78.

Stewart, S. A., Dykxhoorn, D. M., Palliser, D. et al. (2003) Lentivirus-delivered stable gene silencing by RNAi in primary cells. *RNA*, 9, 493–501.

Tarnok, A. and Gerstner, A. O. (2003) Immunophenotyping using a laser scanning cytometer. *Curr Protoc Cytom*, Chapter 6, Unit 6.13, pp. 1–15.

Toomre, D. and Pawley, J. B. (2006) Disk-scanning confocal microscopy. In Pawley, J. B. (Ed.) *Handbook of Biological Confocal Microscopy*, 3rd edn., pp. 221–239, New York, Springer Science + Business Media.

Vermeulen, L., Sprick, M. R., Kemper, K. et al. (2008) Cancer stem cells—Old concepts, new insights. *Cell Death Differ*, 15, 947–958.

Vogt, A., Cholewinski, A., Shen, X. et al. (2009) Automated image-based phenotypic analysis in zebrafish embryos. *Dev Dyn*, 238, 656–663.

Wang, J. and Xie, X. (2007) Development of a quantitative, cell-based, high-content screening assay for epidermal growth factor receptor modulators. *Acta Pharmacol Sin*, 28, 1698–1704.

Wheeler, D. B., Bailey, S. N., Guertin, D. A. et al. (2004) RNAi living-cell microarrays for loss-of-function screens in Drosophila melanogaster cells. *Nat Methods*, 1, 127–132.

Wheeler, D. B., Carpenter, A. E., and Sabatini, D. M. (2005) Cell microarrays and RNA interference chip away at gene function. *Nat Genet*, 37 (Suppl), S25–S30.

Zanella, F., Rosado, A., Blanco, F. et al. (2007) An HTS approach to screen for antagonists of the nuclear export machinery using high content cell-based assays. *Assay Drug Dev Technol*, 5, 333–341.

Zanella, F., Rosado, A., Garcia, B. et al. (2008) Chemical genetic analysis of FOXO nuclear-cytoplasmic shuttling by using image-based cell screening. *Chembiochem*, 9, 2229–2237.

Zhang, Y., Zhou, X., Degterev, A. et al. (2007a) Automated neurite extraction using dynamic programming for high-throughput screening of neuron-based assays. *Neuroimage*, 35, 1502–1515.

Zhang, Y., Zhou, X., Degterev, A. et al. (2007b) A novel tracing algorithm for high throughput imaging Screening of neuron-based assays. *J Neurosci Methods*, 160, 149–162.

Ziauddin, J. and Sabatini, D. M. (2001) Microarrays of cells expressing defined cDNAs. *Nature*, 411, 107–110.

Zuba-Surma, E. K., Kucia, M., Abdel-Latif, A. et al. (2007) The ImageStream System: A key step to a new era in imaging. *Folia Histochem Cytobiol*, 45, 279–290.

15

Optical Manipulation, Photonic Devices, and Their Use in Microscopy

G. Cojoc
C. Liberale
R. Tallerico
A. Puija
M. Moretti
F. Mecarini
G. Das
P. Candeloro
F. De Angelis
E. Di Fabrizio

Magna Graecia University

15.1 Introduction 15-1
15.2 Fluorescence Microscopy Combined with Optical Tweezers 15-2
   Trap Stability • Optical Tweezers Setup • Optical Tweezers Calibration and Force Measurement • Fluorescence Microscopy in Beads and DNA Optical Manipulation • Confocal Microscopy Combined with Optical Tweezers
15.3 Fiber-Optic Tweezers 15-12
   Calculations • Probe Fabrication and Experimental Results
15.4 Plasmonic Devices 15-18
   Device Fabrication • Theoretical Simulation • Experimental Setup and Material Deposition • Raman Scattering Measurements
15.5 Summary 15-28
Acknowledgment 15-28
References 15-28

15.1 Introduction

The use of fluorescence microscopy, especially devoted to biology, medicine, and biophysics studies, is well known, and the techniques and related methodologies are commonly used in day-to-day laboratory activities. In the last few years, new techniques, very often complementary to fluorescence, have become popular in laboratory practices. We suggest the use of optical manipulation, among many others, which is commonly referred to as optical tweezers microscopy, and enhanced spectroscopies, such as surface-enhanced Raman spectroscopy (SERS), through the generation of surface plasmon polariton (SPP). In both cases, the two methodologies allow the investigation of matter at single molecule levels. In future, the combination of these techniques with fluorescence microscopy will probably have a strong impact on the knowledge of molecular events inside the cell, the tissue, and also, possibly, in *in vivo* investigations. This chapter aims only at introducing the subject and does not provide an exhaustive overview.

In 1970, a paper titled "Acceleration and trapping of particles by radiation pressure," written by Arthur Ashkin (1970), started the experimental activity on optical trapping. The possibility of exerting mechanical forces by photon, through radiation pressure and angular momentum exchange, was well recognized theoretically and experimentally since the beginning of modern optics and the discovery of Maxwell equations. Nevertheless, the interest around this phenomenon remained limited due to the weakness of such forces when their effects were considered on macroscopic objects. It was the experimental skill and the scientific awareness of Ashkin that brought the optical trapping from a phenomenon of unpractical interest into the realm of a mature and relevant cross-disciplinary research field, on par with biology, physics, and material science. Since 1970, a strong activity was developed worldwide whose wideness and importance can be appreciated by reading the review paper written by Ashkin (2000).

15.2 Fluorescence Microscopy Combined with Optical Tweezers

Optical forces generated by electromagnetic radiation on microscopic particles can be described in terms of momentum exchange between light and matter. By assuming light to be a flow of photons carrying momentum, and considering for simplicity the case of geometrical optics, when the particle's dimensions are much bigger than the radiation wavelength, it is possible to explain graphically the action of optical forces on a dielectric spherical particle with a refractive index higher than the surrounding medium (Figure 15.1) (Ashkin 1992). As an effect of refraction at an interface between two media of different refractive indices, the momentum carried by a ray changes direction. Since the total momentum is conserved, the difference between the initial momentum and the final momentum is transferred to the medium. Figure 15.1 illustrates how the refraction and the reflection of a single optical ray induce forces on a spherical object. Let vector **p** represent the momentum of the incident light and **p′** that of the emergent light. The momentum **Δp** transferred to the sphere is obtained from **p**=**p′**+**Δp**. The difference **Δp** between **p** and **p′** provides the direction of the force **F** acting on the sphere by the transmitted laser beam, **F** being parallel to **Δp**.

Considering a focused laser beam, formed by a set of different rays, the total force is obtained by adding the forces exerted by each of the rays. Optical forces generated by rays' refractions and reflections on a particle are usually decomposed on "scattering forces" and "gradient forces," depending on the direction parallel or perpendicular, respectively, with respect to the incoming rays' directions (Ashkin 1992). Generally speaking, scattering forces contribute to push the particle in the direction of beam propagation, whereas gradient forces contribute to push the particle in the direction of the beam intensity gradient. In particular, the relative refractive index of the object compared to the surrounding medium, $n=n_{object}/n_{medium}$, affects the direction of the gradient force: if the relative refractive index is greater than unity, the gradient force attracts the particle toward the region of highest beam intensity, whereas in the opposite situation, the particle is pushed out of the beam.

In the case of laser beams focalized by high numerical aperture (NA) lenses, gradient forces are strong enough to counterbalance the action of scattering forces so that an equilibrium point can be created in space when the particle will be trapped (Figure 15.2).

Detailed expressions for the gradient and scattering forces can be derived according to different regimes determined by the ratio between the spherical object diameter, d, and the radiation wavelength, λ. Depending on this ratio there are three different regimes: the Rayleigh ($d\ll\lambda$) (Harada and Asakura 1996) regime, the resonance ($\lambda\sim d$) regime (Tlusty et al. 1998, Rohrbach and Stelzer 2001, Lock 2004, Mansuripur et al. 2005), and the Mie regime ($d\gg\lambda$) (Ashkin 1992).

For the sake of completeness, we give quantitative formulas in the ray optics regime referring to particles of micrometer scale. The ray optics also works in the Rayleigh limit with particles of nanometer scale as described later. Light can be described as the flux of photons with momentum and energy

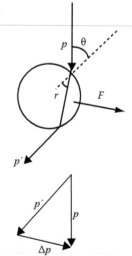

FIGURE 15.1 Direction of the optical force on the sphere as a consequence of momentum conservation in ray refraction.

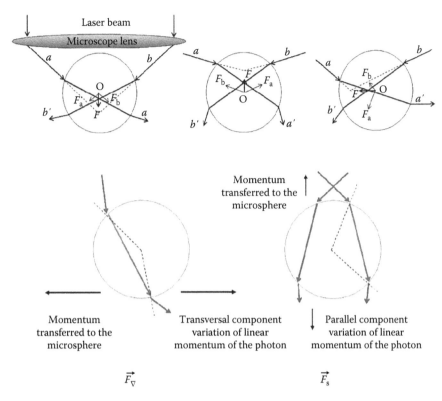

FIGURE 15.2   Restoring forces acting on the spherical particle for small displacements from the trap equilibrium point.

$$|\vec{p}| = \frac{h}{\lambda} \quad \text{and} \quad E = h\nu$$

Thus, we can write $E = pc_m$ where $c_m = c/n$ is the speed of light in a material with index of refraction n. If we introduce the Poynting vector $\mathbf{S} = \mathbf{E} \times \mathbf{H}$ that represents the energy flux density, we can write

$$\vec{S}\,dA = d\dot{E}\hat{n}_s = \frac{c}{n}d\dot{\vec{p}} = \frac{c}{n}d\vec{F}$$

where
  dA is an element of area
  $\hat{n}_s$ is a vector of direction of $\mathbf{S}$

Thus, if we want to know the force $\mathbf{F}$ of light on an object O we have to integrate

$$\vec{F} = \frac{n}{c}\iint_O \left\{ \vec{S}_{in} - \vec{S}_{out} \right\} dA$$

where $\{\mathbf{S}_{in} - \mathbf{S}_{out}\}$ is the net energy flux density of the incoming and outgoing rays. A ray of power $S_{in}\,dA$ that hits a dielectric object O at an angle of incidence θ causes an infinite number of refracted rays of decreasing power. If you sum the forces of all refracted rays by using the above equation, in the limit of Mie scattering (the size of the object $O \gg \lambda$) we obtain the total force that can be written in two components:

$$F_s = (S_{in}dA)\frac{n}{c}\left\{1 + R\cdot\cos 2\theta - \frac{T^2\left[\cos(2\theta - 2r) + R\cos 2\theta\right]}{1 + R^2 + 2R\cos 2r}\right\}$$

$$F_g = (S_{in}dA)\frac{n}{c}\left\{R\cdot\cos 2\theta - \frac{T^2\left[\sin(2\theta - 2r) + R\sin 2\theta\right]}{1 + R^2 + 2R\cos 2r}\right\}$$

where
   θ is the angle of incidence and r is the angle of refraction inside of the object
   the quantities R and T are the Fresnel reflection and transmission coefficients at the surface
   F_s is the scattering force that points into the direction of the incident light
   F_g is the gradient force perpendicular to it (Figure 15.2)

Summing over all incident rays for example of a *TEM00*-mode laser beam gives the total force on the sphere. Qualitatively you find that you have two kinds of forces: a resulting scattering force that tries to push away the sphere and a resulting gradient force that pulls the sphere back to the focus along the gradient of intensity. So, the sphere is kept in the focus of the laser beam in all three dimensions if the gradient force overcomes the scattering one. Practically we also have to overcome the random thermodynamic forces to hold a particle and the frictional forces to move it. It can be shown that the ratio of the radius of the *TEM00* beam and the aperture and the angle q are important parameters for the gradient forces. So, filling the aperture and having a big angle q cause a major force. We can also find similar forces for very small particles. For a sphere of radius a and dielectric constant ε_s, we can describe the force in the Rayleigh limit where $a \ll 1$ by $F = F_\nabla + F_s$ with

$$F_s = \frac{8}{3}\pi\left((kd)^4\right)d^2\frac{\sqrt{\varepsilon_0}}{c}\left(\frac{\varepsilon - \varepsilon_0}{\varepsilon + 2\varepsilon_0}\right)^2 S$$

$$F_\nabla = 2\pi d^3\frac{\sqrt{\varepsilon_0}}{c}\left(\frac{\varepsilon - \varepsilon_0}{\varepsilon + 2\varepsilon_0}\right)\nabla|S|$$

   F_∇ is a gradient force that pulls the sphere to the point of highest intensity along the gradient of $|S|$
   F_s is a scattering force that tries to push the sphere along S (Figure 15.2)
   The other quantities are ε, the dielectric constant of particle; ε_0, the dielectric constant of medium, and k, the wave number.

However, in most situations this force is so much smaller than other forces acting on macroscopic objects that there is no noticeable effect. For example, we can calculate the force due to the change in momentum of light reflecting off of a mirror. In this case, $S_{out} = -S_{in}$, so $F = 2(n/c)\iint(S_{in})dA$. The integral represents the total power of the light, which is usually expressed in watts. In the simple case of 100% reflection, the force is thus $F = 2(n/c)W$, where W is the intensity of the light in watts. If all of the intensity of a 60 W light bulb were focused onto a mirror, the force due to radiation pressure would be 4×10^{-7} N. If a 1 kg mirror exerts a force on a scale of 9.8 N, the additional weight due to radiation pressure is clearly negligible. Objects for which this radiation pressure would be significant would have to weigh less than 1 μg. In optical tweezers experiments, the radiation pressure is provided by laser light, while the objects to be manipulated are generally very small. Micron-sized polystyrene spheres of uniform diameter are easily obtained and can be trapped using the forces described earlier.

In the intermediate regime, where neither Mie nor Rayleigh regime can be fulfilled, a numerical analysis is necessary in order to effectively take into account the specific geometry. In any case, trapping is still possible and no substantial new phenomena occur.

15.2.1 Trap Stability

Another important consideration that remains is the trap stability. Due to the thermal motion, the *Brownian motion*, in order to have a stable trap, the trapping force has to be bigger than that due to the thermal motion. A necessary condition for stability is $F_{grad}/F_{scatt} \geq 1$.

A necessary and sufficient condition for stability is that the potential well U of the gradient force is much larger than the kinetic energy of the Brownian motion of particle, that is,

$$\frac{2\pi n_2 a^3 (m^2 - 1)}{c(m^2 + 2)} \frac{2P}{\pi w_0^2} \geq 10 K_B T$$

This inequality allows to estimate the trap stability knowing the size a of the particle, its relative refractive index m, the laser power P and its focus size (w_0), and the liquid temperature.

In practical conditions, a power of a few mW is enough to have a stable trap of polystyrene bead of a few micron size.

15.2.2 Optical Tweezers Setup

The most basic optical tweezers setup includes a laser beam and a high NA microscope objective. A dielectric particle near the focus will experience a force due to the transfer of momentum from the scattering of incident photons and will be pushed toward the center of the beam, if the particle's index of refraction is higher than that of the surrounding medium (Ashkin 1992).

The optical trap results from the fact that the objects that are trapped in the focus of the laser beam experience a restoring force toward the center of the trap if they try to leave the high intensity volume.

In our optical tweezers setup (see Figure 15.3) at the BioNEM (Bio-Nanotechnology & Engineering for Medicine) laboratory, Catanzaro, Italy, we use an inverted microscope with infinity-corrected optics, *Nikon* ECLIPSE TE 2000-U. The laser source is a single-mode CW ytterbium fiber laser (YLM-5 from IPG Photonics) that is emitting at 1064 nm, linear polarized, and collimated. The output beam diameter is approximately 10 mm, then it is expanded (2×) by the lenses L1 and L2 with focal lengths $f_1 = -100$ mm and $f_2 = 200$ mm, and is subsequently sent onto a LC-SLM (Liquid Crystals-Spatial Light Modulator) (Hamamatsu X8267–15).

The SLM is an electrically addressed phase modulator using an optical image–transmitting element to couple an optically addressed parallel aligned nematic liquid crystal-spatial light modulator (PAL-SLM) with an intensity modulator. The SLM exhibits high efficiency (40%) and a maximum phase shift higher than 2π is provided at 1064 nm, over an active area of 20 mm×20 mm (768×768 pixels). We control it with a computer (PC) sending different diffractive optical elements (DOEs) which are calculated in real time with help of a software that we developed using LabVIEW (National Instruments). For calculation of the phases, we used an approach based on the spherical waves propagation and superposition approach (Di Fabrizio et al. 2003). The same computer is used also for image acquisition, using a 2 Mpix CCD camera from *Nikon*.

A telescope with 0.5× magnification, made of two plano-convex lenses (L3 and L4) with focal lengths $f_3 = 400$ mm and $f_4 = 200$ mm, is used to image the calculated DOE on the back focal plane (BFP) of the microscope objective.

The beam is finally directed (reflected) into the objective (Nikon Plan Apochromat 60×, 1.4 NA, oil immersion, or Nikon Plan Apochromat 100×, 1.3 NA, oil immersion) with an IR dichroic mirror, DM. The mirror allows the visible light used for illumination to pass through.

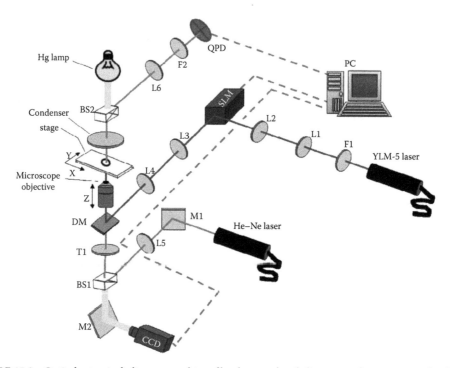

FIGURE 15.3 Optical setup including an ytterbium fiber laser and an helium–neon laser, an expander (lens L1 and L2), LC-SLM, a telescope after the SLM (lens L3 and L4), a dichroic mirror—DM, a microscope objective, a specimen stage, a condenser, a Hg lamp, a CCD camera, and a computer (PC). The tube lens—TL, is used for infinite correction.

The particle movements in the optical trap can be monitored by the BFP interferometry method. A low power laser beam generated by a 633 nm He–Ne (JDSU 1125/P) is sent to the sample collinear with the trapping beam. The microscope condenser (NA 0.5) collects the rays emerging from the sample plane and, if there are source scatterers in or near the focus, scattered light interferes with the non-scattered light. As a result, an interference pattern is registered in the condenser BFP, which is strongly dependent on the relative position of the trapped object and the laser focus. To track particle movements, the BFP interference pattern is imaged, using a plano-convex lens (L6) with focal length $f=50$ mm, onto a four-quadrant photodiode (QPD) (Hamamatsu PD C5460SPL) connected to an high-speed amplifier giving a 850 kHz detection bandwidth.

A short pass dichroic filter (F2) is used to prevent the trapping laser to reach the QPD.

The analog signals are digitized employing an analog-to-digital acquisition board (DAQ) (National Instruments, PCI-6133). The board has a maximum sampling rate of 2.5 M samples/s per channel and 14-bit resolution. Signal acquisition is controlled via a software interface written in LabVIEW.

15.2.3 Optical Tweezers Calibration and Force Measurement

Optical tweezers can be converted into a sensitive micromechanical force transducer that is capable of exerting and/or measuring forces in the pN range. This requires monitoring of the trapped object position with high spatial (nm range) and temporal (ms range) resolutions.

The starting point for this application is indeed that, in a small region around the trapping point, optical forces can be considered to linearly vary with the trapped particle displacement in tune with Hook's law $\vec{F} = -k \cdot \vec{x}$, where k is trap stiffness and x is the displacement. To calculate the applied forces it is then necessary to get a precise measurement of the trapped object displacement from the equilibrium position in the optical trap.

One of the simplest position detection schemes is the BFP detection method, which relies on the interference between forward-scattered light from the trapped object and unscattered light (Allersma et al. 1998). The interference signal can be monitored using QPD positioned at a plane conjugate with the BFP of the microscope condenser. This scheme requires precise calibration of the QPD response to the trapped particle displacement and this is usually done exploiting the statistical analysis of the Brownian motion of the particle in the optical trap.

When a bead of known radius is trapped, the physics of Brownian motion in a harmonic potential is described by a known power spectrum for the thermal fluctuations of a trapped object given by

$$S^x = \frac{k_B T}{\gamma \pi^2 \left(f^2 + f_c^2 \right)} \tag{15.1}$$

which is a Lorentzian, where
  k_B is the Boltzmann constant
  T is the absolute temperature
  f_c is a roll-off frequency given by

$$f_c = \frac{k}{2\pi\gamma} \tag{15.2}$$

where
  k is the trap stiffness
  γ is the drag coefficient given by

$$\gamma = 6\eta\pi r \tag{15.3}$$

with
  η is the medium viscosity
  r is the radius of the particle

This power spectrum can be fit (Svoboda and Block 1994, Gittes et al. 1997), and the trap stiffness (k) can be calculated if the drag coefficient of the particle is known.

As an example, a polystyrene bead with 1 μm diameter was trapped and the QPD output signal, representing the agitation due to the Brownian motion of the trapped bead, was acquired. Figure 15.4 shows a plot of a measured signal amplitude. Each particle tracking was acquired for 2 s with a sampling rate of 200 kHz and this measurement was repeated at least five times for the same trapped bead. This acquisition sequence was then performed at different laser powers.

Each of the data sets were Fourier transformed (DFT). The squared amplitudes of the various transformed data were averaged to obtain a power spectrum for each of the different laser powers. Figure 15.5 shows the power spectra corresponding to the measurement of Figure 15.4. The corner frequency was then determined by fitting the power spectrum to a Lorentzian curve as $f_c = 851$ (Hz), and this value was used to obtain the trap stiffness at the selected power of the trapping laser beam (in this case, 16 mW at the sample).

Using expressions (15.2) and (15.3)

$$k = 12\pi^2 r f_c \eta = 12 \times \pi^2 \times 0.5 \times 10^{-6} \, (m) \times 851 \, (Hz) \times 8.9 \times 10^{-4} \, (Pa \cdot s) \Leftrightarrow$$

$$k = 44.9 \, pN/\mu m$$

where dynamic viscosity is that of water at 300 K, $\eta = 8.9 \times 10^{-4}$ Pa·s, and $r = 0.5$ μm.

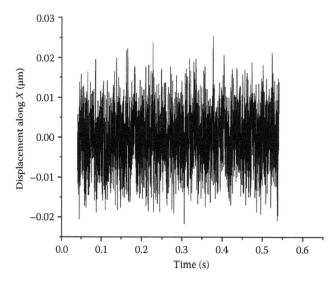

FIGURE 15.4   A plot of the position sensitive detector voltage output corresponding to the amplitude of displacement (in one dimension, i.e., X-axis) of a trapped bead relative to the center of the trap.

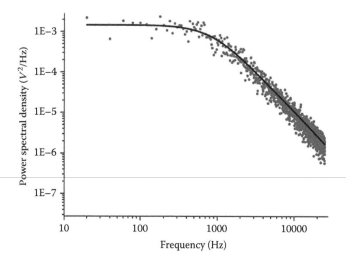

FIGURE 15.5   Power spectral density of the trapped bead movements due to thermal motion is displayed as the dotted line. The line represents the Lorentzian fit.

As an alternative to the power spectrum method for the trap stiffness calibration, the autocorrelation method can be used (Meller et al. 1998). Considering the Brownian motion of a particle in the harmonic potential of the optical trap, the autocorrelation function for displacement in the x direction $g(t) = \langle x(\tau + t)x(\tau) \rangle$ is determined as

$$g(t) = \frac{k_\mathrm{B}T}{k}\exp(-\omega t) \tag{15.4}$$

where ω is the exponential decay factor given by

$$\omega = \frac{k}{\gamma}$$

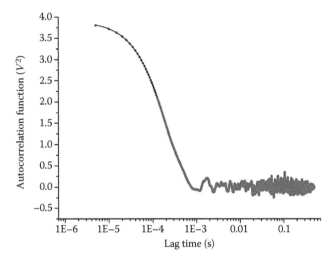

FIGURE 15.6 Semilog-plot of the position-versus-time autocorrelation for the same sample as being used in Figure 15.3. The black line represents the fit according to Equation 15.6.

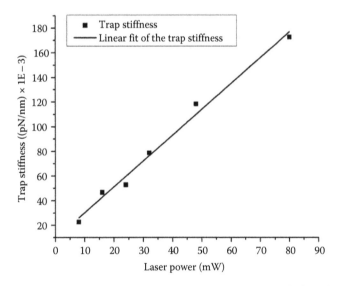

FIGURE 15.7 Measurement of the stiffness of the optical trap along the *x*-axis at different laser powers.

Therefore, the trap stiffness *k* can be retrieved starting from the same set of particle displacement measurements acquired for the power spectrum method. Indeed, after the calculation of autocorrelation function for the measured data (see Figure 15.6) and fitting with Equation (15.4), we can retrieve the exponential decay factor ($\omega = 4926$ rad/s in this case) and finally obtain $k = 41.3$ pN/μm.

Then the stiffness of the optical trap for increasing laser powers (from 8 to 80 mW) is retrieved, and a linear fit of the stiffness values along the *x*-axis is shown in Figure 15.7. It can be observed, as expected from theory, a linear dependence between the stiffness and the laser power. Several papers can be found in the literature, where force measurements are applied to extract quantitative informations on the internal structure of cells (Cojoc et al. 2007) and the biochemistry of single-molecule reactions (Bustamante et al. 2000).

15.2.4 Fluorescence Microscopy in Beads and DNA Optical Manipulation

Recently DNA has not only been regarded as a biological molecule but also as an engineering material. In fact, due to its nanometric structure (2 nm diameter) and its "intelligent" building ability thanks to base pairing, DNA is of considerable interest to nanotechnology. Laser trapping of particles linked to long DNA molecules (e.g., lambda DNA) is quite often used as a manipulating tool in microscope imaging (Perkins et al. 1995, Smith et al. 1996, Brower-Toland et al. 2002, Handa et al. 2005).

Phage lambda DNA (GenBank/EMBL accession numbers J02459, M17233, M24325, V00636, X00906) is a linear double-stranded (48502 nucleotides) chain with 12 base pairs (bp) single-stranded complementary 5'-ends isolated from the temperate *Escherichia coli* bacteriophage Lambda.

A very reliable method to link nucleic acid molecules to many substrates is the non-covalent streptavidin/biotin affinity linkage. In our laboratory, the current protocol for bead–DNA linkage is slightly modified from Perkins et al. (1995). Two μm streptavidin-coated polystyrene beads (Polysciences Ltd.) are linked to a lambda DNA with biotin-modified 5'-ends to obtain one DNA molecule for each bead. The protocol starts with the modification at one end (sticky end) of the lambda DNA with a complementary synthetic oligonucleotide carrying a biotin and a phosphate on each end (5'-P-GGGCGGGCGACCT-bio-3'). Ten μl of 17 nM lambda DNA solution are incubated with 10 μl of 1 μM oligonucleotide solution in a suitable buffer and heated at 65°C for 5 min to denature Lambda DNA sticky ends. The solution is allowed to reach room temperature in the heating block and the ligating enzyme, T4 DNA ligase, is added. Ligation occurs overnight at 16°C. Purification by desalting spin columns follows to eliminate excess oligonucleotide that could compromise beads linkage step. Electrophoresis in 0.7% agarose gel at 30 mA is carried out overnight to assess the integrity of DNA chain (a polymer of such a length is prone to rupture and/or entanglement) and for the quantification of recovered material after desalting. In the next step, beads are incubated overnight at 1:10 bead/DNA molecule ration in phosphate buffer (pH 7.4) with gentle shacking. We found this ratio proper to link only one DNA to each bead. After 3 h settlement and further precipitation in microcentrifuge at 1000×*g* for 20 min, the bead/DNA complex is suspended again in a phosphate buffer solution containing 8% (w/w) glycerol at different NaCl molarities depending on polymer contour length desired (Yoshikawa 2001).

A wide-field fluorescence system using an Hg lamp for excitation is combined to our optical tweezers setup, as shown in Figure 15.8. The excitation band can be selected, using interchangeable filter sets.

Images of Figure 15.9a through f show an experiment where the optical trap is used to manipulate a lambda DNA (indicated with arrows) linked to a bead, which is attached to the cover glass surface. The sequence shows that the laser traps the DNA and elongates a portion of the chain till the laser beam is no longer able to trap the DNA because the elastic force is stronger than the maximum optical force. In Figure 15.9f, it can be observed that lambda DNA escapes from the trap. In this experiment, the lambda DNA is stained with propidium iodide, an intercalating dye whose binding stoichiometry was determined to be one dye molecule per four to five DNA base pairs (Borsali et al. 1998).

Notice in Figure 15.9g through i that some entangled DNA segments not trapped by laser beam maintain a linear configuration.

15.2.5 Confocal Microscopy Combined with Optical Tweezers

A confocal imaging system is also combined with our optical manipulation setup as shown in Figure 15.10. Integrating confocal imaging and optical trapping requires specific measures to avoid the axial

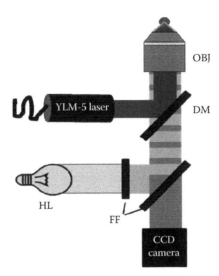

FIGURE 15.8 Schematic of optical manipulation setup combined with fluorescence microscopy. OBJ—microscope objective, DM—dichroic mirror, FF—fluorescence filter, and HL—halogen lamp.

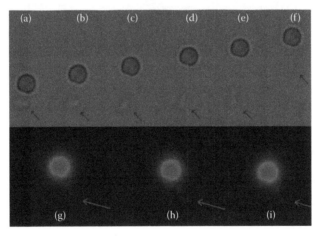

FIGURE 15.9 Lambda DNA linked to beads. Parts (a)–(i) represent fluorescence microscopy images. Arrows indicate lambda DNA filaments.

translation of the trapped object following the microscope objective axial scan in 3D imaging. This is done in our setup by changing the trap position using different phase maps on the SLM while maintaining the same plane (dashed line in Figure 15.10) for the confocal image acquisition. A specific software which permits suitable synchronization between the confocal imaging system and the SLM has been developed. An example of 3D imaging of a suspended and optically trapped cell is shown in Figure 15.11. Imaging is performed using a two-photon microscopy system based on a Tsunami Spectra-Physics laser source and a Nikon C1 scanning head. In the image, FITC-conjugated antibodies stain the MHC class I membrane glycoproteins of a DHL-4 lymphoma B cell.

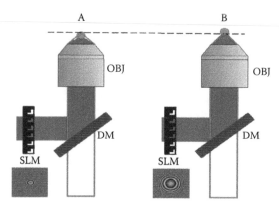

FIGURE 15.10 Schematic figure of the confocal and trapping combination setup. Beam coming from the left is the trapping laser, and beam coming from the bottom is the laser used for confocal imaging. DM—dichroic mirror, OBJ microscope objective, and SLM—spatial light modulator.

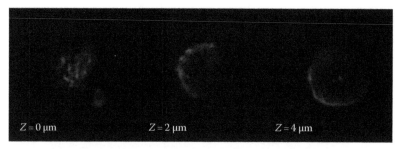

FIGURE 15.11 Selective section imaging of different planes of a DHL-4 lymphoma B–suspended cell using a combination of holographic optical tweezers and two-photon microscopy. FITC-conjugated antibodies stain the MHC class I membrane glycoproteins of the cell.

15.3 Fiber-Optic Tweezers

As described in the previous sections, the standard optical tweezers scheme utilizes one freely propagating laser beam that is tightly focused inside the medium under investigation by means of a suitably modified microscope having a high NA objective. In several situations, this scheme can give rise to serious impairments that could be overcome by an alternative approach in which the laser beam is carried by an optical fiber. As a first point, the use of standard laser tweezers in turbid biological media presents significant challenges, since it is difficult to achieve the tight focusing necessary for optical trapping. The limitation is more severe in thick samples, where the spherical aberration of the focusing beam further weakens the optical trapping forces (Xin-Cheng et al. 2001). Moreover, in the standard tweezers apparatus, the objective used for trapping is usually exploited to observe the trapped sample: this prevents to obtain three-dimensional confocal images of the sample (unless using more complex solutions, as described earlier), and, because of the high NA of this objective, the field of view inside the sample cell is very small. On the contrary, considering a fiber-produced trap, an independent objective, even with lower magnification, can be used, thus providing much more flexibility and a larger field of view. Therefore the fiber-optical approach requires a very simple setup and makes possible the trapping of microparticles in many different environments and also ensures an easy access from almost any direction to the trapped sample. This could be very important, e.g. in-vacuum, where the exploitation of optical trapping is made very difficult, or even prohibited, by the complicated and bulky structure of the microscope-based setup. As a final consideration, it is very important to outline that optical-fiber tweezers can be easily implemented as a probe for endoscopy. This feature could open new frontiers for optical tweezers utilization, like "in situ" analysis or controlled drug delivery. Although the usefulness of optical-fiber tweezers was widely recognized in the literature, no fully satisfactory proposal has been formulated for many years. The main issue to be solved to obtain fiber-based optical tweezers arises from the fact that the maximum NA that can be obtained using an optical fiber, especially considering the small refractive index difference between the fiber and the surrounding medium (water in biological applications), is usually quite low. As a consequence, even after using micro-lenses on top of the fibers (Taguchi et al. 1997, Hu et al. 2004) the gradient forces generated in the focal region are not sufficiently strong enough to accomplish the trapping by counterbalancing the axial scattering forces. On the other hand, even in presence of an extremely tight focusing, the focal point is very close to the fiber tip, making the tweezers extremely unpractical for applications (Liu et al. 2006). The first reported 3D-trap (Constable et al. 1993, Taguchi et al. 2000), based on a fiber-optic setup, achieved stable trapping of small dielectric spheres using two opposing fiber probes at the same time. This solution is quite unpractical due to the critical alignment of the two fibers. In a more recent work (Taylor and Hnatovsky 2003) 3D-trapping with a single-fiber probe is reported, but the trapping mechanism is not entirely optical and so this solution can be used only in specific situations.

Recently our group finally succeeded in conceiving and realizing a single-optical-fiber-based OT (Liberale et al. 2007). This is based on a novel scheme that promises to give new potentiality to such a type of device: the aim of our project is, indeed, the development of a new and powerful tool for trapping, manipulation, and analysis of micro-samples based on optical fibers.

The approach used successfully combines total-internal-reflection (TIR) (instead of refraction) at the interface between fibers and surrounding medium to achieve focusing with high NA and the exploitation of optical fibers in which light propagates through a core structure having an annular geometry. The working principle is represented in Figure 15.12.

By properly shaping the fiber end at an angle θ as shown in Figure 15.12a, the light beam carried in the core (gray annulus in Figure 15.12b) experiences TIR at the fiber/medium interface. The beam is first deflected through the fiber cladding, and then out of the fiber, converging in a point positioned along the probe axis, thus producing a "focalization effect" corresponding to that obtained using an objective with an equivalent NA (in the ray-optics approximation) given by $NA_{eq} = n_F \sin(2\theta)$, where n_F is the fiber refractive index. The same working principle also holds considering a bundle of fibers as shown in

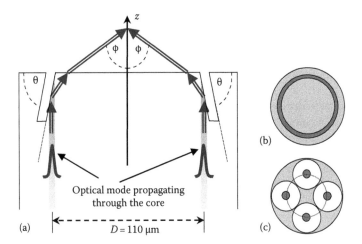

FIGURE 15.12 Scheme of the fiber tweezers. (a) Cross section of the annular-core fiber: the optical beam experiences reflection and refraction at the fiber/medium interface. The θ angle determines the equivalent NA of the fiber probe. (b) Annular-core fiber. (c) Optical fiber bundle. The fiber cores are represented by the dark gray area. (From Minzioni P. et al., *IEEE J. Sel. Top. Quantum Electron.*, 14, 151, 2008. With permission: © 2008 IEEE.)

Figure 15.12c: if the size of the structure and that of the fibers composing the bundle are properly chosen, a fairly symmetrical annular distribution of fibers is obtained. The light emitted by each fiber core can be properly deflected in such a way that all the beams converge in the same point, along the bundle axis, providing the equivalent focusing effect.

In the proposed scheme, the trapping is intrinsically favored by the beam propagation geometry: considering a strongly focused Gaussian beam in the ray optics approximation, only the external (and strongly tilted) rays are mainly responsible for the gradient force in the axial direction. Conversely, the central rays contribute essentially to the scattering force, yielding a negligible contribution to the axial gradient force (Ashkin 1992). In the scheme reported in Figure 15.12a, the trap is built up only by rays slanted with respect to the axis, thus weakening the impact of the scattering force and relaxing the constraints on the NA_{eq} required to achieve efficient trapping. A similar technique has also been used in standard OT to increase their axial-trapping force (O'Neil and Padgett 2001). This approach provides 3D trapping at a large distance from the fiber end by using a single fibre. Furthermore, both the NA_{eq} and the trapping distance can be easily tuned by changing the angle θ and the annulus diameter. As an example, taking $n_F = 1.45$ and $n_M = 1.33$ as the refractive index of the surrounding medium (water), by cutting the fiber surfaces at an angle θ slightly beyond the critical angle for TIR ($\theta_c = 66.5°$), the obtained focusing leads to the equivalent trapping effect of an optical system with $NA_{eq} = 1.06$, a value very close to that of the typical objectives used in bulk optical trapping arrangements. Referring to Figure 15.12a, if the diameter of the core annulus is, as an example, 110 μm, the trapping position is about 35 μm away from the fiber end, thus allowing a high degree of freedom in sample imaging, analysis, and manipulation.

15.3.1 Calculations

The effectiveness and the expected performances of the structures have been numerically assessed in the Mie regime, which holds when the size of the trapped particle is much larger than the wavelength of the trapping beam. In such a case, the optical forces can be calculated using a ray optics approach (Ashkin 1992). The first step for the calculation of the forces is the evaluation of the intensity spatial distribution, given by the superposition of the Gaussian beams emitted by the fibers composing the bundle. The optical forces are then calculated by decomposing the strongly focused beam into individual rays. The power associated to each ray is determined by the corresponding amplitude in the far field, while their angular orientation, with respect to the propagation direction, is given by the gradient of the optical phase (Liberale et al. 2007, Minzioni et al. 2008). Optical forces have been calculated for a polystyrene sphere (refractive index $n_P = 1.59$) with 10 μm diameter immersed in water and have been carried out both for the annular-core and the fiber-bundle configurations (Minzioni et al. 2008, Bragheri et al. 2008). Geometrical parameters used in calculations are the following: the diameter of the fiber cores annulus is $R = 55$ μm, the fibers have a mode field diameter (MFD) = 8 μm, and the cutting angle is $\theta = 66°$. A classical representation of the spatial force distribution is given by means of the dimensionless quantity Q, defined as a function of the total force (F_T), the total optical power (P), the medium refractive index (n_M), and the speed of light in vacuum (c): $Q(x, y, z) = cF_T(x, y, z)/n_M P$. The particle is trapped at the position where the force is equal to zero ($Q = 0$). It can be seen in Figure 15.13 that in the case of a four-fiber bundle, the optical trap is formed at a distance of about 27 μm from the bundle end-face. The Q parameter is shown in the yz plane; due to the symmetric arrangement of the four fibers, identical results can be obtained in the xz plane.

The obtained Q values are comparable to those reported for standard microscope-based tweezers. As expected, at larger θ angles the Q values decrease due to the NA_{eq} reduction, whereas the trapping position progressively moves far away from the fiber end surface. A further increase in NA_{eq} could be obtained by covering the shaped fiber surfaces with a metallic coating that would guarantee TIR for angles smaller than the critical one ($\theta < \theta_c$).

This approach has some advantages with respect to the one based on the annular-core fiber, because it allows an independent control on the different portions of the optical beam emitted by the structure,

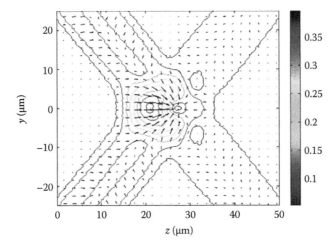

FIGURE 15.13 (See color insert following page 15-30.) Contour plot of the Q factor calculated for the fiber-bundle structure with geometrical parameters: bundle diameter $R=55$ μm, beam diameter $w=8$ μm, and cutting angle $\theta=66°$. Both parameters are shown in the yz plane: due to the TOFT symmetry, identical results have been obtained in the xz plane. (From Bragheri F. et al., *Opt. Express*, 16, 17647, 2008. With permission.)

and also offers the opportunity to "specialize" each of the used fibers for different purposes (e.g., creation of multiple traps, optical analysis, and optical signal collection).

15.3.2 Probe Fabrication and Experimental Results

The experimental demonstration of the proposed all-fiber OT was performed starting from a four-fiber bundle. The fiber-bundle structures have been fabricated by encapsulating the fibers in a fused-silica capillary coated by a polyimide layer (TSP200350, Polymicro Technologies LLC.) and by fixing them through an epoxy resin. The capillary had an inner diameter of 200 μm and an outer diameter of 360 μm. The fiber end surfaces were then polished to obtain a sub-micrometric roughness. Specialty fibers with 80 μm cladding have been used (RC HI 1060, Corning Inc.) in order to keep the overall probe diameter as small as possible, resulting in a miniaturized probe with a size smaller than a syringe needle. The fibers were single mode at 1070 nm, exhibiting a mode field diameter of about 6.5 μm. The fibers end-faces were micro-structured through focused ion beam (FIB) milling (Figure 15.14a through f) that allows a controlled nano-machining of the surfaces through a beam of accelerated gallium ions. The FIB facility is assisted by a scanning electron microscope (SEM) for the sample imaging (NOVA 600i–Fei Company). As a first step, the fiber bundles end-faces are coated by a 20 nm thick gold layer through vapor deposition, in order to avoid the electrostatic charge of the sample during the FIB operation. The bundle is then inserted in the FIB vacuum chamber through a specifically designed support. The bundle axis is tilted with respect to the ion beam direction by about 20° (the angle complementary to θ). The micromachining is obtained by directly milling a slanted hole on each fiber surface in correspondence of the core regions with a 20 nA ion current.

The beam is scanned over a trapezium-like shape area (height 10 μm, area 150 μm²) in order to obtain a large aperture, thus enabling the escape of the sputtered material from the hole. The fraction of material that gets redeposited on the hole surface is removed by a second run performed by scanning a reduced area (Figure 15.15). Finally, the remaining gold layer deposited on the top of the probe is removed, using FIB, to avoid the back reflection of the optical beams used for trapping experiments. In this way, the core regions at the fiber surfaces are properly shaped, as required by the scheme in Figure 15.12a, to obtain TIR at the fiber core/water interface. The images (Figure 15.14f) of the micro-machined probe, taken at the SEM, show that

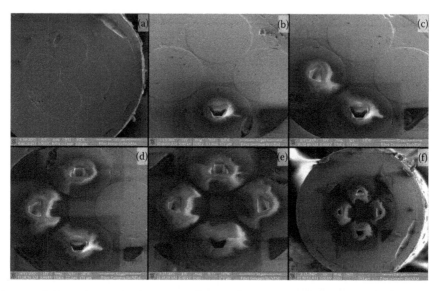

FIGURE 15.14 Different stages of probe micromachining. The symmetry and surface quality is first verified, (a). After this check, the probe is rotated and one fiber is drilled at the proper angle, (b). The probe is subsequently rotated by 90° steps and the procedure is repeated on the remaining fibers, (c) through (e). The images highlight the strong reproducibility of the procedure and the high quality of the surfaces. The final result on the probe is shown in (f). (Reprinted by permission of Macmillan Publishers Ltd: Liberale C. et al., *Nat. Photon.*, 1, 723, 2007. Copyright 2007.)

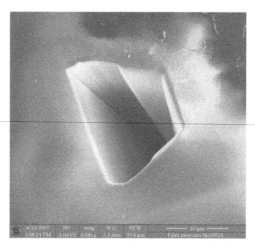

FIGURE 15.15 Close-up image of one hole. From this picture, it is possible to appreciate the flatness of the realized surfaces. (From Minzioni P. et al., *IEEE J. Sel. Top. Quantum Electron.*, 14, 151, 2008. With permission: © 2008 IEEE.)

the overall structure presents a very good symmetry and an excellent quality of the surfaces. The θ angles realized in the fabricated structures are about 70°, giving $NA_{eq} \sim 0.93$. The quality of fabricated probes has been assessed by coupling an Yb-doped fiber laser emitting at 1070 nm into the four fibers composing the device. The probe fibers were connected to the laser through a 1×4 fiber-optic coupler and the optical power carried by each fiber is about 7 mW, yielding to an extremely compact and stable setup. After verifying that all the four beams were converging in the same position using a 160× objective (Figure 15.16), the trapping effectiveness has been tested in the same conditions considered in the numerical simulation by depositing on a coverslip a water suspension of polystyrene spheres having a diameter of 10 μm. The

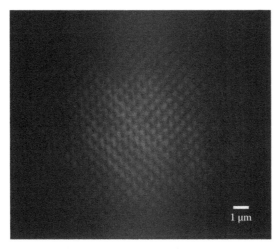

FIGURE 15.16   CCD image of the transverse plane corresponding to the crossing point for the deflected beams from the four-fiber bundle. The image clearly shows the interference pattern.

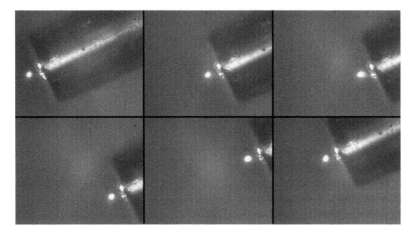

FIGURE 15.17   Sequence of six images showing a polystyrene particle trapped by the TOFT. The images are magnified through a microscope with a 10× objective and recorded by a CCD camera. The trapped sphere scatters the light coming from the fiber probe in the infrared range, which is then captured by the CCD. The trapped particle is moved below the microscope objective: it is first moved to the right (top row), and it is subsequently moved up and to the left (bottom row). Thanks to the use of a low magnification objective, the TOFT end can be clearly seen. The particle appears as a bright point because it strongly scatters the incoming infrared radiation used for trapping.

probe-end was immersed in the suspension and was viewed through a standard microscope using 10×- or 20×-objectives. The probe, mounted on micro-translators, could be easily moved inside the suspension in order to drag the particles that were attracted and trapped by the probe as soon as they entered the range of action of the optical force. In Figure 15.17, a sequence of frames collected by a CCD camera while the probe with two trapped particles was moved inside the field of view of the microscope is shown. The experimental achievement of optical trapping is the most important validation for this new approach.

A final experiment was performed in order to demonstrate simultaneous particle trapping and optical analysis. The probe-end was immersed in a water suspension of 10 μm diameter fluorescent beads (Fluoresbrite® Yellow Green Microspheres Polysciences, Inc.). The experimental setup is shown

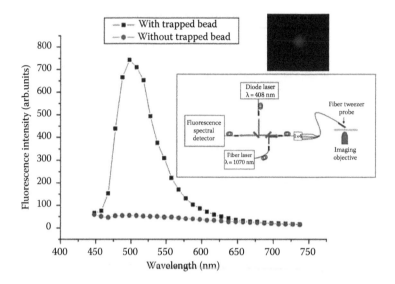

FIGURE 15.18 Fluorescence spectra measured from the fiber probe bundle with (circles) and without (squares) the trapped fluorescent bead. In the inset, a simple scheme of the experimental setup is reported. The laser source emitting at 1070 nm is an ytterbium-doped fiber laser, whereas the radiation at 408 nm used to excite the fluorescent bead is obtained through a laser diode. (Reprinted by permission of Macmillan Publishers Ltd: Liberale C. et al., *Nat. Photon.*, 1, 723, 2007. Copyright 2007.)

in the inset of Figure 15.18. By using two dichroic mirrors, both trapping and fluorescence excitation radiations ($\lambda = 1070$ nm and $\lambda = 408$ nm) were coupled into the fiber probe. The fluorescence signal emitted by the trapped bead was collected by the probe itself, transmitted by the dichroic mirrors, and then detected through a spectrometer with 10 nm spectral resolution (32 PMTs Nikon spectral detector). The optical spectra measured with and without a fluorescent bead trapped by the fiber tweezer are reported in Figure 15.18. The optical analysis function is very effective thanks to the short distance between the bead and the probe-end that acts both as source of the excitation and signal collectors.

The performances could be sensibly improved by designing special bundles including dedicated large mode area fibers or even introducing plasmons-based sensors nanostructured on top of one fiber (see Section 15.3). Furthermore, simple modifications of the probe structure can lead to the formation of multiple traps or to micromanipulation functions like translation, scanning, mechanical squeezing, and oscillation.

15.4 Plasmonic Devices

Recently, a lot of interest is growing for single-molecule detection for characterization and identification of not only organic molecules (Nie and Emory 1997) but also of biological substances such as bacteria (Jarvis and Goodacre 2004), spores (Alexander et al. 2003) and protein (Xu et al. 1999). There are many techniques such as nanosphere lithography, electron-beam lithography, and metal colloidal in order to fabricate a nanostructure for plasmonic enhancement (fluorescence and Raman signal) necessary to study single molecules of biological interest. When the molecules are in close proximity to nanometer-sized metallic structures, the interaction of light with periodic structure causes the generation of plasmons. An enhancement of optical signal can be observed due to the excitation of plasmons on the metallic surface (De Angelis et al. 2008b).

Here, we present a novel device designed for the generation of an SPP in a tapered nanolens made of noble metals, such as gold and silver, combined with a photonic crystal (PC) cavity acting as an efficient coupler between the external optical source and the nanoantenna. Our nanolithography techniques of choice rely on two powerful fabrication methods: FIB milling and chemical vapor deposition (CVD) induced by focused electron beam. We aim at achieving unambiguous chemical information of molecular species with a spatial resolution down to 15 nm in label-free conditions. This avoids unwanted targeting effects at molecular level as well as experimental difficulties related to the use of a near-field configuration. A practical drawback in using Raman scattering is due to the low cross section value compared to first-order spectroscopies. Nevertheless, in our case, optical sensitivity is recovered in favor of Raman spectroscopy exploiting the surface-scattering field enhancement induced by SPP generation. Noticeably, this device can be exploited in fluorescence application for single-molecule detection.

15.4.1 Device Fabrication

The device is fabricated in two steps: (a) FIB milling (FEI Novalab 600) is used to fabricate PC cavity on silicon nitride membrane (100 nm thick); (b) electron beam–induced CVD is used to fabricate the nanoantenna at the center of PC cavity. A thin film of gold is sputtered on the membrane in order to avoid the charging effect due to milling process. For the milling process, the ion beam current and the acceleration voltage are set to be 50 pA and 30 kV, respectively. After the PC fabrication, a nanoantenna is grown in the center of cavity exploiting a gas precursor that contains platinum–carbon polymer ($(CH_3)_3Pt(C_pCH_3)$). The final device, SENSe (surface-enhanced nanosensor), is shown in Figure 15.19. The x-ray microanalysis performed on the nanoantenna confirms that it is composed of a mixture of carbon and platinum, as shown in Figure 15.20 (De Angelis et al. 2008a). Another thin

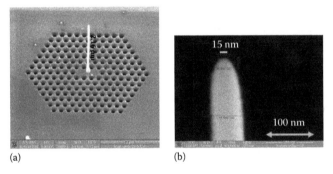

(a)　　　　　　　　　　　　　　(b)

FIGURE 15.19 (a) SEM image of the whole device including the photonic cavity and, at its center, the plasmonic nanoantenna. The latter is 2.5 μm high, and its size gradually decreases from 90 nm in diameter at bottom down to 15 nm radius of curvature at the tip; (b) SEM details of the nanoantenna tip and its radius of curvature.

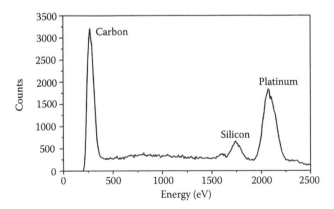

FIGURE 15.20 X-ray microanalysis performed on fabricated pillars. The peaks related to platinum and carbon coming from the pillars, and silicon from the substrate, are clearly identified. (From De Angelis, F. et al., *SPIE Proc. Plasmonics: Metallic Nanostructures and Their Optical Properties*, 7032, 2008. With permission.)

film of gold has been deposited on the surface of the device (25 nm thick), and finally the whole gold layer has been removed from the PC (but not from the pillar surface) using FIB milling. In this way, a gold nanowire has been obtained. The pillar height is 2.5 µm, the base is 80 nm large, and the radius of curvature of the tip is about 15 nm, or less.

15.4.2 Theoretical Simulation

The PC structure containing a L3-type cavity is designed (Akahane et al. 2003, Andreani and Gerace 2006) by a guided-mode expansion method in order to have a cavity mode in the green spectral region. Its spectral response is simulated by using both guided-mode expansion method and finite difference time domain (FDTD) code (Roden and Gedney 2000, Taflove and Hagness 2005) by considering only TE-like modes for which the triangular lattice has a photonic band gap in all directions. The algorithm is supplemented with the convolution perfectly matched layers (CPML)-absorbing boundary conditions and Drude–Lorentz model for gold (plasma frequency $\hbar\omega_p = 8$ eV and broadening $\Gamma = 0.015\omega_p$) that models the optical response of the nanoantenna material. The 3D simulation domain is split into $224 \times 276 \times 107$ cells and the volume of each cell is subdivided into 15.6 nm $\times 15.6$ nm $\times 11.1$ nm. This cell size allows a simulation of a tip radius of curvature of 20 nm. Such a spatial resolution gives rise to the step in time of 0.248×10^{-16} s. A modulated Gaussian pulse covering the spectral range 2.2–2.45 eV is originated from a point source situated at the center of the cavity just below the basis of the metallic nanoantenna. The detector is situated just above the PC cavity or above the tip of the nanoantenna depending on simulation conditions. The simulation is carried out in the presence of nanoantenna with and without the PC cavity. When the nanoantenna is absent, we observe a strong peak at 2.31 eV corresponding to the fundamental TE cavity mode. However, when the nanoantenna is added, the peak is strongly suppressed indicating that the energy of the PC cavity mode is converted in the SPP mode of the nanoantenna. The square module of the e.m. field $(E^2 = E_x^2 + E_y^2)$ in the membrane plane is calculated in the presence of the nanoantenna, without (Figure 15.21a) and with (Figure 15.21b) the PC cavity. By comparison of the two panels, the energy concentration action operated by the PC cavity on the nanoantenna is readily apparent (notice the different color scales in the two panels). The intensity ratio, evaluated at the same exciting power, is one order of magnitude in favor of the nanoantenna on the PC cavity. The simulation allows an evaluation of the electrical field enhancement factor (the ratio between the incoming electrical field at the bottom of the nanoantenna and the electrical field at the tip) to be about 100. For an ideal cone

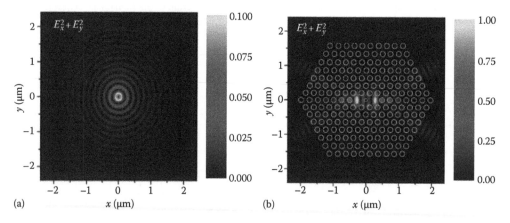

FIGURE 15.21 (See color insert following page 15-30.) FDTD calculation of nanoantenna optical response. Color scales of panels are in arbitrary units. The numerical values reported in the color scale are in the correct relative intensity values. (a) Spatial $E_x^2 + E_y^2$ field profile of the plasmonic nanoantenna on a silicon nitride membrane calculated at $\lambda = 514$ nm. (b) Spatial $E_x^2 + E_y^2$ field profile of the overall device calculated at $\lambda = 514$ nm. (Reprinted with permission from De Angelis F. et al., *Nano Lett.*, 8, 2321, 2008. Copyright 2008 by American Chemical Society.)

geometry the enhancement factor can be higher at the tip apex (Stockman 2004, radius of curvature of the tip about 2 nm); however, in this case the main limitation is related to the radius of curvature of the fabricated tip, which is about 10–15 nm (see Figure 15.19). Since we simulate a radius of curvature of 20 nm, our calculated enhancement factor is expected to be close to the one of the fabricated devices, rather than the ideal device (Stockman 2004). According to FDTD modeling, the plasmonic field is localized around the tip with a decaying length comparable to its radius of curvature.

15.4.3 Experimental Setup and Material Deposition

Microprobe Raman (Renishaw inVia microRaman) spectra were excited by Ar^+ visible laser with 514 nm laser line in transmission configuration through 150× (NA—0.95). The laser power was varied in the range of 0.18–1.8 mW whereas the accumulation time is varying in the range of 150–500 s. Block diagram of micro-Raman setup is shown in Figure 15.22. As it is illustrated, the incident laser path of commercial Raman microscope is modified for transmission configuration to achieve the optimized sample illumination geometry configured in the theoretical simulation described earlier. To characterize the device, four different materials have been deposited on sample surface: hexamethyldisilazane (HMDS), benzenethiol (BTH), quantum dot (QD), and silica nanoparticle (SiO_x). The sample is immersed in HMDS polymer for 10 min and dried with a nitrogen flux. Since, HMDS is a highly volatile molecule it forms a monolayer by chemisorptions on metal surface. In the case of BTH, immediately after taking out from solution, the substrate is washed with water so that the excess BTH will be washed out and a monolayer of BTH will be self-assembled on gold surface due to the covalent binding of thiol group with gold. Later on, the HMDS and the BTH are selectively removed from PC cavity using FIB milling.

The deposition of SiO_x nanoparticles on the vertex is performed by utilizing the precursor gas TEOS (tetraethylorthosilicate). The SiO_x nanoparticle is selectively deposited in a specific position by focusing and positioning the electron beam, with a nanometer precision with the aid of dedicated software. The electron beam is "on" for a few seconds in the presence of a stationary flow of precursor gas. SiO_x is

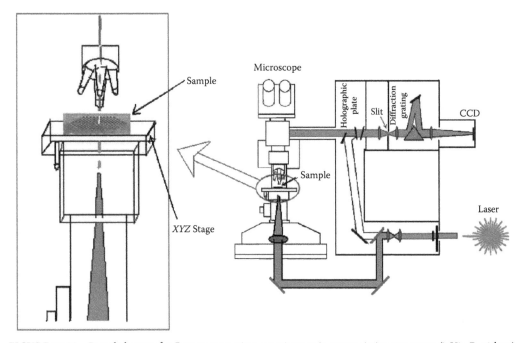

FIGURE 15.22 Detailed setup for Raman scattering experiments in transmission geometry (inVia Renishaw). (Reprinted with permission from De Angelis F. et al., *Nano Lett.*, 8, 2321, 2008. Copyright 2008 by American Chemical Society.)

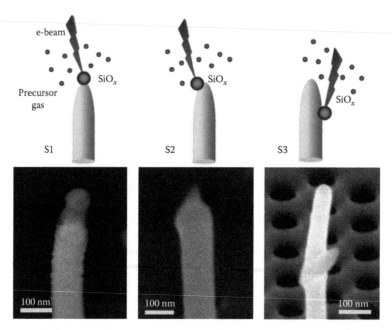

FIGURE 15.23   SiO$_x$ deposition on the nanoantenna. Schematic and SEM images of the SiO$_x$ nanoparticle deposition at three different positions on the nanoantenna, named S1, S2, and S3, corresponding to the tip apex, near the apex, and along the nanoantenna wall, respectively. (Reprinted with permission from De Angelis F. et al., *Nano Lett.*, 8, 2321, 2008. Copyright 2008 by American Chemical Society.)

deposited, for each sample, at three different positions named S1, S2, and S3, as shown in Figure 15.23. Furthermore, the experimental demonstration of high sensitivity reached by the SENSe device is proved by measuring the Raman signal coming from a single QD. This is accomplished by the use of a nanomanipulator, SEM assisted, whose positioning accuracy in the most critical direction is 0.25 nm. In Figure 15.24, SEM images are arranged in a series with progressive magnification (from top left to bottom right) to show the deposition technique of QDs from the manipulator tip to the nanoantenna. The tip of the manipulator has been loaded with QDs by mechanically plowing it in a bulk QD sample (first panel). The deposition of a single QD from the nanomanipulator tip to the nanoantenna tip is not simply due to the mechanical friction but is electrostatically assisted under the SEM beam. We observed that the deposition occurs in two different ways: (a) the nanomanipulator tip and the nanoantenna tip are in close contact and are instantaneously welded by electrostatic forces. By pulling the nanomanipulator tip away from the nanoantenna tip, the QD remains attached to the nanoantenna tip; (b) The nanomanipulator tip and the nanoantenna tip are close but not in contact (200 nm distance). Thanks to the electrical charging of the sample, there is a growing electrical field (estimated field is about 106 V/cm) that is strong enough to be able to transfer the QDs from the nanomanipulator tip to the nanoantenna tip. The manipulator allows the deposition of a single QD (QD ITK amino PEG quantum dots ZnS/CdSe Invitrogen, 15–20 nm diameter) on the nanoantenna tip. The apparent increase in the radius of curvature of the tip is due to carbon codeposition during the operation time of QD manipulation and deposition.

15.4.4 Raman Scattering Measurements

Various Raman measurements have been performed on nanoantenna covered with HMDS or not (bare nanoantenna) in the range between 500 and 3200 cm^{-1} keeping the accumulation time (acc. time) of 150 s. Raman spectrum carried out on bare device (Figure 15.25) clearly illustrates the C–C vibrational band to be at around 1570 cm^{-1}, coming from residual carbon present on the vertex of nanoantenna, as it

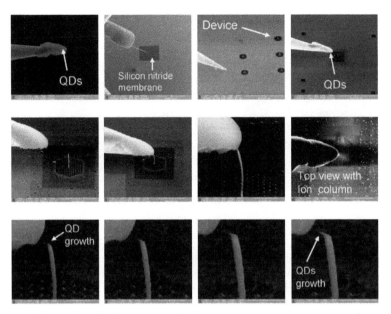

FIGURE 15.24 Single quantum dot deposition on the nanoantenna. SEM images arranged in progressive magnification (from top left to bottom right) to show the deposition technique of QDs from the manipulator tip to the nanoantenna. (Reprinted with permission from De Angelis F. et al., *Nano Lett.*, 8, 2321, 2008. Copyright 2008 by American Chemical Society.)

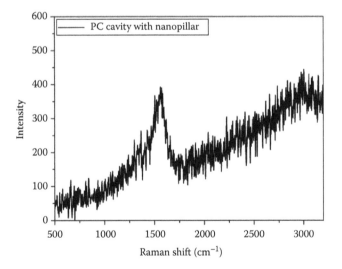

FIGURE 15.25 Raman spectra performed on baretapered nanopillar, showing C–C-related vibrational band. (From De Angelis, F. et al., *SPIE Proc. Plasmonics: Metallic Nanostructures and Their Optical Properties*, 7032, 2008. With permission.)

has been observed from x-ray microanalysis (Figure 15.20) (De Angelis et al. 2008a). The measurements have also been performed on sole nanoantenna (without PC, not shown in figure), showing no Raman band, centered at around 1570 cm^{-1}: it could be due to the fact that high scattering and diffraction losses avoid an efficient coupling of the radiation with the nanoantenna.

Though, the accumulation time and the laser power are raised up to 500 s and 1.8 mW, no Raman band is observed for sole nanoantenna structure, showing that, in the absence of PC cavity, the coupling

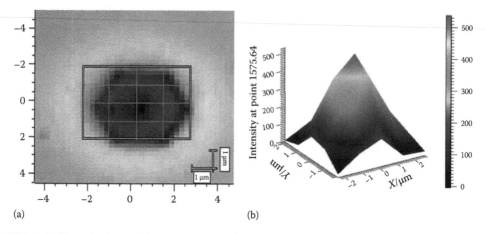

(a)   (b)

FIGURE 15.26 (See color insert following page 15-30.) Micro-Raman mapping measurement. (a) *XY*-mapping area where the Raman measurement is performed, (b) mapping analysis for the Raman band centered at 1575.64 cm^{-1} attributed to the C–C-related vibrational band. (From De Angelis, F. et al., *SPIE Proc. Plasmonics: Metallic Nanostructures and Their Optical Properties*, 7032, 2008. With permission.)

between the antenna and the incident laser is inefficient. Raman mapping is also performed on bare device in confocal mode. *XY* mapping area is shown in Figure 15.26a, and the Raman measurements on each point are performed at the *XY* surface area (De Angelis et al. 2008a). Intensity mapping analysis corresponding to the Raman shift of 1575.64 cm^{-1} is shown in Figure 15.26b. Figure clearly shows that as the measurement point on *XY* mapping surface moves from center (nanoantenna) toward outside the structure, the C–C band decreases smoothly.

Raman measurements are also performed on Si wafer (backscattering configuration), silicon nitride membrane (backscattering geometry), PC cavity without nanoantenna (transmission mode), and PC cavity with nanoantenna (transmission mode); all are deposited with HMDS polymer in order to investigate and to evaluate the device functionality for few/single molecule detection. Raman spectra for HMDS-deposited various substrates are shown in Figure 15.27 (De Angelis et al. 2008a). Peaks in the range of 1300–1700 cm^{-1} are attributed to the sp$_2$ a-C vibrational band whereas the band at around 3080 cm^{-1} is related to the C–H$_x$ stretching vibration. Figure 15.27 clearly shows that in order to achieve a detectable Raman spectra from HMDS polymer deposited over the device, much lower laser excitation power and accumulation time are needed with respect to silicon wafer and silicon nitride membrane. SERS enhancement from PC cavity with nanopillar enables us in the remarkable detection of HMDS compound. By assuming that HMDS molecular surface area equals 0.4 nm^2, the number of molecules present on the tip of the nanopillar is estimated to be around 200. While comparing the Raman spectra obtained on silicon nitride and on device, and considering a laser beam diameter of about one micron, a Raman scattering enhancement of 3×10^6 can be calculated.

In order to verify that the Raman signal of HMDS is coming only from the tip of the nanoantenna, a set of measurements changing the *z* position of the sample have been performed (Figure 15.28). At first the microscope focus has been adjusted on the sample surface, and a Raman spectrum has been recorded (*z*=0 in figure scale). In this configuration, the tip of the nanoantenna is just a little bit out of the focus depth, which is about 1 μm. Then, the sample plane has been progressively moved down and a set of Raman spectra have been measured without adjusting the focus of the microscope (*z* from −1 to −5 μm of figure scale). As can be clearly seen in Figure 15.28, the maximum value of signal intensity of the peak at 3083 cm^{-1} has been obtained when the stage is about 3 μm below the unchanged focal plane, and the tip of the nanoantenna is correctly positioned in the focal plane itself. It means that the main part of Raman signal of HMDS is coming from the tip of the nanoantenna, confirming that the device is properly working, as predicted from theoretical calculation (Section 15.3.2).

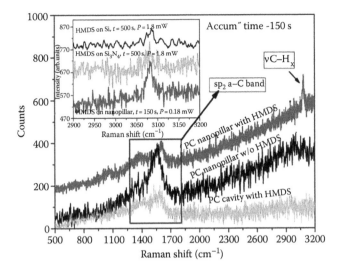

FIGURE 15.27 Raman spectra for bare PC cavity with nanopillar (black), without HMDS polymer, PC cavity with (dark gray) and without nanopillar (light gray) covered by HMDS monolayer in the range between 500 and 3200 cm⁻¹. All the measurements are performed by keeping all the parameters same. In the inset of figure, HMDS deposited on Si wafer (black), on silicon nitride membrane (light gray), and on PC cavity nanopillar (dark gray). (From De Angelis, F. et al., *SPIE Proc. Plasmonics: Metallic Nanostructures and Their Optical Properties*, 7032, 2008. With permission.)

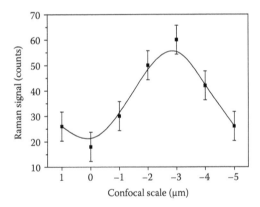

FIGURE 15.28 Raman scattering intensities of the peak at 3083 cm⁻¹ of HMDS deposited on the device for different *z* positions of the sample stage. The maximum value of the signal is obtained when the stage is about 3 μm below the "zero" position (focus adjusted on sample surface): it means that the signal is coming from the tip of the nanoantenna.

15.4.4.1 Selectively Deposited SiO$_x$ Nanoparticles

Raman measurements were carried out for SiO$_x$ in the range between 250 and 900 cm⁻¹ for bulk and nanodeposited SiO$_x$ at three different selected positions (S1, S2, and S3) on nanoantenna, keeping the laser power always 0.18 mW (the deposition technique has been described in Section 15.3.3, Figure 15.23). The corresponding Raman measurements are reported in Figure 15.29. The experimental results show that the maximum signal is obtained when the SiO$_x$ nanoparticle is deposited at the tip center (red line) of the nanoantenna and it rapidly vanishes as soon as the particle is deposited away from the tip (blue line). From SEM investigations, we estimate that the scattering volumes of SiO$_x$ in S1 and S2 are comparable (radius of SiO$_x$ sphere is about 35 nm). The decrease in Raman scattering intensity is the

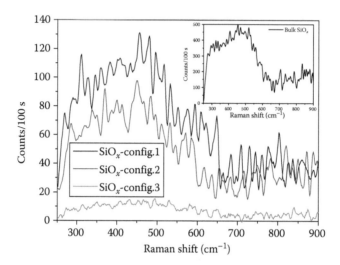

FIGURE 15.29   Raman scattering spectra ($\lambda_{exc}=514$ nm, $P=0.18$ mW, $T_{int}=100$ s) of a SiO$_x$ nanoparticle deposited on the nanoantenna at positions S1, S2, and S3. In the inset, Raman spectrum of bulk SiO$_x$ is also shown.

experimental proof that the SPP intensity probed at S2 is lower with respect to S1, as expected from the calculated field localization. Notice that even if at S3 the scattering volume is about 4 times larger, the Raman intensity is almost at background level (green line). This is expected from the negligible field intensity along the straight portion of the nanoantenna. This experimental condition is equivalent to having a single isolated particle where no enhancement is present. Raman spectrum of bulk SiO$_x$ is also shown in the inset of Figure 15.29.

15.4.4.2 Selectively Deposited CdSe Quantum Dots

Figure 15.24 shows the selective deposition of CdSe (QD ITK amino PEG quantum dots ZnS/CdSe, 15–20 nm diameter) QDs on the vertex of nanoantenna. Raman measurements are carried out for CdSe QDs, deposited on the tip of nano-optic device, in the range of 500–3700 cm^{-1}. Various bands are observed for the substance deposited on the device tip and from the unwilling layer of carbon deposited on the tip during the manipulation of CdSe QD. The presence of bands around 1592 cm^{-1} and shoulder at around 1354 cm^{-1} are ascribed to C–C vibrational modes. In Figure 15.30, the Raman scattering spectrum of the single QD is compared to a QD bulk assembly of cylindrical shape (1.2 μm diameter and 200 nm thickness, not shown), keeping the identical spectroscopic conditions (laser power, substrate material, sample preparation).

   The Raman intensity peak details from the single QD on the nanoantenna matches the spectroscopic information from the bulk sample, but the Raman spectra from single QD show more clearly the av-NH$_2$ asymmetric stretching vibration at 3580 cm^{-1} (Figure 15.30, inset). This vibration is in general weak compared to the CH$_x$ one and its presence in the spectrum is due to the tip local field enhancement. A single QD allows an estimate of the number of NH$_2$ molecules contributing to this Raman vibration. In fact, from the supplier specifications (www.invitrogen.com), the QD has an inner core of 2.5 nm radius, on which PEG (poly(ethylene glycol)) molecules are bound with NH$_2$-terminated groups, thus resulting in an average single QD diameter of about 18 nm. A conservative estimation can be done assuming very high close packing of PEG–NH$_2$ groups, that is, 1 anchorage/nm^2. With this assumption, the total number of NH$_2$ groups detected is equal to 80 and the Raman scattering signal enhancement is calculated to be about 10^5. When using the anchorage specification given by the QD supplier, the number of amino groups detected is estimated to be 10.

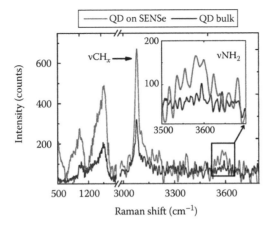

FIGURE 15.30 Raman scattering spectra (λ_{exc}=514 nm, P=0.18 mW, T_{int}=150 s) taken from the single QD (light gray line) as compared to that of a QD bulk sample (dark gray line). In the inset, the asymmetric stretching vibration, $a\nu$-NH$_2$, at 3580 cm^{-1} is reported in the range between 3400 and 3800 cm^{-1}.

15.4.4.3 Benzenethiol Self-Assembled Monolayer

Raman measurements are carried out in the range between 500 and 3500 cm^{-1} for both bulk BTH and monolayer-deposited BTH on device, as shown in Figure 15.31. The agreement with the characteristic peaks is evident. The additional peaks on the spectrum are found at 1353.8, 1591.8, and 2709.5 cm^{-1}, which are attributed to amorphous carbon D-band, G-band, and second-order D-band, respectively. All experiments presented here are designed to check both the plasmonic localization field and the device sensitivity at level of a few molecules in SERS conditions. In the first experiments (QD and silica), we experimentally demonstrated, through Raman scattering and controlled deposition at the nanometer level, that the SPP field is strongly localized in a region comparable to the radius of curvature of the nanoantenna tip. This demonstration allows us to infer that also for the BTH monolayer, the intense Raman signals are due to a strong localization and enhancement caused by the efficient generation of an SPP in the nanoantenna. The field enhancement on the nanoantenna tip is strong enough to allow the detection of a monolayer of molecules deposited on that area, similar to the case of HMDS reported in Figure 15.27.

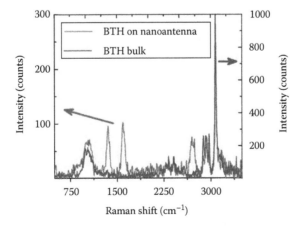

FIGURE 15.31 Confocal Raman scattering on benzenethiol. Raman scattering measurements (λ_{exc}=514 nm, P=0.18 mW, T_{int}=100 s) on benzenethiol monolayer on the nanoantenna (light gray line) as compared to the bulk spectrum (dark gray line).

15.5 Summary

Optical tweezers and photonic devices based on plasmons are less popular in the community of fluorescence microscopy. We foresee a common future and integration of these fields. The possibility of doing force spectroscopy, fluorescence analysis, and Raman spectroscopy at the same time and at the level of single molecule on cell, tissue, and possibly *in vivo* will become realistic in a few years and will be a prolific field for young researchers who will attempt to innovate in these fascinating fields.

Acknowledgment

Authors thank for support from the EU Project NanoScale Contract n.214566 (FP7-NMP-2007-SMALL-1).

References

Akahane, Y., Asano, T., Song, B. S., and Noda, S. 2003. Photonics: Tuning holes in photonic crystal nano cavities. *Nature* 425: 944–947.

Alexander, T.A., Pellegrino, P.M., and Gillespie, J.B. 2003. Near-infrared surface enhanced Raman scattering mediated detection of single optically trapped bacterial spores. *Applied Spectroscopy* 57: 1340–1345.

Allersma, M.W., Gittes, F., and Schmidt, C.F. 1998. Two-dimensional tracking of NCD motility by back focal plane interferometry. *Biophysics Journal* 74: 1074–1085.

Andreani, L.C. and Gerace, D. 2006. Photonic-crystal slabs with a triangular lattice of triangular holes investigated using a guided-mode expansion method. *Physical Review B* 73: 235114–235130.

Ashkin, A. 1992. Forces of a single beam gradient laser trap on a dielectric sphere in the ray optics regime. *Biophysical Journal* 61: 569–582.

Ashkin, A. 2000. History of optical trapping and manipulation of small-neutral particle, atoms, and molecules. *IEEE Journal of Selected Topics in Quantum Electronics* 6: 841–856.

Borsali, R., Nguyen, H., and Pecora, R. 1998. Small-angle neutron scattering and dynamic light scattering from a polyelectrolyte solution: DNA. *Macromolecules* 31: 1548–1555.

Bragheri, F., Minzioni, P., Liberale, C. et al. 2008. Design and optimization of a reflection-based fiber-optic tweezer. *Optics Express* 16: 17647–17653.

Brower-Toland, B.D., Smith, C.L., Yeh, R.C. et al. 2002. Mechanical disruption of individual nucleosomes reveals a reversible multistage release of DNA. *Proceedings of the National Academy of Sciences of the United States of America* 99: 1960–1965.

Bustamante, C., Macosko, J.C., and Wuite, G.J. 2000. Grabbing the cat by the tail: Manipulating molecules one by one. *Nature Reviews Molecular Cell Biology* 1: 130–136.

Cojoc, D., Difato, F., Ferrari, E. et al. 2007. Properties of the force exerted by filopodia and Lamellipodia and the involvement of cytoskeletal component. *PLoS ONE* 2(10): e1072, on line October 2007, www.plosone.org.

Constable, A., Kim, J., Mervis, J. et al. 1993. Demonstration of a fiber-optical light-force trap. *Optics Letters* 18: 1867–1869.

De Angelis, F., Das, G., Patrini, M. et al. 2008a. Novel plasmonic nanodevice for few/single molecule detection. *SPIE Proceedings* 7032: 70320T-1–70320T-8.

De Angelis, F., Patrini, M., Das, G. et al. 2008a. A hybrid plasmonic-photonic nanodevice for label-free detection of a few molecules. *Nano Letters*, 8: 2321–2327.

Di Fabrizio, E.M., Cojoc, D., Cabrini, S. et al. 2003. Nano-optical elements fabricated by x-ray lithography. *SPIE Proceedings* 5225: 113–125.

Gittes, F., Schnurr, B., Olmsted, P.D. et al. 1997. Microscopic viscoelasticity: Shear moduli of soft materials determined from thermal fluctuations. *Physical Review Letters* 79: 3286–3289.

Handa, N., Bianco, P.R., Baskin, J.R., and Kowalczykowski, S.C. 2005. Direct visualization of RecBCD. Movement reveals cotranslocation of the RecD motor after χ recognition. *Molecular Cell* 17: 745–750.

Harada, Y. and Asakura, T. 1996. Radiation forces on a dielectric sphere in the Rayleigh scattering regime. *Optics Communications* 124: 529–541.

Hu, Z., Wang, J., and Liang, J. 2004. Manipulation and arrangement of biological and dielectric particles by a lensed fiber probe. *Optics Express* 12: 4123–4128.

Jarvis, R.M. and Goodacre, R. 2004. Discrimination of bacteria using surface enhanced Raman spectroscopy. *Analytical Chemistry* 76: 40–47.

Liberale, C., Minzioni, P., Bragheri, F. et al. 2007. Miniaturized all-fiber probe for three dimensional optical trapping, manipulation and analysis. *Nature Photonics* 1: 723–727.

Liu, Z., Guo, C., Yang, J., and Yuan, L. 2006. Tapered fiber optical tweezers for microscopic particle trapping: Fabrication and application. *Optics Express* 14: 12511–12516.

Lock, J.A. 2004. Calculation of the radiation trapping force for the laser tweezers by use of generalized Lorenz–Mie theory. I. Localized model description of an on-axis tightly focused laser beam with spherical aberration. *Applied Optics* 43: 2532–2544.

Mansuripur, M., Zakharian, A.R., and Moloney, J. 2005. Radiation pressure and the distribution of electromagnetic force in dielectric media. *Optics Express* 13: 2312–2336.

Meller, A., Bar-Ziv, R., Tlusty, T. et al. 1998. Localized dynamic light scattering: A new approach to dynamic measurements in optical microscopy. *Biophysical Journal* 74: 1541–1548.

Minzioni P., Bragheri F., Liberale C. et al. 2008. A novel approach to fiber-optic tweezers: Numerical analysis of the trapping efficiency. *IEEE Journal of Selected Topics in Quantum Electronics* 14: 151–157.

Nie, S. and Emory, S.R. 1997. Probing single molecules and single nanoparticles by surface-enhanced Raman scattering. *Science* 275: 1102–1106.

O'Neil A.T. and Padgett M.J. 2001. Axial and lateral trapping efficiency of Laguerre–Gaussian modes in inverted optical tweezers. *Optics Communications* 193: 45–50.

Perkins, T.T., Smith, D.E., Larson, R.G., and Chu, S. 1995. Stretching of a single tethered polymer in a uniform flow. *Science* 268: 83–87.

Rohrbach, A. and Stelzer, E.H.K. 2001. Optical trapping of dielectric particles in arbitrary fields. *Journal of Optical Society of America* 18: 839–853.

Roden, J.A. and Gedney, S.D. 2000. Convolutional PML (CPML): An efficient FDTD implementation of the CFS-PML for arbitrary media. *Microwave and Optical Technology Letters* 27: 334–339.

Smith, S.B., Cui, Y., and Bustamante, C. 1996. Overstretching B-DNA: The elastic response of individual double-stranded and single-stranded DNA molecules. *Science* 271: 795–799.

Stockman, M.I. 2004. Nanofocusing of optical energy in tapered plasmonic waveguides. *Physical Review Letters* 93: 137404–137408.

Svoboda, K. and Block, S.M. 1994. Biological applications of optical forces. *Annual Review of Biophysics and Biomolecular Structure* 23: 247–285.

Taflove, A. and Hagness, S.C. 2005. *Computational Electrodynamics: The Finite-Difference Time-Domain Method*, 3rd edn. Artech House: Boston, MA.

Taguchi, K., Atsuta, K., Nakata, T., and Ikeda, M. 2000. Levitation of a microscopic object using plural optical fibers. *Optics Communications* 176: 43–47.

Taguchi, K., Ueno, H., Hiramatsu, T., and Ikeda, M. 1997. Optical trapping of dielectric particle and biological cell using optical fibre. *Electronics Letters* 33: 1413–1414.

Taylor, R.S. and Hnatovsky, C. 2003. Particle trapping in 3D using a single fiber probe with an annular light distribution. *Optics Express* 11: 2775–2782.

Tlusty, T., Meller, A., and Bar-Ziv, R. 1998. Optical gradient forces of strongly localized fields. *Physical Review Letters*, 81: 1738–1741.

Xin-Cheng, Y., Zhao-Lin, L., Hong-Lian, G. et al. 2001. Effects of spherical aberration on optical trapping forces for Rayleigh particles. *Chinese Physics Letters* 18: 432–434.

Xu, H., Bjerneld, E.J., Kall, M., and Borjesson, L. 1999. Spectroscopy of single hemoglobin molecules by surface enhanced Raman scattering. *Physical Review Letters* 83: 4357–4360.

Yoshikawa, K. 2001. Controlling the higher-order structure of giant DNA molecules. *Advanced Drug Delivery Reviews* 52(3): 235–244.

16

Optical Tweezers Microscopy: Piconewton Forces in Cell and Molecular Biology

Francesco Difato
*Italian Institute
of Technology*

Enrico Ferrari
National Research Council

Rajesh Shahapure
*International School
for Advanced Studies*

Vincent Torre
*International School
for Advanced Studies*

Dan Cojoc
National Research Council

16.1 Introduction 16-1
16.2 Optical Trapping Principle and Setups......................................16-3
16.3 Optical Trap Calibration...16-6
16.4 Optical Tweezers versus Fluorescence Microscopy.................16-8
16.5 Optical Tweezers in Biology...16-10
References 16-14

16.1 Introduction

Optical microscopes, from simple lens optical systems to advanced fluorescence instruments, play a fundamental role in medicine and biology as they permit the observation of living systems in their native environment with a low level of structural and functional perturbation.[1]

The spatial resolution in normal microscopes is diffraction limited to about 200 nm laterally and 500 nm axially, but the size of the features of interest in cell biology (organelles, molecules, macromolecules) is much less.[2] Therefore, the enhancement of the spatial resolution represents a very important research issue for scientists worldwide. The advent, in the last 15 years, of viable physical concepts for overcoming the limiting role of diffraction set off a quest that has led to readily applicable and widely accessible fluorescence microscopes with a nanoscale spatial resolution.[3,4]

On the other hand, all living cells face mechanical forces that are converted into biochemical signals and integrated into the cellular responses (mechanotransduction). Therefore, the development of new techniques for exerting mechanical stresses on cells and for observing their responses is crucial to clarify the molecule- and cell-level structures that may participate in mechanotransduction.[5] Forces can be exerted on a cell by a variety of experimental techniques. If the force is in the right range of magnitude, it is capable of eliciting a biological response from the cell. For instance, in the case of a fluid shear, the critical level of stress for a variety of biological responses has been observed to be of about 1 Pa. Integrated over the entire apical surface of a vascular endothelial cell (about 1000 μm^2), this produces a total force of 1 nN. Other experiments with forces applied via tethered beads also exhibit a threshold value of about 1 nN. If the total applied force is balanced solely by the forces in focal adhesions that occupy only 1% of the basal surface area, then the stress on the focal adhesions amplifies 100-fold to 100 Pa. On the other hand, a stress of 1 Pa on a focal adhesion, which might activate a local biological response of the cell, requires an applied force of about 10 pN only. The level of force needed to produce a significant conformational change in the force-transmitting proteins can also be estimated. The force that causes the bond between two proteins to rupture establishes an upper bound. Several studies have measured the fibronectin/integrin bond strength,[6] producing estimates in the range of 30–100 pN. Forces as low as 3–5 pN have also been shown to be sufficient to unfold certain subdomains in fibronectin.[7] If an external force is capable of producing a significant change in intracellular biochemical reaction rates, then the effect of force on protein conformation must exceed that associated with thermal fluctuations. Given that the thermal energy, kT, is ~4 pN.nm, and considering conformational changes with a characteristic length scale of 1–10 nm, the corresponding force levels would fall in the range of 0.4–4 pN.[5] This happens to coincide with the magnitude of force that can be produced by a single myosin molecule,[8] consistent with the theory that active cellular contraction can induce cell signaling. Therefore, all this would suggest that the critical values of force exerted on a single molecule fall within the pN range.

Interestingly, this is also the range of forces exerted on the particle in an optical trap by the radiation pressure of light. This relatively young technique[9] provides the non-mechanical manipulation of biological particles such as viruses, living cells, and subcellular organelles.

Optical tweezers are now being used in the investigation of an increasing number of biochemical and biophysical processes, from the mechanical properties of biological polymers to the multitude of molecular machines that drive the internal dynamics of the cell.[10]

Optical tweezers enable the control of the spatial organization of samples to perform sorting and/or to induce specific interactions between sample particles at an arbitrary location and time.[11]

An optical trap can be calibrated to perform force spectroscopy measurements. The optical tweezers system in this configuration is sometimes called the photonic force microscope (PFM).[12]

The acronym, PFM, has been introduced in analogy with the atomic force microscope. The probe, i.e., the cantilever tip for the AFM, is replaced by the trapped bead in the PFM. From the point of view of force measurement, the main difference between the AFM and the PFM is the stiffness of the probe. The stiffness for an optically trapped probe is usually much lower (one to two order of magnitudes) than the mechanical cantilever probe of the AFM. This makes the PFM complementary to the AFM for force measurements in cell biology.

Optical tweezers microscopy is compatible with fluorescence-imaging techniques. Therefore, it is possible to apply localized mechanical and chemical stimuli on cells while following changes in cell shape and organization.[13] The combination of chemical and mechanical stimulation is useful and relevant in cell biology, since cells test the extracellular matrix rigidity during their differentiation[14] and they take up a polarized organization in a culture dish, which influences their structure and function (in fact tissue-specific architecture and cell–cell communication are lost on the 2D arrangement of the culture dish[15]). Many laboratories are now studying different types of 3D scaffolds to grow cell cultures in a three-dimensional tissue-like fashion.[16]

In this context, a better understanding of the mechanisms by which cells compute mechanical transductions during differentiation in tissue development is necessary, i.e., how contact inhibition regulates cell proliferation and how mechanical tensions regulate single cell and global tissue shape.[17]

In the late nineteenth century, Julius Wolff proposed the idea that bone is deposited and modeled in response to mechanical stress. Intrinsic mechanical properties of the cell microenvironment influence cell function both *in vitro* and *in vivo*; therefore, Wolff's law may be extended to the development of any kind of tissue throughout the body.[14] In Ingberg's tensegrity model, the cellular organization is explained as a scaffold of tensed and compressed cables, i.e., the cytoskeleton, which defines the cell compartmentalization. Structural changes in the equilibrium of such cables influence the cell behavior.[18] It is known, for instance, that cells expand where the extra-cellular matrix (ECM) is stiffer, and to test the stiffness of the external environment, a cell performs contraction on its binding sites,[19] which are clustered at special protruding structures formed by a cytoskeleton components arrangement.[20]

The high sensitivity of the PFM has permitted the measurement of forces in the piconewton range, which is relevant in cell biology for the quantification of mechanical properties of cell membranes, DNA molecules, and filamentous proteins.[21] Force measurements represent an additional information in the multidimensional dataset, which can be obtained using an optical microscope, and, moreover, a new point of view in biology studies indicates that it can be applied to understand how physical forces within the cell interact to form a stable architecture,[22] how cells resist physical stresses, and how changes in the cytoskeleton initiate biochemical reactions. In addition, the PFM can help in explaining the mechanical rules that cause molecules to assemble.

An interesting example is the study of the shell of a virus, which consists of several subunits of proteins, to understand how it can self-assemble starting from an apparent chaotic sequence of collisions.[23] Another useful technique implemented in optical microscopy is the localized laser-based dynamic light scattering (DLS). DLS has been used to monitor changes in the Brownian motion of virus-like particles (VLPs) in solution,[24] giving information related to particle size and diffusion properties.[23,25] Viruses represent a simple or primitive self-replicating organism since they have genes and evolve by natural selection. Consequently, they are exploited as the simplest biological model to study essential characteristic of living organisms[26] and the understanding of their self-assembling represents a challenging task for biophysicists. Moreover, altered mechanical characteristics of tissues or a single cell compartment may either correlate with or play an important role in the onset of pathology.[27] For example, changes in the compliance of blood vessels are associated with atherosclerosis, and changes of the mechanical properties of malignant cells[28] can be characterized by a change of the Brownian motion fluctuation in an optical trap.[29]

The use of the PFM and DLS to probe cells for quantitative biophysical parameters can represent an alternative descriptive marker for the onset and evolution of pathology.

Furthermore, optical tweezers open the field for a single molecule manipulation and its biophysical characterization. With the appropriate chemistry, a bead can be attached to a single molecule as a handle allowing the application of forces on a single molecule.[30] In conclusion, optical tweezers microscopy can be used for a wide range of experiments, from a single molecule level to the cell level and helps to understand their organization up to the tissue level.

16.2 Optical Trapping Principle and Setups

The first demonstration that light radiation pressure induces forces on microparticles suspended in fluid was given by Ashkin in the early 1970s.[9] He observed that microparticles were confined to the laser optical axis and pushed in the direction of propagation. Using two counter propagating laser beams the first three-dimensional trapping of a particle was demonstrated. Later, in 1986, Ashkin and his colleagues demonstrated the single-beam gradient force optical trap using a single laser beam tightly focused by a high numerical aperture (NA) objective,[31,32] which is at the base of most of the optical trapping setups of today. Interestingly, the optical tweezers technique reported in this paper was thought of as proof of the concept for atom trapping,[33] that was extensively investigated in that period at Bell laboratories and demonstrated in the same year.

Light carries both linear and angular momentum and can thus exert forces and torques on matter. Optical tweezers exploit this fundamental property to trap objects in a potential well formed by light. Optical traps involve the balance of two types of optical forces: scattering forces that push objects along the direction of propagation of the light, and gradient forces that pull objects along the spatial gradient of the light intensity.[32]

When gradient optical forces exceed those from the scattering, an object is attracted to the point of the highest intensity formed by the focused light and can be stably trapped at this position in all three dimensions. Actually, since more particles usually interact in the laser beam, a third force, the binding force, which represents the self-consistent interaction between the multiple particles and the incident wave, is present.[32]

The trapping force, F, is proportional to the power of the laser, W, and the refractive index of the fluid, n_m:

$$F = Q \frac{n_m W}{c} \tag{16.1}$$

where
  c is the velocity of light
  Q is a dimensionless coefficient expressing the efficiency of the trap and depending on a series of factors as the material and the shape of the particle[32]

In order to get a feeling of the level of the forces induced by the laser radiation pressure, let us consider the following simplified example. Given a spherical particle (bead) of diameter $d = 1$ μm, which totally reflects the incident laser beam of power $W = 1$ mW, we have to estimate the force, F_p, exerted on the particle and the acceleration, a, of the particle.[34] Considering the linear momentum of a photon, p, the number, N, of photons/second carried in the laser beam and the variation of the momentum by total reflection, the force, F_p, is:

$$F_p = \frac{dP}{dt} = N\,2p = 2W/c = 10 \text{ pN} \tag{16.2}$$

where
  P is the momentum of the laser beam
  c the velocity of light

This force is very small, but since the particle is also small (its mass, $m \sim 10^{-12}$ g), the acceleration, $a = 1000\,g$ (where g is the gravitational acceleration) is very big. Even if we consider losses in the reflection, and a force of only $F_p = 1$ pN, the acceleration of the particle $a = 100\,g$ is still big.

A detailed investigation of the trapping mechanism and the forces can be approached considering the size, d, of the trapped particle with respect to the wavelength, λ, of the trapping laser. There are three cases:

  1. $d \gg \lambda$
  2. $d \ll \lambda$
  3. $d \sim \lambda$

The ray optics approach can be applied for the first case, assuming the laser beam is described by rays that refract and reflect at the particle–fluid interface and considering the changes of the linear momentum associated to these laser rays.[34]

If $d \ll \lambda$ (case 2), the particle in the laser focus can be thought of as a dipole in the electromagnetic field and the radiation forces can be derived considering the Rayleigh scattering regime.[35]

When the size of the particle is about the laser wavelength (case 3), the above mentioned approaches with their approximations cannot be applied, and a rigorous theory of the electromagnetic field governed by Maxwell's equations should be considered. The force of the electromagnetic field (e-m filed) when impinging on the particle can be computed either via a direct application of the Lorenz force and the bound/free current/charges within the volume of changes or via the Maxwell stress tensor. The second approach has the advantage of computation efficiency since the e-m fields need to be evaluated only on a surface enclosing the object, while the first method requires the evaluation of the fields within the whole volume. However, the disadvantage of the second method is that the polarizability of the object within its volume is not computed.[36] Different techniques to compute the forces in these regime are reported.[37–42]

From the point of view of the optical setup, a single trap optical tweezers system is a relatively simple optical architecture to adapt to an optical microscope. The trapping laser beam is collimated and expanded to slightly overfill the pupil aperture of the microscope objective. The laser beam is coupled to the optical axis of the microscope through an appropriate dichroic mirror.[43] A high NA objective (NA > 1) is required to obtain a high gradient of the intensity in the focal plane. Oil immersion objectives with NA as high as 1.4 and even higher for TIRF objectives can be used but the drawback consists in the presence of important spherical aberrations. Therefore, water immersion objectives with NA up to 1.2 are preferred. To change the position of the trapped object, two mirrors can be used to deflect the laser beam. However, this implies a reduced efficiency of the trap when the trapping position is moved off axis. An alternative is to move the sample cell and keep the particle fixed in the focal point. This allows a change in the relative position of the particle to the surrounding environment with no change in the optical path of the beam and, hence, keeps the optical properties of the trap unaffected. Scanning the beam in the sample is more cumbersome—the mirrors and the pupil aperture of the objective need to be optically conjugated by a telescope. Such a system has the drawback that the trap position can be controlled only in the x-, y-directions. To move the trap in the z-direction the stage should be moved. Another approach is to change the distance between the lenses of the beam expander, (the collimation of the beam at the entrance of the microscope objective is varied) and consequently, a change in the equilibrium between the scattering and the gradient force produces a shift of the relative position of the trapped object with respect to the objective focus.[44]

To increase the number of trapping spots in the sample more than one laser can be used. Alternatively, a single beam can be split by a polarizing beam splitter cube and the second beam can be slightly shifted with respect to the first.[45] Such a solution allows a maximum of two traps for each laser beam. Many laboratories in the last decade have been improving their optical setups to obtain multitrapping in the sample and to increase the manipulation capabilities. Two kinds of approaches have been practically used. One is based on acousto-optical devices[46] that are able to fast steer the laser beam in different positions which time-share the light power, so that trapped objects do not feel the scanning of the laser. Another lies on diffractive optical elements that reshape the Gaussian beam of the laser in an arbitrarily shaped intensity profile to obtain the desired trapping configuration on the sample.[47] The first approach permits a change in the number and the position of spots at high speed, but trapping spots are created only in a planar arrangement. The second method requires higher computational skills, but allows a three-dimensional array of trapping spots, or more complex shapes of trapping spots[48] to be obtained: e.g., Bessel beams are "non-diffracting" and therefore they are propagation invariant beams when they propagate trough opaque obstacle.[49] Laguerre–Gauss beams can transfer their angular orbital momentum onto trapped particles or they can trap low refractive index probes as bubbles.[50] Anisotropic beam shapes create a potential which influence molecular dipole diffusion and distribution and they could influence cell growth.[51] Recently, this technique based on diffractive optical elements has been transferred to microfabricated solid supports, i.e., presenting high NA Fresnel lenses built on the coverslip, to trap particles without even the need of a microscope objective.[52]

Once the optical tweezers are built, they can be calibrated, as discussed in the next section, to obtain the stiffness of the optical trap and to perform force measurements. To calibrate the optical tweezers system it is necessary to measure with high accuracy the displacement of the trapped probe from the equilibrium position in the trap. One method is to use video tracking. Optically resolved probes can be tracked by simple brightfield imaging applying centroid algorithms, while subresolved probes can be tracked if they emit fluorescent light.[53] In such a setup, no additional hardware is required and particle position measurements can be obtained with accuracy of the order of 10 nm and a bandwidth of few kilohertz. Video tracking is a simple method to calibrate the stiffness of a single- or multi-trap configuration. Furthermore, when the detection path of the system is modified to project a video hologram (and not an image) to the CCD, tracking of the probe reaches a nanometer resolution and its refractive index can be measured with high accuracy.[54]

Another popular method to track the probe position is based on light interferometry measurements.[55] This configuration measures the interference between the light scattered by the trapped object and the unscattered light passing through the focus. The interference pattern is collected at the back focal plane of the condenser and projected to a position sensitive detector.[56] This detector can be either a position-sensitive detector (PSD) that is able to track arbitrary shape objects in the x-, y-direction, or a quadrant photodiode (QPD) that is able to track a symmetric probe in three directions (x, y, z). In these kinds of measurements, both types of detectors permit a combination of a sub-nanometer resolution with a temporal bandwidth of hundreds of kHz. Another important issue in the setup is the noise. The PFM is sensible to any kind of environmental noise: thermal, acoustic, or convective. Therefore, the sensitivity of the system can be improved by closing up the setup into an acoustic chamber, which also reduces the convective noise, and controls the environmental temperature. Furthermore, feedback control on the output intensity of the laser or feedback control in the stage position to reduce the mechanical drift of the sample, permit to reach a subnanometer displacement resolution and a femtonewton force resolution.[57,58]

16.3 Optical Trap Calibration

The optical manipulation setup started to be used as force transducers after 1989, when Block et al.,[12] made the first calibrated measurement of the compliance of bacterial flagella using the tweezers to force bacteria to rotate (they were tethered to a microscope cover-glass by their flagellum).

To calibrate the optical force of the system it is necessary to take into account all external forces acting on the trapped probe. Starting from the Ornstein–Uhlenbeck[59] equation to describe the trajectory of an object:

$$m\hat{x}(t) = -\gamma\dot{x}(t) + F_{\text{Brownian}} \tag{16.3}$$

where
  m is the inertial mass of the particle
  while the Stokes friction $-\gamma\dot{x}(t)$ and the random Brownian force represent the surrounding medium
    influence on the object trajectory

The Brownian forces acting on the trapped probe can be modeled as: $\sqrt{2D}\,\gamma h(t)$, where D is the diffusion coefficient, γ is the friction coefficient, and $h(t)$ is an independent white Gaussian random process.

In an optical trap, the electric field of the laser radiation produces forces on charged particles. Consequently, the motion of the induced dipole, arising from the electronic polarizability of the trapped particle, is also affected by the potential well of the laser radiation. From the distribution histogram of the particle position in the trapping volume it can be shown that the potential is harmonic. When the potential is harmonic, such a distribution should be Gaussian.[7] This permits the optical forces to be modeled as a spring: $F_{opt} = -kx$, where k is the optical stiffness and x is the probe displacement from the equilibrium position.[34] Thus Equation (16.3) becomes

$$m\hat{x}(t) = -\gamma\dot{x}(t) + F_{Brownian} + F_{opt} \qquad (16.4)$$

In this equation, the inertial term can be neglected since the low Reynolds number conditions apply.

Once the equation of motion of a trapped object is modeled, the optical stiffness of the system can be obtained by measuring the probe displacements in the trap. Different methods have been implemented to calibrate the optical tweezers and they are based on some *a priori* information. The tolerance on such parameters produce the accuracy of the calibration method applied.

The drag force method can be applied by video or laser interferometric tracking. The drag coefficient, including surface proximity corrections, should be known *a priori*. In this method, a force is generated on the trapped probe by a fluid flow, and the trapped bead response is measured as the distance between the new equilibrium position and the trap center. Although the lateral displacement of the object due to the drag force is relatively easy to measure by video tracking, axial displacements represent a more difficult task. To measure the probe position in the z-direction by video tracking, the bead image can be calibrated[60] by its fluorescence intensity if acquired in a confocal setup, or by the diameter of the airy disk pattern in a widefield microscope.[53]

Other methods rely on the Brownian motion detection of the trapped probe and a simple video tracking does not have the required spatial and temporal resolution. Therefore, interferometric measurements are applied.[56]

In the equipartition method, the kinetic energy of the trapped object is assumed to be equal to the thermal energy:

$$\frac{1}{2}k_B T = \frac{1}{2}k\left\langle x^2 \right\rangle \qquad (16.5)$$

where
   $k_B T$ is the thermal energy
   k the optical stiffness

In such an equation, the temperature T is the *a priori* parameter. In addition, a separate calibration of the detector transduction from volts to nanometers has to be performed separately.

In the power spectrum method such calibration of the detector is not required and Boltzmann statistics is directly applied to the power spectrum of the position fluctuations of the probe, measured in volts by the PSD or the QPD. In such a method, the fit with a Lorentzian function is applied to the power spectrum (measured in volt²/Hz) to obtain the optical stiffness and the sensitivity of the detector (in volt/nm). The temperature and the viscosity of the surrounding medium have to be known *a priori*.

The power spectrum method is most often applied because it permits a fast calibration of the system. PSDs and QPDs permit the measurement of position fluctuation with a temporal bandwidth of hundreds of kHz, and therefore, few seconds of recording are enough to obtain sufficient data points. This means that optical tweezers can be calibrated easily at the environmental condition in the sample.

In fact, it is well known that the optical properties of the sample affect the optical characteristic of the setup, as its point spread function (PSF), and therefore its trapping efficiency.

The drawback of the method is that it relies on two parameters that have to be known *a priori* and therefore they decrease the accuracy.

Interferometric recordings make it possible to resolve the Brownian motion and to observe the hydrodynamic effect on medium viscosity.[59] Therefore, they were taken into account in data analysis to obtain a better estimate of the optical stiffness of the system. A mathematical modeling of the optical system may also improve the performances. Hence, the low-pass filtering effect of the detector,[61] the shape of the trapping potential at the focus spot, and the condenser NA influence on the instrument sensitivity were modeled[62,63] and the local-shift variance sensitivity of the QPD was measured.[64] Such studies represented an important contribution to obtain system calibration and to achieve high sensitivity.[65]

Therefore, the accuracy limit in the calibration of the system is due to the *a priori* parameters that need to be included in the data analysis. This has been overcome by a new calibration strategy where two methods are combined to obtain the experimental parameters of the probe without the need of *a priori* information. The method relies on the power spectral measurements of the thermal motion of the probe during a sinusoidal motion of a translation stage. This method allows the extrapolation of the parameters that are used as *a priori* information in the procedures described above, with accuracy within 3%.[57] This calibration inaccuracy enables to resolve surface forces which can be important issue in case of biological interfaces as membranes. Furthermore, the method has been used to calibrate the detector sensitivity for particles tracking inside the cell,[66,67] which could represent the first step toward the challenging field of intracellular force measurements.

16.4 Optical Tweezers versus Fluorescence Microscopy

Lukosz's principle[68] states that resolution can be increased at the expense of the field of view.

Fluorescence microscopy makes it possible to observe only the compartment marked in the sample, and therefore, provides a better contrast with respect to the overall specimen. In addition, technological improvements in optical microscopy further reduce the field of view, e.g., in confocal architecture, that confers a higher spatial resolution in the two-dimensional field of view and allows access to the third dimension of the sample. Much confocal architecture has been implemented[69]: in the one photon confocal microscope, the field of view is decreased in the detection path, while in the two-photon microscope such a confinement is obtained in the excitation process. Both types of confocal microscopes make it possible to observe a small volume (the detection volume is in the order of femtoliters) and to supply the possibility of observing single fluorescent molecules.[70] Many efforts have been undertaken to reduce the spatial dimensions of the PSF of the confocal systems to overcome the diffraction limit and to develop the nano-scope[3,71] for the application of molecular imaging to living systems.

The relatively limited spatial resolution of optical microscopy is counter balanced by the multidimensionality of the data acquired. For example, the time lapse imaging of a 3D volume allows the exploitation of a fourth dimension. Moreover, the wavelength, λ, enables multilabeling and a discrimination of different entities in the sample. Spectral properties of fluorochromes such as the lifetime of the excited state, give information on the environmental condition of the molecule.[72] Changes in the intracellular environment, related to the cell metabolism, can influence the structure of the fluorescent molecules and their optical properties.[73] Therefore, cell physiology can be studied at the molecular level. This is the so-called lifetime dimension. Furthermore, the Second Harmonic Generation, occurring when an intense laser beam passes through a polarizable material with non-centrosymmetric molecular organization, gives another possible dimension for optical microscopy measurements.[74]

The photonic force microscope, adds one more controlled or measurable[75] variable to optical microscopy: the force. Therefore, optical tweezers represent a further resolution improvement for the optical microscope: force measurements permit the polymerization steps of single filaments to be followed where the added monomer raises the length of the polymer of some nanometers,[76] or, the Brownian motion analysis can be applied to resolve 3D structures with resolution on the order of tens of nanometers.[77]

Laser tweezers are highly compatible with optical microscopy techniques because they exploit the same optical path.[43] However, it induces photodamage that arises from the exposure of the trapped sample to a high intensity of light.[78] To overcome this problem, it is possible to decrease the intensity reaching the sample, if a better trapping efficiency of the system is reached. For example, by adaptive optics methods it is possible to recover the nonideal behavior of the objective lens to obtain unblurred focus spots and a better spatial confinement of photons.[79]

Such methods to improve the trapping efficiency are similar to the deconvolution processing of microscopy images to improve the experimental spatial resolution of microscopes.[80,81] In such techniques, it is necessary to characterize the optical system by its optical point-spread function[2] (OPSF). In this respect, theoretical modeling can characterize the image formation process of the microscope, but only experimental measurements of the PSF can quantify the limitations of the real system.[82] Indeed, the experimental OPSF presents shape asymmetry due to spherical aberrations introduced by optical elements, while the theoretical OPSF is symmetric and accounts only for the resolution limits of an ideal imaging system.[83] The disadvantage of the experimental OPSF is that it could be corrupted by noise, otherwise, deconvolution with the theoretical OPSF offers only a qualitative enhancement of the image, because the introduced artifacts cannot be quantified.

In photonic force microscopy the Brownian motion of the trapped probe follows the shape of the potential well created by the focused laser which represents the force point-spread function (FPSF) of the optical tweezers (Figure 16.1). If the trapped probe is tracked for enough time it is possible to measure the trapping volume with a nanometer resolution and therefore define the spherical aberration of the system with the same accuracy.[84,85] It has already been proved that spherical aberrations produce an enlargement of the OPSF[86,87]; and that the recovery of such spherical aberrations by adaptive optics produces a smaller trapping volume, which represents less Brownian noise in force spectroscopy measurements and a better trapping efficiency.[79] Techniques such as photo-activated localization microscopy (PALM) and stochastic optical reconstruction microscopy (STORM) apply a fit between the PSF of the system and the point-like features of the image to obtain the center of the mass of the labeled molecule at a 20 nm resolution.[71]

In this context, the trapping efficiency is used as a quantitative parameter to estimate the optical spherical aberrations recovering by hardware deconvolution through the adaptive optics method.[88] This recovering could be also quantified by an experimental fluorescent PSF acquisition, which can then be used in a software deconvolution algorithm.[89]

Therefore, the correlation of the force and the fluorescence intensity data could improve the calculation of an experimental OPSF; while a combination of hardware and software deconvolution can be applied to recover both types of errors in an optical system: degradation due to the process of image formation usually denoted as blurring, and degradation introduced by the recording process usually

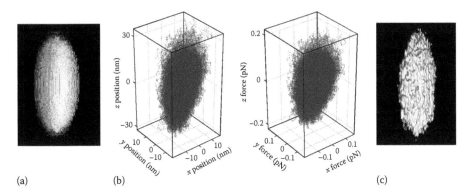

(a)   (b)   (c)

FIGURE 16.1 (See color insert following page 15-30.) OPSF and FPSF. (a) Three-dimensional rendering of an OPSF of a confocal microscope (objective 100× NA 1.4 oil immersion) measured with 64 nm diameter latex beads immersed in oil. Field of view 0.8 × 1 μm. (b) Three-dimensional scatter plots of 1 μm diameter silica bead, trapped with a CW infrared laser (1064 nm wavelength, objective 100× NA 1.4 oil immersion) in water. Scatter plots are computed from the *x*, *y*, *z* trace, in nm and pN, low pass filtered at 20 kHz and acquired at 10 kHz. (c) Three-dimensional rendering of a FPSF representing the volume observed from the trapped object. Field of view 50 × 70 μm.

denoted as noise.[89] In turn, this can be exploited to obtain a higher single-molecule localization accuracy of a STORM microscope.[90]

Another way to decrease phototoxicity is through the use of a laser source in the wavelength's window for which the absorption of the biological matter is low. The typical wavelength used for biological applications is in the infrared range for which it has been shown by means of viability tests, both on bacteria and eukaryotic cells, that the damage versus exposure time is reduced.[78,91] Since the best wavelength range for trapping is the same as that used in two-photon excitation, some groups started to use a femto-second pulsed laser source to obtain trapping, ablation, and fluorescence excitation by using only one beam.[92,93]

In conclusion, by the optical tweezers technique, it is possible to trap particles starting from a single molecule, to living cells and to confine them to a subdiffraction limited volume that can be observed with an extended area detector.[94] Therefore, in principle, it is possible with optical tweezers to reduce the field of view of the system without the need of a pinhole in front of the detector, or an infrared femto-second pulsed laser to confine the excitation to a 3D volume. For example, such a possibility can be exploited in fluorescence techniques as fluorescence correlation spectroscopy,[95] or for single-molecule studies: Bustamante was the first scientist who used such a technique to manipulate single molecules of DNA.[96] After his seminal paper, many works followed on the subject of single molecule biophysical characterization[21] and the application of single molecule manipulation in combination with fluorescence techniques.[97,98]

16.5 Optical Tweezers in Biology

Cells need to organize their internal space and to interact with their environment.

The network filaments, i.e., the cytoskeleton, represent connections to distribute tensile forces through the cytoplasm to the nucleus as a signaling pathway.[99] Transmembrane receptor, as integrins, function as mechanoreceptors and provide a preferred path to transfer mechanical forces across the cell surface.[100] During motion, cells organize integrins distribution in clusters at special complexes called focal adhesions.[20] Forces transmitted to the cytoskeleton from these sites produce a stress in associated cytoskeleton molecules which can be converted into gene expression, protein recruitment at local sites, and dynamic cytoskeleton architecture modulation.[101,102] Thus cell shape and stability are the effects of a mechanical force balance between the cytoskeleton traction forces and the extracellular substrate local rigidity.[19,99]

A regulation of the assembly processes dynamically modulates the mechanical properties of the cytoskeleton network, locally and globally to balance the compression and the tension force, to extend cell protrusions such as lamellipodia and filopodia, or to propel intracellular vesicles or pathogens through the cytoplasm.[103]

For cells moving through tissues, the resistive force comes from the surrounding extracellular matrix. For cells *in vitro*, the major force against the extension arises from a tension in the plasma membrane which modulates the shape of the extruding processes.[104,105]

In force spectroscopy measurements, the membrane elasticity is quantified by the tether extraction of the membrane-embedded beads[106] and piconewton tensions are applied to the phospholipids bi-layer to stimulate receptor membrane trafficking.[107,108]

An important factor in membrane resistance is the number of membrane–cytoskeleton adhesion components. It has been shown that the force needed to extract a membrane tether can be modified by alterations of the cytoskeleton.[107,109] It was also demonstrated that mechanical stimuli, which induce membrane tension changes, can regulate membrane traffic through exocytosis and endocytosis.[108] These results are in agreement with the observations of the normal behavior of the cell; for example, during endocytosis, clathrins displace the membrane-associated cytoskeleton to decrease membrane tension and to cause membrane engulfment.[105]

In the case of membrane protrusion, the displacement of the normal cytoskeletal components is required to create a gap for the insertion of new monomers at the filament tips.[110]

Apparent membrane tension, which is in the order of tens of pN, have been measured from the force exerted on membrane tethers.[106] An analysis of the tether-formation processes is important because it allows the calculation of the bending rigidity of membranes and the apparent surface tension, and therefore to obtain a theoretical model of the force produced or the energy dissipated during motion. Furthermore, optical tweezers can also be used to track membrane proteins without the need of a fluorescent marker at the unraveled resolution.[111] For example, receptor tracking permitted a distinction between membrane fluidity in living and dead cells.[112]

The membrane deforms passively in response to a cytoskeleton reorganization.

Since actin filaments are rather fragile, mechanical properties of a single actin fiber are difficult to measure. Microtubules are much stiffer, and force generation by this polymer has been measured by observing how a single microtubule buckles when polymerizing against a wall. This permitted the polymerization steps of single filaments to be followed and demonstrated that a microtubule growing at the "plus end" develops a force of a few piconewtons and that polymerization velocity decreases exponentially as a function of the load force.[45] In addition, microtubules are long filamentous protein structures that alternate between periods of elongation and shortening in a process termed dynamic instability.[76] The average time a microtubule spends in an elongation phase, known as the catastrophe time, is regulated from the biochemical machinery of the cell and a comparison of catastrophe times for microtubules growing freely or in front of an obstacle leads to the conclusion that force reduces the catastrophe time by limiting the rate of the tubulin addition.

Although the actin filament is less stiff than microtubules, it represents an essential component in cellular motility: it provides protrusive forces to extend cell protrusions such as lamellipodia and filopodia, or propels intracellular vesicles or pathogens through the cytoplasm.[113] The forces generated by the elongation of a few parallel-growing actin filaments against a rigid barrier, mimicking the geometry of filopodial protrusion, was measured,[114,115] and it was shown that a growth of approximately eight actin parallel-growing filaments can be stalled by relatively small applied load forces on the order of 1 pN. These results suggest that force generation by small actin bundles is limited, and therefore, living cells must use actin-associated factors to generate a substantial force to overcome membrane resistance[116] and to produce stable protrusions at the leading edge of the cell during locomotion.[117]

Another important role that the cytoskeleton plays in cell physiology is the intracellular transport of various cargos. In intracellular transport two aspects have to be analyzed: one is how cargos use the network and the second is how the cell changes the structure of the networks to assist cargo motion. Pathogens have been used as simplified model systems to study eukaryotic cell motility. *In vitro* monitoring of their movement in cell extracts saw the conversion of a complex cell biology problem into a biochemically affordable one, and opened the way to design a minimal motility medium to study the biochemical mechanism of an actin dynamics control.[118]

One of the most important recent discoveries about the actin propulsion is the proof that the actin tail is attached to the surface of the pathogens.[119] Noireaux et al.[120] used an optical trap to measure the force required to separate the bacterial cell from the actin tail, which turned out to be greater than 10 pN. Finally, Cameron et al.[121] used electron microscopy to observe that the actin filaments of the branching network are transiently attached to the surface of the bead. This leads to a more widely accepted model: the elastic tethered ratchet model.[122] in which attached fibers are in tension and resist the forward motion of the bacterium/bead, while dissociated fibers are in compression, and generate the force of propulsion.

Recently, atomic force microscopy (AFM) measurements have succeeded in stalling actin network growth and have revealed that two or more stable growth velocities can exist at a single load value. These results demonstrate that a single force–velocity relationship does not completely describe the behavior of the system which depends on the loading history and not only on the instantaneous load.[123] Another experiment reports the effort to stop the motion of *Listeria* with an optical trap.[114,124] It has been shown that the trap could temporarily stop its motion, and then the bacterium could escape due to an increase in the force supplied by the tail, like in an autocatalytic model.[125] An autocatalytic model is also sustained by measurements with fluorescent actin, where it has been shown that the fluorescence intensity

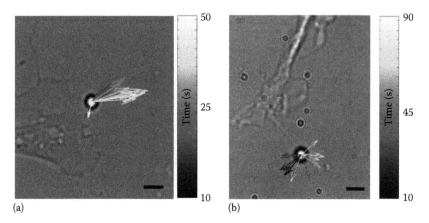

(a)   (b)

FIGURE 16.2 (See color insert following page 15-30.)   Force measurement and cell motility. Vectorial representation of the force exerted by the cell on obstacle (1 μm diameter silica bead). Different colors of the vector represent the different times at which the cell exerts the force on the trapped probe. In (a) lamellipodium increases the force exerted on the obstacle and tries to avoid it, as suggested from the time evolution of the length and the direction of the force. In (b) the random succession of the force vector direction in time represents the exploratory motion of a filopodium which senses the environment. Bars 2 μm.

in the actin tail increases during the stationary period of the bacteria motion, suggesting that the actin density is building up to overcome the opposing forces.[126]

Optical tweezers can be also utilized to measure and to analyze the force operated by cells[127] in an exploratory motion and therefore to understand how cells operate in mechanical transduction. With optical microscopy, it is possible to follow the motion of the cell, but force measurements permit to quantify how many molecular motors are involved in a movement, and therefore, to also quantify how much of the molecular machinery of the cell is recruited in such an action. This means that thanks to force measurement, it is possible to distinguish the random motion from the well-organized shift of the cell (Figure 16.2).[128]

It has been observed that filopodia, which explore the environment by rapidly moving in all directions, modulate their activity by changing the duration of the collision with a bead in response to a different stiffness of the load. This could represent a sensing of the obstacle force and is in agreement with measurements on a single microtubule catastrophic time modulation.[76]

Lamellipodia, which follow the pathway analyzed by filopodia, showed a more complex behavior in response to the obstacle: sometimes, they entirely retracted, at other times they moved around it to progress forward or they removed the obstacle by lifting it and giving it back (Figure 16.3). Lamellipodia have a more differentiated structure and are thought to exert a force with variable directions in space. Therefore, multi-tweezers measurements have been applied[129] to understand their overall organization and to measure complex forces exerted during the growth cone motion.

These kinds of measurements can be used to understand the response of the cells to molecular cues or mechanic-chemical stimuli[130] localized in the trapping volume, or to understand the mechanisms and to identify which molecules are involved in the different steps of motor planning.

Single biological particles, usually used as simple biological models, have also been trapped: Tobacco mosaic virus, *Escherichia coli* bacteria, and manipulation of particles within the cytoplasm of cells.[131] Optical tweezers also became a well-established technique for single-molecule studies: Bustamante et al. were able to follow the DNA encapsidation steps of the bacteriophage and to measure the force delivered by the capsid molecular motors.[132] A study of single molecular motors is one of the most attractive applications of optical tweezers: the forces generated by single motor molecules such as kinesin and myosin has been quantified[133,134] or steps of RNA polymerase along DNA filaments have been measured.[135]

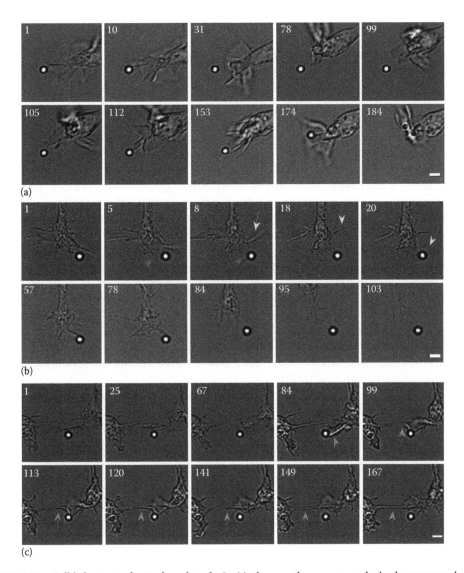

FIGURE 16.3 Cell behavior in front of an obstacle. In (a), the growth cone senses the load, retracts and grows again toward the obstacle to remove it. In (b), the growth cone senses the load, and completely retracts. White and grey arrows indicate the growing filopodia which retract and bend toward the obstacle. In (c), a bead is trapped between two connected growth cones. Lamellipodium of the growth cone on the right side is attracted from the load, but then avoids the obstacle. Grey arrows indicate how the connection between the two growth cones is reinforced when the lamellipodium of the growth cone on the right side stops to push the trapped probe and steers toward the connection between the two neurites. Bars 2 μm.

A complementary system to light manipulation tools that is fast evolving is laser dissection.[136] By laser scissors it is possible to cut living biological samples with a subcellular resolution: the cytoskeleton organization could be disturbed while leaving the cells unaffected in any other respect.[137] Optical tweezers and laser dissection tools are also used to control the location and the timing of optical uncaging in chemical stimulation experiments,[138] or for optoporating the cell membrane.[139,140] These opportunities encourage the use of optical tools in probing and quantifying biophysical cellular parameters, and open a new field of nanosurgery.[92,141]

References

1. Diaspro, A. et al. Multi-photon excitation microscopy. *Biomed. Eng. Online* **5**, 36 (2006).
2. Diaspro, A., Annunziata, S., Raimondo, M., and Robello, M. Three-dimensional optical behaviour of a confocal microscope with single illumination and detection pinhole through imaging of sub-resolution beads. *Microsc. Res. Tech.* **45**, 130–131 (1999).
3. Hell, S.W. Far-field optical nanoscopy. *Science* **316**, 1153–1158 (2007).
4. Hell, S.W. Microscopy and its focal switch. *Nat. Methods* **6**, 24–32 (2009).
5. Huang, H., Kamm, R.D., and Lee, R.T. Cell mechanics and mechanotransduction: Pathways, probes, and physiology. *Am. J. Physiol. Cell Physiol.* **287**, C1–C11 (2004).
6. Lahenkari, P.P. and Horton, M.A. Single integrin molecule adhesion forces in intact cells measured by atomic force microscopy. *Biochem. Biophys. Res. Commun.* **259**, 645–650 (1999).
7. Erikson, H.P. Reversible unfolding of fibronectin type III and immunoglobulin domains provides the structural basis for stretch and elasticity of titin and fibronectin. *Proc. Natl. Acad. Sci. USA* **91**, 10114–10118 (1994).
8. Finer, J.T., Simmons, R.M., and Spudich, J.A. Single myosin molecule mechanics: Piconewton forces and nanometre steps. *Nature* **368**, 113–119 (1994).
9. Ashkin, A. Acceleration and trapping of particles by radiation pressure. *Phys. Rev. Lett.* **24**, 156–159 (1970).
10. Moffitt, J.R., Chemla, Y.R., Smith, S.B., and Bustamante, C. Recent advances in optical tweezers. *Annu. Rev. Biochem.* **77**, 205–228 (2008).
11. Macdonald, M.P. et al. Creation and manipulation of three-dimensional optically trapped structures. *Science* **296**, 1101–1103 (2002).
12. Block, S.M., Blair, D.F., and Berg, H.C. Compliance of bacterial flagella measured with optical tweezers. *Nature* **338**, 514–518 (1989).
13. Neuman, K.C. and Block, S.M. Optical trapping. *Rev. Sci. Instrum.* **75**, 2787–2809 (2004).
14. Peyton, S.R., Ghajar, C.M., Khatiwala, C.B., and Putnam, A.J. The emergence of ECM mechanics and cytoskeletal tension as important regulators of cell function. *Cell Biochem. Biophys.* **47**, 300–320 (2007).
15. Pampaloni, F., Reynaud, E.G., and Stelzer, E.H. The third dimension bridges the gap between cell culture and live tissue. *Nat. Rev. Mol. Cell Biol.* **8**, 839–845 (2007).
16. Bakunts, K., Gillum, N., Karabekian, Z., and Sarvazyan, N. Formation of cardiac fibers in Matrigel matrix. *Biotechniques* **44**, 341–348 (2008).
17. Riveline, D. et al. Focal contacts as mechanosensors: Externally applied local mechanical force induces growth of focal contacts by an mDia1-dependent and ROCK-independent mechanism. *J. Cell Biol.* **153**, 1175–1185 (2001).
18. Ingber, D.E. and Tensegrity, I. Cell structure and hierarchical systems biology. *J. Cell Sci.* **116**, 1157–1153 (2003).
19. Giannone, G. et al. Lamellipodial actin mechanically links myosin activity with adhesion-site formation. *Cell* **128**, 561–575 (2007).
20. Galbraith, C.G., Yamada, K.M., and Galbraith, J.A. Polymerizing actin fibers position integrins primed to probe for adhesion sites. *Science* **315**, 992–995 (2007).
21. Bustamante, C., Bryant, Z., and Smith, S.B. Ten years of tension: Single-molecule DNA mechanics. *Nature* **421**, 423–427 (2003).
22. Coirault, C., Pourny, J.C., Lambert, F., and Lecarpentier, Y. Optical tweezers in biology and medicine. *Med. Sci. (Paris)* **19**, 364–367 (2003).
23. Santos, N.C. and Castanho, M.A. Teaching light scattering spectroscopy: The dimension and shape of tobacco mosaic virus. *Biophys. J.* **71**, 1641–1650 (1996).
24. Pattenden, L.K., Middelberg, A.P.J., Niebert, M., and Lipin, D.I. Towards the preparative and large-scale precision manufacture of virus-like particles. *Trends Biotechnol.* **23**, 523–529 (2005).

25. Meller, A. et al. Localized dynamic light scattering: A new approach to dynamic measurements in optical microscopy. *Biophys. J.* **74**, 1541–1548 (1998).

26. Flint, S.J., Enquist, L.W., Krug, R.M., Racaniello, V.R., and Skalka, A.M. *Principles of Virology. Molecular Biology, Pathogenesis, and Control*. ASM Press, Washington, DC, 804 pp. (2000).

27. Berrier, A.L. and Yamada, K.M. Cell-matrix adhesion. *J. Cell. Physiol.* **213**, 565–573 (2007).

28. Guck, J. et al. Optical deformability as an inherent cell marker for testing malignant transformation and metastatic competence. *Biophys. J.* **88**, 3689–3698 (2005).

29. De Luca, A.C. et al. Real-time actin-cytoskeleton depolymerization detection in a single cell using optical tweezers. *Opt. Express* **15**, 7922–7932 (2007).

30. Bustamante, C., Smith, S.B., Liphardt, J., and Smith, D. Single-molecule studies of DNA mechanics. *Curr. Opin. Struct. Biol.* **10**, 279–285 (2000).

31. Ashkin, A., Dziedzic, J.M., Bjorkholm, J.E., and Chu, S. Observation of a single-beam gradient force optical trap for dielectric particles. *Opt. Lett.* **11**, 288–290 (1986).

32. Burns, M.M., Fournier, J.M., and Golovchenko, J.A. Optical binding. *Phys. Rev. Lett.* **63**, 1233–1236 (1989).

33. Ashkin, A. *Optical Trapping and Manipulation of Neutral Particles Using Lasers: A Reprint Volume with Commentaries*. World Scientific Publishing Company, Hackensack, NJ, 940 pp. (2006).

34. Ashkin, A. Forces of a single-beam gradient laser trap on a dielectric sphere in the ray optics regime. *Methods Cell Biol.* **55**, 1–27 (1998).

35. Harada, Y. and Asakura, T. Radiation forces on a dielectric sphere in the Rayleigh scattering regime. *Opt. Commun.* **124**, 529–541 (1996).

36. MIT Center for Electromagnetic Theory and Applications, http://web.mit.edu/~ceta/obt/fund-forces.html (2009).

37. Mansuripur, M. Radiation pressure and the linear momentum of the electromagnetic field. *Opt. Express* **12**, 5375–5401 (2004).

38. Kemp, B.A., Grzegorczyk, T.M., and Kong, J.A. Ab initio study of the radiation pressure on dielectric and magnetic media. *Opt. Express* **13**, 9280–9291 (2005).

39. Gordon, J.P. Radiation forces and momenta in dielectric media. *Phys. Rev. A* **8**, 14–21 (1973).

40. Loudon, R. Theory of the radiation pressure on dielectric surfaces. *J. Mod. Opt.* **49**, 812–836 (2002).

41. Obukhov, Y.N. and Hehl, F.W. Electromagnetic energy-momentum and forces in matter. *Phys. Lett. A* **311**, 277–284 (2009).

42. Rohrbach, A. and Stelzer, E.H. Trapping forces, force constants, and potential depths for dielectric spheres in the presence of spherical aberrations. *Appl. Opt.* **41**, 2494–2507 (2002).

43. Lee, W.M., Reece, P.J., Marchington, R.F., Metzger, N.K., and Dholakia, K. Construction and calibration of an optical trap on a fluorescence optical microscope. *Nat. Protoc.* **2**, 3226–3238 (2007).

44. Kress, H., Stelzer, E.H., Griffiths, G., and Rohrbach, A. Control of relative radiation pressure in optical traps: Application to phagocytic membrane binding studies. *Phys. Rev. E. Stat. Nonlin. Soft. Matter Phys.* **71**, 061927 (2005).

45. Kikumoto, M., Kurachi, M., Tosa, V., and Tashiro, H. Flexural rigidity of individual microtubules measured by a buckling force with optical traps. *Biophys. J.* **90**, 1687–1696 (2006).

46. Visscher, K., Brakenhoff, G.J., and Krol, J.J. Micromanipulation by "multiple" optical traps created by a single fast scanning trap integrated with the bilateral confocal scanning laser microscope. *Cytometry* **14**, 105–114 (1993).

47. Dufresne, E.R. and Grier, D.G. Optical tweezer arrays and optical substrates created with diffractive optics. *Rev. Sci. Instrum.* **69**, 1974 (1998).

48. Cojoc, D. et al. Dynamic multiple optical trapping by means of diffractive optical elements. *Microelectron. Eng.* **73–74**, 927–932 (2009).

49. McGloin, D., Garces-Chavez, V., and Dholakia, K. Interfering Bessel beams for optical micromanipulation. *Opt. Lett.* **28**, 657–659 (2003).

50. Garbin, V. et al. Optical micro-manipulation using Laguerre–Gaussian beams. *Jpn. J. Appl. Phys.* **44**, 5773 (2005).

51. Carnegie, D.J., Stevenson, D.J., Mazilu, M., Gunn-Moore, F., and Dholakia, K. Guided neuronal growth using optical line traps. *Opt. Express* **16**, 10507–10517 (2008).

52. Schonbrun, E. and Crozier, K.B. Spring constant modulation in a zone plate tweezer using linear polarization. *Opt. Lett.* **33**, 2017–2019 (2008).

53. Speidel, M., Jonas, A., and Florin, E.L. Three-dimensional tracking of fluorescent nanoparticles with subnanometer precision by use of off-focus imaging. *Opt. Lett.* **28**, 69–71 (2003).

54. Lee, S.H. et al. Characterizing and tracking single colloidal particles with video holographic microscopy. *Opt. Express* **15**, 18275–18282 (2007).

55. Keen, S., Leach, J., Gibson, G., and Padgett, M.J. Comparison of a high-speed camera and a quadrant detector for measuring displacements in optical tweezers. *J. Opt. A: Pure Appl. Opt.* **9**, S264–S266 (2007).

56. Gittes, F. and Schmidt, C.F. Interference model for back-focal-plane displacement detection in optical tweezers. *Opt. Lett.* **23**, 7–9 (1998).

57. Schaffer, E., Norrelykke, S.F., and Howard, J. Surface forces and drag coefficients of microspheres near a plane surface measured with optical tweezers. *Langmuir* **23**, 3654–3665 (2007).

58. Carter, A.R. et al. Stabilization of an optical microscope to 0.1 nm in three dimensions. *Appl. Opt.* **46**, 421–427 (2007).

59. Lukic, B. et al. Motion of a colloidal particle in an optical trap. *Phys. Rev. E: Stat. Nonlin. Soft Matter Phys.* **76**, 011112 (2007).

60. Dreyer, J.K., Berg-Sorensen, K., and Oddershede, L. Improved axial position detection in optical tweezers measurements. *Appl. Opt.* **43**, 1991–1995 (2004).

61. Berg-Sorensen, K., Peterman, E.J.G., Weber, T., Schmidt, C.F., and Flyvbjerg, H. Power spectrum analysis for optical tweezers. II: Laser wavelength dependence of parasitic filtering, and how to achieve high bandwidth. *Rev. Sci. Instrum.* 77, 063106–063110 (2006).

62. Rohrbach, A. Stiffness of optical traps: Quantitative agreement between experiment and electro-magnetic theory. *Phys. Rev. Lett.* **95**, 168102 (2005).

63. Rohrbach, A., Kress, H., and Stelzer, E.H. Three-dimensional tracking of small spheres in focused laser beams: Influence of the detection angular aperture. *Opt. Lett.* **28**, 411–413 (2003).

64. Tischer, C., Pralle, A., and Florin, E.L. Determination and correction of position detection non-linearity in single particle tracking and three-dimensional scanning probe microscopy. *Microsc. Microanal.* **10**, 425–434 (2004).

65. Rohrbach, A. Switching and measuring a force of 25 femtoNewtons with an optical trap. *Opt. Express* **13**, 9695–9701 (2005).

66. Sacconi, L., Tolic-Norrelykke, I.M., Stringari, C., Antolini, R., and Pavone, F.S. Optical micromanipu-lations inside yeast cells. *Appl. Opt.* **44**, 2001–2007 (2005).

67. Desai, K.V. et al. Agnostic particle tracking for three-dimensional motion of cellular granules and membrane-tethered bead dynamics. *Biophys. J.* **94**, 2374–2384 (2008).

68. Lukosz, W. Optical systems with resolving powers exceeding the classical limit. *J. Opt. Soc. Am.* **56**, 1463–1471 (1966).

69. Diaspro, A. *Confocal and Two-Photon Microscopy: Foundations, Applications and Advances.* Wiley-Liss, New York, 567 pp. (2002).

70. Lakowicz, J.R. *Principles of Fluorescence Spectroscopy*, 3rd Edn. Springer, p. 954 (2006).

71. Betzig, E. et al. Imaging intracellular fluorescent proteins at nanometer resolution. *Science* **313**, 1642–1645 (2006).

72. Periasamy, A., Elangovan, M., Elliott, E., and Brautigan, D.L. Fluorescence lifetime imaging (FLIM) of green fluorescent fusion proteins in living cells. *Methods Mol. Biol.* **183**, 89–100 (2002).

73. Bastiaens, P.I. and Squire, A. Fluorescence lifetime imaging microscopy: Spatial resolution of bio-chemical processes in the cell. *Trends Cell Biol.* **9**, 48–52 (1999).

74. Gauderon, R., Lukins, P.B., and Sheppard, C.J. Three-dimensional second-harmonic generation imaging with femtosecond laser pulses. *Opt. Lett.* **23**, 1209–1211 (1998).

75. Svoboda, K. and Block, S.M. Biological applications of optical forces. *Annu. Rev. Biophys. Biomol. Struct.* **23**, 247–285 (1994).

76. Laan, L., Husson, J., Munteanu, E.L., Kerssemakers, J.W., and Dogterom, M. Force-generation and dynamic instability of microtubule bundles. *Proc. Natl. Acad. Sci. USA* **105**, 8920–8925 (2008).

77. Tischer, C. et al. Three-dimensional thermal noise imaging. *Appl. Phys. Lett.* **79**, 3878–3880 (2001).

78. Neuman, K.C., Chadd, E.H., Liou, G.F., Bergman, K., and Block, S.M. Characterization of photo-damage to *Escherichia coli* in optical traps. *Biophys. J.* 77, 2856–2863 (1999).

79. Wulff, K.D. et al. Aberration correction in holographic optical tweezers. *Opt. Express* **14**, 4169–4174 (2006).

80. Vicidomini, G., Mondal, P.P., and Diaspro, A. Fuzzy logic and maximum a posteriori-based image restoration for confocal microscopy. *Opt. Lett.* **31**, 3582–3584 (2006).

81. Difato, F. et al. Improvement in volume estimation from confocal sections after image deconvolution. *Microsc. Res. Tech.* **64**, 151–155 (2004).

82. Diaspro, A., Corosu, M., Ramoino, P., and Robello, M. Adapting a compact confocal microscope system to a two-photon excitation fluorescence imaging architecture. *Microsc. Res. Tech.* **47**, 196–205 (1999).

83. Periasamy, A., Skoglund, P., Noakes, C., and Keller, R. An evaluation of two-photon excitation versus confocal and digital deconvolution fluorescence microscopy imaging in *Xenopus* morphogenesis. *Microsc. Res. Tech.* **47**, 172–181 (1999).

84. Tatarkova, S.A., Sibbett, W., and Dholakia, K. Brownian particle in an optical potential of the washboard type. *Phys. Rev. Lett.* **91**, 038101 (2003).

85. Lukic, B. et al. Direct observation of nondiffusive motion of a Brownian particle. *Phys. Rev. Lett.* **95**, 160601 (2005).

86. Egner, A., Andresen, V., and Hell, S.W. Comparison of the axial resolution of practical Nipkow-disk confocal fluorescence microscopy with that of multifocal multiphoton microscopy: Theory and experiment. *J. Microsc.* **206**, 24–32 (2002).

87. Hell, S., Reiner, G., Cremer, C., and Stelzer, E.H.K. Aberrations in confocal fluorescence microscopy induced by mismatches in refractive-index. *J. Microsc.* **169**, 391–405 (1993).

88. Booth, M.J. Adaptive optics in microscopy. *Philos. Trans. A: Math. Phys. Eng. Sci.* **365**, 2829–2843 (2007).

89. Bertero, M. and Boccaccio, P. *Introduction to Inverse Problems in Imaging*. IOP publishing, Bristol, UK (1998).

90. Kano, H., Hans, T.M., van der Voort, Martin Schrader, Geert, M.P., van Kempen, and Hell, S.W. Avalanche photodiode detection with object scanning and image restoration provides 2–4 fold resolution increase in two-photon fluorescence microscopy. *Bioimaging* **4**, 187–197 (1996).

91. Ericsson, M., Hanstorp, D., Hagberg, P., Enger, J., and Nystrom, T. Sorting out bacterial viability with optical tweezers. *J. Bacteriol.* **182**, 5551–5555 (2000).

92. Sacconi, L. et al. In vivo multiphoton nanosurgery on cortical neurons. *J. Biomed. Opt.* **12**, 050502 (2007).

93. Visscher, K. and Brakenhoff, G.J. Single beam optical trapping integrated in a confocal microscope for biological applications. *Cytometry* **12**, 486–491 (1991).

94. Wilson, T. and Sheppard, C.J.R. *Theory and Practice of Scanning Optical Microscopy*. Academic Press, London, UK, 213 pp. (1984).

95. Meng, F. and Ma, H. Fluorescence correlation spectroscopy analysis of diffusion in a laser gradient field: A numerical approach. *J. Phys. Chem. B* **109**, 5580–5585 (2005).

96. Bustamante, C., Marko, J.F., Siggia, E.D., and Smith, S. Entropic elasticity of lambda-phage DNA. *Science* **265**, 1599–1600 (1994).

97. Lang, M.J., Fordyce, P.M., Engh, A.M., Neuman, K.C., and Block, S.M. Simultaneous, coincident optical trapping and single-molecule fluorescence. *Nat. Meth.* **1**, 133–139 (2004).

98. Capitanio, M., Maggi, D., Vanzi, F., and Pavone, F.S. FIONA in the trap: The advantages of combining optical tweezers and fluorescence. *J. Opt. A: Pure Appl. Opt.* **9**, S157–S163 (2007).

99. Wang Y.-L. and Discher, D.E. *Cell Mechanics.* Elsevier, New York, 496 pp. (2007).

100. Cai, Y. et al. Nonmuscle myosin IIA-dependent force inhibits cell spreading and drives F-actin flow. *Biophys. J.* **91**, 3907–3920 (2006).

101. Palazzo, A.F., Eng, C.H., Schlaepfer, D.D., Marcantonio, E.E., and Gundersen, G.G. Localized stabilization of microtubules by integrin- and FAK-facilitated Rho signaling. *Science* **303**, 836–839 (2004).

102. Adams, J.C. et al. Cell-matrix adhesions differentially regulate fascin phosphorylation. *Mol. Biol. Cell* **10**, 4177–4190 (1999).

103. Carlsson, A.E. Growth of branched actin networks against obstacles. *Biophys. J.* **81**, 1907–1923 (2001).

104. Liu, A.P. and Fletcher, D.A. Actin polymerization serves as a membrane domain switch in model lipid bilayers. *Biophys. J.* **91**, 4064–4070 (2006).

105. Raucher, D. and Sheetz, M.P. Cell spreading and lamellipodial extension rate is regulated by membrane tension. *J. Cell Biol.* **148**, 127–136 (2000).

106. Dai, J. and Sheetz, M.P. Cell membrane mechanics. *Meth. Cell Biol.* **55**, 157–171 (1998).

107. Sheetz, M.P. and Dai, J. Modulation of membrane dynamics and cell motility by membrane tension. *Trends Cell Biol.* **6**, 85–89 (1996).

108. Apodaca, G. Modulation of membrane traffic by mechanical stimuli. *Am. J. Physiol. Renal Physiol.* **282**, F179–F190 (2002).

109. Hochmuth, R.M., Shao, J.Y., Dai, J.W., and Sheetz, M.P. Deformation and flow of membrane into tethers extracted from neuronal growth cones. *Biophys. J.* **70**, 358–369 (1996).

110. Mogilner, A. and Oster, G. Polymer motors: Pushing out the front and pulling up the back. *Curr. Biol.* **13**, R721–R733 (2003).

111. Pralle, A. and Florin, E.L. Cellular membranes studied by photonic force microscopy. *Meth. Cell Biol.* **68**, 193–212 (2002).

112. Tolic-Norrelykke, I.M., Munteanu, E.L., Thon, G., Oddershede, L., and Berg-Sorensen, K. Anomalous diffusion in living yeast cells. *Phys. Rev. Lett.* **93**, 078102 (2004).

113. Carlier, M.F., Le Clainche, C., Wiesner, S., and Pantaloni, D. Actin-based motility: From molecules to movement. *Bioessays* **25**, 336–345 (2003).

114. Footer, M.J., Kerssemakers, J.W.J., Theriot, J.A., and Dogterom, M. Direct measurement of force generation by actin filament polymerization using an optical trap. *Proc. Natl. Acad. Sci. USA* **104**, 2181–2186 (2007).

115. Kovar, D.R. and Pollard, T.D. Insertional assembly of actin filament barbed ends in association with formins produces piconewton forces. *Proc. Natl. Acad. Sci. USA* **101**, 14725–14730 (2004).

116. Shaevitz, J.W. and Fletcher, D.A. Load fluctuations drive actin network growth. *Proc. Natl. Acad. Sci. USA* **104**, 15688–15692 (2007).

117. Dent, E.W. and Kalil, K. Axon branching requires interactions between dynamic microtubules and actin filaments. *J. Neurosci.* **21**, 9757–9769 (2001).

118. Carlier, M.F.T., Loisel, T.P., Boujemaa, R., and Pantaloni, D. Reconstitution of actin-based motility of *Listeria* and *Shigella* using pure proteins. *Biophys. J.* **78**, 240A (2000).

119. McGrath, J.L. et al. The force–velocity relationship for the actin-based motility of *Listeria monocytogenes. Curr. Biol.* **13**, 329–332 (2003).

120. Noireaux, V. et al. Growing an actin gel on spherical surfaces. *Biophys. J.* **78**, 1643–1654 (2000).

121. Cameron, L.A., Svitkina, T.M., Vignjevic, D., Theriot, J.A., and Borisy, G.G. Dendritic organization of actin comet tails. *Curr. Biol.* **11**, 130–135 (2001).

122. Mogilner, A. and Oster, G. Force generation by actin polymerization. II: The elastic ratchet and tethered filaments. *Biophys. J.* **84**, 1591–1605 (2003).

123. Parekh, S.H., Chaudhuri, O., Theriot, J.A., and Fletcher, D.A. Loading history determines the velocity of actin-network growth. *Nat. Cell Biol.* **7**, 1219–1223 (2005).

124. Theriot, J.A. Actin polymerization and the propulsion of *Listeria-monocytogenes. Biophys. J.* **66**, A352 (1994).

125. Gerbal, F. et al. Measurement of the elasticity of the actin tail of *Listeria monocytogenes. Eur. Biophys. J. Biophys. Lett.* **29**, 134–140 (2000).

126. Theriot, J.A. and Fung, D.C. *Listeria monocytogenes*-based assays for actin assembly factors. *Meth. Enzymol.* **298**, 114–122 (1998).

127. Kress, H. et al. Filopodia act as phagocytic tentacles and pull with discrete steps and a load-dependent velocity. *Proc. Natl. Acad. Sci. USA* **104**, 11633–11638 (2007).

128. Cojoc, D. et al. Properties of the force exerted by filopodia and lamellipodia and the involvement of cytoskeletal components. *PLoS ONE* **2**, e1072 (2007).

129. Allioux-Guerin, M. et al. Spatio-temporal analysis of cell response to a rigidity gradient: A quantitative study by multiple optical tweezers. *Biophys. J.* **96**, 238–247 (2008).

130. Sheetz, M.P., Sable, J.E., and Dobereiner, H.G. Continuous membrane–cytoskeleton adhesion requires continuous accommodation to lipid and cytoskeleton dynamics. *Annu. Rev. Biophys. Biomol. Struct.* **35**, 417–434 (2006).

131. Ashkin, A. and Dziedzic, J.M. Optical trapping and manipulation of viruses and bacteria. *Science* **235**, 1517–1520 (1987).

132. Bustamante, C., Macosko, J.C., and Wuite, G.J.L. Grabbing the cat by the tail: Manipulating molecules one by one. *Nat. Rev. Mol. Cell Biol.* **1**, 130–136 (2000).

133. Jeney, S., Florin, E.L., and Horber, J.K. Use of photonic force microscopy to study single-motor-molecule mechanics. *Methods Mol. Biol.* **164**, 91–108 (2001).

134. Molloy, J.E. and Padgett, M.J. Lights, action: Optical tweezers. *Contemp. Phys.* **43**, 241–258 (2002).

135. Wang, M.D. et al. Force and velocity measured for single molecules of RNA polymerase. *Science* **282**, 902–907 (1998).

136. Michael Conn, P. *Laser Capture Microscopy and Microdissection.* Academic Press, San Diego, CA, 775 pp. (2002).

137. Colombelli, J., Reynaud, E.G., and Stelzer, E.H. Investigating relaxation processes in cells and developing organisms: From cell ablation to cytoskeleton nanosurgery. *Methods Cell Biol.* **82**, 267–291 (2007).

138. Stracke, F., Rieman, I., and Konig, K. Optical nanoinjection of macromolecules into vital cells. *J. Photochem. Photobiol. B* **81**, 136–142 (2005).

139. Steubing, R.W., Cheng, S., Wright, W.H., Numajiri, Y., and Berns, M.W. Laser induced cell fusion in combination with optical tweezers: The laser cell fusion trap. *Cytometry* **12**, 505–510 (1991).

140. Schneckenburger, H., Hendinger, A., Sailer, R., Strauss, W.S., and Schmitt, M. Laser-assisted optoporation of single cells. *J. Biomed. Opt.* **7**, 410–416 (2002).

141. Berns, M.W. et al. Laser microsurgery in cell and developmental biology. *Science* **213**, 505–513 (1981).

17

In Vivo Spectroscopic Imaging of Biological Membranes and Surface Imaging for High-Throughput Screening

Jo L. Richens
University of Nottingham

Peter Weightman
University of Liverpool

Bill L. Barnes
University of Exeter

Paul O'Shea
University of Nottingham

17.1   Introduction 17-1
17.2   Background: Major Cell Biological Aims That Require
         Imaging Solutions 17-3
17.3   Surface Imaging Techniques 17-5
17.4   Reflection Anisotropy Spectroscopy and Reflection
         Anisotropy Microscopy...17-6
17.5   Plasmonics...17-7
17.6   Conclusions: Ultrafast, Ultrasensitive, Super-Resolution,
         Label-Free Imaging Applications...17-10
Acknowledgments...17-11
References 17-11

17.1 Introduction

In recent years, "imaging" and particularly "biological imaging" has almost become a science in its own right. Fundamentally, however, it is just a set of tools directed toward a specific measurement problem that includes spatial discrimination over a defined coordinate system. The emphasis on spatial discrimination therefore, essentially defines the measurement as imaging. This is not to underestimate the enormous importance and impact that different imaging approaches have made on biological and biomedical research. In fact, the reason that "imaging" occupies this lauded position resides in the spectacular success and impact such techniques have brought to life science research. Imaging as an experimental endeavor has three essential requirements or components that must be present to ensure that most information is available for making some assessment of a biological process. The first deals with the acquisition of an imaging signal itself using, for example, a CCD camera or a photomultiplier output from a scanned sample. This aspect of the imaging process, customarily, is what most would associate with the imaging. There are two further requirements, however, that may augment the information content of an imaging experiment. Thus, a suitable optical probe can "target" a feature to "light it up." Examples of these features will be indicated later in this chapter (see, e.g., Figures 17.1 and 17.2) but perhaps the most well known are the fluorescent proteins such as GFP and intracellular calcium ion indicators such as Fura.

While the use of optical probes is clearly an important factor in contributing to the imaging process (unless the imaging modality does not require a probe and is thus deemed "label-free"), the data-processing algorithms that extract relevant signal data from often complex backgrounds are also extremely important. The latter therefore, represent image-processing procedures and are often not given the credit they really merit as the emphasis is usually on the hardware fabrication part of the imaging science.

Biological and biomedical research have presented many significant challenges to the acquisition of imaging information. These problems mostly reside in the need to acquire signals that are noisy or at extremely low levels. However, in the case of cell biological research directed at single cells (see Figure 17.1) as opposed to the multicellular imaging directed at larger structures, some problems have seemed insurmountable as they apparently require that the laws of physics be breached. Thus, it is often necessary to interrogate structures that have diameters of, or are in regions that are well below the resolving power of standard optical microscopes. This is the well-known diffraction limit and it presents a formidable challenge to reaching the necessary spatial resolution of visible light–based imaging modalities. Fortunately, a number of highly innovative and ingenious solutions to these problems have become available. These are the so-called superresolution techniques (reviewed in O'Shea et al. 2008), and of late, also called nanoscopy directed at single living cells some of which are outlined in other chapters in this book.

There are of course, other requirements for imaging technologies that yet other chapters in this book deal with. The distinct emphasis in this chapter, however, will be to adopt a much more forward-looking "work in progress" viewpoint rather than a retrospective review. We outline some exciting developments in the interactions of visible and UV radiation with biological surfaces, and particularly biological membranes, that offer novel information and lend themselves to spatial imaging. Our aim therefore, is not just to visualize the topography of biological material but to identify and characterize molecular structures with different functionalities on and within biological membranes. We will outline, in very simple terms, the basic physical mechanisms that can be used within an imaging regime. These approaches are quite different to other techniques that involve symmetry-breaking events such as those that reside in the interfacial properties of biological materials such as second-harmonic imaging covered in Chapter 9. Similarly, there are numerous other surface-oriented techniques that yield useful

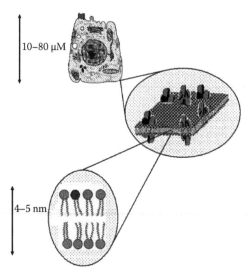

FIGURE 17.1   Living cell indicating a fluorescent phospholipid membrane probe (solid circle) and approximate length scales (for details see Wall et al. 1995b, Cladera et al. 2001).

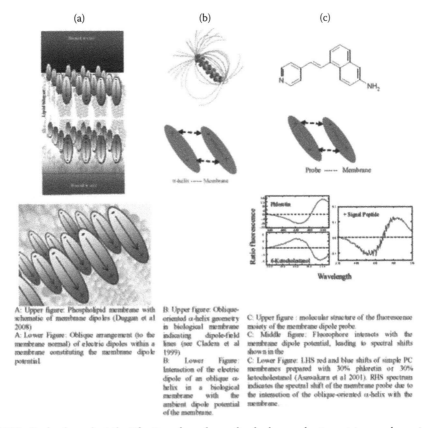

A: Upper figure: Phospholipid membrane with schematic of membrane dipoles (Duggan et al 2008)
A: Lower Figure: Oblique arrangement (to the membrane normal) of electric dipoles within a membrane constituting the membrane dipole potential

B: Upper figure: Oblique-oriented α-helix geometry in biological membrane indicating dipole-field lines (see Cladera et al 1999)
B: Lower Figure: Interaction of the electric dipole of an oblique α-helix in a biological membrane with the ambient dipole potential of the membrane.

C: Upper figure : molecular structure of the fluorescence moiety of the membrane dipole probe.
C: Middle figure: Fluorophore interacts with the membrane dipole potential, leading to spectral shifts shown in the
C: Lower Figure: LHS red and blue shifts of simple PC membranes prepared with 30% phloretin or 30% ketocholestanol (Asawakarn et al 2001). RHS spectrum indicates the spectral shift of the membrane probe due to the interaction of the oblique-oriented α-helix with the membrane.

FIGURE 17.2 Probe-dependent identification of membrane dipole changes due to protein–membrane interaction.

information. Several of the diffraction techniques (such as low-angle x-ray or neutron diffraction), for example, have made an impact on membrane biology. In this chapter, however, we seek to establish how novel approaches may yield structural information on living cellular systems as well as (e.g.) in microarrays for a high-throughput analysis of intermolecular interactions.

17.2 Background: Major Cell Biological Aims That Require Imaging Solutions

In many ways, contemporary cell biological research relies heavily on the many different manifestations of imaging technologies. These technologies range from microarray spot readers to live cell imaging techniques with many being based on signals from specific labels (probe) that are often fluorescent through to label-free techniques typically based on surface plasmon resonance. The former, based on measuring specific molecular interactions with confirmation that they have occurred or not, is usually the simple goal.

The latter are designed typically to render biological macromolecular topography but not usually with molecular and geometrical detail. However, for biological imaging science to maintain a research momentum, it is necessary that the molecular structure be identified and particularly those changes of structure that can be resolved over time periods relevant to cellular interactions and signaling processes. The inspection of molecular structure (and structural change) is highly accessible, however, using any of a number of well-established spectroscopic techniques. According to the nature of the structure to be resolved, it is possible to define a frequency of electromagnetic radiation that may be used diagnostically. These include UV/Vis absorption for electronic transitions (1.5–5.5 eV), IR for vibrational modes, and numerous other dichroic and polarization approaches that identify asymmetric structures such as α-helix geometry in proteins (see Figure 17.2). If these are brought to bear within a microscope arrangement, then the same property can be resolved with a spatial resolution. Clearly, these applications applied within an imaging regime are not as simple as in the glib way they are stated above. For many reasons, not least because techniques that purport to illuminate molecular structure are typically "label-free," EM absorption methods directed toward large ensembles of molecules are often in an isotropic environment. For an imaging application, however, discrimination in recognizing a particular structure from a background of several (actually very many) similar structures on a relatively small surface area would be extremely difficult. Thus for example, a circular dichroism in the UV offers a very useful spectral signature for protein secondary structures. It would be very helpful if we could obtain such information of proteins on a living cell surface. Unfortunately, a CD "image" of a single living cell would not only suffer from a very low total dichroic signal to background noise (i.e., the total CD signature would be vanishingly small even with very small path lengths) but it would also be compromised with the nearly impossible task of discriminating the "desired" signal of a given protein from the total signal arising from the cell. All this, of course, would not get a near diffraction-limited resolution and with all the problems combined it can be determined that no CD-imaging instrument for cellular studies is really viable at present. There are reports, however, of interesting approaches to make CD-imaging measurements on very much simpler systems (spatially) that possess large signals such as biaxial crystals of 1,8-dihydroxyanthraquinone (Claborn et al. 2003). Application to cellular imaging, however, is not likely to be viable for quite some time yet.

In view of these forgoing comments, we explored the possibilities that other spectroscopic modalities may be more appropriate for revealing molecular structural changes due to (e.g.) cell signaling events. We considered both label-dependent and label-free approaches each with their own merits and problems. The kinds of changes we seek to study include for example, changes of the α-helix geometry as a result of conformational changes during the course of a ligand-induced signaling event at a membrane receptor. Thus Figure 17.2 indicates that helical geometry may change during the course of the interaction of a specific ligand that binds to a membrane receptor. This was identified using a membrane fluorophore that responds to the membrane dipole potential and may be applied in a suspension of cells or in artificial membranes vesicles i.e., both with no spatial information (see Cladera and O'Shea 1998, Asawakarn et al. 2001), as well as in an imaging regime (O'Shea 2003, 2005, Duggan et al. 2008).

We developed the application of classes of fluorophores, as illustrated in Figure 17.2, that reports the magnitude of the membrane dipole potential and exhibits changes during the course of ligand binding to a receptor (Asawakarn et al. 2001) or the insertion of a peptide into the membrane that enters either fully folded or folds into an α-helix as it enters the membrane (Cladera and O'Shea 1998, Golding et al. 1996).

Following up our work with the modulation of peptide–membrane interactions by the membrane dipole potential, it became apparent that the membrane dipole potential is quite different in magnitude in membrane microdomain structures known generically as rafts (Simons and Vaz 2004). Thus, this seemed to offer a new mechanism of biological control if protein structures were to be modulated depending on whether they were located in membrane microdomains or elsewhere in the membrane (O'Shea 2003).

Typically, membrane rafts are "detergent-insoluble," "viscous patches," about 40–150 nm in diameter in the plasma membranes of all eukaryotic cells. The importance of the raft paradigm is underlined by the fact that in recent years, it has been one of the most spectacularly productive areas of cell biology in general and membrane research in particular (Simons and Vaz 2004, O'Shea 2005). Rafts exhibit a number of features such as the recruitment of GPI-linked proteins and are linked to cell signaling and vesicular uptake including endo/exocytosis and transcytosis as well as to infection (Cladera et al. 2001).

Over the last few years we have developed an hypothesis that augments the accepted view of the function of rafts (i.e., sequestration of reactants/proteins, etc.) in that they also modulate receptor activity by virtue of their elevated membrane dipole potential (e.g., reviewed in O'Shea 2005) and briefly described above (see also Figure 17.2).

It is essential, therefore, to study these nanoscopic cellular structures more systematically and functionally. Some of the outstanding key questions, include: (1) Do rafts in cells really exist? (2) If so, is assembly and disassembly solely a function of the thermodynamic phase behavior of the membrane lipids, i.e., the more cholesterol/Sphingomyelin, the more rafts, etc. (see, e.g., Richardson et al. 2007) or are there additional control mechanisms? (3) How large are they (i.e., what is their diameter)? (4) How many of them are there? (5) How quickly do they "turnover" (what is their lifetime in the membrane)? (6) Are all rafts the same? Based on an imaging analysis study of the number density and sizes of membrane microdomains, we published the first physical model of how rafts may be formed and disassembled based on a statistical mechanics of the membrane lipids that underlies their observed equilibrium thermodynamic organization in model membranes (Richardson et al. 2007). Recent data however, indicates that the cellular systems are much more complicated and we have evidence that not all rafts are actually the same. This whole area of cell biology, therefore, really requires a much more thorough inspection, and progress will only take place if better spatial information becomes available. Some quite beautiful examples of imaging tools relevant to these problems are described in this book. In Section 17.3 we describe how these biological problems may be addressed using other kinds of imaging approaches based on some techniques directed toward surface analysis and thus when they are also combined with microscopy they are especially appropriate for studies of biological membranes.

17.3 Surface Imaging Techniques

There are a number of techniques that lend themselves to imaging surfaces appropriate to studies of cell membranes. These include several rather simple approaches that we have developed to illuminate changes of membrane dipoles as the result of protein binding and insertion (Cladera and O'Shea 1998), ligand binding of membrane receptors (Asawakarn et al. 2001) as well as changes of the surface electrostatic surface potential in model membranes (Wall et al. 1995a) and cell membranes (Wall et al. 1995b). A rigorous physical description of each of these membrane potentials can be found in O'Shea (2004). Some of the imaging applications designed to illuminate these membrane properties from a functional point of view that includes figures and images illustrating the level of information available are outlined by Duggan et al. (2008).

In addition to this work, we have also developed the use of two-photon surface wave microscopy applied to membranes (Goh et al. 2005). We described how a scanning optical microscope that uses coverslips upon which supported membranes are deposited, can be modified to support the propagation of surface waves capable of large field enhancements. We outlined the beam conditioning necessary to ensure efficient use of the illumination for signal capture. A two-photon surface-wave fluorescent excitation was demonstrated on fluorescent nanospheres, exhibiting a point-spread function width of ≈220 nm at an illumination wavelength of 925 nm. The potential of nonlinear surface-wave excitation for both fluorescence and harmonic imaging microscopy is also an interesting application of these approaches. Some of these areas are discussed however, in Chapter 9, so we will not dwell further on this possible application. We illustrated these with examples in which membrane rafts could be visualized and analyzed along the lines we described in Richardson et al. (2007).

Clearly, these are all based on a requirement for membrane labeling (i.e., proteins or lipids, etc.) with appropriate fluorescent probes. There are other approaches, however, that appear to yield some of the biological information we desire without the need to label membranes and a discussion of these form the remainder of this chapter.

17.4 Reflection Anisotropy Spectroscopy and Reflection Anisotropy Microscopy

The first analytical technique that we consider is known as reflection anisotropy spectroscopy (RAS) and it involves the measurement of the difference in the reflectivity in two orthogonal directions of normal incident linearly polarized light. In contrast to most surface optical techniques, the signal is not dominated by contributions from the bulk (see also the later section on plasmonics—particle vs surface plasmons). Most applications to date have involved a cubic substrate with the cancellation of the bulk signal by symmetry. For molecules that adopt an ordered structure on the surface, then, irrespective of the symmetry of the substrate, RAS is sensitive to the molecular alignment. It is this later quantity that we seek and particularly so as it changes during a membrane process involving ligand-mediated signaling mechanisms in many eukaryotic biological systems (e.g., Cladera et al. 1999, Asawakarn et al. 2001).

Early manifestations of RAS directed toward semiconductor surfaces were necessarily undertaken under ultra high vacuum or in a gaseous milieu. Fortunately, more recent developments enable measurements in liquids, particularly in the aqueous environment necessary to support cells and model membrane structures. Thus, RAS has now been established as a highly sensitive and reproducible method for identifying molecular interactions at surfaces (Smith et al. 2003, 2004). It has been used to probe the interactions of nucleic acid bases (Weightman et al. 2006, Smith et al. 2009) and amino acids (LeParc et al. 2006) at Au(110)/electrolyte interfaces and to monitor the adsorption of single and double stranded (ss and ds) DNA at Au(110)/electrolyte interfaces (Cuquerella et al. 2007, Mansley et al. 2008). The latter work showed that both ss-DNA and ds-DNA bind to the gold surface and that RAS could detect the bound molecules.

Quite recently, Messiha et al. (2008) demonstrated that RAS offers real-time measurements of a conformational change in proteins induced by electron transfer reactions. A bacterial electron transferring flavoprotein (ETF) was modified so it could become adsorbed onto an Au(110) electrode. This facilitated a reversible electron transfer to and from the protein cofactor in the absence of mediators (i.e., these are used typically to deliver electrons from an electrode to redox proteins in solution). Reversible changes of the RAS signature were observed with this protein that are interpreted as arising from conformational changes accompanying the transfer of electrons to and from the redox site within the protein. This is quite extraordinary as this means the RAS technique is able to determine the binding (reduction) and the loss (oxidation) of an electron to the protein on the gold surface in a time-resolved manner. The rapid acquisition of such a signal means that it may be possible to study protein dynamics on surfaces although the photon budget of the RAS in an imaging microscope configuration (see, e.g., Figure 17.3b)

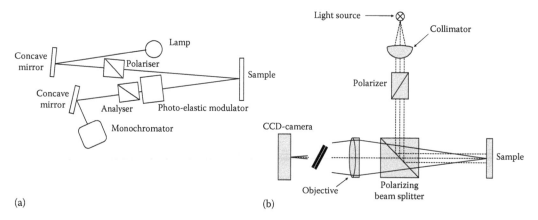

(a)   (b)

FIGURE 17.3 Optical arrangements for (a) RAS spectrometry and (b) RAM incorporating a scanning capability for spatial imaging of RAS.

may well be restrictive to what can be obtained currently. RA spectrometers presently cover the range 1.5–5.5 eV but one of the present authors (PW) has just described the development of an RAS instrument that, in combination with a low energy instrument, provides an RA spectra from 1.5 to 7.0 eV (Mansley et al. 2009). The overall aim of this is to enable RAS studies of biomolecules on Au/electrolyte interfaces where the optical transitions are most intense.

Our optical arrangement for a spectrometer with an RAS capability is illustrated in Figure 17.3a. Following this, it is not difficult to include a scanning facility in which images can be acquired as illustrated in Figure 17.3b. Reflection anisotropy microscopy (RAM), therefore, represents an optical RAS imaging probe of surfaces. We do not claim priority for the latter as Ertl and coworkers (Haas et al. 1996) developed a working instrument more than a decade ago. These authors report a RAM instrument based on the sensitivity of the plane of polarization of reflected light to the electronic nature of the surface and hence responds to the chemical nature of a surface. RAM achieves surface sensitivity in a similar way to RAS by using a near normal incidence and a reflection of plane polarized light from the surface of a cubic crystal so that the response of the bulk cancels by symmetry. As originally developed, RAM uses laser illumination so it is essentially a single frequency probe. This is a disadvantage since it cannot be used to probe specific molecules by being tuned to a particular molecular transition. As a microscope, it cannot give a spatial resolution better than the wavelength of light that is used but may well benefit from combining it with a structured light module (e.g., Gustafson 2005). Thus, RAM could be used to follow dynamic processes on surfaces in liquid environments with a spatial detection possibility.

17.5 Plasmonics

Surface plasmons are surface electromagnetic waves that propagate along the interface between a noble metal surface and an external medium such as a liquid. Since the wave exists on the boundary of the metal and the external liquid medium, these oscillations are very sensitive to many kinds of change at this boundary. Plasmons essentially, comprise light that is trapped at the surface of a metal through the interaction of the electromagnetic field with the free conduction electrons in the metal surface.

Although the phenomenon of SPR was reported by Turbadar in 1959, it was not until the 1990s that a commercial sensor was developed by Biacore (now owned by GE Healthcare PLC). SPR responds to changes on a metal surface conditioned to "sense" the binding of analytes. This has been used for specific protein detection and is now exploited routinely as a generic label-free sensor. For commercial instruments the detection units are measured in resonance units (RUs), where one RU corresponds to a change in 10^{-40} in the angle of the intensity minimum. For an average protein (i.e., based on molecular mass) this change corresponds to the addition of around 1 pg/mm^2 onto the sensor surface, although there are many variables. So, a more exact conversion factor between the RU and the surface concentration depends on the properties of the sensor surface and the interacting molecules, etc.

There are many more possibilities for the application of plasmonics than by simply measuring the coarse addition of mass to a flat SPR-active sensor surface. If we consider the nature of the plasmon field at the surface for example, we can exploit this to yield much greater sensitivities, but there is also the potential to utilize these fields to excite fluorescence as well as to image a surface. Some of these approaches have been taken with some success by Lackowicz and co-workers (see, e.g., Lakowicz et al. 2006). In particular, by preparing corrugated surfaces it has proved possible to obtain very high sensitivities.

We describe below some developments that offer great promise with the plasmonic imaging of biological structures. In our laboratories, for example, we have investigated the nature of the interaction of (e.g.) a protein with a metal surface (and particularly with a metal coated biological membrane—e.g., Figure 17.4) exhibiting plasmonic behavior. Thus, the question we consider is the explicit nature of the interaction as for all label-free measurements it is essential to be able to understand the information that such signals impart. Thus, we gave consideration to the possibility that the plasmon field may interact in other ways with the protein other than as a simple change of local refractive index indicating that a binding

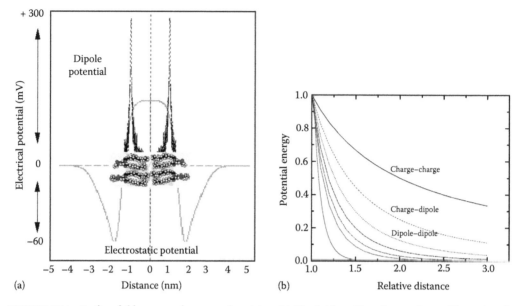

FIGURE 17.4 Surface fields on membranes and proteins. (a) The field profiles of a membrane illustrating the directional magnitude and the range of the electrostatic potential and the dipole fields associated with a biological membrane. (From the Cell Biophysics Web site, see http://www.nottigham.ac.uk/%7Emazpso/NewFiles/theory.htm) (b) Illustration of the variety of intermolecular interactions between a protein surface and a metal surface. (Personal communication and modified from unpublished work of T. Laue.)

event had taken place. In other words, we addressed questions associated with the electronic interaction of a protein with a plasmonic surface. Some of the results of such analyses are outlined in Lunt et al. (2008, 2009). It seems clear, however, that changes of electric potential at a metal (e.g., gold) surface or particle may affect the behavior of free electrons in the metal. In the simplest terms, an applied potential may essentially be taken as a charging of the interfacial capacitance of the metal in contact with a liquid electrolyte; the behavior of such surfaces has been modeled and measured comprehensively within the science of electrochemistry (see, e.g., O'Shea 2004, 2005). Such charging however, may lead to changes in the density of free electrons and so modulate the complex refractive index and change the spectrum. More sophisticated analyses include screening effects (e.g., Thomas-Fermi) and how the accumulation of a poly-charged protein may influence the solvent environment at the plasmonic surface. The latter has been modeled at length also and indicates that the dielectric constant of the surface in the locality of the protein may change dramatically. Such behavior is shown schematically in Figure 17.4 and described fully in O'Shea (2004). These latter effects may not be so dominant for surface plasmons as their decay length is rather large compared to the size of a bound protein but for particle plasmonics the effects could be dramatic. Some further features of this latter phenomenon are discussed below.

The strength of the optical field associated with surface plasmons falls off exponentially with distance away from the metal surface but there are a number of interesting properties of this field that can be exploited not least within imaging modes.

Three factors make surface plasmons ideally suited to imaging at surfaces (Barnes 2006). First, the light is localized in the vicinity of the surface (~200 nm). Second, the strength of the optical field is enhanced near the surface (typically around 10-fold). Third, the character of surface plasmons depends in a sensitive way on the nature of the material in the region adjacent to the metal surface. An additional bonus is that the wavelength of the surface plasmon is less than that of light of the same frequency. Zhang et al. (2004) demonstrated that microscopy using surface plasmons in conjunction with a solid immersion lens could be used for microscopy.

The resolution that may be achieved by this technique offers a useful but limited improvement over conventional approaches (see, e.g., Goh et al. 2005). While there was some hope that this might be overcome using a wavevector enhancing scheme (Smolyaninov et al. 2005), it now appears that the approach may not be viable (Drezet et al. 2007). Other more exotic alternatives have been proposed. Pendry (2000), for example, suggested using negative index materials to make a perfect lens. The usual diffraction limit arises from the fact that the optical near-field associated with an object is not transferred into the far field by our usual lenses. In the perfect lens, the near field components of the optical field are enhanced so that they may successfully be transmitted into the far field. The perfect lens is made from a slab of material that supports surface modes, surface plasmons, and it is these modes that can be used to enhance the strength of the near-field. These fascinating ideas have some experimental backing (Fang et al. 2005) but we are a very long way from providing a practical useful imaging technology. We and others (Sentenac et al. 2006, Liu et al. 2008) are also developing an alternative approach based on spatially structuring the light (Gustafsson 2005). Our approach is based on the use of the localized surface plasmon modes associated with metallic nanoparticles. The key advantage of using nanoparticles rather than a continuous metal surface is that the optical field associated with the nanoparticles is much more closely tied to the metal surface. Thus, as illustrated in Figure 17.5 this extends to around ~20 nm, rather than ~200 nm as in the case of a plasmonic surface. This state of affairs is illustrated schematically in Figure 17.5, in which the field strength is shown as a function of distance from the metal surface for the surface plasmon mode of a planar metal film and for a metallic nanoparticle. For materials such as a biological membrane placed adjacent to the metal surface, the membrane is in a more intense contact (i.e., with a much better overlap) with the field of the nanoparticle-based plasmon than that associated with the planar interface. In a similar manner to our two-photon surface wave studies (Goh et al. 2005) therefore, we anticipate that a membrane supported on a nanoparticle array will exhibit a much greater signal sensitivity.

Models of the degree of field localization are illustrated in Figure 17.6. Triangular metallic nanoparticles may be produced using a technique known as nanosphere lithography (NSL). The particles are formed by depositing metal through a template made from an ordered array of plastic beads using a self-assembly approach (Hulteen and van Duyne 1995). The NSL technique shows that particle arrays being considered for plasmon assisted imaging may be made using relatively low-cost fabrication procedures (Murray et al. 2006).

Presently, we can only speculate on where the use of plasmons may lead in imaging. It is possible to use plasmon modes to confine optical fields down to ~10 nm length scales. An array of particles might be used to build up a spatial map but imaging requires that we are able to distinguish between the responses of different particles. This may be achieved by employing rod-like particles and using a combination of particle orientation and polarization of the light to achieve the required selectivity. Another approach might be to develop a surface enhanced coherent anti-stokes Raman spectroscopy microscopy that employs arrays of nanoparticles. The advantage here would be that of potentially gaining access to compositional information.

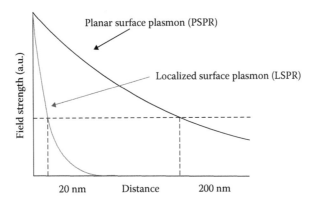

FIGURE 17.5 Plasmon fields associated with particles and surfaces.

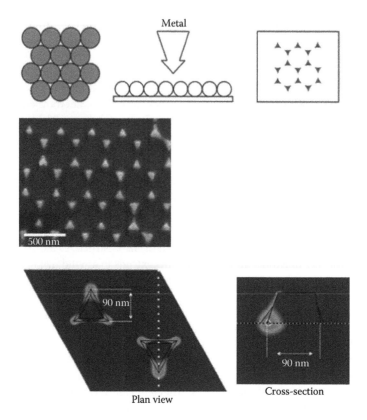

FIGURE 17.6   Field localization of nanopyramids. (From Murray, A.W. and Barnes, W.L., *Adv. Mater.*, 19, 3771, 2007. With permission.)

Other recent innovations in nanoparticle technologies coupled to various excited modes include some recent work by Shegai et al. (2008) who utilized novel properties of nanoparticles to obtain a control of light polarization. Thus, it is now possible to control the direction of light propagation by means of surface plasmon propagation in metal nanostructures, enhanced transmission through nanoholes in optically thin metal films, and light beaming with a subwavelength hole array. The control of the intensity and the spectrum of light mainly involves the enhancement of the local electromagnetic field by a localized surface plasmon resonance and coupling. The novelty of the work reported in Shegai et al. (2008), however is interesting to us as this may be a way forward to combine nanoparticle technologies together with RAS for a highly sensitive RAM system. The future holds great promise for many of these imaging modalities.

17.6 Conclusions: Ultrafast, Ultrasensitive, Super-Resolution, Label-Free Imaging Applications

There is no doubt that nanoparticle based imaging holds much promise with many attributes that can be exploited to deliver better information of the activities taking place on a cell surface. For the moment however, this field is in its infancy but with several other rather spectacular advances in nanoparticle-based screening tending to dominate the research efforts of both basic science and industrial laboratories. However, even for these array-based analytical procedures we utilize SPR imaging to identify molecular interactions. The virtue is that we can interrogate many interactions with great sensitivity, very quickly, and without any requirement for any kind of sophistication such as chemical derivatization of chromophoric conjugation in the sample preparation.

We are in the process, with the latter project, of developing analytical instruments that will contribute as platform technologies for systems biology and for biomedical screening. Richens et al. (2009) for example, outlines how patterns of molecular markers in serum may be screened to indicate (or not) the presence of a disease condition. Thus a number of diseases exhibit elevated concentrations of a single type of molecular marker (e.g., prostate specific antigen—PSA—as a marker of prostate cancer). Even so, there are also many false positives with this approach. There are many other disease conditions, however, that do not benefit from such a simple diagnostic test and which are typically diagnosed using rather subjective approaches. There appears a strong possibility that such conditions (e.g., LSE, COPD Alzheimer's disease), however, may be identifiable from patterns of molecular markers in biological fluids (such as CSF, serum, etc.) rather than from a sole marker (Richens et al. 2009). Our goal, therefore, is to identify such markers from their complex molecular backgrounds in a label-free, ultrasensitive, and high-throughput manner. This is a challenging task but early success does indeed offer a way forward offering new tools for both "systems biology" and even single cell proteomics as well as for medical diagnosis.

The other chapters, as well as our considerations above, indicate that nanoscopy imaging in its many guises will consolidate its role as a cornerstone of biological and biomedical research. The superb sensitivity that the imaging modalities described in the accompanying chapters in this book offer, will ensure that major biological impacts in understanding are possible by illuminating the subtleties of molecular behavior over smaller and smaller spatial dimensions. Attainment of these objectives is essential if we are to reach our goal of having a predictive and a quantitative understanding of cell biology in the future.

Acknowledgments

We are grateful to the RCUK/EPSRC (grant no. EP/C52389X/1) and the EPSRC (grant no. EP/C534697/1) for financial support that enabled some of this work to progress. We are also grateful to our colleagues in the Cell Biophysics Group and the Institute of Biophysics, Imaging & Optical Science, University of Nottingham for helpful discussions.

References

Asawakarn, T., Cladera, J., and O'Shea, P. (2001) Effects of the membrane dipole potential on the interaction of Saquinavir with phospholipid membranes and plasma membrane receptors of $CaCO_2$ cells. *Journal of Biological Chemistry*, **276**, 38457–38463.

Barnes, W. L. (2006) Surface plasmon–polariton length scales: A route to sub-wavelength optics. *Journal of Optics A: Pure and Applied Optics*, **8**, S87–S93.

Claborn, K., Puklin-Faucher, E., Kurimoto, M., Kaminsky, W., and Kahr, B. (2003) Circular dichroism imaging microscopy: Application to enantiomorphous twinning in biaxial crystals of 1,8-Dihydroxyanthraquinone. *Journal of the American Chemical Society*, **125**, 14825–14831.

Cladera, J., Martin, I., Ruysschaert, J.-M., and O'Shea, P. (1999) Characterization of the sequence of interactions of the fusion domain of the simian immunodeficiency virus with membranes; role of the membrane dipole potential. *Journal of Biological Chemistry*, **274**, 29951–29959.

Cladera, J., Martin, I., and O'Shea, P. (2001). The fusion domain of HIV gp41 interacts specifically with heparan sulfate on the T-lymphocyte cell surface. *EMBO Journal* **20**, 19–26.

Cuquerella, M. C., Smith, C. I., Fernig, D. G., Edwards, C., and Weightman, P. (2007) Adsorption of calf thymus DNA on Au(110) studied by reflection anisotropy spectroscopy. *Langmuir*, **23**, 2078–2082.

Drezet, A., Hohenau, A., and Krenn, J. R. (2007) Comment on "Far-field optical microscopy with a nanometer-scale resolution based on the in-plane image magnification by surface plasmon polaritons". *Physical Review Letters*, **98**, 209703.

Duggan, J., Jamal, G., Tilley, M., Davis, B., McKenzie, G., Vere, K., Somekh, M. G., O'Shea, P., and Harris, H. (2008) Functional imaging of microdomains in cell membranes. *European Biophysics Journal*, **37**, 1279–1289.

Fang, N., Lee, H., Sun, C., and Zhang, X. (2005) Sub-diffraction-limited optical imaging with a silver superlens. *Science*, **308**, 534–537.

Goh, J., Somekh, M., See, C. W., Pitter, M. C., Vere, K.-A., and O'Shea, P. (2005) Two-photon fluorescence surface wave microscopy. *Journal of Microscopy*, **220**, 168–175.

Golding, C., Senior, S., Wilson, M. T., and O'Shea, P. (1996) Time resolution of binding and membrane insertion of a mitochondrial signal peptide: Correlation with structural changes and evidence for cooperativity. *Biochemistry* **20**, 10931–10937.

Gustafsson, M. G. L. (2005) Nonlinear structured-illumination microscopy: Wide-field fluorescence imaging with theoretically unlimited resolution. *Proceedings of the National Academy of Sciences of the United States of America*, **102**, 13081–13086.

Haas, G., Franz, R. U., Rotermund, H. H., Tromp, R. M., and Ertl, G. (1996) Imaging surface reactions with light. *Surface Science*, **352–354**, 1003–1006.

Hulteen, J. C. and van Duyne, R. P. (1995) Nanosphere lithography—A materials general fabrication process for periodic particle array surfaces. *Journal of Vacuum Science & Technology A—Vacuum Surfaces and Films*, **13**, 1553–1558.

Lakowicz, J. R., Chowdhury, M. H., Ray, K., Zhang, J., Fu, Y., Badugu, R., Sabanayagam, C. R., Nowaczyk, K., Szmacinski, H., Aslan, K., and Geddes, C. D. (2006) Plasmon-controlled fluorescence: A new detection technology *Plasmonics in Biology and Medicine III* (Vo-Dinh, T., Lakowicz, J. R., and Gryczynski, Z., Eds.). *Proceedings of the SPIE*, **6099**, 34–48.

LeParc, R., Smith, C. I., Cuquerella, M. C., Williams, R. L., Fernig, D. G., Edwards, C., Martin, D. S., and Weightman, P. (2006) Reflection anisotropy spectroscopy study of the adsorption of sulfur-containing amino acids at the Au(110)/electrolyte interface. *Langmuir*, **22**, 3413–3420.

Liu, S., Chunag, C.-J., See, C. W., Zoriniants, G., Barnes, W. L., and Somekh, M. G. (2008) Double grating structured light microscopy using plasmonic nanoparticle arrays. *Optics Letters*, **34**, 1255–1257.

Lunt, E. A. M., Pitter, M. C., Somekh, M. G., and O'Shea, P. (2008) Studying protein binding to conjugated gold nanospheres; application of Mie light scattering to reaction kinetics. *Journal of Nanoscience and Nanotechnology*, **8**, 4335–4340.

Lunt, E.A.M., Pitter, M. C., and O'Shea, P. (2009) Quantitative studies of the interactions of metalloproteins with gold nanoparticles; identification of dominant properties of the protein that underlie the spectral changes. *Langmuir*, **25**, 10100–10106.

Mansley, C. P., Smith, C. I., Cuquerella, M. C., Farrell, T., Fernig, D. G., Edwards, C., and Weightman, P. (2008) Ordered structures of DNA on Au(110). *Physica Status Solidi C*, **5**, 2582–2586.

Mansley, C. P., Farrell, T., Smith, C. I., Harrison, P., Bowfield, A., and Weightman, P. (2009) A new UV reflection anisotropy spectrometer and its application to the Au(110) electrolyte surface. *Journal of Physics D: Applied Physics*, **42**, 115303 1–5.

Messiha, H. L., Smith, C. I., Scrutton, N. S., and Weightman, P. (2008) Evidence for protein conformational change at a Au(110)/protein interface. *Europhysics Letters* **83**, 18004 1–5.

Murray, A. W. and Barnes, W. L. (2007) Plasmonic materials. *Advanced Materials* **19**, 3771–3782.

Murray, A. W., Suckling, J. R., and Barnes, W. L. (2006) Overlayers on silver nanotriangles: Field confinement and spectral position of localized surface plasmon resonances. *Nano Letters*, **6**, 1772–1777.

O'Shea, P. (2003) Intermolecular interactions with/within cell membranes and the trinity of membrane potentials; kinetics and imaging. *Biochemical Society Transactions*, **31**, 990–996.

O'Shea, P. (2004) Membrane potentials; measurement, occurrence and roles in cellular function, in *Bioelectrochemistry of Membranes* (Walz, D. et al., Eds.). Birkhauser Verlag, Switzerland, pp. 23–59.

O'Shea, P. (2005) Physical landscapes in biological membranes. *Philosophical Transactions of the Royal Society: Mathematical, Physical and Engineering Sciences*, **363**, 575–588.

O'Shea, P., Somekh, M. G., and Barnes, W. L. (2008) Shedding light on life; visualising nature's complexity. *Physics World*, **21**, 29–34.

Pendry, J. B. (2000) Negative refraction makes a perfect lens. *Physical Review Letters*, **85**, 3966–3969.

Richardson, G., Cummings, L. J., Harris, H., and O'Shea, P. (2007) Towards a mathematical model of the assembly and disassembly of membrane microdomains: Comparison with experimental models. *Biophysical Journal*, **92**, 4145–4156.

Richens, J. L., Urbanowicz, R. A., Lunt, E. A., Metcalf, R., Corne, J., Fairclough, L., and O'Shea, P. (2009) Systems biology coupled with label-free high-throughput detection as a novel approach for diagnosis of chronic obstructive pulmonary disease. *Respiratory Research*, **22**, 10–29.

Sentenac, A., Chaumet, P. C., and Belkebir, K. (2006) Beyond the rayleigh criterion: Grating assisted far-field optical diffraction tomography. *Physical Review Letters*, **97**, 243901.

Shegai, T., Li, Z., Dadosh, T., Zhang, Z., Xu, H., and Haran, G. (2008) Managing light polarization via plasmon–molecule interactions within an asymmetric metal nanoparticle trimer. *Proceedings of the National Academy of Sciences of the United States of America*, **105**, 16448–16453.

Simons, K. and Vaz, W. L. C. (2004) Model systems, lipid rafts and cell membranes. *Annual Reviews of Biophysics and Biomolecular Structure*, **33**, 269–295.

Smith, C. I., Maunder, A. J., Lucas, C. A., Nichols, R. J., and Weightman, P. (2003) Adsorption of pyridine on Au(110) as measured by reflection anisotropy spectroscopy. *Journal of the Electrochemical Society*, **150**, E233.

Smith, C. I., Dolan, G. J., Farrell, T., Maunder, A. J., Fernig, D. G., Edwards, C., and Weightman, P. (2004) The adsorption of bipyridine molecules on Au(110) as measured by reflection anisotropy spectroscopy. *Journal of Physics: Condensed Matter*, **16**, S4385.

Smith, C. I., Bowfield, A., Dolan, G. J., Cuquerella, M. C., Mansley, C. P., Fernig, D. G., Edwards, C., and Weightman, P. (2009) Determination of the structure of adenine monolayers adsorbed at Au(110)/electrolyte interfaces using reflection anisotropy spectroscopy. *Journal of Chemical Physics*, **130**, 044702.

Smolyaninov, I. I., Elliott, J., Zayats, A. V., and Davis, C. C. (2005) Far-field optical microscopy with a nanometer-scale resolution based on the in-plane image magnification by surface plasmon polaritons. *Physical Review Letters*, **94**, 4.

Wall, J., Golding, C. A., Van Veen, M., and O'Shea, P. (1995a) The use of fluorescein phosphatidylethanolamine (FPE) as a real-time probe for peptide-membrane interactions. *Molecular Membrane Biology*, **12**,183–192.

Wall, J., Ayoub, F., and O'Shea, P. (1995b) Interactions of macromolecules with the mammalian cell surface. *Journal of Cell Science*, **108**, 2673–2682.

Weightman, P., Dolan, G. J., Smith, C. I., Cuquerella, M. C., Almond, N. J., Farrell, T., Fernig, D. G., Edwards, C., and Martin, D. S. (2006) Orientation of ordered structures of cytosine and cytidine 5′-monophosphate adsorbed at Au(110)/liquid interfaces. *Physical Review Letters*, **96**, 086102, 1–4.

Zhang, J., See, C. W., Somekh, M. G., Pitter, M. C., and Liu, S. G. (2004) Wide-field surface plasmon microscopy with solid immersion excitation. *Applied Physics Letters*, **85**, 5451–5453.

18

Near-Field Optical Microscopy: Insight on the Nanometer-Scale Organization of the Cell Membrane

18.1 Near-Field Scanning Optical Microscopy...................................**18**-1
Modern Microscopy Pushes the Resolution • A Scan through
History • The Near-Field Approach • The NSOM Head and the
Feedback Mechanism • Getting an NSOM Image Requires the
Right Probe • A Bio-NSOM

18.2 NSOM in the Biological Domain...**18**-11
Artificial Lipid Mono/Bilayers as Test Models • Protein
Organization on the Cell Membrane

18.3 Future Perspectives in Near-Field Optical Microscopy..........**18**-17
Exploiting the Power of Optical Nanoantennas

Acknowledgments...**18**-21

References **18**-21

Davide Normanno

Thomas van Zanten

María F. García-Parajo

*Institute for Bioengineering
of Catalonia*

18.1 Near-Field Scanning Optical Microscopy

Nowadays, the near-field optical microscopy community is facing new and exciting challenges as discussed throughout this chapter. However, we would like to begin by considering two events that took place more than a century ago but still maintain their relevance. It was precisely in 1873, when Ernest Karl Abbé stated a rigorous criteria, the well-known *diffraction limit*, that establishes the maximal resolving power that can be obtained in optical microscopy (Abbé 1873). Only 1 year earlier, Claude Monet painted "Impression, soleil levant" opening officially the school of impressionism, which revolutionized the concept of painting. The defined sign of the brush on the canvas was passed over by a new concept: color patches were demanded to create the atmosphere and the impression of a sunrise at Le Havre harbor. If we look at the coincidence of these two events in our recent history, both of them propose the same concept: we cannot observe and describe the reality with sharp boundaries, we can only achieve a blurred sight on the world. In the case of optical microscopy, the limit of our observation capabilities has been fixed by the diffraction limit for more than one century until the mid-1980s when the near-field scanning optical microscope (NSOM) was realized and the diffraction limit was broken for the first time.

The implications in biology for achieving such super-resolution are tremendous. It is now possible to observe phenomena at the molecular scale, not only visualizing cells but also their details and the organization of individual molecules. In fact, finally the nanometer-scale cellular "building blocks" can be studied in their natural environment, thus preserving their functions, while obtaining a high spatial resolution combined together with a wealth of information on structural, dynamical, and chemical properties of the specimens.

In this chapter, we first describe the principle of NSOM going through the development of the technique. In Section 18.2, we will review the most outstanding achievements obtained in biology using NSOM, while in Section 18.3 we highlight new frontiers in near-field optical microscopy, which are now pushing the NSOM community to new and exciting challenges.

18.1.1 Modern Microscopy Pushes the Resolution

The current broad search for super-resolution microscopes should not be surprising. In fact, an increasing resolution of imaging devices has driven science and technology for many centuries. Seminal studies about optics were already carried out in the eleventh century by Ibn al-Haytham, a Persian polymath, who is regarded as the father of optics for his influential *Kitab al-Manazir* (*Book of Optics*). The invention of the compound microscope during the last decade of 1500, thanks to Hans Lippershey, and subsequent improvements in the nineteenth century to optical microscopy by Robert Hooke, Anthony van Leeuwenhoek, and Abbé have revolutionized all aspects of science, and especially biology: it became possible, for instance, to see bacteria and blood cells. Since then, Nobel Prizes have also been awarded several times for the development of new imaging techniques: in 1903 to Richard Zsigmondy for the slit optical ultramicroscope; in 1953 to Frits Zernike for the phase-contrast microscope; in 1986 to Ernst Ruska for the electron microscope, and to Gerd Binnig and Heinrich Rohrer for the scanning tunneling microscope (STM). Nevertheless, despite so many achievements, the stimulus to develop super-resolution microscopy techniques has never been lost and during the last 30 years new powerful imaging techniques have been proposed and implemented.

Regarding optical microscopy, the photonics community accepts that the quickly fading (nonpropagating) evanescent field, which also originates from scattering when a sample is illuminated with light, is essential to achieve subwavelength resolutions. This idea is exactly what NSOM exploits: irradiate in the near-field and/or collect the near-field radiation by placing a probe in close proximity to the specimen (i.e., a few nanometers away from the sample). By the way, as we will discuss, achieving subdiffraction limit images is only one of the capabilities of the NSOM.

Today, as a consequence of the increasing complexity of biological requirements, subdiffraction images with single-molecule sensitivity are achievable in the near-field with NSOM, as well as in the far-field (Hell 2007) thanks to techniques like stimulated emission depletion (STED) microscopy (Hell and Wichmann 1994, Schmidt et al. 2008), photo activation localization microscopy (PALM) (Hess et al. 2006, Betzig et al. 2006), and stochastic optical reconstruction microscopy (STORM) (Rust et al. 2006, Huang et al. 2008), which have been proposed, respectively, in 1994 (STED) and 2006 (PALM/STORM). The appearance of these new subdiffraction techniques and the severe technical difficulties of NSOM have caused a slowdown in the evolution of near-field microscopy after an initial broad enthusiasm in the 1990s (Edidin 2001). However, as we will see, new added capabilities achievable with near-field microscopy are now again driving the attention of researchers. In fact, STED, PALM, and STORM rely on fluorescence and on specific photo-physical properties of fluorescent markers. On the contrary, NSOM is a truly optical technique and encompasses all possible contrast mechanisms furnished by light (absorption, transmission, scattering, interference, and fluorescence), which enlarges the applications of the NSOM.

18.1.2 A Scan through History

The idea of employing near-field optics to achieve images beyond the diffraction limit was initially formulated by an Irish scientist, Edward H. Synge, who described a method "which makes the attainment of a resolution of 0.01 μm and even beyond" possible (Synge 1928). Synge, encouraged by Albert Einstein, published his ideas and provided a very clear and clever proposal on how to realize such an instrument. The suggested implementation involved the realization of an illumination source with dimension *a* much smaller than the wavelength of the light (λ), and then to place such a subwavelength source close to the sample, which is then scanned. In such a way, an image with subwavelength resolution will be obtained. The original suggestion of Synge was to create a minute aperture (of the order of 100 nm) in a metallic screen. Later, to use a strong light source behind the screen to obtain, through the aperture, a very small light source. The tiny spot of light created in this way would then be used as local illumination source for the imaging of biological specimens. In order to achieve near-field illumination, Synge pointed out that it is necessary to keep the aperture at a distance (*d*) from the sample much shorter than the aperture dimension itself ($d \ll a$). The light transmitted through the sample would be later recorded point by point with a sensitive detector to form the image. However, at the time, several technical difficulties (including the realization of the aperture and the nanometer control of the distance between aperture and sample) made Synge discard his initial idea and he never tried to realize it (Synge 1931). In 1956, O'Keefe proposed a similar concept and he failed again due to the lack of technology (O'Keefe 1956). The first experimental realization of Synge's proposal came more than 40 years after the initial suggestion. In 1972, Ash and Nicholls realized such a scheme operating in the microwave region and they reported a resolution of the order of one-sixtieth of the wavelength (Ash and Nicholls 1972). In the mid-1980s, exploiting the technical progress after the recent introduction of the STM (Binnig and Rohrer 1982), Synge's scheme was realized with visible light. In 1984, Pohl and coworkers (Pohl et al. 1984) at the IBM research laboratories in Zurich and, independently, Lewis and his group (Lewis et al. 1984) at Cornell University realized the first NSOM prototype. In both instruments, the key innovation was the fabrication and implementation of subwavelength optical sources with sufficient output power. The idea of realizing a small hole in a metallic screen was replaced by the realization of an aperture at the apex of a sharp transparent probe tip coated with metal (Betzig et al. 1986, Dürig et al. 1986). Finally, two major issues toward the consolidation of NSOM as the first subdiffraction optical microscope were implemented by Betzig and Trautman at AT & T Bell Laboratories in the early 1990s. First, the development of a (shear-force) feedback system; and second, the utilization of aluminum-coated, tapered, single-mode optical fibers as nanosources (Betzig and Trautman 1992).

18.1.3 The Near-Field Approach

In far-field optical microscopy, the diffraction limit implies that the minimum distance Δ*x* required to independently resolve two distinct objects is dependent on the wavelength λ of the light used to observe the specimen, and by the condenser and objective lens system, through their refractive indices *n* and angle of acceptance* α:

$$\Delta x = \frac{\lambda}{2n \sin \alpha},$$ (18.1)

this implies that Δ*x* exceeds 300 nm in the case of visible light.

---

* Usually, the quantity $n \sin \alpha$ is referred to as the numerical aperture (NA) of an optical system. In order to increase the optical resolution of an imaging technique, it is worth increasing NA either by a large index of refraction *n* of the surrounding medium or with a large angle of acceptance α.

When an object, such as microscopic specimen, is illuminated with a (monochromatic) plane wave, the transmitted or reflected light, scattered by the object in a characteristic way, is collected by a lens and projected onto a detector to form the image. Usually, for convenience and practicality, the detector is placed in the far-field, i.e., multiple wavelengths away from the sample, so that the far-field component of the light, which propagates in an unconfined way, is the only component used to generate the image. On the other hand, the interaction between the imaging light and the specimen also generates a near-field component, which consists of a nonpropagating (evanescent) field existing only near the object at distances less than the wavelength of the light. Because the near-field decays exponentially within a distance less than the wavelength, it cannot usually be collected by the lens; thus, it is not detected (Goodman 1968). This effect leads to the well-known Abbé's diffraction limit.

By detecting the near-field component before it undergoes diffraction, NSOM allows nondiffraction-limited high-resolution optical imaging. This is achieved in NSOM by placing a probe tip in close proximity to the sample in order to illuminate and/or detect in the near-field. In NSOMs, the resolution Δx no longer depends on λ but instead on the aperture diameter a of the probe. Near-field optical microscopy relies on a confined flux of photons between the local probe and the sample surface. Besides that, the overall instrument design can vary significantly depending on the requirements of the particular applications. Although the illumination mode (see Figure 18.1) was first introduced in NSOM and is still the most employed in near-field optical microscopy, the reciprocity theorem suggests the possibility of exchanging the positions of sources and detectors: one may collect the light in the near-field in order to get the information contained in the evanescent waves. This NSOM variant was developed a few years after the illumination mode set up (Betzig et al. 1987). The idea was to reverse the light path in the illumination mode NSOM: the sample is illuminated through a classical microscope objective and the detection in the near-field is done with the tip of the optical fiber. For an exhaustive discussion about image formation in collection-mode NSOM, see Greffet and Carminati (1997). Another completely different type of NSOM is the so-called apertureless NSOM (aNSOM) introduced in the early 1990s (Specht et al. 1992, Inouye and Kawata 1994, Zenhausern et al. 1995). In this case, the advantage is to place both source and detector in the far-field while bringing only a small scattering tip into the near-field. The tip acts as a small probe that locally perturbs the electromagnetic field. The evanescent field arising from the interaction between the radiation and the specimen is converted to propagating radiation by the tip. It is then detected in the far-field at the same frequency of the incident light, for a recent review on aNSOM see Novotny and Stranick (2006). Although aNSOMs could present a higher signal-to-noise ratio, potential topographic artifacts complicate the interpretation of the signal recorded for the image formation (Hecht et al. 1997, Gucciardi et al. 2007). Even if DNA patterns have been successfully resolved with aNOSM (Akhremitchev et al. 2002), this particular technique is not (yet) commonly employed in near-field optical microscopy for biological applications. Finally, as we will see in

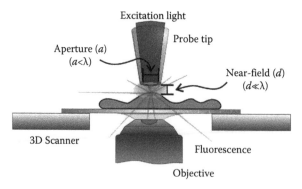

FIGURE 18.1   Schematic of aperture-type NSOM. A probe is placed in the near-field of the specimen (i.e., the tip-to-specimen separation d is of the order of few nanometers) and used to illuminate the sample. To a first approximation, resolution is defined by the size of the aperture a and not by the wavelength λ of the excitation light.

the last section of this chapter, instead of using a pointed probe as a local scatterer one can use it as a local light source. Under proper polarization and excitation conditions and probe geometries, the tip can locally enhance the electromagnetic field and behave as a local excitation source (Denk and Pohl 1991, Novotny et al. 1997). This approach enables simultaneous spectral and (subdiffraction) spatial measurements and is commonly referred to as tip-enhanced near-field optical microscopy. It has to be emphasized that the field enhancement and the scattering efficiency are interrelated phenomena. Hence, a near-field optical microscope working efficiently in local excitation mode also works efficiently in the local scattering mode. Finding conditions for a strong local-field enhancement benefits both measurement modalities.

Some of the possible near-field configurations are depicted in Figure 18.2. Despite the several possible approaches, it is worth noting that artifacts, particularly when imaging rough surfaces, may arise because of asymmetrical illumination or detection (Wu 2006), and only a few NSOM configurations have been commercialized.*

18.1.4 The NSOM Head and the Feedback Mechanism

The heart of all scanning probe microscopy techniques is the probe head and the scanning system. Their design and function determine the attainable scan resolution. Scanners, usually piezoelectric actuators, must have low noise (small position fluctuations) and precision positioning capability (typically less than 1 nm) and are commonly used to move the sample under the tip, which remains coaxial with the microscope objective. The required precision in the control of the distance between the sample and the probe implies that the instrument rests on a vibration-isolating optical table, or suspended in order to minimize mechanical vibrations. For most NSOM applications, it is necessary to precisely maintain the probe above the specimen surface but to prevent it from contacting the surface. In the case of contact, in fact, either the damage of the probe or of the specimen or debris accumulation on the tip are highly likely. The engagement of a constant gap between the probe and the specimen is best met by employing a

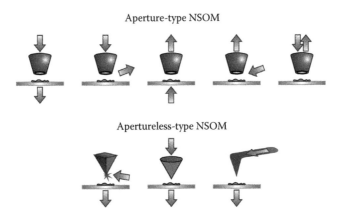

FIGURE 18.2 Different NSOM configurations. Upper panel: Aperture-type NSOMs may work both in reflection and transmission modes either with coaxial or asymmetric illumination and detection. Lower panel: Apertureless type of NSOM configuration is also possible. In these cases, excitation light can be either locally scattered by a tip or exit out from an uncoated taper or a bent optical fiber.

---

* At the moment, only a few companies produce and commercialize near-field optical microscopes. These companies are Olympus (Japan), WITec GmbH (Germany), Nanonics Imaging Ltd. (United States), Jasco (United Kingdom), and Omicron NanoTechnology GmbH (Germany). The configurations commercialized always rely on symmetric illumination-detection pathways, both in transmission and reflection modes.

real-time feedback control system. In this way, accurate optical signal levels are obtained together with simultaneous topography of the sample surface.

Several different techniques have been implemented so far to monitor the vertical position of the probe tip. First NSOMs relied on electron-tunneling feedback (Pohl et al. 1984, Dürig et al. 1986), later extended to photon tunneling in the photon scanning tunneling microscopy (PSTM) (Reddick et al. 1989, Courjon et al. 1989). Today, the majority of NSOMs utilize two distinct feedback methods, which have analogous sensitivity and performance and are similar to noncontact AFM. These techniques are called shear-force feedback (Betzig et al. 1992, Toledo-Crow et al. 1992) and tapping mode feedback (Shalom et al. 1992, Tsai and Lu 1998). This latter method implies the use of bent tips (see also Figure 18.2), which are usually characterized by lower optical throughput, technical difficulties in the realization, and higher mechanical vibrations. However, straight probes, typically employed in the shear-force feedback, are easier to fabricate, and have lower cost and higher throughput. On the contrary, the shear-force feedback technique works best when specimens have relatively low surface contrast but require longer scan times. To date, the most commonly employed NSOM configuration relies on shear-force feedback based on the use of quartz tuning forks (Karraï and Grober 1995a,b). In this approach, the NSOM tip is glued onto one of the arms of the tuning fork (see Figure 18.3). The crystal, which is a sort of piezoelectric diapason, is externally driven with a dither piezo at its resonance frequency.* In this way, the probe is oscillated laterally (with an oscillation amplitude of a few nanometers) over the sample. In typical operation, as the oscillating probe approaches the specimen surface, the amplitude, phase, and frequency of oscillation change due to dissipative and adiabatic forces present at the tip end. The probe oscillation damping due to tip–specimen interaction increases nonlinearly with decreasing tip-to-pecimen separation. During the final approach, when the distance between tip and sample becomes smaller than 20 nm, short-range interaction forces between the tip and the sample become detectable. Due to the action of shear forces, the resonance frequency is detuned with respect to the driving oscillation generating a decrease in the oscillation amplitude and a phase shift. The origin of the shear force is not fully understood and different mechanisms, such as viscous damping of water films or electrostatic (Van der Waals) forces, have been invoked to play a role. Although a clear picture of the mechanisms acting in the shear force is still pending, the use of pure phase feedback (Ruiter et al. 1997) or a combination of amplitude and phase feedback (Pfeffer et al. 1997) furnishes a signal reacting fast enough to perform feedback. Basically, the piezoelectric potential is acquired from the electrodes on

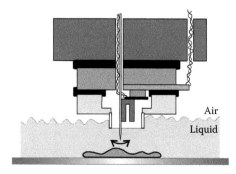

FIGURE 18.3 The NSOM scanning head. Schematic of the NSOM configuration for shear-force feedback. The NSOM probe is oscillated parallel to the surface via a tuning fork. Typically, the peak resonance, the amplitude, and the phase of the oscillation are found to change upon approach of the tip to the specimen surface and used as input to the feedback loop to engage the tip at a fixed distance from the sample. (Drawing not to scale.)

---

* The tuning fork can be either driven internally (electrically) or externally by a dither piezo to which the fork is rigidly attached. The mode of oscillation of the tuning fork depends on the type of excitation. If the fork is directly driven (electrically), the arms vibrate in opposite directions, whereas external mechanical excitation produces oscillations in which both arms of the tuning fork move in the same direction (Rensen 2002).

the quartz and properly amplified. The signal is subsequently fed into a lock-in amplifier and referenced to the driving signal of the oscillating tuning fork. The output from the lock-in amplifier (amplitude, phase, or a combination of amplitude and phase) is then compared to a user-specified reference signal in the control loop to maintain the probe at a fixed distance over the specimen obtaining the topography of the sample and avoiding the tip–sample contact.

When the tip is oscillated in order to sense the distance to the sample, the quality of the mechanical oscillation is crucial. The quality of a mechanical oscillator is given by a dimensionless parameter called the quality factor, or Q factor, or simply Q.[*] It is generally beneficial to maximize the Q of the probe in order to achieve greater stability and more sensitive tip height regulation. The lower the Q of the oscillating probe, the lower the signal-to-noise ratio, which results in a correspondingly lower quality topographic information. When quartz tuning forks are utilized, their very high Q factors and the corresponding high gain provide the system with high sensitivity to small forces, typically on the order of 100 pN. In addition, to reduce the interaction force[†] between the probe and the sample, the tip is usually oscillated ideally with an amplitude of the order of 1–5 nm.

Shear-force imaging with straight probes, however, is usually very difficult to perform in liquid conditions. In fact, the additional viscous damping effect of the fluid causes a dramatic decrease in the probe oscillation amplitude. Considering the relevance for biological applications to work in liquid conditions, many efforts are now devoted to improve NSOM operation in liquid. A possible solution might be the use of tuning forks with as high as possible resonance frequencies. The tuning forks routinely employed are extracted from clock oscillators[‡] and their resonance frequency is ~32 kHz, but devices with resonances ranging from 10 kHz to several tens of megaHertz are also available. Nevertheless, it has been recently shown that it is possible to image soft samples in liquid conditions with conventional tuning forks (Gheber et al. 1998), also by taking advantage of a diving bell concept (see Figure 18.3) to keep the tuning fork system in air (Koopman et al. 2003).

18.1.5 Getting an NSOM Image Requires the Right Probe

In NSOM probes not only is the Q factor crucial, the optical properties and the throughput efficiency of the tip are also fundamental and many efforts have been directed to the microfabrication of near-field optical probes. Despite that, only a few different types of tips are commercially available and most NSOM users still rely on self-made probes. The most widely used NSOM probes, especially in the case of biological applications, are straight probes based on tapered optical fiber with an aperture at the apex. Nevertheless, other different geometries have also been proposed. For instance, pyramidal tips with an aperture at the apex have been used to increase the resolution in first NSOMs (van Hulst et al. 1993) while triangular tips have been introduced very recently and present the advantage of a precise determination of molecule orientations (Molenda et al. 2005).

The realization of NSOM straight probes based on optical fibers is achieved by first creating a transparent taper with a sharp apex. Then, an aluminum coating is deposited on the walls of the probe cone in order to confine the light inside the tapered region. Aluminum is commonly preferred to other opaque materials because of its very small penetration depth, which implies a high reflectivity. Two different strategies are commonly used to produce tapered optical fibers with sharp tips: chemical etching

---

[*] Q is defined as the oscillator's resonance frequency f_R divided by its resonance width Δf. Δf is the width of the resonance peak at the points where the amplitude is equivalent to the peak amplitude divided by the square root of 2, or approximately 70.7% of the peak amplitude (Karraï and Grober 1995a).

[†] The interaction force F between the tip and the sample can be approximated as $F = k_D \cdot x$, where x is the oscillation amplitude and k_D is an effective spring constant that is related to the tuning fork spring constant k through the relation $k_D \propto k/\sqrt{Q}$, where Q is the quality factor. Thus to minimize F, the Q factor should be as high as possible, whereas the lateral oscillation x should be as small as possible.

[‡] Normally these forks have a Q of 10,000 when encapsulated, while a bare tuning fork has a quality factor of about 2000–3000 which further reduces to about 1000–1500 once the tip is glued to the fork.

(Hoffmann et al. 1995) and the "heating and pulling" method (Betzig et al. 1991). Chemical etching consists of dipping the optical fiber into a HF solution with a supernatant organic solvent: the tip formation occurs at the interface between HF and solvent around the fiber. The undergoing phenomenon is the relation between the meniscus height and the remaining fiber diameter. Achieving reproducibility on a large-scale production of NSOM probes is, without doubt, an interesting aspect of chemical etching. In addition, chemical etching permits to control the taper angle by varying the organic solvent used in the process. On the other hand, the microscopic roughness of the taper surface after the etching process limits the optical properties of these probes. A few years ago, this limitation has been partially overcome with the introduction of the so-called tube-etching method (Lambelet et al. 1998). The heating and pulling method is able to produce very smooth glass surfaces on the taper allowing a high-quality aluminum deposition on the tip. Unfortunately, with this latter method it is difficult to obtain large cone angles leading to lower values of the transmission coefficient (Hecht et al. 2000). Basically, the heating and pulling method consists in locally heating (with a CO_2 laser or a filament) an optical fiber that is then pulled apart and finally fractured to form two facets at the apex. By controlling temperature, pulling, and heating it is possible to achieve control on the shape of the resulting fiber.

Whatever technique is used to form the tapered fiber (see Burgos et al. 2003a) for a comparison between the two different methods of taper realization), the aperture is realized during the evaporation process. In fact, as depicted in Figure 18.4a, the aluminum evaporation takes place in a slightly tilted geometry and the deposition rate at the apex is smaller than on the side. Thus, a kind of shadow effect leads to the formation of the aperture at the apex. Nevertheless, it is worth noticing that nowadays it is more common to see "sculptured" tips obtained by precise control of focused ion beam (FIB) milling (Veerman et al. 1998). FIB milling enables not only an accurate control of the aperture size, as shown in Figure 18.4b, but also provides a means of carving specific shapes with nanometer-scale precision (see also Figure 18.12).

The size a of the NSOM probe tip aperture determines (to a large extent) the resolution of the NSOM and is a primary factor in achieving high-resolution images: smaller apertures provide higher resolution. However, reducing the aperture diameter decreases the number of photons exiting the aperture. This, in turn, lowers the signal levels at the detector and decreases the signal-to-noise ratio, which imposes a conflicting limit on the resolution improvement that can be achieved by further reduction of the aperture. Thus, a sufficient diameter to provide the desired optical throughput must be maintained. The throughput can be significantly increased by using probes with large cone angle (Hecht et al. 2000), or by modifying the tip physical characteristics, for example, by changing the heating and pulling parameters, or the etching variables, or by varying the angle of the fiber during metal evaporation.

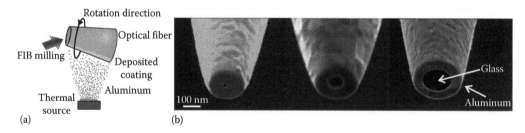

FIGURE 18.4 Near-field optical probes. (a) Sketch of the evaporation process. Evaporation is performed in order to deposit an opaque layer on the taper of the optical fiber to achieve transmission through the aperture. Aluminum is the most commonly employed metal for this purpose. Evaporation, due to the tilted geometry, already leads to the formation of an aperture at the apex of the taper. FIB is commonly used to precisely shape the tip end. (Drawing not to scale.) (b) SEM images of aluminum-coated FIB-milled NSOM probes with different aperture diameters at the apex.

The tip optical throughput has been experimentally shown to be proportional to the diameter of the aperture, a, divided by the wavelength of light, λ, raised to the fourth power (Massey 1984):

$$\text{Optical throughput} \sim \left(\frac{a}{\lambda}\right)^4 ; \tag{18.2}$$

nevertheless, previous calculations have proposed a throughput efficiency of the order of 10^{-6} (Bethe 1944). To have an idea of the effective throughput, let us consider an aperture a of the order of 100 nm and let us suppose to use red light (from a He:Ne laser for instance, $\lambda = 633$ nm) to obtain our image. In these conditions, the maximal optical throughput is of the order of 6×10^{-4}. In reality, probes and especially commercial ones have a much lower throughput efficiency, with typical values of the order of 10^{-6} (Valaskovic et al. 1995). Furthermore, in illumination mode NSOM, due to thermal effects, it is not possible to compensate the low throughput of the fiber probes by increasing the power of the light: absorption in the taper region will literally fuse the optical fiber. In fact, the light coupled into the core of the fiber when it reaches the tapered region of the probe begins to propagate through the cladding to the metal coating. Because the metal has a nonzero absorption coefficient, photon absorption occurs in the metal layer, and heat is generated. The optical signal transmitted through the probe is limited by the ability of the tapered region to physically tolerate and to dissipate this heat.* Among the strategies for reducing probe warming are increasing the thickness of the deposited metal layer and increasing the cone angle of the tapered tip. Interestingly, in certain NSOM applications, controlled tip heating could even be used to have a beneficial side effect (Erickson and Dunn 2005, Gucciardi et al. 2005).

18.1.6 A Bio-NSOM

For biological applications, the most widely used configuration is an aperture-type NSOM, incorporated into an inverted optical microscope, with near-field excitation and far-field detection (see Figure 18.5). This scheme preserves most of the conventional imaging modes (confocal microscopy, for instance), which remain available in combination with the near-field approach. Light that is emitted by the aperture locally interacts with the sample. It may be absorbed, phase shifted, or used to locally excite fluorescent markers, depending on the sample and the contrast mechanisms employed. In any case, light emerging from the imaging zone must be collected with the highest possible efficiency. For this purpose, high NA (oil immersion) microscope objectives are usually employed. The collected light is directed to sensitive detectors, such as avalanche photo-diodes (APD) or photo-multiplier tubes (PMT), via a suitable dichroic mirror for spectral splitting or through a polarizing beam splitter cube for polarization detection. Filters are also commonly used to select the spectral regions of interest removing unwanted spectral components, and inverted optical microscopes are an advantageous solution for light collection, redistribution, and filtering.

In Figure 18.5, the excitation light from one or more laser sources is coupled into the optical fiber. As already described, the tip is maintained in the near-field of the specimen by the feedback system operating in close loop that precisely controls the separation between the probe and the sample. In addition, a 3D scanner (usually based on three piezoelectric actuators) is employed to control the relative positioning of sample and probe. Depending on design and applications, in principle the scanner

---

* The thermal destruction of an aluminum-coated NSOM probe tip has been experimentally shown (Stähelin et al. 1996) to occur at approximately 470°C (considerably lower than the melting temperature of aluminum, 660°C), corresponding to an input-coupled power of about 10 mW. The tip damage most likely results from the different thermal expansion coefficients of the aluminum and the optical fiber, which causes the aluminum coating to separate from the glass. Interactive Java tutorials, illustrating this heating effect as well as other topics related to NSOM operation, can be found in the Microscope Resource Centre Web site of Olympus (http://www.olympusmicro.com/primer/techniques/nearfield/nearfieldhome.html).

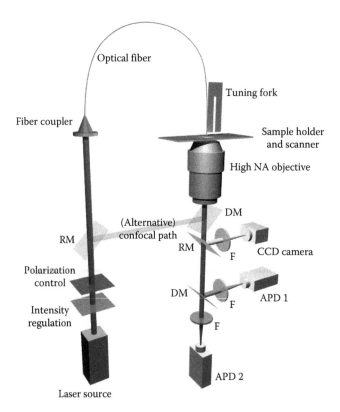

FIGURE 18.5 Rendering of a near-field optical microscope. In this arrangement, the NSOM head with tuning fork, probe, and scanners is placed at the sample location while the objective remains beneath the stage. The system depicted in the figure also implements one or more external laser sources to provide illumination, neutral density filters, and polarization optics to regulate the excitation conditions. When integrated into inverted optical microscopes, NSOM could exploit halogen illumination also to achieve sample inspection via removable mirrors (RM) used to project the bright-field image of the sample onto a charge-coupled device (CCD) camera and also standard confocal microscopy could be guaranteed. Single-molecule sensitivity is provided by two APD, which can be used either to detect two different spectral components or polarization states. Adequate optics, such as dichroic mirrors (DM) and filters (F), are routinely used to select the desired optical path and spectral components. Tip-probe distance is regulated through the feedback loop. Control of the scanners and data acquisition are usually driven through additional electronics and remotely controlled with a computer.

may either move the specimen or the probe. In the case of the scanner locked to the specimen, which is the most employed configuration for biological imaging, the sample is moved in a raster pattern. The image is generated from the signal arising from the tip–specimen interaction under the probe, which is fixed and aligned coaxially with the objective. The size of the area imaged depends uniquely on the coarse of the scanners. During raster scanning, data obtained both from the feedback system and from the optical detectors are simultaneously stored by a computer point by point. Finally, the PC compiles and renders the acquired data and simultaneously furnishes topography and optical image of the specimen.

In the following section, we will concentrate on the biological achievements obtained using NSOM. Such results mainly proceed from the capabilities of visualizing single molecule (Betzig and Chichester 1993) and proteins (Dunn et al. 1995) and from the accurate determination of molecular location and orientation (Veerman et al. 1999, van Hulst et al. 2000), as shown in Figure 18.6.

Let us conclude this section on near-field scanning optical microscopy by mentioning that NSOM is a well-established and unique tool in biology as well as in several other research fields, among which

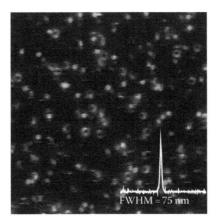

FIGURE 18.6 (See color insert following page 15-30.) Individual Dil molecules embedded in a thin PMMA layer. The color coding indicates the in-plane transition dipole of individual molecules, with green for molecules oriented along the x-direction and red for molecules oriented in the y-direction. Doughnut-shaped features correspond to out-of-plane oriented molecules excited by the strong electric field localized at the boundaries between glass and aluminum. Size of the image: $4.2\,\mu m \times 4.2\,\mu m$ ($2.5\,ms/pixel$). The inset shows the intensity profile of a single molecule with a full width half maximum of 75 nm.

single-molecule spectroscopy (Trautman et al. 1994, Sunney-Xie and Dunn 1994), Raman spectroscopy (Webster et al. 1998), semiconductor heterostructure spectroscopy (Richter et al. 1997, Intonti et al. 2001), and ultrafast (Guenther et al. 1999) and coherent (Toda et al. 2000) spectroscopy, and is also used for nonlinear optical tuning of photonic crystal microcavities (Vignolini et al. 2008).

18.2 NSOM in the Biological Domain

While in France, Claude Monet revolutionized the concept of painting, it was just over the border in Spain that at the end of the nineteenth century Antoni Gaudí began revolutionizing concepts in the field of architecture. Particularly interested in the angles and curvatures he found in nature during his frequent long walks, he started incorporating these into the design of buildings, such as the Sagrada Familia and Casa Milà. Actually, a common feature of his work was the incorporation of organic-shaped designs decorated with a mosaic of variously colored pieces (see Figure 18.7a). This was in sharp contrast to the strict and sober geometrical style seen before him. In recent years, a similar concept-change has been taking place in cell biology: the plasma membrane of the cell is no longer seen as just a geometrically restricted interface, which contains proteins floating randomly around in a lipid–fluid bilayer (Singer and Nicolson 1972). Indeed, evidence such as distinct lipid domain formation (Simons and Ikonen 1997) and protein microclustering (Sieber et al. 2007) point toward the need for a mosaic representation of the plasma membrane (Edidin 2003, Engelman 2005), as illustrated in Figure 18.7b. The size distributions of membrane mosaic pieces are thought to range from 5 nm for lipid shells (Anderson and Jacobson 2002), 5–500 nm for protein clusters (Sharma et al. 2004, Douglass and Vale 2005, de Bakker et al. 2007), to 20–200 nm for lipid domains called lipid rafts (Simons and Toomre 2000, Kusumi et al. 2004, Pike 2006). This organization appears to be of eminent biological importance, as demonstrated by the formation of supramolecular complexes required to sustain cell engagement during the immunological synapse (Grakoui et al. 1999), receptor clustering to enhance pathogen binding and internalization (Cambi et al. 2004), and cell adhesion (Cambi et al. 2006). More intriguing is the potential association of these receptors to lipid rafts to supposedly assist in the formation of larger functional complexes (Simons and Ikonen 1997). However, the driving mechanism for the formation of these nanodomains remains poorly understood. Do different components diffuse stochastically to form larger functional units upon

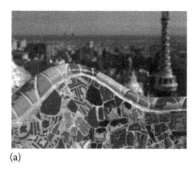
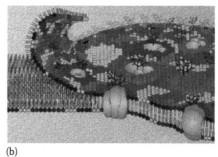

(a)   (b)

FIGURE 18.7 Mosaic-like structures in art and cell biology. (a) Photograph of a part of the mosaic at Parc Güell and (b) schematic representation of the cell membrane. (Schematic reproduced with permission from Eddin, M., *Nat. Rev. Mol. Cell. Biol.*, 4, 414, 2003.)

random encounters, or are they already preorganized into stable domains? So far, these questions have been difficult to assess because both protein and lipid domains are well below the resolution of fluorescence microscopy, and too densely packed to be resolved individually.

Along these rapidly changing developments in biology, NSOM has evolved itself from a technical point of view as discussed in the first part of this chapter. Because NSOM is essentially a surface-sensitive technique, it is ideally suited for cell membrane studies (de Lange et al. 2001, García Parajo 2005). In the following section, we focus on some examples of the use of NSOM in the biological domain. We especially focus on those works that, after the first demonstrations (Hwang et al. 1995, 1998, Enderle et al. 1997), have revealed parts of the still puzzling cell membrane mosaic. Obviously, NSOM is not the only technique that has been used to unravel the membrane mosaic. For some other biophysical reviews, the reader is referred to the other chapters within the book as well as to published reviews (Kusumi and Suzuky 2005, Marguet et al. 2006, Day and Kenworthy 2009). Additionally, NSOM research in the biological domain has not only focused on membrane compartmentalization and as such we refer to other relevant reviews (Dunn 1999, van Hulst et al. 2000, Lewis et al. 2003).

18.2.1 Artificial Lipid Mono/Bilayers as Test Models

The membrane of cells is the interface through which all interactions of the cell with its direct outer world take place. Current estimates consider the plasma membrane to consist of about 500–1000 different kinds of lipids and a multitude of different proteins. It is thought that different types of lipids actively phase-segregate (Mayor and Rao 2004), forming distinct compartments. A basic segregation principle could be dependent on the saturation of the carbohydrate chains, which reflect the phospholipid length and flexibility. Treating the cells with a cold detergent, such as Triton-X, demonstrated the differential solubility of the plasma membrane (Simons and van Meer 1988). One of those fractions, defined as the detergent-resistant membrane, or lipid rafts, is enriched in cholesterol and sphingolipids (Brown and London 2000). These lipid rafts are small domains that possibly act as nucleation sites for signaling platforms involved in many essential cell functions (Simons and Toomre 2000). To test this hypothesis but to avoid the inherent complexity of the cell membrane, model membranes have been used to investigate the segregation behavior of lipids and different proteins in predetermined lipid mixtures. The typical binary or ternary lipid mixtures used to mimic the lipid composition of cell membranes indeed phase-segregate into liquid condensed (LC) and liquid expanded (LE) phases.

By transferring monolayers of a lipid mixture on a substrate using standard Langmuir–Blodgett techniques, Hwang et al. have used NSOM to reveal previously unresolved features of around 50 nm (Hwang et al. 1995). When a higher pressure was used to form the monolayer, the domains of LC phase appeared to decrease in size and an increasingly complex fine web structure of the LE phase arised (Hwang et al. 1995). Cholesterol addition, typically enriching the LC phase, resulted in the formation of elongated thin LC domains. From these morphological changes, it was concluded that cholesterol reduced the line tension between the domains in regions of LC/LE coexistence. Likewise, the addition of the ganglioside GM1, again a LC constituent, affected the monolayer morphology significantly. Moreover, GM1 induced a more pronounced segregation between the LC and LE phases. These results suggested the formation of genuine distinct domains, thus favoring the occurrence of a lipid raft type of phenomenon. The lipids typically enriching the LC phase are significantly more saturated than lipids constituting the LE phase. Thus, when all lipids pack in their subsequent phase, the LE phase will be lower in height. Indeed, by specifically labeling the LE phase, a perfect correlation was found between the topographical and fluorescence signals (Hollars et al. 1998), see Figure 18.8. To extend these findings, Hollars and Dunn used their tapping-mode feedback NSOM set up to additionally obtain compliance information of the lipid monolayer (Hollars and Dunn 1997). Note that because the carbohydrate chains of the lipids from the LC phase are highly saturated, these lipids pack in an ordered fashion as compared to the lipids from the LE phase. As expected, the LC phase was found less compliant than the LE phase (Hollars and Dunn 1997). As such Hollars and Dunn demonstrated the strength of NSOM as compared to fluorescence or scanning probe techniques on their own.

In a more recent work supporting the formation of lipid rafts on model membranes, small amounts of labeled GM1 revealed that GM1 is not homogeneously distributed throughout the LC phase. Instead, they were seen to constitute their own 100–200 nm-sized domains (Burgos et al. 2003b). In fact, upon closer examination the labeled GM1 distribution appeared to be more complex. To better characterize the GM1 behavior, GM1 lipids were labeled with Bodipy. This fluorophore displays a redshift in the emission spectra when present in higher concentrations, thus being able to probe the local lipid density (Dahim et al. 2002). Due to the strong tendency of GM1 to partition in gel or liquid-ordered phases, high concentration GM1 was found in the LC phase, showing the redshifted emission, even while using low deposition pressures (Coban et al. 2007). Nevertheless, a rather large fraction of single Bodipy-GM1 was still found randomly distributed in the LE phase. Upon increasing the deposition pressure toward expected cell membrane pressures, the LC domain phases became smaller and the labeled GM1 appeared to preferentially partition into the LC phase (Coban et al. 2007). To conclude, it appears that sizes as well as segregation-strength of a lipid monolayer is influenced by composition and deposition pressure, both of which the cell is able to locally control.

The use of NSOM to investigate monolayers has been extended toward bilayers (Ianoul et al. 2003) and protein-containing lipid layers (Flanders and Dunn 2002, Sibug Aga and Dunn 2004, Murray et al. 2004, Höppener et al. 2005). The addition of proteins to such lipid phase-segregated model systems will be an important step in understanding how lipid-based interaction can influence protein distribution. Subsequently, monitoring the dynamics will then provide a more complete spatiotemporal

(a)   4 μm   (b)

FIGURE 18.8 NSOM image of a phase-separated lipid monolayer. Topography (a) and near-field optical image (b) of a fluorescently doped lipid monolayer showing the NSOM capability of obtaining topography and fluorescence simultaneously. (Courtesy of Prof. R. C. Dunn.)

map of proteins and lipids in a lipid bilayer. Work in this direction has been performed using atomic force microscopy in combination with fluorescent correlation spectroscopy (Chiantia et al. 2006). The recent proof-of-principle that dynamical studies can also be performed with NSOM (Vobornik et al. 2008) opens up an exciting field that combines high-resolution imaging with ultrafast dynamics. Indeed, the advantage of performing FCS on confined volumes has been recently demonstrated on living cells (Wenger et al. 2007, Eggeling et al. 2009). The incorporation of this approach in NSOM in addition provides surface sensitivity, topography, and multicolor fluorescence cross-correlation spectroscopy.

18.2.2 Protein Organization on the Cell Membrane

Because NSOM is a scanning probe method, optical and topographical information are simultaneously obtained, enabling the direct correlation of protein organization and position on intact biological membranes. Our group has used NSOM to image pathogen recognition receptors with high resolution on cells of the immune system, providing insight into the mechanisms exploited by the cell to ensure proper control of these receptors. By labeling the pathogen recognition receptor DC-SIGN with a specific monoclonal antibody, we found that as much as 80% of DC-SIGN is clustered on the cell membrane of immature dendritic cells (de Bakker et al. 2007), see Figure 18.9. These domains were randomly distributed over the plasma membrane and contained on average 5–20 DC-SIGN molecules (Cambi et al. 2004, de Bakker et al. 2007). The size distribution of the DC-SIGN domains was found to be around 185 nm, with values ranging from 10 to 500 nm. Furthermore, we could also demonstrate a remarkable heterogeneity of the DC-SIGN packing density in these clusters. This suggests that the large spread in DC-SIGN density per cluster likely serves to maximize the chances of DC-SIGN binding to a large variety of viruses and pathogens having different binding affinities (de Bakker et al. 2007). Indeed, the organization of DC-SIGN in nanodomains appeared crucial for the efficient binding and internalization of pathogens (Cambi et al. 2004). In contrast, other receptors such as those from the interleukin family (IL2R and IL15R) were found to have a constant packing density albeit forming domains of different sizes (de Bakker et al. 2008). Although IL2R and IL15R were found to pack at different densities, the

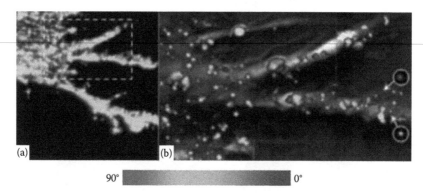

FIGURE 18.9 (See color insert following page 15-30.)   Organization of pathogen receptor DC-SIGN on the cell membrane of immature dendritic cells. (a) Confocal image (20 µm × 20 µm) of a dendritic cell stretched on fibronectin-coated glass expressing DC-SIGN on the membrane. (b) Combined topography (grey) and near-field fluorescence image (color) of the frame (12 µm × 7 µm) highlighted in (a). In both (a) and (b), the fluorescence signal is color coded according to the detected polarization (red for 0° and green for 90°). To illustrate the single-molecule detection sensitivity of the setup, the size as well as the intensity of two individual molecules are slightly enlarged and rescaled in the image [shown in circles in (b)]. Individual molecules are identified by their unique dipole emission (i.e., red and green color coding). The yellow color of most fluorescent spots results from additive emissions from multiple molecules with random in-plane orientation (combination of red and green) in one spot. (From de Bakker, B. I. et al., *Chemphyschem*, 8, 1473, 2007. Wiley-VCH Verlag GmbH & Co. KGaA. Reproduced with permission.)

linear increase in the number of receptors with domain size suggests a general building block type of assembly for these receptors (de Bakker et al. 2008).

Very recently, Chen et al. used NSOM in combination with quantum dots to label the T cell receptor (TCR) of T cells in live animals before and after cell stimulation (Chen et al. 2008b). In the resting state, the TCR complexes are found monomerically organized on the T cell membrane. Upon T cell stimulation, the TCR complexes reorganized and formed 270–390 nm sized domains. Interestingly, these small-sized domains were not only formed but also sustained for days. Additional experiments showed that although unstimulated cells could produce an immune response, stimulated cells produced significant higher levels of cytokines (Chen et al. 2008b). By means of these high-resolution NSOM experiments, it was shown that the TCR reorganization plays a significant role in antigen recognition and cytokine production.

In the case of members of the epidermal growth factor (EGF) receptor tyrosine kinase family, clustering is thought to have a negative effect. Some EGFs, like the erbB2 receptor, are found to be over-expressed in breast cancerous cells. It is thought that this over-expression leads to cluster formation causing the highly oncogenic activation of very potent kinase activity. In fact, by applying NSOM the clustering behavior of EGF receptors was found to be associated with the activation state of the cell (Nagy et al. 1999). Additionally, it was found that EGF cluster sizes increased if the quiescent cells were treated with EGF activators to the same extent as cells over-expressing these EGFs (Nagy et al. 1999). Since activation of the EGF signaling pathways requires extensive interaction between individual members of the EGF family, it is likely that concentrating one of these EGF receptors in clusters increases the likelihood of coclustering of other EGF members. This coclustering would then subsequently increase the EGF signaling efficiency. In other words, a higher local concentration will decrease the lag time for direct inter-receptor contact. This increased colocalization of individual members of the interleukin family has been actually demonstrated by combining dual-color excitation and single molecule detection NSOM (de Bakker et al. 2008). IL2R and IL15R did not interact if their organizations were monomeric. However, in their clustered form, both receptors were found to colocalize significantly suggesting that the clustering of both receptors takes place in the same nanocompartments (de Bakker et al. 2008).

Cell signaling events commonly involve a multitude of spatially segregated proteins and lipids. As such, standard confocal microscopy studies in biology usually involve multiple colors corresponding to multiple specifically labeled proteins. However, inherent to all lens-based techniques are chromatic aberrations that cause multiple wavelengths to never perfectly overlap. On the other hand, NSOM guarantees a perfect overlay between any excitation wavelength, which is an essential requirement to resolve the true nanoscale mosaic on a cell membrane.

In this sense, Enderle et al. used dual-color NSOM to directly measure the association of a host protein (protein 4.1) and parasite proteins (MESA and PfHRP1) in malaria (*Plasmodium falciparum*)-infected erythrocytes (Enderle et al. 1997). As the parasitic proteins interact with the host proteins, 100 nm-sized knob-like topographical features appear on the membrane of the host cell. To investigate the direct interaction of host and parasite proteins, the proteins were specifically labeled and subsequently imaged with NSOM. As expected, the fluorescence from the two labeled parasitic proteins and the labeled host protein were found on the knob-like structures. However, this did not necessarily involve the colocalization of host and parasite proteins (Enderle et al. 1997). Another more recent NSOM work spatially related topographical features to two different lipid species (Chen et al. 2008a). Both GM1 and GM3 were seen to cluster in 40–360 nm domains that distributed randomly on the plasma membrane of epithelial cells. However, upon closer examination it appeared that the GM3 clusters were localized on the peaks of microvillus-like structures (Chen et al. 2008a). In contrast, the majority of the GM1 lipid clusters were found in the valleys or slopes of these topographical protrusions (Chen et al. 2008a). These results highlight the importance of correlating topography and optical information uniquely afforded by NSOM.

In another study trying to relate the association of different components on the cell membrane, Ianoul et al. used NSOM to visualize both β-adrenergic receptors (βAR) and caveolae. High-resolution NSOM imaging revealed that about 12–72 β_2AR proteins aggregate into 140 nm-sized clusters, suggesting loosely packed domains (Ianoul et al. 2005). These clusters were distributed randomly on the cell membrane covering ~1 cluster · μm^{-2}. Consistent with biochemical experiments, differences between distinct βARs were observed with NSOM: β_1AR receptors had an increased and more heterogeneous receptor packing density in 100 nm-sized clusters as well as higher cluster coverage of ~3.1 cluster · μm^{-2}. The authors suggested that the higher expression level and inhomogeneous distribution of β_1AR compared to β_2AR probably reflect slight differences in function between both receptors (Ianoul et al. 2005). Because the caveolar localization of the β_2AR appeared to be essential for physiological signaling, the researchers performed colocalization NSOM experiments. These caveolae are frequently associated to lipid-rafts because of their detergent insolubility as well as the enrichment of cholesterol in these domains. High-resolution NSOM showed that ~15%–20% β_2ARs colocalize in caveolae (Ianoul et al. 2005), as shown in Figure 18.10. The lack of complete colocalization of β_2AR with the caveolae suggested that the diverse functional properties of the β_2AR could arise from its association with multiprotein complexes of different compositions that may not be caveolar in nature (Ianoul et al. 2005). Interestingly the fraction of β_2ARs not colocalizing with the caveolae appeared to be proximal to the caveolae. These results suggested that the β_2AR complexes are pre-assembled in, or near caveolae. More conventionally used techniques such as FRET are unable to report on such a proximity effect at spatial scales >10 nm. On the other extreme, diffraction-limited techniques such as confocal microscopy will not be able to reveal a lack of colocalization if multiple components are located at distances <300 nm. NSOM is capable of bridging the gap between 10 and 300 nm providing valuable information at these important spatial scales.

The next important step ahead is performing NSOM measurements on the plasma membrane of living cells. First preliminary demonstrations of NSOM measurements on living cells have been reported (Kapkiai et al. 2004, Ueda et al. 2007, Longo et al. 2008) although a high-resolution investigation on the membrane of living cells is yet to be demonstrated. Obviously, if the scanning speed is not significantly faster than protein diffusion, the optical signal will be blurred. Nevertheless, the promising demonstration

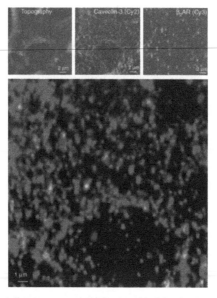

FIGURE 18.10 (See color insert following page 15-30.) Dual color NSOM imaging for colocalization studies. (Top panel) NSOM topography of neanatal cardiac myocytes (right) with fluorescence-labeled caveolin-3 (middle) and β2AR (left). (Bottom panel) NSOM fluorescent overlay image showing colocalization (yellow) of β2AR (red) and caveolin-3 (green) on the nm scale. (Reprinted by permission of Macmillan Publishers Ltd: Ianoul, A., *Nat. Chem. Biol.*, 1, 196, 2005. Copyright 2005.)

of using the subwavelength aperture to perform fluorescence correlation spectroscopy (Vobornik et al. 2008) opens the way for probing dynamics at relevant spatial scales. The ability to probe dynamics of the plasma membrane domains more directly is essential to unravel the driving mechanisms for nanodomains formation and evolution during cell activation. Additionally, multispecies cross-correlation spectroscopy should indicate if certain proteins are diffusing in identical domains. In conclusion, there is a wealth of knowledge in cell biology that still awaits to be uncovered by means of high-resolution techniques. The combination of capabilities that is offered though the use of NSOM makes the technique a worthy and essential asset in the spectra of biophysical techniques that nowadays exist.

18.3 Future Perspectives in Near-Field Optical Microscopy

New exciting ideas are triggering a renewed interest in the NSOM community. On one side, there is the promise of increasing further the NSOM resolution through the use of small optical antennas at the tip apex to achieve field enhancement and resolution of the order of few nanometers (García-Parajo 2008). On the other side, more biologically relevant applications will require routine NSOM imaging on live cells. This, in turn, would probably imply the use of tuning forks resonating at high frequencies in order to achieve fast scanning rates and to minimize the interaction forces between the tip and the living cell. Furthermore, there are also promising technical advancements to be exploited such as the combination of NSOM with other optical techniques such as fluctuation correlation spectroscopy (FCS), and Föster resonance energy transfer (FRET). In this respect, pioneering works have shown the capabilities of detecting FRET simultaneously to NSOM imaging (Vickery and Dunn 1999) and the possibility to perform FCS in the near-field by employing zero-mode optical waveguides (Levene et al. 2003) or NSOM (Vobornik et al. 2008).

There is also a growing interest in incorporating into NSOM other capabilities, which have been recently reported for AFM. In particular, as depicted in Figure 18.11, it will be extremely appealing to implement the molecular recognition technique (TREC) in NSOMs. TREC, which has been introduced by Peter Hinterdorfer a few years ago (Stroh et al. 2004, Hinterdorfer and Dufrêne 2006), involves the use of chemically functionalized tips to identify and localize specific molecules on the cell membrane during scanning (Chtcheglova et al. 2007). Finally, further technical improvements may come from the fast (r)evolution in the optical fiber field. In fact, novel structured optical fibers, such as photonic crystal fibers (Russell 2003, Carlson and Woehl 2008), hopefully could lead to new capabilities for near-field optical microscopy.

18.3.1 Exploiting the Power of Optical Nanoantennas

The idea of employing antennas to achieve subwavelength concentration of light beyond the diffraction limit was first proposed by Oseen in 1922, who demonstrated that an arbitrarily large fraction of the emitted energy can be sent into an arbitrarily small solid angle (Oseen 1922). As noted by Toraldo di Francia "fortunately it appears that the microwave researchers were not very much concerned, or perhaps even acquainted, with the old well-established theorems of wave-optics [on the Abbé/Rayleigh resolution criterion]…" (di Francia 1952), so that, for several decades, the microwave community contemplated the possibility of constructing antennas that beat the diffraction limit in directivity. Again, the lack of technical and manufacturing skills

FIGURE 18.11 Chemically active NSOM probes. One of the more interesting challenges for the near-field community is the implementation of the chemical recognition capability into NSOM. This can be accomplished by specific chemical or biochemical functionalization of the tips that would also allow the localization of specific molecules on the scanned surface.

made the realization of such antennas impossible and the idea was forgotten. Concerning near-field optical microscopy (in the visible region of the electromagnetic spectrum), the underlying concept is to use nanoantennas to locally enhance the electric field at the nanoscale to obtain a smaller illumination source. As we previously mentioned, resolution in near-field microscopy actually depends on the dimensions of the illumination aperture. Consequently, the reduction of these dimensions improves the ultimate resolution of the microscope. Unfortunately, it is not possible to infinitely reduce the aperture size because the optical throughput (and thus the detected signal) will drastically decrease. Using "aperture-type" NSOM, the optical throughput is already of the order of 10^{-4}–10^{-6} for a 70 nm aperture and, in practice, the resolution of NSOM is limited to about 50 nm (Hecht et al. 2000). In order to overcome these limitations, the "nanoantenna" concept has been exploited by the near-field community. Also in this case, researchers first looked back and decided to exploit the field enhancement effects, which arise from antennas when an electromagnetic radiation is resonantly coupled into the (metallic) nanoantenna. The final aim is to get a nanosource of radiation or, in other words, an almost punctual and well-controllable light source, while maintaining a throughput energy density high enough to perform optical microscopy. In a way, filling the gap between conventional illumination and the use of single molecules (Michaelis et al. 2000) and/or single photons (Kim et al. 1999, Lounis and Moerner 2000, Michler et al. 2000) as light sources. Despite earlier discussions on theoretical grounds (Wessel 1985, Denk and Pohl 1991, Novotny et al. 1998, Kawata et al. 1999), only in 1989, the field enhancement effect was first experimentally observed (Fischer and Pohl 1989), while the first high-resolution fluorescent image recorded with a sharply pointed gold tip was reported 10 years later (Sánchez et al. 1999). Field enhancement is stronger in metal tips (Haefliger et al. 2004) because, at optical frequencies, metals are characterized by small penetration depths and by electromagnetic resonances associated with the free electrons. The collective response of the free electrons can greatly enhance the electric field of the incoming radiation. The coupled excitation of electrons and electromagnetic field is generally referred to as a surface plasmon. Besides apertures, metal tips are the most commonly used probes in near-field optical microscopy. They are easy to fabricate and manipulate. The field enhancement at a sharp tip arises from a combination of quasistatic lightning-rod effect and surface-plasmon excitation (Novotny et al. 1997).

In the context of near-field optical microscopy, pointed metal probes are used in localizing and enhancing optical radiation. In principle, these probes fulfill the role of standard optical lenses used in imaging. However, in this context they no longer work as linear elements and are no longer limited by the laws of diffraction. Their function is similar to electromagnetic antennas that convert propagating radiation into a confined zone called the feedgap. In the feedgap, electric circuitry either releases or receives the signal associated with the electromagnetic field. Because of this similarity, pointed metal probes used in near-field optics are also referred to as optical antennas. Nanoantennas have advantages also in terms of lateral resolution in addition to the high field concentrations through the enhancement process. To increase the imaging resolution, the most promising approach is to employ the so-called tip-on-antenna (TOA) probes (see Figure 18.12). The first report concerning imaging with a TOA probe came in 1998 and accomplished a lateral resolution of the order of 150 nm (Matsumoto et al. 1998). Just a few years later, this value was already pushed down to 25 nm (Frey et al. 2002), then to 20 nm when taking advantage of a two-photon absorption process (Sánchez et al. 1999), and finally reached 10 nm (Frey et al. 2004). Nowadays, TOA-imaging resolution is in the range of 10–20 nm (Hillenbrand and Keilmann 2002).

In principle, by making use of electromagnetic resonances associated with surface plasmons, any metal nanostructure can be viewed as an optical antenna (including conventional NSOM probes). Of

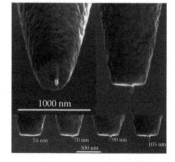

FIGURE 18.12 Tip-on-aperture probes. Upper panel: Different views of a nanoantenna at the apex of an aperture tip. Lower panel: SEM images of nanoantennas of different lengths. (Reprinted with permission from Taminiau, T. H. et al., *Nano Lett.*, 7, 28, 2007a. Copyright 2007 by American Chemical Society.)

course, the efficiency depends on the material composition and the geometry of the structure (Martin et al. 2001). Metal planar nanostructure on glass substrates is one of the most successful method reported up to date. Either resonant stripes (Mühlschlegel et al. 2005) and bow-tie antennas (Schuck et al. 2002), with gap dimension as small as 30 nm, or adjacent gold nanoparticles (Rechberger et al. 2003) have shown local intensity enhancements with factors of the order of 1000 times, providing also a new method for high-harmonic generation (Kim et al. 2008). The extraordinarily enhanced emission and the extremely small excitation volume achieved in these structures would enable the study of single-(bio)molecule interactions. The approach would be extremely interesting in applications where the stoichiometry of the system prevents the visualization of individual molecules with other techniques due to the size of the excitation volume. Exploiting optical antennas to boost field confinement in small areas further increases the limit of detection by orders of magnitude, enabling the monitoring of extremely weak chemical or biological reactions (see Figure 18.13a). Recent promising reports have shown that by exploiting nanoantennas, it is possible to reduce the illumination volume by a factor of 10,000 in FCS experiments (Estrada et al. 2008), and to enhance single molecule fluorescence (Wenger et al. 2008). It may even be possible for these resonant optical antennas to be used for single-molecule Raman spectroscopy. The arrays of optical nanoantennas could readily be used on substrates to which cells are attached. Feedgaps would work as "hot spots," where local FCS could be performed. This technique could be particularly suitable for investigating slowly diffusing supramolecular complexes on the cell surface.

Although planar structures would be highly suitable for monitoring (bio)-chemical reactions and transient dynamic interactions *in vitro* and *in vivo*, real imaging with nanometer resolution should rely on free-standing structures in combination with scanning probe methods. Along these lines, gold nanoparticles glued at the apex of glass tips have been exploited as nanoantennas (Kalkbrenner et al. 2001, Anger et al. 2006, Kühn et al. 2006) and used, for instance, to tune the radiative properties of single quantum dots (Farahani et al. 2005). With regard to biological nanoimaging, single Ca^{2+} channels on erythrocyte plasma membranes have recently been visualized for the first time using a gold nanoparticle-based optical antenna (Höppener and Novotny 2008). Furthermore, these experiments have the additional value of being performed in liquid conditions, thus, opening the door to a multitude of physiologically relevant experiments, including elucidating the complex relationship between the cell membrane and the cell function. Yet, the concept is still far from fully mastered and requires further improvements. In the approach reported by Höppener et al., a substantial amount of background signal is generated by the focused laser beam that is used to irradiate the optical antenna (Höppener and Novotny 2008). Also, although the work is performed at physiological conditions, at present the approach requires the cells to be fixed, which prevents dynamic studies.

A different excitation scheme that suppresses background illumination has been proposed (Frey et al. 2004) and, more recently, applied to antenna concepts (Taminiau et al. 2007a). In these TOA antennas, the main idea is to take advantage of the local illumination properties of aperture-type NSOM to drive the antenna to resonance. The antenna is thus a metallic tip on top of a normal nano aperture. In the initial experiment, reported by Frey et al., the tip was considered mainly as a nanoscale-scattering tip (Frey et al. 2004). Yet, simultaneous topography and fluorescence imaging of Cy-3-labeled DNA with 10 nm resolution was convincingly demonstrated (Frey et al. 2004). Using a similar experimental approach, but exploiting the resonant properties of the metallic tip (tuning the antenna dimensions to the wavelength used*), and using optimal excitation conditions, single-molecule detection with 30 nm resolution and virtually no background was demonstrated (Taminiau et al. 2007a, b).

---

* To this respect, it is worth noticing that at optical frequencies the simple wavelength scaling, i.e., the relationship between the characteristic lengths L of the antenna and the wavelength λ of the incoming (or outgoing) radiation, which usually follows the law $L = $ const. λ, where const is an antenna-design constant, breaks down because the incident radiation is no longer perfectly reflected from the metal's surface. Instead, radiation penetrates into the metal and gives rise to oscillations of the free-electron gas. Hence, at optical frequencies an antenna no longer responds to the external wavelength but to a shorter effective wavelength λ_{eff}, which depends on the material properties (Novotny 2007).

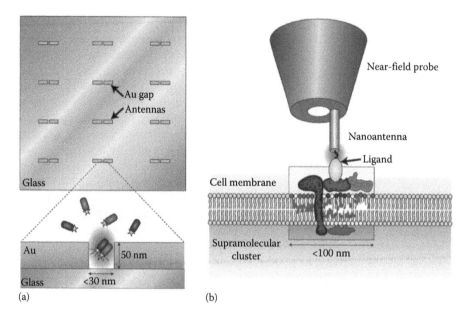

FIGURE 18.13 Potential applications of optical nanoantennas. (a) Sensing molecular interactions at high concentrations using gold gap nanoantennas. The reduced observation volume combined with the field enhancement enables the detection of weak interactions at the individual level. (b) Imaging molecular interactions within a dense supramolecular cluster on the cell membrane. The ligand attached to the nanoantenna activates the cell receptor and triggers cellular signaling. Topography, biochemical recognition, and fluorescence images at the nanometer scale are simultaneously obtained. (Drawings not to scale.) (Reprinted by permission of Macmillan Publishers Ltd: Garcia-Parajo, M. F., *Nat. Phaton.*, 2, 201, 2008. Copyright 2008.)

It is well recognized that nanometer-sized metal objects influence the emission pattern of individual quantum emitters. For instance, in 2000, Gersen et al. demonstrated that the proximity to the metallic aperture of a near-field probe was sufficient to shift the angular emission of single molecules from side to side (Gersen et al. 2000). These changes suggested that local optical fields could redirect the emission of individual molecules. The redirection of the angular emission was dependent on the molecule dipole orientation and the dominant emission pattern was still determined by the molecule itself. To truly redirect molecular emission requires coupling the original molecule to a second, much more efficient radiating system. In the microwave and radiowave fields (which are deeply inspiring the nanoantenna researches), efficient and directional electromagnetic radiation is often mediated by antennas. The same concept has been very recently translated into the nanoscale and into the optical domain, and the control over the nanoantenna emission direction has been finally achieved (Taminiau et al. 2008).

Within the biological domain, the use of nanoantennas would have an enormous impact on super-high-resolution imaging (see Figure 18.13b). Investigating multiple components on the native cell membrane at physiologically relevant packing densities, and correlating membrane topology (also with nanometer resolution owing to the sharp tip) with structural information, such as lipid rafts, membrane ruffles, or caveolae, are only a few examples of the many topics that can be addressed. Finally, the nanoantenna can be functionalized with specific chemical groups or ligands in a similar way as it is done in atomic force microscopy (Stroh et al. 2004, Hinterdorter and Dufrêne 2000). This would make it possible to simultaneously record topography, biochemical recognition, and fluorescence at the nanometer scale, and all at the individual-molecule level (see Figures 18.11 and 18.13b). Moreover, other configurations using metallic tips, such as infrared spectroscopy NSOM (Knoll and Keilmann

1999, Brehm et al. 2006), tip-enhanced Raman spectroscopy (Stöckle et al. 2000, Hartschuh et al. 2003, Pettinger et al. 2004), or coherent anti-stokes Raman scattering (CARS) (Ichimura et al. 2004), which provide chemical recognition by means of molecular vibrational "fingerprints" would benefit from free-standing nanoantennas. As already suggested, sharper tips should enable the recognition of chemical composition and the molecular conformation of single proteins or even their subunits (Brehm et al. 2006). The rapid development of nano-optics technology means that a bright future in biological nano imaging and (bio)chemical sensing lies ahead of us.

Acknowledgments

We are indebted to Dr. Olga Esteban, Ruth Diez Ahedo, and Gert-Jan Bakker (Single-Molecule BioNanoPhotonics group at IBEC) for discussions and continuous support. In addition, we appreciated discussions and references suggested by Silvia Vignolini and Francesca Intonti (LENS, Italy). Davide Normanno also acknowledges Silvia Baracani (Switch Project, Italy) for discussion about art, society, and science. Financial support from NEST-ADVENTURE EC: Bio-Light-Touch (to Davide Normanno and María García-Parajo) and ImmunoNanoMap (to Thomas van Zanten and María García-Parajo) is gratefully acknowledged.

References

Abbé, E. (1873), Beiträge zur theorie des mikroskops und der mikroskopischen wahrnehmung, *Arch. Mikrosc. Anat. Entwicklungsmech.* **9**, 413–468.

Akhremitchev, B. B., Sun, Y., Stebounova, L., and Walker, G. C. (2002), Monolayer-sensitive infrared imaging of DNA stripes using apertureless near-field microscopy, *Langmuir* **18**, 5325–5328.

Anderson, R. G. and Jacobson, K. (2002), A role for lipid shells in targeting proteins to caveolae, rafts, and other lipid domains, *Science* **296**, 1821–1825.

Anger, P., Bharadwaj, P., and Novotny, L. (2006), Enhancement and quenching of single-molecule fluorescence, *Phys. Rev. Lett.* **96**, 113002(4).

Ash, E. A. and Nicholls, G. (1972), Super-resolution aperture scanning microscope, *Nature* (London) **237**, 510–512.

Bethe, H. A. (1944), Theory of diffraction by small holes, *Phys. Rev.* **66**, 163–182.

Betzig, E. and Chichester, R. J. (1993), Single molecules observed by near-field scanning optical microscopy, *Science* **262**, 1422–1425.

Betzig, E. and Trautman, J. K. (1992), Near-field optics: Microscopy, spectroscopy, and surface modification beyond the diffraction limit, *Science* **257**, 189–195.

Betzig, E., Lewis, A., Harootunian, A., Isaacson, M., and Kratschmer, E. (1986), Near field scanning optical microscopy (NSOM): Development and biophysical applications, *Biophys. J.* **49**, 269–279.

Betzig, E., Isaacson, M., and Lewis, A. (1987), Collection mode near-field scanning optical microscopy, *Appl. Phys. Lett.* **51**, 2088–2090.

Betzig, E., Trautman, J. K., Harris, T. D., Weiner, J. S., and Kostelak, R. L. (1991), Breaking the diffraction barrier: Optical microscopy on a nanometric scale, *Science* **251**, 1468–1470.

Betzig, E., Finn, P. L., and Weiner, J. S. (1992), Combined shear force and near-field scanning optical microscopy, *Appl. Phys. Lett.* **60**, 2484–2486.

Betzig, E., Patterson, G. H., Sougrat, R. et al. (2006), Imaging intracellular fluorescent proteins at nanometer resolution, *Science* **313**, 1642–1645.

Binnig, G. and Rohrer, H. (1982), Scanning tunneling microscopy, *Helv. Phys. Acta* **55**, 726–735.

Brehm, M., Taubner, T., Hillenbrand, R., and Keilmann, F. (2006), Infrared spectroscopic mapping of single nanoparticles and viruses at nanoscale resolution, *Nano Lett.* **6**, 1307–1310.

Brown, D. A. and London, E. (2000), Structure and function of sphingolipid- and cholesterol-rich membrane rafts, *J. Biol. Chem.* **275**, 17221–17224.

Burgos, P., Lu, Z., Ianoul, A. et al. (2003a), Near-field scanning optical microscopy probes: A comparison of pulled and double-etched bent NSOM probes for fluorescence imaging of biological samples, *J. Microsc.* **211**, 37–47.

Burgos, P., Yuan, C., Viriot, M.-L., and Johnston, L. J. (2003b), Two-color near-field fluorescence microscopy studies of microdomains ("Rafts") in model membranes, *Langmuir* **19**, 8002–8009.

Cambi, A., de Lange, F., van Maarseveen, N. M. et al. (2004), Microdomains of the C-type lectin DC-SIGN are portals for virus entry into dendritic cells, *J. Cell Biol.* **164**, 145–155.

Cambi, A., Joosten, B., Koopman, M. et al. (2006), Organization of the integrin LFA-1 in nanoclusters regulates its activity, *Mol. Biol. Cell* **17**, 4270–4281.

Carlson, C. A. and Woehl, J. C. (2008), Fabrication of optical tips from photonic crystal fibers, *Rev. Sci. Instrum.* **79**, 103707.

Chen, Y., Qin, J., and Chen, Z. (2008a), Fluorescence-topographic NSOM directly visualizes peak-valley polarities of GM1/GM3 rafts in cell membrane fluctuations, *J. Lipid Res.* **49**, 2268–2275.

Chen, Y., Shao, L., Ali, Z., Cai, J., and Chen, Z. W. (2008b), NSOM/QD-based nanoscale immunofluorescence imaging of antigen-specific T-cell receptor responses during an in vivo clonal Vγ2Vδ2 T-cell expansion, *Blood* **111**, 4220–4232.

Chiantia, S., Kahya, N., Ries, J., and Schwille, P. (2006), Effects of ceramide on liquid-ordered domains investigated by simultaneous AFM and FCS, *Biophys. J.* **90**, 4500–4508.

Chtcheglova, L. A., Waschke, J., Wildling, L., Drenckhahn, D., and Hinterdorfer, P. (2007), Nano-scale dynamic recognition imaging on vascular endothelial cells, *Biophys. J.* **93**, L11–L13.

Coban, O., Burger, M., Laliberte, M., Ianoul, A., and Johnston, L. J. (2007), Ganglioside partitioning and aggregation in phase-separated monolayers characterized by bodipy GM1 monomer/dimer emission, *Langmuir* **23**, 6704–6711.

Courjon, D., Sarayeddine, K., and Spajer, M. (1989), Scanning tunneling optical microscopy, *Opt. Commun.* **71**, 23–28.

Dahim, M., Mizuno, N., Li, X. M., Momsen, W. E., Momsen, M. M., and Brockman, H. L. (2002), Physical and photophysical characterization of a BODIPY phosphatidylcholine as a membrane probe, *Biophys. J.* **83**, 1511–1524.

Day, C. A. and Kenworthy, A. K. (2009), Tracking microdomain dynamics in cell membranes, *Biochim. Biophys. Acta* **1788**, 245–253.

de Bakker, B. I., de Lange, F., Cambi, A. et al. (2007), Nanoscale organization of the pathogen receptor DC-SIGN mapped by single-molecule high-resolution fluorescence microscopy, *Chemphyschem* **8**, 1473–1480.

de Bakker, B. I., Bodnár, A., van Dijk, E. P. et al. (2008), Nanometer-scale organization of the alpha subunits of the receptors for IL2 and IL15 in human T lymphoma cells, *J. Cell Sci.* **121**, 627–633.

de Lange, F., Cambi, A., Huijbens, R. et al. (2001), Cell biology beyond the diffraction limit: Near-field scanning optical microscopy, *J. Cell Sci.* **114**, 4153–4160.

Denk, W. and Pohl, D. W. (1991), Near-field optics: Microscopy with nanometer-size fields, *J. Vac. Sci. Technol. B* **9**, 510–513.

di Francia, G. T. (1952), Super-gain antennas and optical resolving power, *Suppl. Nuovo Cim.* **9**, 426–438.

Douglass, A. D. and Vale, R. D. (2005), Single-molecule microscopy reveals plasma membrane microdomains created by protein–protein networks that exclude or trap signaling molecules in T cells, *Cell* **121**, 937–950.

Dunn, R. C. (1999), Near-field scanning optical microscopy, *Chem. Rev.* **99**, 2891–2927.

Dunn, R. C., Allen, E. V., Joyce, S. A., Anderson, G. A., and Sunney-Xic, X. (1995), Near-field fluorescent imaging of single proteins, *Ultramicroscopy* **57**, 113–117.

Dürig, U., Pohl, D., and Rohner, F. (1986), Near-field optical-scanning microscopy, *J. Appl. Phys.* **59**, 3318–3327.

Edidin, M. (2001), Near-field scanning optical microscopy, a siren call to biology, *Traffic* **2**, 797–803.

Edidin, M. (2003), Lipids on the frontier: A century of cell-membrane bilayers, *Nat. Rev. Mol. Cell Biol.* **4**, 414–418.

Eggeling, C., Ringemann, C., Medda, R. et al. (2009), Direct observation of the nanoscale dynamics of membrane lipids in a living cell, *Nature* **457**, 1159–1163.

Enderle, T., Ha, T., Ogletree, D. F., Chemla, D. S., Magowan, C., and Weiss, S. (1997), Membrane specific mapping and colocalization of malarial and host skeletal proteins in the *Plasmodium falciparum* infected erythrocyte by dual-color near-field scanning optical microscopy, *Proc. Natl. Acad. Sci. USA* **94**, 520–525.

Engelman, D. M. (2005), Membranes are more mosaic than fluid, *Nature* **438**, 578–580.

Erickson, E. S. and Dunn, R. C. (2005), Sample heating in near-field scanning optical microscopy, *Appl. Phys. Lett.* **87**, 201102(3).

Estrada, L. C., Aramendía, P. F., and Martínez, O. E. (2008), 10,000 times volume reduction for fluorescence correlation spectroscopy using nano-antennas, *Opt. Express* **16**, 20597–20602.

Farahani, J. N., Pohl, D. W., Eisler, H.-J., and Hecht, B. (2005), Single quantum dot coupled to a scanning optical antenna: A tunable superemitter, *Phys. Rev. Lett.* **95**, 017402(4).

Fischer, U. C. and Pohl, D. W. (1989), Observation of single-particle plasmons by near-field optical microscopy, *Phys. Rev. Lett.* **62**, 458–461.

Flanders, B. N. and Dunn, R. C. (2002), A near-field microscopy study of submicron domain structure in a model lung surfactant monolayer, *Ultramicroscopy* **91**, 245–251.

Frey, H. G., Keilmann, F., Kriele, A., and Guckenberger, R. (2002), Enhancing the resolution of scanning near-field optical microscopy by a metal tip grown on an aperture probe, *Appl. Phys. Lett.* **81**, 5030–5032.

Frey, H. G., Witt, S., Felderer, K., and Guckenberger, R. (2004), High-resolution imaging of single fluorescent molecules with the optical near-field of a metal tip, *Phys. Rev. Lett.* **93**, 200801(4).

García-Parajo, M. F. (2008), Optical antennas focus in on biology, *Nat. Photon.* **2**, 201–203.

García-Parajo, M. F., de Bakker, B. I., Koopman, M. et al. (2005), Near-field fluorescence microscopy: An optical nanotool to study protein organization at the cell membrane, *Nanobiotech.* **1**, 113–120.

Gersen, H., García-Parajo, M. F., Novotny, L., Veerman, J.-A., Kuipers, L., and van Hulst, N. F. (2000), Influencing the angular emission of a single molecule, *Phys. Rev. Lett.* **85**, 5312–5315.

Gheber, L. A., Hwang, J., and Edidin, M. (1998), Design and optimization of a near-field scanning optical microscope for imaging biological samples in liquid, *Appl. Opt.* **37**, 3574–3581.

Goodman, J. W. (1968), *Introduction to Fourier Optics*, Physical and Quantum Electronics Series. McGraw-Hill, San Francisco, CA.

Grakoui, A., Bromley, S. K., Sumen, C. et al. (1999), The immunological synapse: A molecular machine controlling T cell activation, *Science* **285**, 221–227.

Greffet, J.-J. and Carminati, R. (1997), Image formation in near-field optics, *Progr. Surf. Sci.* **56**, 133–237.

Gucciardi, P. G., Patané, S., Ambrosio, A. et al. (2005), Observation of tip-to-sample heat transfer in near-field optical microscopy using metal-coated fiber probes, *Appl. Phys. Lett.* **86**, 203109(3).

Gucciardi, P. G., Bachelier, G., Allegrini, M. et al. (2007), Artifacts identification in apertureless near-field optical microscopy, *J. Appl. Phys.* **101**, 064303(8).

Guenther, T., Emiliani, V., Intonti, F. et al. (1999), Femtosecond near-field spectroscopy of a single GaAs quantum wire, *Appl. Phys. Lett.* **75**, 3500–3502.

Haefliger, D., Plitzko, J. M., and Hillenbrand, R. (2004), Contrast and scattering efficiency of scattering-type near-field optical probes, *Appl. Phys. Lett.* **85**, 4466–4468.

Hartschuh, A., Sánchez, E. J., Sunney-Xie, X., and Novotny, L. (2003), High-resolution near-field Raman microscopy of single-walled carbon nanotubes, *Phys. Rev. Lett.* **90**, 095503(4).

Hecht, B., Bielefeldt, H., Inouye, Y., Pohl, D. W., and Novotny, L. (1997), Facts and artifacts in near-field optical microscopy, *J. Appl. Phys.* **81**, 2492–2498.

Hecht, B., Sick, B., Wild, U. P. et al. (2000), Scanning near-field optical microscopy with aperture probes: Fundamentals and applications, *J. Chem. Phys.* **112**, 7761–7774.

Hell, S. W. (2007), Far-field optical nanoscopy, *Science* **316**, 1153–1158.

Hell, S. W. and Wichmann, J. (1994), Breaking the diffraction resolution limit by stimulated emission: Stimulated-emission-depletion fluorescence microscopy, *Opt. Lett.* **19**, 780–782.

Hess, S. T., Girirajan, T. P. K., and Masony, M. D. (2006), Ultra-high resolution imaging by fluorescence photoactivation localization microscopy, *Biophys. J.* **91**, 4258–4272.

Hillenbrand, R. and Keilmann, F. (2002), Material-specific mapping of metal/semiconductor/dielectric nanosystems at 10 nm resolution by backscattering near-field optical microscopy, *Appl. Phys. Lett.* **80**, 25–27.

Hinterdorfer, P. and Dufrêne, Y. F. (2006), Detection and localization of single molecular recognition events using atomic force microscopy, *Nat. Methods* **3**, 347–355.

Hoffmann, P., Dutoit, B., and Salathé, R.-P. (1995), Comparison of mechanically drawn and protection layer chemically etched optical fiber tips, *Ultramicroscopy* **61**, 165–170.

Hollars, C. W. and Dunn, R. C. (1997), Submicron fluorescence, topography, and compliance measurements of phase-separated lipid monolayers using tapping-mode near-field scanning optical microscopy, *J. Phys. Chem. B* **101**, 6313–6317.

Hollars, C. W. and Dunn, R. C. (1998), Submicron structure in 1-α-dipalmitoylphosphatidylcholine monolayers and bilayers probed with confocal, atomic force, and near-field microscopy, *Biophys. J.* **75**, 342–353.

Höppener, C. and Novotny, L. (2008), Antenna-based optical imaging of single Ca^{2+} transmembrane proteins in liquids, *Nano Lett.* **8**, 642–646.

Höppener, C., Siebrasse, J. P., Peters, R., Kubitscheck, U., and Naber, A. (2005), High-resolution near-field optical imaging of single nuclear pore complexes under physiological conditions, *Biophys. J.* **88**, 3681–3688.

Huang, B., Wang, W. Q., Bates, M., and Zhuang, X. (2008), Three-dimensional super-resolution imaging by stochastic optical reconstruction microscopy, *Science* **319**, 810–813.

Hwang, J., Tamm, L. K., Böhm, C., Ramalingam, T. S., Betzig, E., and Edidin, M. (1995), Nanoscale complexity of phospholipid monolayers investigated by near-field scanning optical microscopy, *Science* **270**, 610–614.

Hwang, J., Gheber, L. A., Margolis, L., and Edidin, M. (1998), Domains in cell plasma membranes investigated by near-field scanning optical microscopy, *Biophys. J.* **74**, 2184–2190.

Ianoul, A., Burgos, P., Lu, Z., Taylor, R. S., and Johnston, L. J. (2003), Phase separation in supported phospholipid bilayers visualized by near-field scanning optical microscopy in aqueous solution, *Langmuir* **19**, 9246–9254.

Ianoul, A., Grant, D. D., Rouleau, Y., Bani-Yaghoub, M., Johnston, L. J., and Pezacki, J. P. (2005), Imaging nanometer domains of β-adrenergic receptor complexes on the surface of cardiac myocytes, *Nat. Chem. Biol.* **1**, 196–202.

Ichimura, T., Hayazawa, N., Hashimoto, M., Inouye, Y., and Kawata, S. (2004), Tip-enhanced coherent anti-Stokes Raman scattering for vibrational nanoimaging, *Phys. Rev. Lett.* **92**, 220801(4).

Inouye, Y. and Kawata, S. (1994), Near-field scanning optical microscope with a metallic probe tip, *Opt. Lett.* **19**, 159–161.

Intonti, F., Emiliani, V., Lienau, C., Elsaesser, T., Nötzel, R., and Ploog, K. H. (2001), Near-field optical spectroscopy of localized excitons in a single GaAs quantum wire, *Phys. Rev. B* **63**, 075313(5).

Kalkbrenner, T., Ramstein, M., Mlynek, J., and Sandoghdar, V. (2001), A single gold particle as a probe for apertureless scanning near-field optical microscopy, *J. Microsc.* **202**, 72–76.

Kapkiai, L. K., Moore-Nichols, D., Carnell, J., Krogmeier, J. R., and Dunn, R. C. (2004), Hybrid near-field scanning optical microscopy tips for live cell measurements, *Appl. Phys. Lett.* **84**, 3750–3752.

Karraï, K. and Grober, R. D. (1995a), Piezo-electric tip-sample distance control for near field optical microscopes, *Appl. Phys. Lett.* **66**, 1842–1844.

Karraï, K. and Grober, R. D. (1995b), Piezo-electric tuning fork tip-sample distance control for near field optical microscopes, *Ultramicroscopy* **61**, 197–205.

Kawata, Y., Xu, C., and Denk, W. (1999), Feasibility of molecular-resolution fluorescence near-field microscopy using multi-photon absorption and field enhancement near a sharp tip, *J. Appl. Phys.* **85**, 1294–1301.

Kim, J., Benson, O., Kan, H., and Yamamoto, Y. (1999), A single-photon turnstile device, *Nature (London)* **397**, 500–503.

Kim, S., Jin, J., Kim, Y.-J., Park, I.-Y., Kim, Y., and Kim, S.-W. (2008), High-harmonic generation by resonant plasmon field enhancement, *Nature (London)* **453**, 757–760.

Knoll, B. and Keilmann, F. (1999), Near-field probing of vibrational absorption for chemical microscopy, *Nature (London)* **399**, 134–137.

Koopman, M., de Bakker, B. I., García-Parajo, M. F., and van Hulst, N. F. (2003), Shear force imaging of soft samples in liquid using a diving bell concept, *Appl. Phys. Lett.* **83**, 5083–5085.

Kühn, S., Håkanson, U., Rogobete, L., and Sandoghdar, V. (2006), Enhancement of single-molecule fluorescence using a gold nanoparticle as an optical nanoantenna, *Phys. Rev. Lett.* **97**, 017402(4).

Kusumi, A. and Suzuki, K. (2005), Toward understanding the dynamics of membrane-raft-based molecular interactions, *Biochim. Biophys. Acta* **1746**, 234–251.

Kusumi, A., Koyama-Honda, I., and Suzuki, K. (2004), Molecular dynamics and interactions for creation of stimulation-induced stabilized rafts from small unstable steady-state rafts, *Traffic* **5**, 213–230.

Lambelet, P., Sayah, A., Pfeffer, M., Philipona, C., and Marquis-Weible, F. (1998), Chemically etched fiber tips for near-field optical microscopy: A process for smoother tips, *Appl. Opt.* **37**, 7289–7292.

Levene, M. J., Korlach, J., Turner, S. W., Foquet, M., Craighead, H. G., and Webb, W. W. (2003), Zero-mode waveguides for single-molecule analysis at high concentrations, *Science* **299**, 682–686.

Lewis, A., Isaacson, M., Harootunian, A., and Muray, A. (1984), Development of a 500 Å spatial resolution light microscope I. Light is efficiently transmitted through λ/16 diameter apertures, *Ultramicroscopy* **13**, 227–231.

Lewis, A., Taha, H., Strinkovski, A. et al. (2003), Near-field optics from subwave-length illumination to nanometric shadowing, *Nat. Biotechnol.* **21**, 1378–1386.

Longo, G., Girasole, M., and Cricenti, A. (2008), Implementation of a bimorph-based aperture tapping-SNOM with an incubator to study the evolution of cultured living cells, *J. Microsc.* **229**, 433–439.

Lounis, B. and Moerner, W. E. (2000), Single photons on demand from a single-molecule at room temperature, *Nature (London)* **407**, 491–493.

Marguet, D., Lenne, P. F., Rigneault, H., and He, T. H. (2006), Dynamics in the plasma membrane: How to combine fluidity and order, *EMBO J.* **25**, 3446–3457.

Martin, Y. C., Hamann, H. F., and Wickramasinghe, H. K. (2001), Strength of the electric field in aperture-less near-field optical microscopy, *J. Appl. Phys.* **89**, 5774–5778.

Massey, G. A. (1984), Microscopy and pattern generation with scanned evanescent waves, *Appl. Opt.* **23**, 658–660.

Matsumoto, T., Ichimura, T., Yatsui, T., Kourogi, M., Saiki, T., and Ohtsu, M. (1998), Fabrication of a near-field optical fiber probe with a nanometric metallized protrusion, *Opt. Rev.* **5**, 369–373.

Mayor, S. and Rao, M. (2004), Rafts: Scale-dependent, active lipid organization at the cell surface, *Traffic* **5**, 231–240.

Michaelis, J., Hettich, C., Mlynek, J., and Sandoghdar, V. (2000), Optical microscopy using a single-molecule light source, *Nature (London)* **405**, 325–328.

Michler, P., Kiraz, A., Becher, C. et al. (2000), A quantum dot single-photon turnstile device, *Science* **290**, 2282–2285.

Molenda, D., des Francs, G. C., Fischer, U. C., Rau, N., and Naber, A. (2005), High-resolution mapping of the optical near-field components at a triangular nano-aperture, *Opt. Express* **13**, 10688–10696.

Mühlschlegel, P., Eisler, H.-J., Martin, O. J., Hecht, B., and Pohl, D. W. (2005), Resonant optical antennas, *Science* **308**, 1607–1609.

Murray, J., Cuccia, L., Ianoul, A., Cheetham, J., and Johnston, L. J. (2004), Imaging the selective binding of synapsin to anionic membrane domains, *Chembiochem* **5**, 1489–1494.

Nagy, P., Jenei, A., Kirsch, A. K., Szöllősi, J., Damjanovich, S., and Jovin, T. M. (1999), Activation-dependent clustering of the erbB2 receptor tyrosine kinase detected by scanning near-field optical microscopy, *J. Cell Sci.* **112**, 1733–1741.

Novotny, L. (2007), Effective wavelength scaling for optical antennas, *Phys. Rev. Lett.* **98**, 266802(4).

Novotny, L. and Stranick, S. J. (2006), Near-field optical microscopy and spectroscopy with pointed probes, *Annu. Rev. Phys. Chem.* **57**, 303–331.

Novotny, L., Bian, R. X., and Sunney-Xie, X. (1997), Theory of nanometric optical tweezers, *Phys. Rev. Lett.* **79**, 645–648.

Novotny, L., Sánchez, E. J., and Sunney-Xie, X. (1998), Near-field optical imaging using metal tips illuminated by higher-order Hermite–Gaussian beams, *Ultramicroscopy* **71**, 21–29.

O'Keefe, J. A. (1956), Resolving power of visible light, *Opt. Soc. Am.* **46**, 359.

Oseen, C. W. (1922), Einstein's pinprick radiation and Maxwell's equations, *Ann. Phys. (Leipz.)* **69**, 202–204.

Pettinger, B., Ren, B., Picardi, G., Schuster, R., and Ertl, G. (2004), Nanoscale probing of adsorbed species by tip-enhanced Raman spectroscopy, *Phys. Rev. Lett.* **92**, 096101(4).

Pfeffer, M., Lambelet, P., and Marquis-Weible, F. (1997), Shear-force detection based on an external cavity laser interferometer for a compact scanning near field optical microscope, *Rev. Sci. Instrum.* **68**, 4478–4482.

Pike, L. J. (2006), Rafts defined: A report on the keystone symposium on lipid rafts and cell function, *J. Lipid Res.* **47**, 1597–1598.

Pohl, D. W., Denk, W., and Lanz, M. (1984), Optical stethoscopy: Image recording with resolution λ/20, *Appl. Phys. Lett.* **44**, 651–653.

Rechberger, W., Hohenau, A., Leitner, A., Krenn, J. R., Lamprecht, B., and Aussenegg, F. R. (2003), Optical properties of two interacting gold nanoparticles, *Opt. Commun.* **220**, 137–141.

Reddick, R., Warmack, R., and Ferrell, T. (1989), New form of scanning optical microscopy, *Phys. Rev. B* **39**, 767–770.

Rensen, W. H. (2002), Tuning fork tunes: Exploring new scanning probe techniques, PhD thesis, University of Twente, Twente.

Richter, A., Behme, G., Süptitz, M. et al. (1997), Real-space transfer and trapping of carriers into single GaAs quantum wires studied by near-field optical spectroscopy, *Phys. Rev. Lett.* **79**, 2145–2148.

Ruiter, A. G., Veerman, J. A., van der Werf, K. O., and van Hulst, N. F. (1997). Dynamic behavior of tuning fork shear-force feedback, *Appl. Phys. Lett.* **71**, 28–30.

Russell, P. (2003), Photonic crystal fibers, *Science* **299**, 358–362.

Rust, M. J., Bates, M., and Zhuang, X. (2006), Sub-diffraction-limit imaging by stochastic optical reconstruction microscopy (STORM), *Nat. Methods* **3**, 793–795.

Sanchez, E. J., Novotny, L., and Sunney-Xie, X. (1999), Near-field fluorescence microscopy based on two-photon excitation with metal tips, *Phys. Rev. Lett.* **82**, 4014–4017.

Schmidt, R., Wurm, C. A., Jakobs, S., Engelhardt, J., Egner, A., and Hell, S. W. (2008), Spherical nanosized focal spot unravels the interior of cells, *Nat. Methods* **5**, 539–544.

Schuck, P. J., Fromm, D. P., Sundaramurthy, A., Kino, G. S., and Moerner, W. E. (2002), Improving the mismatch between light and nanoscale objects with gold bowtie nanoantennas, *Phys. Rev. Lett.* **94**, 017402(4).

Shalom, S., Lieberman, K., Lewis, A., and Cohen, S. R. (1992), A micropipette force probe suitable for near-field scanning optical microscopy, *Rev. Sci. Instrum.* **63**, 4061–4065.

Sharma, P., Varma, R., Sarasij, R. C. et al. (2004), Nanoscale organization of multiple GPI-anchored proteins in living cell membranes, *Cell* **116**, 577–589.

Sibug-Aga, R. and Dunn, R. C. (2004), High-resolution studies of lung surfactant collapse, *Photochem. Photobiol.* **80**, 471–476.

Sieber, J. J., Willig, K. I., Kutzner, C. et al. (2007), Anatomy and dynamics of a supramolecular membrane protein cluster, *Science* **317**, 1072–1076.

Simons, K. and Ikonen, E. (1997), Functional rafts in cell membranes, *Nature (London)* **387**, 569–572.

Simons, K. and Toomre, D. (2000), Lipid rafts and signal transduction, *Nat. Rev. Mol. Cell Biol.* **1**, 31–39.

Simons, K. and van Meer, G. (1988), Lipid sorting in epithelial cells, *Biochemistry* **27**, 6197–6202.

Singer, S. J. and Nicolson, G. L. (1972), The fluid mosaic model of the structure of cell membranes, *Science* **175**, 720–731.

Specht, M., Pedarnig, J., Heckl, W., and Hänsch, T. (1992), Scanning plasmon near-field microscope, *Phys. Rev. Lett.* **68**, 476–479.

Stähelin, M., Bopp, M. A., Tarrach, G., Meixner, A. J., and Zschokke-Gränacher, I. (1996), Temperature profile of fiber tips used in scanning near-field optical microscopy, *Appl. Phys. Lett.* **68**, 2603–2605.

Stöckle, R. M., Suh, Y. D., Deckert, V., and Zenobi, R. (2000), Nanoscale chemical analysis by tip-enhanced Raman spectroscopy, *Chem. Phys. Lett.* **318**, 131–136.

Stroh, C., Wang, H., Bash, R. et al. (2004), Single-molecule recognition imaging microscopy, *Proc. Natl. Acad. Sci. USA* **101**, 12503–12507.

Sunney-Xie, X. and Dunn, R. C. (1994), Probing single molecule dynamics, *Science* **265**, 361–364.

Synge, E. H. (1928), A suggested method for extending microscopic resolution into the ultra-microscopic region, *Phil. Mag.* **6**, 356–362.

Synge, E. H. (1931), A microscopic method, *Phil. Mag.* **11**, 65–80.

Taminiau, T. H., Moerland, R. J., Segerink, F. B., Kuipers, L., and van Hulst, N. F. (2007a), λ/4 resonance of an optical monopole antenna probed by single molecule fluorescence, *Nano Lett.* **7**, 28–33.

Taminiau, T. H., Segerink, F. B., Moerland, R. J., Kuipers, L., and van Hulst, N. F. (2007b), Near-field driving of a optical monopole antenna, *J. Opt. A: Pure Appl. Opt.* **9**, S315–S321.

Taminiau, T. H., Stefani, F. D., Segerink, F. B., and van Hulst, N. F. (2008), Optical antennas direct single-molecule emission, *Nat. Photon.* **2**, 234–237.

Toda, Y., Sugimoto, T., Nishioka, M., and Arakawa, Y. (2000), Near-field coherent excitation spectroscopy of InGaAs/GaAs self-assembled quantum dots, *Appl. Phys. Lett.* **76**, 3887–3889.

Toledo-Crow, R., Yang, P. C., Chen, Y., and Vaez-Iravani, M. (1992), Near-field differential scanning optical microscope with atomic force regulation, *Appl. Phys. Lett.* **60**, 2957–2959.

Trautman, J. K., Macklin, J. J., Brus, L. E., and Betzig, E. (1994), Near-field spectroscopy of single molecules at room temperature, *Nature (London)* **369**, 40–42.

Tsai, D. P. and Lu, Y. Y. (1998), Tapping-mode tuning fork force sensing for near-field scanning optical microscopy, *Appl. Phys. Lett.* **73**, 2724–2726.

Ueda, A., Niwa, O., Maruyama, K., Shindo, Y., Oka, K., and Suzuki, K. (2007), Neurite imaging of living PC12 cells with scanning electrochemical/near-field optical/atomic force microscopy, *Angew. Chem. Int. Ed. Engl.* **46**, 8238–8241.

Valaskovic, G. A., Holton, M., and Morrison, G. H. (1995), Parameter control, characterization, and optimization in the fabrication of optical fiber near-field probes, *Appl. Opt.* **34**, 1215–1228.

van Hulst, N. F., Moers, M. H., Noordman, O. F., Tack, R. G., Segerink, F. B., and Bölger, B. (1993), Near-field optical microscope using a silicon-nitride probe, *Appl. Phys. Lett.* **62**, 461–463.

van Hulst, N. F., Veerman, J.-A., García-Parajo, M. F., and L. Kuipers (2000), Analysis of individual (macro) molecules and proteins using near-field optics, *J. Chem. Phys.* **112**, 7799–7810.

Veerman, J.-A., Otter, A. M., Kuipers, L., and van Hulst, N. F. (1998), High definition aperture probes for near-field optical microscopy fabricated by focused ion beam milling, *Appl. Phys. Lett.* **72**, 3115–3117.

Veerman, J.-A., García-Parajo, M. F., Kuipers, L., and van Hulst, N. F. (1999), Single molecule mapping of the optical field distribution of probes for near-field microscopy, *J. Microsc.* **194**, 477–482.

Vickery, S. A. and Dunn, R. C. (1999), Scanning near-field fluorescence resonance energy transfer microscopy, *Biophys. J.* **76**, 1812–1818.

Vignolini, S., Intonti, F., Balet, L. et al. (2008), Nonlinear optical tuning of photonic crystal microcavities by near-field probe, *Appl. Phys. Lett.* **93**, 023124.

Vobornik, D., Banks, D. S., Lu, Z., Fradin, C., Taylor, R., and Johnston, L. J. (2008), Fluorescence correlation spectroscopy with sub-diffraction-limited resolution using near-field optical probes, *Appl. Phys. Lett.* **93**, 163904.

Webster, S., Batchelder, D. N., and Smith, D. A. (1998), Submicron resolution measurement of stress in silicon by near-field Raman spectroscopy, *Appl. Phys. Lett.* **72**, 1478–1480.

Wenger, J., Conchonaud, F., Dintinger, J. et al. (2007), Diffusion analysis within single nanometric apertures reveals the ultrafine cell membrane organization, *Biophys. J.* **92**, 913–919.

Wenger, J., Gérard, D., Dintinger, J. et al. (2008), Emission and excitation contributions to enhanced single molecule fluorescence by gold nanometric apertures, *Opt. Express* **16**, 3008–3020.

Wessel, J. (1985), Surface-enhanced optical microscopy, *J. Opt. Soc. Am. B* **2**, 1538–1540.

Wu, S. (2006), Review of near-field microscopy, *Front. Phys. China* **3**, 263–274.

Zenhausern, F., Martin, Y., and Wickramasinghe, H. K. (1995), Scanning interferometric apertureless microscopy: Optical imaging at 10 Angstrom resolution, *Science* **269**, 1083–1085.

Index

A

Acceptor-sensitized emission, 13-4
Acousto optical beam splitter (AOBS), 7-7
Acousto-optical modulator (AOM), 2-4, 2-9
Acousto-optical tunable filter (AOTF), 5-2–5-4
β-Adrenergic receptors (βAR), 18-15–18-16
Aequorea victoria, 10-2, 10-17
Atomic force microscopy (AFM), 16-2, 16-11
Autocorrelation function (ACF), 6-3–6-5, 6-9–6-12,
    15-8–15-9
Avalanche photodiodes (APD), 6-17

B

Back focal plane (BFP), 15-5–15-7
Benzenethiol (BTH), 15-21
Bessel beams, 16-5
Bioluminescence resonance energy transfer
    (BRET), 13-15
Brownian motion, 15-5, 15-7–15-8

C

CFP-EYFP ratiometric indicators, 10-14
Charge couple device (CCD)
    camera, 5-2–5-4, 12-18
    fluorescence correlation spectroscopy (FCS), 6-24
    parallelized scanning, 14-10
Chemical vapor deposition (CVD), 15-18–15-19
Complementary metal–oxide–semiconductor (CMOS)
    detectors, 6-17
Confocal laser scanning microscopy (CLSM), 3-5, 6-19
Convolution perfectly matched layers (CPML), 15-20
CroH/Cro proton exchange, 10-8
Cross-correlation function (CCF), 6-2–6-5
Cuvette-based spectrofluorometry, 12-11
Cyan fluorescent protein (CFP), 10-14, 12-7

D

Data acquisition
    activation and readout laser alignment, 11-10–11-11
    data recording, 11-12
    flowchart, protocol, 11-16
    sample region selection, 11-11–11-12
Data analysis
    background subtraction, 11-12–11-13
    flowchart, protocol, 11-16
    image rendering, 11-14–11-15
    optical parameters, 11-12
    thresholds and molecules, 11-13–11-14
Dendra2-Actin, fixed cell imaging, 11-17
Dichroic mirror (DC), 1-6–1-7
Diffraction-limited spot, 11-1
Diffractive optical elements (DOEs), 15-5
Drude–Lorentz model, 15-20
Dual laser excitation fluorescence cross-correlation
    spectroscopy (DLE-FCCS), 6-21
Dynamic light scattering (DLS), 16-3
Dynamic saturation optical microscopy (DSOM), 2-3

E

Ecliptic pHlourins (EcGFP), 10-11–10-12
Electron microscopy (EM), 11-3
Electron multiplying charge-coupled device cameras
    (EMCCD), 6-17, 6-24
Electron transferring flavoprotein (ETF), 17-6
Electro-optic modulator (EOM), 7-7
Emission filter wheels, 14-6
Excitation–emission lifetime matrix, 12-18
Excited state proton transfer (ESPT), 10-5,
    10-12, 10-18
Extended depth of field (EDF)
    Airy beam acceleration
        Airy wave packet, 4-7–4-8
        illumination PSF, 4-8–4-9
        lateral intensity distribution, 4-7–4-8
        Schrödinger equation, 4-7
    annular apertures, 4-9–4-10
    applications, 4-14–4-15
    Bessel beam excitation
        axicons, 4-4
        FWHM resolution, 4-5
        multiplex Bessel beam excitation, 4-6–4-7
        two-photon Bessel beam excitation, 4-6

exocytosis and endocytosis, 4-1
extended focus PSF
    Ewald sphere, 4-4
    interference pattern, 4-3–4-4
    plane wave propagation, 4-3
    pupil plane apertures, 4-4–4-5
mRSI, 4-14
scanning interferometric EDF, 4-11–4-13
    constructive and destructive channels,
        4-11–4-12
    inverted field distribution, 4-11
    partial image inversion, 4-11–4-12
    signal-to-noise level, 4-13
software-based EDF, 4-2–4-3
stereo image possessing, 4-1
wave-front coding, 4-10–4-11
widefield interferometric EDF, 4-13
Extra-cellular matrix (ECM), 16-3
EYFP, 10-13–10-14

F

Far-field fluorescence microscopy
    Abbe-limit, 3-2–3-3
    confocal laser scanning 4Pi-microscopy
        excitation and detection pinhole, 3-4
        experimental realization, 3-6
        fluorescence excitation, 3-5
        hologram, 3-5
        optical resolution, 3-5
        point-by-point scanning, 3-4
        superresolution perspective, focusing, 3-6–3-7
        tandem scanning reflected light
            microscope, 3-4
    data registration and evaluation software
        data acquisition, 3-18
        data segmentation, 3-20–3-21
        parameters, 3-19
        photon numbers, 3-20
        position determination, 3-21–3-22
    emGFP-tagged tubulin molecule distribution,
        3-23–3-24
    GFP-labeled histone distribution, 3-24–3-26
    light optical analysis of biostructures by enhanced
        resolution (LOBSTER), 3-3
    nanostructure analysis, 3-28–3-29
    point-spread-function engineering, 3-3
    single molecule counting, 3-29–3-31
    specimen preparation, 3-22–3-23
    spectrally assigned localization microscopy
        (SALM)
        colocalization, 3-8
        fluorescence emission spectra, 3-9
        fluorescence lifetimes, 3-16–3-17
        fluorophore molecule, 3-17
        mutual Euclidean distances, 3-9
        optical isolation, 3-8, 3-10

        photoswitchable fluorochromes, 3-15
        proof-of-principle SPDM experiments,
            3-15–3-16
        reversible photobleaching, 3-17–3-18
        $SPDM_{Phymod}$ technique, 3-18
        spectral signature, 3-8–3-9
        stochastic labeling schemes, 3-16
        virtual SPDM I, 3-10–3-13
        virtual SPDM II, 3-13–3-15
    stimulated emission depletion microscopy, 3-7
    structured illumination microscopy, 3-7–3-8
    three-dimensional (3D) nanoimaging
        objective lenses, 3-26, 3-38
        optical resolution, 3-27–3-28
        subwavelength-sized fluorescence object, 3-27
    vertical SMI microscope, 3-18–3-19
    in vivo imaging, 3-29
F64L/S65T/T203Y mutant, 10-12
Fluorescein arsenical helix binder (FlAsH), 6-18
Fluorescence correlation spectroscopy (FCS)
    advanced computation, 6-24–6-25
    CCD-based FCS, 6-24
    confocal illumination, 6-18–6-19
    correlation
        autocorrelation function (ACF), 6-3–6-4
        characteristics, 6-5–6-6
        correlated, anticorrelated, and uncorrelated
            variables, 6-2–6-3
        cross-correlation function (CCF), 6-2–6-4
        diffusion propagator, 6-8
        fluorescence fluctuation correlation
            function, 6-7
        fluorescent probes, 6-4
        Gaussian laser beam profile, 6-8
        models used, 6-9–6-12
        statistical properties, 6-5
    correlation function calculation, 6-9, 6-13
    correlator schemes, 6-13–6-14
    fluorescence correlation microscopy (FCM), 6-19
    fluorescence cross-correlation spectroscopy
        (FCCS), 6-20–6-23
    instrumentation
        detection mode, 6-16–6-17
        excitation mode, 6-16
        fluorophores, 6-17–6-18
        light source, 6-15–6-16
    protein–protein interactions, 6-26
    scanning fluorescence correlation spectroscopy
        (SFCS), 6-23–6-24
    single molecule spectroscopy, 6-2
    temporal and spatial resolutions, 6-1
    total internal reflection (TIR), 6-20
    two-photon excitation (TPE), 6-19
Fluorescence cross-correlation spectroscopy (FCCS)
    dual-color labeling, 6-20
    dual laser excitation (DLE), 6-21
    Forster resonance energy transfer (FRET), 6-21

multiple focal spot excitation, 6-23
pulsed interleaved excitation (PIE), 6-22–6-23
single wavelength (SW), 6-22
total internal reflection (TIR), 6-23
two-photon excitation (TPE), 6-22
Fluorescence intensity measurements
fixed emission wavelengths
acceptor and donor photobleaching, 13-8
apparent FRET efficiencies, 13-7–13-8
arbitrary wavelength, 13-6
excitation and emission spectra, 13-6
spectral cross talk and bleed-through, 13-7
multiple emission wavelengths
apparent FRET efficiency, 13-10
composite spectrum, 13-9
emission spectra, 13-8
fluorescence species, 13-9–13-10
fluorescence spectra, 13-8–13-9
single-molecule/molecular complex sensitivity,
13-10–13-11
Fluorescence-labeled cellular nanostructures. *see*
Far-field fluorescence microscopy
Fluorescence lifetime imaging (FLIM), 13-14–13-15
Fluorescence loss in photobleaching (FLIP), 8-2
Fluorescence photoactivation localization microscopy
(FPALM)
artifacts, suboptimal conditions, 11-19–11-20
calibration, 11-19
data analysis
background subtraction, 11-12–11-13
image rendering, 11-14–11-15
optical parameters, 11-12
thresholds and molecules, 11-13–11-14
definitions, 11-1–11-2
electron microscopy (EM), 11-3
flowchart, protocol, 11-15–11-18
light exposure, 11-20
light microscopy, 11-2–11-3
localization precision, 11-5–11-7
materials
experimental setup, 11-9–11-10
filters, 11-8–11-9
photoactivatable fluorophores, 11-7
protein expression *vs.* antibody labeling, 11-7–11-8
protocols, background minimization, 11-8
microscope position stability and drift, 11-21
near-field scanning optical microscopy
(NSOM), 11-3
setup and data acquisition
activation and readout laser alignment,
11-10–11-11
data recording, 11-12
sample region selection, 11-11–11-12
single molecule localization and reconstruction
microscopy, 11-4
super-resolution microscopy techniques, 11-3–11-4
troubleshooting, 11-18–11-19

two-photon microscopy, 11-3
widefield microscopy resolution, 11-2
Fluorescence quantum yield (Φ), 11-2
Fluorescence recovery after photobleaching (FRAP),
7-9–7-10, 8-2, 12-15
Fluorescence resonance energy transfer. *See* Förster
resonance energy transfer
Focal adhesion kinase (FAK), 12-10
Focused ion beam (FIB) milling, 15-15, 15-18–15-19
Force point-spread function (FPSF), 16-9
Förster resonance energy transfer (FRET)
channel multiplexing, 12-21
consequences, 12-5–12-6
dimeric complexes
acceptors and donor spectroscopic
properties, 13-2
acceptor-sensitized emission, 13-4
donor quenching, 13-3
excitation rate constants, 13-4
quantum yields, 13-2–13-3
RET efficiency, 13-3
disadvantages, 12-19–12-20
distance, 12-3–12-4
donor emission and acceptor excitation
spectrum, 12-3
fluorescence cross-correlation spectroscopy
(FCCS), 6-21
fluorescence intensity measurements
fixed emission wavelengths, 13-6–13-8
multiple emission wavelengths, 13-8–13-10
single-molecule/molecular complex sensitivity,
13-10–13-11
fluorescence lifetime imaging microscopy (FLIM)
donor lifetime, 12-21
homodyning and heterodyning
approaches, 12-18
intensity-modulated excitation signal, 12-17
lifetime detection, 12-15
molar extinction coefficient, 12-16
multi-channel plate (MCP), 12-18
multiple exponential fitting, 12-17
quantum yield, 12-15
STED, 12-19
time-and frequency-domain techniques,
12-16–12-17
fluorometric-imaging plate reader (FLIPR)
instrument, 12-3
green fluorescent protein (GFP), 12-1
intermolecular FRET
direct excitation/emission detection, 12-11
donor/acceptor and acceptor/donor ratios, 12-12
molar extinction coefficient, 12-13
normalization, 12-11–12-12
wide-field *vs.* confocal microscope, 12-13
intramolecular FRET
cyan fluorescent protein (CFP) donor, 12-7
emission and excitation spectra tail, 12-7

focal adhesion kinase (FAK), 12-10
immunofluorescence labeling, 12-11
innocent bystanders, 12-10
ratiometric detection schemes, 12-8–12-9
sensitized emission, 12-6
yellow fluorescent protein (YFP) acceptor, 12-7
labeling and sensing paradigms, 12-22–12-23
molecular machines, 12-1–12-2
multibeam scanning, 12-21
non-radiative process, 13-1
oligomeric complexes, 13-4–13-5
orientation, 12-4–12-5
photobleaching, 12-14–12-15, 12-22
polarization, 12-2–12-3
protein structure in vivo
apparent FRET efficiency, 13-12–13-14
Gaussian functions, 13-12
pixel-level spectrally resolved FRET, 13-14–13-15
Ste2p-GFP$_2$ and Ste2p-YFP fusion proteins, 13-11–13-12
spectrofluorometry, 12-2
theranostics, 12-21
Fourier transforms, 2-10–2-11, 15-7
Full-width at half maximum (FWHM), 11-2

G

Gaussian beam, 15-14
Gaussian laser beam profile, 6-8
GFPuv, 10-13
Green fluorescent proteins (GFPs), 7-1, 7-5, 7-9, 8-8
Aequorea victoria, 10-2, 10-17
biosynthetic cargo mislocalization, 10-1
eukaryotic cells, 10-2
Förster resonance energy transfer (FRET), 12-1
high-resolution quantitative fluorescence imaging microscopy, 10-2
mitochondria, 10-1
multidimensional fluorescence microscopy, 14-1
pH sensing
one-site model, 10-6–10-7
pH-dependent optical properties, 10-4–10-6
protonation states, chromophore, 10-4
two-site model, 10-7–10-8
pHi measurement
confocal microimaging, 10-14–10-15
two-photon microimaging, 10-16–10-17
protein structure and folding, 10-3–10-4
protonation and resolution, 10-8–10-9
ratiometric fluorescent indicators, 10-9–10-11
in vivo application, 10-11–10-14

H

Hexamethyldisilazane (HMDS), 15-21, 15-24–15-25
High power continuous wave laser, 7-3

High-resolution quantitative fluorescence imaging microscopy, 10-2
High-throughput/content microscopy
confocal high-content screening microscopy
CCD-based detection, 14-10
laser scanning confocal microscopy, 14-9
selective plane illumination microscopy (SPIM), 14-11
spatial parallelization, 14-9
spectral resolution, 14-10
flow cytometry, 14-11
image screening microscopy
academic resources, 14-3, 14-5
acquisition software, 14-5
commercial and open source solutions, 14-3–14-4
computational autofocusing, 14-8
depth of field, 14-7
image degradation, 14-7
intensity fluctuation, 14-5
laser-based autofocus systems, 14-7–14-8
liquid-handling systems, 14-3
multiband filters, 14-6
object recognition, 14-8
phototoxic side effects, 14-6
retroviral immobilization, 14-4
robotized acquisition system, 14-3
spatial resolution, 14-6, 14-8
spatial sampling, 14-9
laser-scanning cytometry, 14-12–14-13
Homogeneous time-resolved fluorescence (HTRF), 12-16
Hook's law, 15-6

I

Image reconstruction, 14-9
Image rendering
data analysis, 11-14–11-15
flowchart, protocol, 11-16
Interferometric image inversion
scanning interferometric EDF, 4-11–4-13
constructive and destructive channels, 4-11–4-12
inverted field distribution, 4-11
partial image inversion, 4-11–4-12
signal-to-noise level, 4-13
widefield interferometric EDF, 4-13
Intracellular pH indicators (pH$_i$), GFPs
Aequorea victoria, 10-2, 10-17
biosynthetic cargo mislocalization, 10-1
eukaryotic cells, 10-2
high-resolution quantitative fluorescence imaging microscopy, 10-2
mitochondria, 10-1
pH sensing
one-site model, 10-6–10-7
pH-dependent optical properties, 10-4–10-6

protonation states, chromophore, 10-4
two-site model, 10-7–10-8
pHi measurement
confocal microimaging, 10-14–10-15
two-photon microimaging, 10-16–10-17
protein structure and folding, 10-3–10-4
protonation and resolution, 10-8–10-9
ratiometric fluorescent indicators, 10-9–10-11
in vivo application, 10-11–10-14
In vivo spectroscopic imaging
CD image, 17-4
equilibrium thermodynamic organization, 17-5
image acquisition, 17-1–17-2
living cellular systems, 17-2–17-3
molecular structure, 17-3–17-4
nanoscopic cellular structures, 17-5
nanoscopy, 17-2
peptide–membrane interaction, 17-4
plasmonics
field localization, 17-9–17-10
field strength, 17-9
light propagation, 17-10
membrane and protein fields, 17-8
nanosphere lithography (NSL), 17-9
resonance units (RUs), 17-7
probe-dependent identification, 17-3–17-4
reflection anisotropy microscopy (RAM), 17-7
reflection anisotropy spectroscopy (RAS), 17-6–17-7
spatial discrimination, 17-1
super-resolution and label-free imaging applications, 17-10–17-11
surface imaging techniques, 17-5
surface plasmon resonance (SPR), 17-3, 17-7
ultrafast and ultrasensitive imaging applications, 17-10–17-11

K

Kinetic resolution, 10-8

L

Laguerre–Gauss beams, 16-5
Light microscopy, 11-2–11-3
Lorentzian curve, 15-7

M

Maxwell equations, 15-1
Mean square displacement (MSD) analysis
anomalous diffusion, 5-10
definition, 5-9
diffusion coefficient, 5-11–5-12
first time lag, 5-9
free 2-D diffusion, 5-10–5-11

3-D image acquisition, 5-12
intracellular transport, 5-13
second time lag, 5-9–5-10
variance, 5-11
Mie scattering, 15-3
Monochromators (MC), 1-6
Multiband dichroic filters, 14-6
Multidimensional fluorescence microscopy
acquisition protocol, 14-3
automated image analysis
cell cycle progression analysis, 14-13–14-14
data annotation and browsing, 14-13
histopathological analysis, 14-14
hyperspectral analysis, 14-15
living cells analysis, 14-15
signal quantification, 14-13
subcellular analysis, 14-15
biomedical and oncological research applications
automated cell movement tracking, 14-16
fluorescence-based assays, 14-15
G-protein-coupled receptor, 14-15
high content–automated approach, 14-16–14-17
laser-scanning cytometry, 14-16
cancer stem cells, 14-2
far-field nanoscopy, 14-17
high-throughput/content microscopy
confocal high-content screening microscopy, 14-9–14-11
flow cytometry, 14-11
image screening microscopy, 14-3–14-9
laser-scanning cytometry, 14-12–14-13
mosaic reconstruction, tissue slice, 14-2–14-3
statistical resolution, 14-2, 14-17
in vivo intracellular tomography, 14-1
Multiple focal spot excitation FCCS, 6-23
Murine Achilles tendon
longitudinal section, 9-9–9-10
3D reconstruction, 9-8–9-9
transversal section, 9-6–9-8, 9-10
Myosin muscular fibers, 9-7–9-8

N

Nanosphere lithography (NSL), 17-9
Near-field scanning optical microscopy (NSOM)
aperture-type NSOM, 18-4
bio-NSOM
artificial lipid mono/bilayers, 18-12–18-14
DiI molecules, 18-10–18-11
inverted optical microscope, 18-9
lipid–fluid bilayer, 18-11
lipid rafts, 18-11
mosaic representation, plasma membrane, 18-11–18-12
photonic crystal microcavities, 18-11

protein organization, cell membrane,
18-14–18-17
rendering, 18-9–18-10
tip–specimen interaction, 18-10
diffraction limit, 18-1–18-2
far-field optical microscopy, 18-3–18-4
FPALM, 11-3
head and feedback mechanism
electron-tunneling feedback, 18-6
lock-in amplifier, 18-7
piezoelectric actuators, 18-5
probe oscillation damping, 18-6
quality factor, 18-7
shear-force feedback, 18-7–18-8
illumination source, 18-3
image, probes
chemical etching, 18-7–18-8
heating and pulling method, 18-8
optical throughput, 18-9
self-made and straight probes, 18-7
tube-etching method, 18-8
local excitation and scattering mode, 18-5
minimum distance, 18-3
molecular recognition technique (TREC), 18-17
nanometer-scale cellular building blocks, 18-2
optical nanoantennas
chemical recognition, 18-21
Cy-3-labeled DNA, 18-19
feedgap, 18-18–18-19
field enhancement, 18-18
illumination aperture, 18-18
microwave and radiowave fields, 18-20
potential applications, 18-19–18-20
waveoptics theorem, 18-17
subwavelength resolution, 18-3
super-resolution, 18-2
tip-enhanced near-field optical microscopy, 18-5
Nitrilotriacetate (NTA) labeling, 6-18
Nonlinear fluorescence imaging, saturated
excitation (SAX)
acousto-optical modulator (AOM), 2-4, 2-9
discussion and perspectives, 2-13, 2-15
fluorescence microscopy resolution improvement
dynamic saturation optical microscopy
(DSOM), 2-3
fluorescent labels, 2-1
nonlinearity, 2-2
photoswitchable proteins, 2-3
stimulated-emission depletion (STED), 2-1–2-2
focused scanning laser, 2-4
high-resolution saturated fluorescence microscopy,
2-12–2-13
imaging-mode resolution, 2-3
imaging properties and point spread function
confocal fluorescence microscopy, 2-7
excitation intensity, 2-5
fluorescence intensity, demodulated response, 2-6

harmonic demodulation, 2-4
harmonic demodulation frequencies, 2-7–2-8
illumination and the detection distribution, 2-6
intensity profiles, 2-7–2-8
lock-in amplifier, 2-8
rhodamine-6G molecules, saturation
response, 2-5
second harmonic generation (SHG)
microscopy, 2-6
signal-to-noise ratio, 2-8–2-9
saturation nonlinearity effect
Fourier transforms, 2-10–2-11
harmonic demodulation, 2-10, 2-12
lock-in amplifier, 2-9
modulation and demodulation frequency,
2-10–2-11
onset of saturation and second-order
dependence, 2-9–2-10
photomultiplier tube, 2-12
signal-to-noise ratio, 2-10
time-modulated intensity, 2-9
single channel detector, 2-4
Nonratiometric pH indicators, 10-11
Nyquist theorem, 14-9

O

One-photon excitation (OPE), 6-22
Optical cellular biosensors, 12-15
Optical point-spread function (OPSF), 16-9
Optical projection tomography (OPT), 4-14–4-15
Optical-transfer function (OTF), 4-11–4-12
Optical tweezers
applications, biology
actin filaments, 16-11
force measurement and cell motility, 16-12
lamellipodia, 16-12–16-13
laser dissection, 16-13
membrane–cytoskeleton adhesion
components, 16-10
network filaments, 16-10
tether-formation processes, 16-11
atomic force microscopy (AFM), 16-2
dynamic light scattering (DLS), 16-3
fiber-optic tweezers
equivalent focusing effect, 15-14
force calculations, 15-14–15-15
probe fabrication and experimental results,
15-15–15-18
scattering and gradient forces, 15-14
in situ analysis, 15-13
three-dimensional confocal images, 15-12
total-internal-reflection (TIR), 15-13
fluorescence microscopy
beads and DNA, 15-10
calibration and force measurement, 15-6–15-10
confocal architecture, 16-8

confocal microscopy, 15-10–15-12
femto-second pulsed laser, 16-10
fluorescence correlation spectroscopy, 16-10
force point-spread function (FPSF), 16-9
light force, 15-3
Mie scattering, 15-3
optical forces, 15-2
optical point-spread function (OPSF), 16-9
optical setup, 15-5–15-6
photonic force microscope, 16-8–16-9
Rayleigh limit, 15-3, 15-4
restoring forces, 15-2–15-3
scattering and gradient forces, 15-2
*TEM*00-mode laser beam, 15-4
total force, 15-3–15-4
trap stability, 15-5
optical trap calibration
drag force method, 16-7
equipartition method, 16-7
laser radiation, 16-6
low-pass filtering effect, 16-8
Ornstein–Uhlenbeck equation, 16-6
optical trapping principle and setups
Bessel and Laguerre–Gauss beams, 16-5
electromagnetic field, 16-5
gradient and scattering forces, 16-4
laser beam, 16-3, 16-5
position sensitive detector, 16-6
trapping force, 16-4
video tracking, 16-6
particle acceleration and trapping, 15-1
photonic force microscope (PFM), 16-2–16-3
spatial resolution, 16-1
vascular endothelial cell, 16-2
virus-like particles (VLPs), 16-3
Ornstein–Uhlenbeck equation, 16-6

P

PA-GFP-HA, live cell imaging, 11-17–11-18
Phase filter (PF), 1-3–1-4
Phosphate buffered saline (PBS), 11-17
Photoactivatable fluorophores, 11-2
Photoactivation (PA), 7-9–7-11
Photo activation localization microscopy (PALM), 18-2
Photoactivation quantum yield (Φ_{PA})
Photobleaching minimization
Ca^{++}-bound species, 8-11
controlled light exposure microscopy (CLEM)
basic principles, 8-14
bleaching curve analysis, 8-15
experimental realization, 8-14–8-15
feedback circuit operations, 8-15
vs. non-CLEM, tobacco, 8-15–8-16
dark state relaxation
GFP and Atto532, 8-18
high flux illumination densities, 8-16

illumination pause, 8-18
intense excitation, 8-16
saturation, 8-17
S_1 state molecules, pathways, 8-16–8-17
definition, 8-9
differential rate equations, 8-7–8-8
eigen state, Hamiltonian, 8-21
fluorescence emission intensity, 8-10
FRAP and FLIP, 8-2
GFP and fluorescein, 8-8
imaging system resolution, 8-2
near-and far-field excitation, 8-1
nonlinear photo switching, 8-11
out-of-focus photodamage, 8-10–8-11
Perrin–Jablonski diagram, 8-9–8-10
perturbation free Hamiltonian, 8-22
photodynamic interaction, 8-9
phototoxicity limit, 8-7
quantum light microscopy
fluorescent signal, 8-20–8-21
Heisenberg uncertainty principle, 8-21
high-order photon interaction, 8-18
macroscopic superposition states, 8-19
Mandel Q parameter, 8-20
photon number distribution, 8-19–8-20
Poisson distribution, 8-20–8-21
quantum radiation source, 8-18, 8-21
s-parameterized characteristic function, 8-19
Wigner function, 8-19–8-20
rhodamine 6G molecules, 8-10
Schrodinger notation, 8-22
signal-to-noise ratio (SNR), 8-2, 8-7
single-and multi-photon excitation
cross section, 8-5–8-7
first-order coefficient, 8-4
perturbation-free time-independent Hamiltonian, 8-4
second-order coefficient, 8-4–8-5
two-photon excitation process, 8-3
zero-order coefficient, 8-4
steady-state solution, 8-8
system function, 8-22
transition probability, 8-23
triplet state depletion method
absorption cross section, 8-11
differential rate equations, 8-12
fluorescein and GFP, 8-12–8-13
molecular energy diagram, 8-11–8-12
normalized population, 8-13
reactive oxygen species, 8-14
steady-state solution, 8-12
two photon excitation (TPE), 8-3
UV lasers, 8-1
Photobleaching quantum yield (Φ_B), 11-2
Photomultiplier tubes (PMT), 6-17
Photonic force microscope (PFM), 16-2–16-3
Photoswitchable fluorophores, 11-2

Phototonic crystal fibers (PCF), 1-6
Plasmonic devices
    benzenethiol (BTH), 15-21
    device fabrication, 15-19–15-20
    FIB milling and CVD, 15-18
    finite difference time domain (FDTD) code, 15-20
    guided-mode expansion method, 15-20
    nanomanipulator and nanoantenna tip, 15-22
    quantum dot (QD), 15-21–15-22
    radius of curvature, 15-21
    Raman scattering measurements
        benzenethiol self-assembled monolayer, 15-27
        C–C vibrational band, 15-22–15-23
        HMDS polymer, 15-24–15-25
        Raman mapping, 15-24
        selectively deposited CdSe quantum dots,
            15-26–15-27
        selectively deposited SiO_x nanoparticles,
            15-25–15-26
        signal intensity, 15-24–15-25
        silicon wafer and silicon nitride membrane,
            15-24
    scanning electron microscope (SEM) images,
        15-22–15-23
    silica nanoparticle (SiO_x), 15-21–15-22
    single-molecule detection, 15-18–15-19
Pointillist microscopy. *see* Single molecule localization
    and reconstruction microscopy
Point-spread function (PSF), 4-3–4-4, 11-2
Positionsensitive detector (PSD), 16-6
Promyelocytic leukemia protein (PML), 10-13
Protein organization, cell membrane
    βAR and caveolae, 18-15–18-16
    dual color NSOM imaging, 18-16
    epidermal growth factor (EGF), 18-15
    fluorescence correlation spectroscopy, 18-17
    multispecies cross-correlation spectroscopy,
        18-17
    pathogen recognition receptor, DC-SIGN, 18-14
    T cell receptor, 18-15
Protein structure in vivo
    apparent FRET efficiency
        image blur and constrains, 13-12
        oligomeric complexes, 13-13–13-14
        parallelogram-shaped tetramer, 13-14
        tetrameric complex, 13-13
    Gaussian functions, 13-12, 13-14
    pixel-level spectrally resolved FRET, 13-14–13-15
    Ste2p-GFP_2 and Ste2p-YFP fusion proteins,
        13-11–13-12
Pulsed interleaved excitation fluorescence cross-
    correlation spectroscopy (PIE-FCCS),
    6-22–6-23

Q

Quadrant photodiode (QPD), 15-6–15-7, 16-6

R

$1/e^2$ Radius (r_0), 11-2
Ratiometric pH indicators
    cyan fluorescent protein (CFP), 10-14
    deGFPs, 10-12
    E^1GFP, 10-13
    E^2GFP, 10-12–10-13
    GFpH and YFpH sensors, 10-13–10-14
    mtAlpHi, 10-14
    ratiometric pHlourin (RaGFP), 10-12
Readout activation, 11-2
Reflection anisotropy microscopy (RAM), 17-7
Reflection anisotropy spectroscopy (RAS), 17-6–17-7
Resolution (R_0), 11-2
Reversible photobleaching microscopy (RPM), 3-2
Reversible saturable fluorescent transitions
    (RESOLFT), 1-5
Rotational shearing interferometry coherence
    imaging, 4-14

S

Sapphire laser systems, 1-6
Saturated excitation (SAX), 2-1–2-2
Scanning fluorescence correlation spectroscopy
    (SFCS), 6-23–6-24
Second harmonic generation (SHG)
    biological observations, tissues
        3D reconstruction, Achilles tendon, 9-8–9-9
        Gouy phase anomaly, 9-9
        human trabecular bone fragment, 9-11
        longitudinal section, murine Achilles tendon,
            9-9–9-10
        myosin muscular fibers, 9-7–9-8
        peritumoral murine blood vessels, 9-8–9-9
        punctuate segmental collagen, 9-10
        titanium–sapphire lasers, 9-7
        transversal section, murine Achilles tendon,
            9-8, 9-10
        zebrafish tail, 9-8
    cross-section, 9-3
    excitation beam focusing, 9-4
    Gouy shift, 9-3
    homogeneous molecular distribution, 9-4
    imaging modes and microscope design
        backward phase matching, 9-4
        BSHG and PSHG, 9-6
        FSHG signal, 9-5
        nonlinear coherent scattering process, 9-4
        optical scheme, 9-5
        transversal section, murine Achilles tendon,
            9-6–9-7
    molecular harmonic up-conversion, 9-2
    nonlinear microscopy, 9-1
    SAX microscopy, 2-6
    second-and third-order response, 9-2

Sensitized acceptor emission, 12-4–12-5
Short-pulsed lasers, 7-3–7-4
Simulated Raman scattering (SRS), 1-6, 1-10
Single molecule localization and reconstruction
        microscopy, 11-4
Single particle tracking (SPT)
    applications, 5-13
    extracellular and intracellular matrices, 5-1
    fluorescence correlation spectroscopy (FCS), 5-2
    fluorescence recovery after photobleaching
        (FRAP), 5-1
    instrumentation
        achromat lens, 5-3
        epi-fluorescence microscope, 5-2
        fluorescence band-pass filters, 5-4
        notch filter, 5-2–5-3
        widefield illumination, 5-2
    particle localization, image processing
        centroid, 5-7–5-8
        diffraction-limited spot, 5-4
        Gaussian distribution, 5-6
        2-D Gaussian fit, 5-7–5-8
        image acquisition, 5-5
        intensity thresholding, 5-6–5-7
        low-pass filter, 5-5–5-6
        multiple particle tracking, 5-5
        particle position, 5-4–5-5
        photobleaching, 5-7
        unsharp filter, 5-5
    trajectory analysis, MSD
        anomalous diffusion, 5-10
        diffusion coefficient, 5-11–5-12
        first time lag, 5-9
        free 2-D diffusion, 5-10–5-11
        3-D image acquisition, 5-12
        intracellular transport, 5-13
        second time lag, 5-9–5-10
        variance, 5-11
    trajectory construction, 5-8–5-9
Single wavelength fluorescence cross-correlation
        spectroscopy (SW-FCCS), 6-22
Spatial light modulator (SLM), 15-5, 15-11
SPDM$_{Phymod}$ technique, 3-2, 3-18
Spectrally assigned localization microscopy (SALM)
    colocalization, 3-8
    fluorescence emission spectra, 3-9
    fluorescence lifetimes, 3-16–3-17
    fluorophore molecule, 3-17
    mutual Euclidean distances, 3-9
    optical isolation, 3-8, 3-10
    photoswitchable fluorochromes, 3-15
    proof-of-principle SPDM experiments, 3-15–3-16
    reversible photobleaching, 3-17–3-18
    SPDM$_{Phymod}$ technique, 3-18
    spectral signature, 3-8–3-9
    stochastic labeling schemes, 3-16
    virtual SPDM I

        cumulative frequency, 3-11
        3D distance errors, 3-10
        epifluorescence imaging, 3-13
        FWHM, 3-12
        localization errors, 3-11–3-12
        numerical modeling, 3-12
    virtual SPDM II
        concentric ring structure, 3-13–3-14
        variable localization accuracy and molecule
            density, 3-14–3-15
Spectral precision distance/position determination
        microscopy (SPDM), 3-1
Spontaneous activation, 11-2
Stimulated emission depletion (STED) microscopy
    diffraction limit, 1-1
    far-field fluorescence nanoscopy
        electric field, phase filter, 1-3
        Jablonski diagram, 1-2
        point-spread function (PSF), 1-4
        radius, molecule, 1-2
        random photoswitching, 1-5
        resolution gain, 1-3
        reversible saturable fluorescent transitions
            (RESOLFT), 1-5
        vortex phase plates, 1-4
    light sources
        broadband laser source, 1-7
        colocalization, 1-9
        confocal measurements, 1-10–1-11
        CW lasers, 1-5
        immunostained neurofilaments and tubulin
            fibers, 1-7–1-8
        phototonic crystal fibers (PCF), 1-6
        picosecond-pulsed diode lasers, 1-10
        quantitative fluorescence suppression, 1-5
        resolution enhancement, 1-7
        simulated Raman scattering (SRS), 1-6, 1-10
        Stokes line, 1-10
        supercontinuum laser source and comb
            spectrum, 1-6
        superresolution, 1-9
    near-field scanning microscopy (NSOM), 1-1
    nonlinear fluorescence imaging, saturated
        excitation (SAX), 2-1–2-2
    super-resolution microscopes, 18-2
Stochastic FRET, 13-15
Stochastic optical reconstruction microscopy
        (STORM), 18-2
Superecliptic pHlourins (sEcGFP), 10-11
Super-resolution microscopy, 11-3–11-4
Surface plasmon resonance (SPR), 17-3, 17-7
Surface plasmons, 17-7–17-9

T

T cell receptor, 18-15
*TEM*00-mode laser beam, 15-4

Thermodynamic equilibrium, 10-8
Three-dimensional (3D) nanoimaging
    objective lenses, 3-26, 3-38
    optical resolution, 3-27–3-28
    subwavelength-sized fluorescence object, 3-27
Time-correlated single photon counting (TCSPC),
    6-22, 12-17
Ti:sapphire laser, 6-16, 7-6
Titanium–sapphire lasers, 9-7
Total internal reflection fluorescence correlation
    spectroscopy (TIR-FCS), 6-20
Total internal reflection fluorescence cross-correlation
    spectroscopy (TIR-FCCS), 6-23
Two-photon excitation fluorescence correlation
    spectroscopy (TPE-FCS), 6-19
Two-photon excitation fluorescence cross-correlation
    spectroscopy (TPE-FCCS), 6-22
Two-photon excitation (2PE) microscopy, fluorescence
    imaging
    advantages, excitation localization
        diffusion coefficient, 7-10
        FRAP and photoactivation, 7-9–7-10
        HeLa cells, 7-10–7-11
        high spatial confinement, 7-8
        photobleaching process, 7-9–7-10
        sub femtoliter volume, 7-9
    applications, 7-8–7-9
    confocal microscopy, 7-2
    design considerations, 7-6–7-7
    fluorescence intensity, 7-3
    fluorescent probes, 7-5–7-6
    high power continuous wave laser, 7-3
    mode-locked lasers, 7-1
    nonlinear process, 7-2
    short-pulsed lasers, 7-3–7-4
    single-photon excitation (1PE), 7-1–7-2
    two-photon point-spread function, 7-4–7-5
Two-photon excitation (TPE)-pulsed infrared lasers,
    6-16

U

Ultrafast fluorescence upconversion spectroscopy,
    10-12
Ultra-high-resolution microscopy, 11-2

V

Volumetric reconstruction, 14-9

Y

Yb-doped fiber laser, 15-16
Yellow fluorescent protein (YFP) acceptor, 12-7

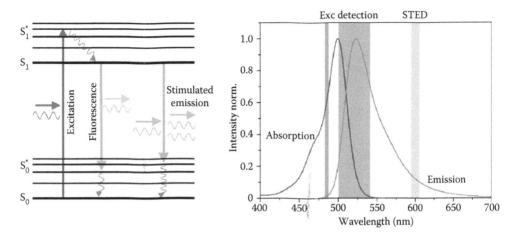

FIGURE 1.1 The Jablonski diagram (left) shows the processes utilized in STED microscopy. The spectrum (right) shows the corresponding wavelength inducing excitation and stimulated emission. As shown, the STED wavelength, and hence photons created via stimulated emission, are excluded from the detection band.

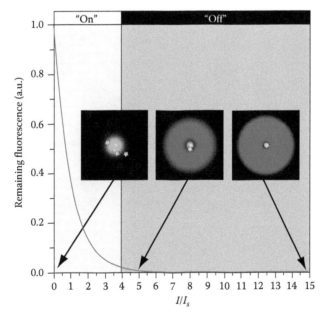

FIGURE 1.2 Fluorescence as a function of STED intensity I (normalized to the saturation intensity I_s which is characteristic of the dye used) and resolution scaling. Beyond a certain value of I/I_s, the probability for fluorescence to occur is negligible for all practical purposes. With increasing power of the STED laser beam the region where this value is reached converges toward the focal center. As a result, nearby features that are illuminated simultaneously by the diffraction-limited excitation focus can be separated because only those features in the center of the focus remain fluorescent.

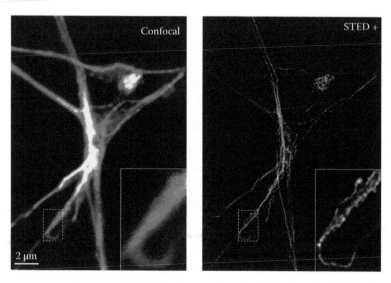

FIGURE 1.6 Fluorescence images of immunostained neurofilaments (light subunit, ATTO 590). Clearly visible, the STED (right) image shows more structural details than the confocal (left) counterpart. The STED image is deconvolved to further increase the contrast. (From Wildanger, D. et al., *Opt. Express.*, 16, 9614, 2008. With permission.)

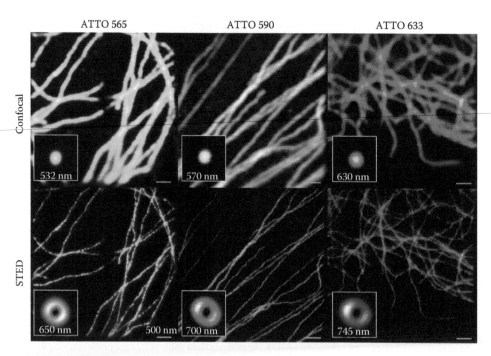

FIGURE 1.7 Tubulin fibers immunostained with ATTO 565, ATTO 590, and ATTO 633. The upper panels show confocal images obtained by employing excitation wavelengths of 532, 570, or 630 nm. The lower panels show the corresponding STED images. Depending on the dye, the central STED wavelength was adjusted to 650, 700, and 750 nm. Obviously, the STED images reveal single fibers where the confocal counterpart is blurred. (From Chi, K.R., *Nat. Meth.*, 6, 15, 2009. With permission.)

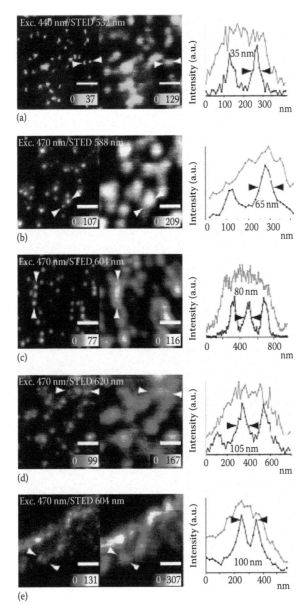

FIGURE 1.9 STED measurements are shown at left, with corresponding confocal measurements of the same site in the sample center, and line-profile measurements, at sites indicated by arrows, at right. Scale bars 500 nm. Excitation and STED wavelengths are indicated. (a) 20–30 nm silica beads labeled with ATTO 425. (b and c) 40 nm beads. (d) 20 nm beads. (e) Neurofilaments labeled with ATTO 532. (From Rankin, B.R. et al., *Opt. Lett.*, 33, 2491, 2008. With permission.)

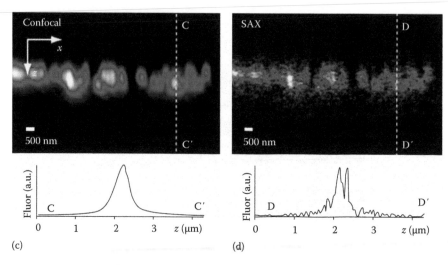

(c)

(d)

FIGURE 2.12 Stained microtubules in HeLa cells. The images show $x-z$ sections, and were taken with a modulation frequency of $\omega = 10\,\text{kHz}$, and (c) demodulated at 2ω, which corresponds to typical confocal fluorescence microscopy, or (d) demodulated at 2ω, where the higher frequencies present in SAX microscopy can be used to increase the imaging resolution.

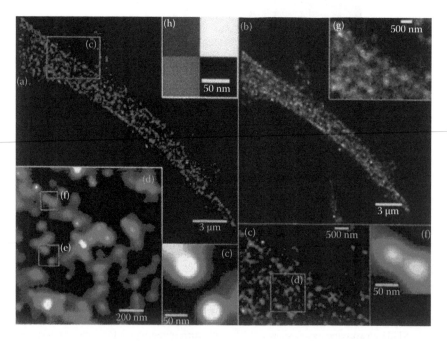

FIGURE 3.10 SPDM$_{\text{Phymod}}$ of emGFP-tagged tubulin molecules in human fibroblasts. (a) A cell protrusion indicating single tubulin molecules, scale bar $3\,\mu\text{m}$; (b) Conventional epifluorescence image calculated from (a); (c) Insert from (a), scale bar $500\,\text{nm}$; (d) Insert from (c), scale bar $200\,\text{nm}$; (e and f) Inserts from (d), scale bar $50\,\text{nm}$; (g) Conventional epifluorescence image calculated from (c); (h) Conventional epifluorescence image calculated from (e); The distance of the two molecules in (e) was estimated to be $106\,\text{nm}$; The distance of the two molecules in (f) was estimated to be $62\,\text{nm}$. The mean localization accuracy determined was $35 \pm 10\,\text{nm}$; total number of tubulin molecules counted in the entire fibroblast cell (optical section of ca. $600\,\text{nm}$ thickness): $N = 42{,}821$, and $N = 1032$ in (a).

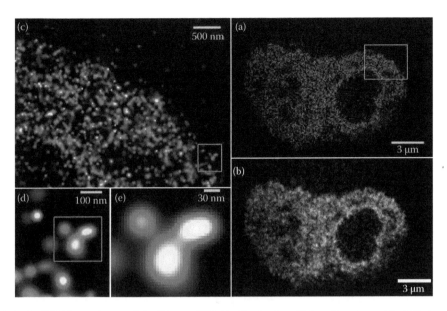

FIGURE 3.12 SPDM$_{Phymod}$ image of the Histone H2B distribution in a HeLa cell nucleus with stably transfected GFP-tagged H2B. (a and b) Entire nucleus, (scale bar 3 µm): (a) SPDM image; (b) epifluorescence image (calculated from the SPDM image by convolution with a PSF with FWHM=200 nm; (c) insert from SPDM image (a) at higher magnification (scale bar 500 nm); (d) insert from (c) at higher magnification (scale bar 100 nm); (e) insert from (d) at higher magnification (scale bar 30 nm) featuring three single histone molecules at distances of about 70–80 nm. The "size" of the molecules corresponds to the localization accuracy. Total number of H2B molecules counted: N=17,409 (optical section of ca. 600 nm thickness).

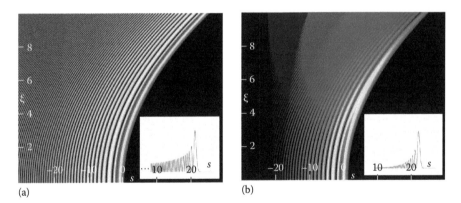

(a)   (b)

FIGURE 4.6 Propagation of 1D Airy wave packets in generalized coordinates ξ and s. While true Airy wave packets are diffractionless, this theoretical solution requires infinite energy and therefore cannot be realized physically (a). Finite energy approximations can be realized but are no longer truly diffractionless (b). In both cases, the maximum of the distribution undergoes constant acceleration, resulting in parabolic trajectories. The trajectories can be extended to negative values of ξ by mirroring with respect to the s-axis. (Reprinted figure with permission from Siviloglou, G.A. et al., *Phys. Rev. Lett.*, 99, 213901, 2007. Copyright 2007 by the American Physical Society.)

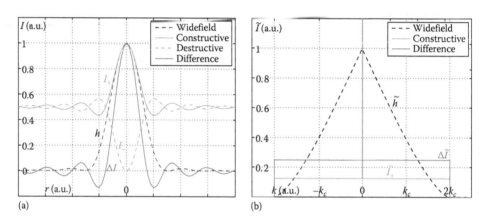

(a) (b)

FIGURE 4.13   (a) From a maximum signal of 100% the constructive channel's point-spread function in interfero-metric detection quickly approaches a constant background of 50%. Subtracting the destructive channel's signal yields a point-spread function which has a better full-width half-maximum resolution than an equivalent widefield point-spread function. (b) From the corresponding optical-transfer functions it can be seen that compared to the widefield case interferometric detection has a much higher transfer efficiency for high spatial frequencies. The lateral point-spread function and the optical-transfer function are independent of defocus, and are therefore ideally suited for extended depth of field imaging. (From Wicker, K. and Heintzmann, R., *Opt. Express*, 15, 12206, 2007. With permission.)

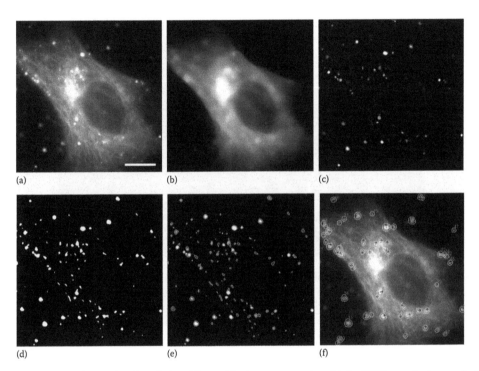

FIGURE 5.2 Particles can be identified and located by image processing. (a) An SPT image is shown of a RPE cell with EGFP-tubulin and YOYO-1 labeled pDNA-polymer complexes of approximately 100 nm diameter. The scale bar is 10 μm. (b) Using a low pass filter (21×21 median filter), the particles can be removed from the image, resulting in an approximate background image containing only the lower spatial frequencies. (c) By subtracting the background image from the original image (unsharp filtering, Equation 5.1), a filtered image is obtained with the particles on a much more homogeneous background. (d) Intensity thresholding can be used to binarize the filtered image. The pixels above the threshold are assigned a value of 1 (displayed in white), while the pixels below the threshold are set to zero (displayed in black). (e) A contour line is calculated for each object in the binary image. (f) Pixels inside a contour belong to a (potential) object of interest. From the pixel values in the original image, object properties can be calculated, such as mean intensity, size, and center location. Based on user-defined criteria, a final selection is made to filter out unwanted objects and retain the objects of interest.

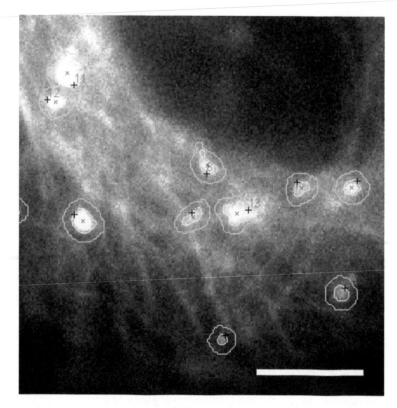

FIGURE 5.4 A magnified subregion of image 2F showing yellow contours around particles. The blue × symbols indicate the centroid position. The green contour is drawn at a user-defined distance (here 3 pixels) from the object contour and allows to calculate a local background value for each particle (the average value of the pixels along this contour). The scale bar is 5 μm.

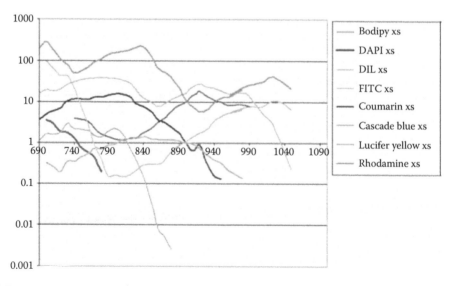

FIGURE 7.4 2PE cross sections for common dyes in the IR range. (After Diaspro, A. et al., *Quart. Rev. Biophys.*, 38, 97, 2005.)

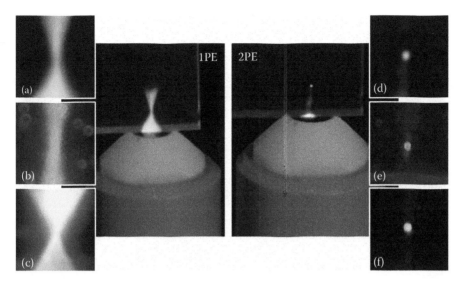

FIGURE 8.1 Experimental excitation profile of three different fluorophores along the axial axis for single- and two-photon excitation.

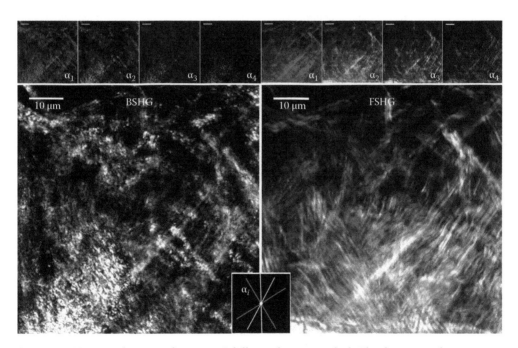

FIGURE 9.4 Transversal section of a murine Achilles tendon 15 mm thick. The objective used is a 63×1.4 NA oil immersion. The sample is acquired in the PSHG mode. α_i angles are related to the orientation of the incoming beam polarization with respect to the sample. At the top inset row there are the images resulting from the acquisition at different angles false color coded. The full color images come from the stitching of the images in the top row acquired in the BSHG and the FSHG modes, respectively. All the scale bars are 10 mm. (From Bianchini, P. and Diaspro, A., *J. Biophoton.*, 1, 443, 2008. Wiley-VCH Verlag GmbH & Co. KGaA. Reproduced with permission.)

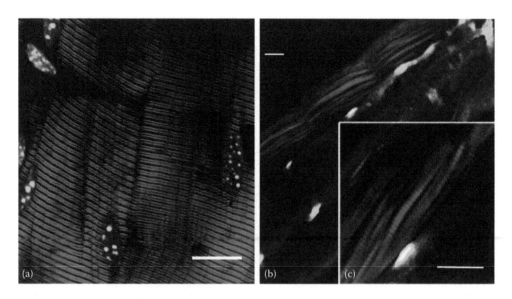

FIGURE 9.5 (a) Mouse muscle fibers (gastrocnemius muscle) single optical section. Infrared excitation at 800 nm. Nuclei are labeled with Hoechst33342 (green) and acquired in the epichannel in the spectral range 440–480 nm. SHG (magenta) from myosin fibers is acquired in the transmission channel. (Sample courtesy of E. Ralston, NIH, Bethesda, MD.) (b, c) Tail of a whole zebrafish. Epithelial blood vessel cells are genetically modified in order to express EGFP (green). Forward SHG comes from the tail muscles (magenta). Panel (c) is a magnification of the muscle fibers to evidence the contrast achieved by SHG imaging. The scale bar is 20 μm.

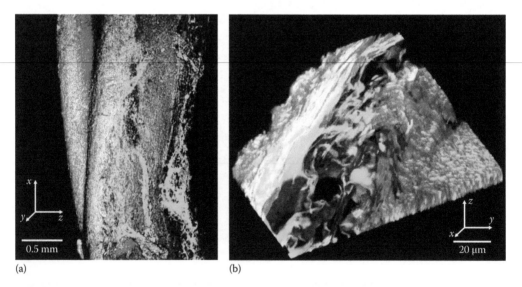

FIGURE 9.7 A 3D reconstruction of the Achilles tendon. The excitation wavelength is 860 nm. Green is the 2PE autofluorescence, magenta and cyan are the backward and forward SHGs, respectively. (a) Entire tendon viewed with 5×0.4 NA objective. (b) Transverse histological slice of the tendon viewed with 63×1.4 NA oil immersion objective. (From Bianchini, P. and Diaspro, A., *J. Biophotonics*, 1, 443, 2008. Wiley-VCH Verlag GmbH & Co. KGaA. Reproduced with permission.)

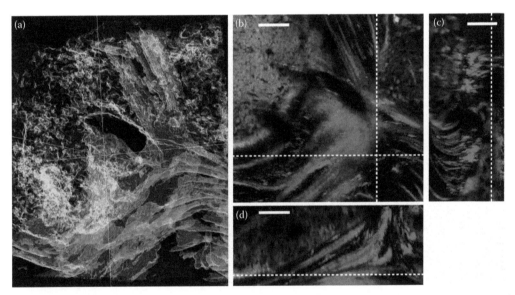

FIGURE 9.10 A human trabecular bone fragment. Panel (a) is the 3-dimensional reconstruction of the z-stack depicted in panels (b), (c), and (d). The 3D render is obtained by the MicroScoBioJ software package (containing image segmentation, registration, restoration, surface rendering, and surface estimation features) (MicroSCoBioJ, http://imagejdocu.tudor.lu/doku.php?id=plugin:stacks:microscobioj:start). Panels (b), (c), and (d) are the orthogonal view of the acquired optical z-sections. The dashed lines indicate which planes of the stack are shown. In particular, (b) is the xy, (c) is the yz, and (d) is the xz plane. For all the panels, green represents the 2PEF of advanced glycation end products (AGEs), blue represents the backward SHG, and red the forward SHG. All the scale bars are 30 μm.

FIGURE 10.1 3D structure of wtGFP as derived from x-ray analysis (Ormo et al. 1996). The frontal part of the barrel is torn open to show the chromophore buried at the protein center.

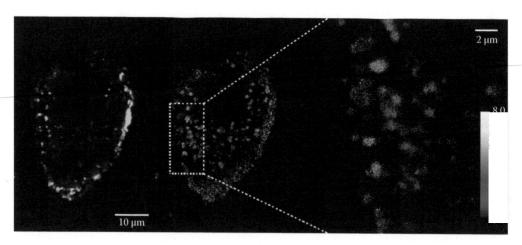

FIGURE 10.4 Real-time monitoring of endocytosis of Tat- E^{I}GFP in HeLa cells (Serresi et al. 2009). Images were taken 4 h after the exposure of cells to Tat-E^{I}GFP. Left image: fluorescence of Tat-E^{I}GFP on the membrane of a cell and internalized in endocytotic vesicles. Central image: a ratiometric pH$_i$ map of the same cell. Right image: a magnified ratiometric pH$_i$ map of the area enclosed in the white dotted rectangle shown in the central image. The pH$_i$ maps are color-coded according to the lookup table placed on the extreme right of the figure. Note the higher pH experienced by Tat-E^{I}GFP on the cell membrane (around neutrality) compared to the one associated with internalized vesicles (5.8–6.5). The progressive acidification upon internalization is a hallmark of the dynamic evolution of organelle contents from endosomes and pinosomes to the degradative lysosomal compartments. (Based on Serresi, M. et al., *Anal. Bioanal. Chem.*, 393, 1123, 2009.)

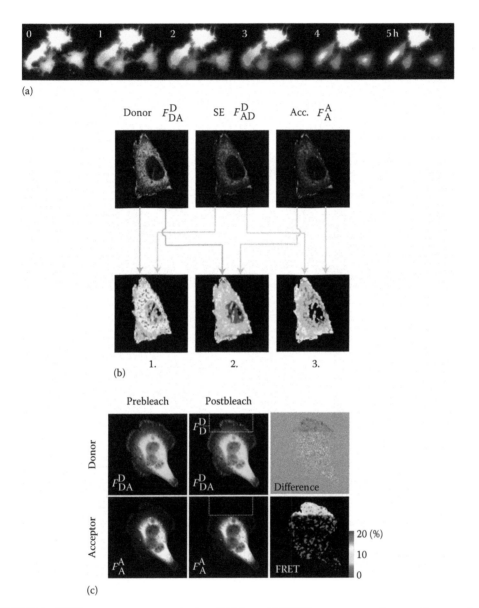

(a)

Donor F_{DA}^{D} SE F_{AD}^{D} Acc. F_{A}^{A}

1.   2.   3.

(b)

Prebleach   Postbleach

Donor

F_{D}^{D}

F_{DA}^{D} F_{DA}^{D} Difference

Acceptor

20 (%)

10

F_{A}^{A} F_{A}^{A} FRET   0

(c)

FIGURE 12.4 FRET measurement examples in cell biology. (a) MCF7 mammary carcinoma cells expressing a CFP–YFP fusion sensor for caspase 3 proteolytic activity. Cleavage of the DEVD recognition peptide releases both fluorophores, at the cost of FRET. Caspase 3 was activated by an induction of apoptosis by incubation with Fas ligand for the indicated time periods. FRET was measured from the ratio of acceptor direct excitation over sensitized emission (Method 3 in Figure 12.3) so that the ratio increases with decreasing FRET, i.e., increasing caspase activity. (b) Conformational activation of focal adhesion kinase (FAK). An intramolecular biosensor for activation-correlated conformational change in the FERM domain of FAK ("FERM sensor") shows higher FRET in the focal adhesions as in the cytoplasm. The three ratio methods presented in the text and in Figure 12.3 are shown. The high intracellular signal in method 1 is most likely caused by substantial autofluorescence in the CFP image. (c) Confirmation of FRET in the FERM sensor by acceptor photobleaching (see Section 12.3).

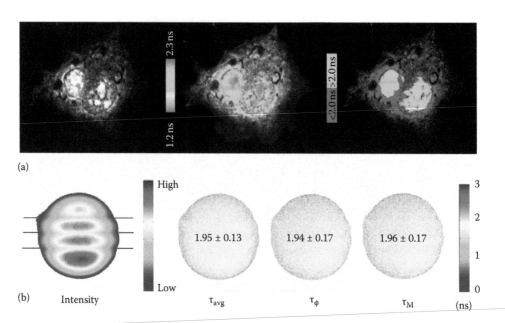

(a)

2.3 ns

1.2 ns

<2.0 ns >2.0 ns

High

3

2

1.95 ± 0.13   1.94 ± 0.17   1.96 ± 0.17

1

0

Low

(b)   Intensity τ_{avg} τ_ϕ τ_M (ns)

FIGURE 12.5   Lifetime detection of FRET. (a) Hela cell co-expressing CFP in the nucleus (especially in the nucle-oli) and a CFP–YFP fusion protein construct in the cytoplasm. Left: Fluorescence intensity distribution of the CFP fluorescence. Middle: 2-photon TCSPC FLIM image of the CFP fluorescence lifetime. Note the high, unperturbed lifetimes in the nucleoli, the slightly lowered lifetimes in the remainder of the nucleus caused by mixing in with the lower lifetime of the fusion construct as it passively enters the nucleus from the cytosol. Right: simple threshold-ing of the lifetime at 2.0 ns suffices to segment the nucleus from the cytosol. (b) Demonstration that lifetimes are independent of fluorophore concentration. GFP, covalently coupled to a sepharose bead, is bleached in a confocal microscope in three lines to locally reduce the concentration of fluorescent species. The lifetime calculated from the phase shift (τ_f) and demodulation (τ_M) in wide-field frequency-domain FLIM is identical over the entire bead. Note that both lifetimes are very similar; this is an indication for the single lifetime nature of GFP, i.e., of the absence of lifetime heterogeneity.

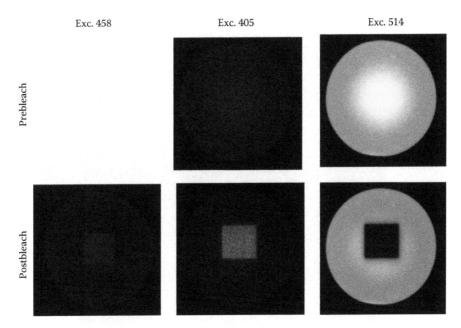

FIGURE 12.6 Photobleaching YFP creates a photoproduct with CFP emission properties. Shown is a sepharose bead covalently conjugated with EYFP. After photobleaching YFP at 514 nm, a photoproduct is generated that emits in the CFP channel when excited at 405 nm. The excitation of this photoproduct is considerably less efficient at 548 nm, which allows the separation of dequenched CFP emission from this artifact.

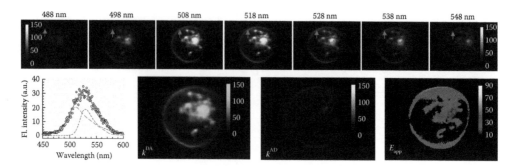

FIGURE 13.3 First row: Microphotographs of yeast cells coexpressing the fusion proteins Ste2p-GFP$_2$ and Ste2p-YFP, obtained at seven selected emission wavelengths with a TPM-SR upon excitation at 800 nm (±15 nm). Second row: Analysis of the images in the top row: First panel from left: Fluorescence spectrum from a single pixel located on the cell membrane and indicated by the red arrows in the first row. Circles, measured intensities; long-dashed lines, decomposed donor-only (GFP$_2$ or Ste2p-GFP$_2$) spectrum; short-dashed lines, acceptor-only (YFP or Ste2p-YFP) spectrum; solid lines, sums of individual spectra. Second panel: donor-only image. Third panel: acceptor-only image. Fourth panel: spatial distribution of the apparent FRET efficiency. (Adapted from Raicu, V. et al., *Nat. Photon.*, 3, 107, 2009. With permission.)

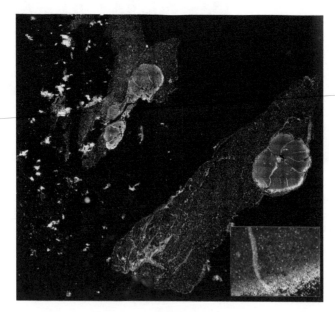

FIGURE 14.1 Mosaic reconstruction of a tissue slice for high content analysis. A mouse ovary tissue slice was fixed and stained for three color parameters analysis (DNA:DAPI. BrdU: FITC anti BrdU. CK5: Cy5 anti CK5). A Scan^R acquisition station assembled on an upright motorized microscope was employed for acquisition. Magnification for each field of view is 10× and the entire scanned area is 2 cm × 2 cm obtained by a mosaic reconstruction of 25 × 30 images. The inset shows the detail of a single acquired field. (Sample kindly provided by Dr. Alessandra Insinga, IEO, Milan, Italy.)

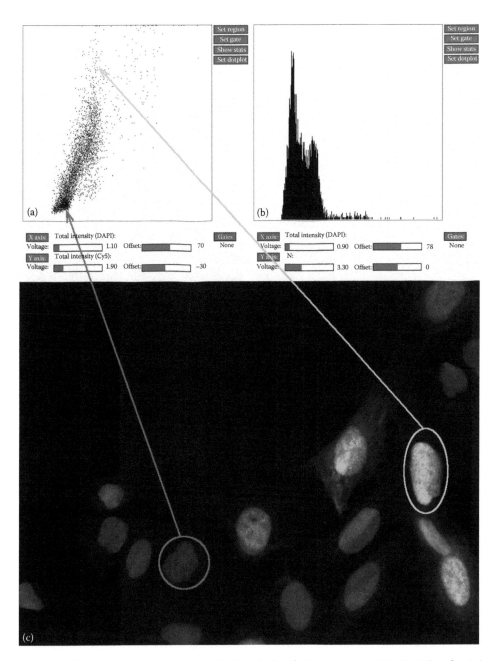

FIGURE 14.2 The image cytometry analysis of cell cycle distribution. Bivariate DNA (DAPI)/cyclin A (anti-Cyclin A rabbit polyclonal antibody + cy3 conjugated goat anti-rabbit secondary antibody) content in exponentially growing U2OS osteosarcoma cell line. Analysis was performed using an ImageJ macro written by the author. In the figure, an example of the link between the shown cytogram in (a) (DNA versus Cyclin A) and a sampled image in (c) reflecting the position of the outlined cells in the graph based on evaluation of their fluorescence intensity in the two channels is shown. The histogram on the right panel (b) shows the DNA distribution of the analyzed cell population. (Cells kindly provided by Dr. Sara Barozzi, IEO, Milan, Italy.)

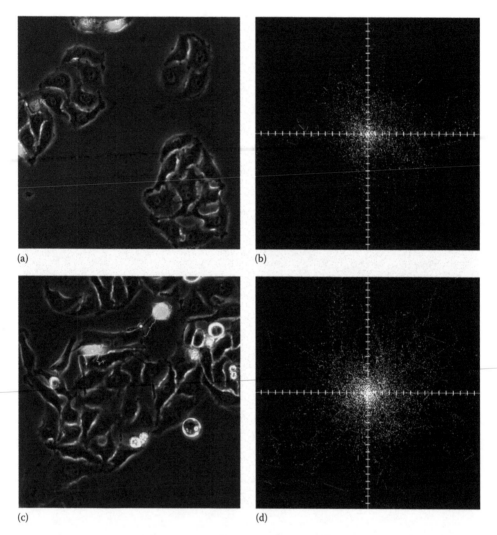

FIGURE 14.3 Automated tracking of cell movements. Migratory phenotype upon overexpression of RAB5 [panels (c), (d)] protein in comparison to wild type HeLa cells [panels (a), (b)] was evaluated by automated GFP fluorescence tracking. Image segmentation for cell identification and reconstruction of trajectories was applied to 49 fields/well, transfected respectively with GFP or GFP + RAB5 expression vectors. "Spider Graphs" in panels (b) (ctrl) and (d) (RAB5) report the tracked movements of the two cell populations. (Cells kindly provided by Dr. Andrea Palamidessi, IFOM, Milan, Italy.)

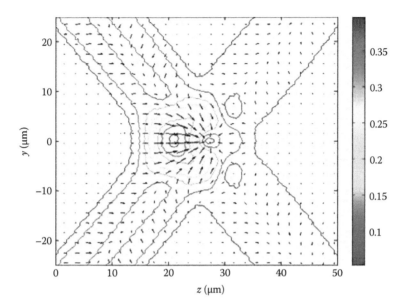

FIGURE 15.13 Contour plot of the Q factor calculated for the fiber-bundle structure with geometrical parameters: bundle diameter $R=55$ μm, beam diameter $w=8$ μm, and cutting angle $\theta=66°$. Both parameters are shown in the yz plane: due to the TOFT symmetry, identical results have been obtained in the xz plane. (From Bragheri F. et al., *Opt. Express*, 16, 17647, 2008. With permission.)

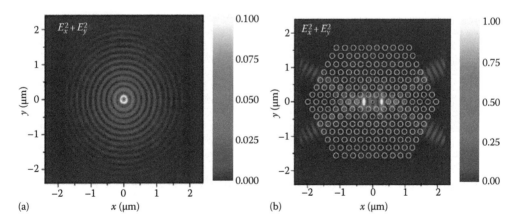

FIGURE 15.21 FDTD calculation of nanoantenna optical response. Color scales of panels are in arbitrary units. The numerical values reported in the color scale are in the correct relative intensity values. (a) Spatial $E_x^2+E_y^2$ field profile of the plasmonic nanoantenna on a silicon nitride membrane calculated at $\lambda=514$ nm. (b) Spatial $E_x^2+E_y^2$ field profile of the overall device calculated at $\lambda=514$ nm. (Reprinted with permission from De Angelis F. et al., *Nano Lett.*, 8, 2321, 2008. Copyright 2008 by American Chemical Society.)

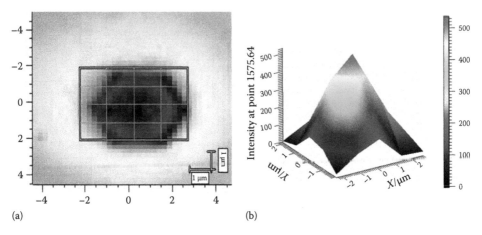

(a)   (b)

FIGURE 15.26 Micro-Raman mapping measurement. (a) *XY*-mapping area where the Raman measurement is performed, (b) mapping analysis for the Raman band centered at 1575.64 cm^{-1} attributed to the C–C-related vibrational band. (From De Angelis, F. et al., *SPIE Proc. Plasmonics: Metallic Nanostructures and Their Optical Properties*, 7032, 2008. With permission.)

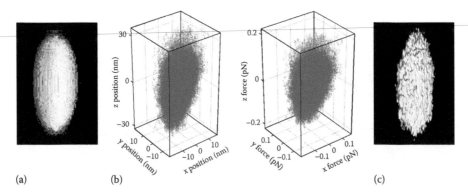

(a)   (b)   (c)

FIGURE 16.1 OPSF and FPSF. (a) Three-dimensional rendering of an OPSF of a confocal microscope (objective 100× NA 1.4 oil immersion) measured with 64 nm diameter latex beads immersed in oil. Field of view 0.8 × 1 μm. (b) Three-dimensional scatter plots of 1 μm diameter silica bead, trapped with a CW infrared laser (1064 nm wavelength, objective 100× NA 1.4 oil immersion) in water. Scatter plots are computed from the *x, y, z* trace, in nm and pN, low pass filtered at 20 kHz and acquired at 10 kHz. (c) Three-dimensional rendering of a FPSF representing the volume observed from the trapped object. Field of view 50 × 70 μm.

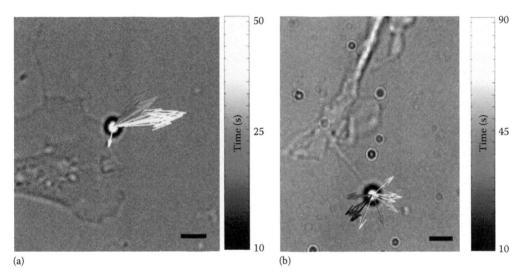

(a)   (b)

FIGURE 16.2 Force measurement and cell motility. Vectorial representation of the force exerted by the cell on obstacle (1 μm diameter silica bead). Different colors of the vector represent the different times at which the cell exerts the force on the trapped probe. In (a) lamellipodium increases the force exerted on the obstacle and tries to avoid it, as suggested from the time evolution of the length and the direction of the force. In (b) the random succession of the force vector direction in time represents the exploratory motion of a filopodium which senses the environment. Bars 2 μm.

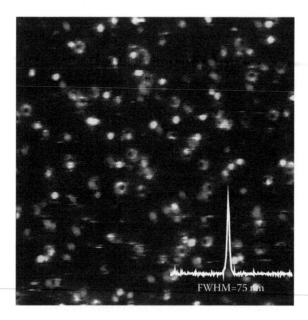

FIGURE 18.6 Individual DiI molecules embedded in a thin PMMA layer. The color coding indicates the in-plane transition dipole of individual molecules, with green for molecules oriented along the *x*-direction and red for molecules oriented in the *y*-direction. Doughnut-shaped features correspond to out-of-plane oriented molecules excited by the strong electric field localized at the boundaries between glass and aluminum. Size of the image: $4.2\,\mu m \times 4.2\,\mu m$ (2.5 ms/pixel). The inset shows the intensity profile of a single molecule with a full width half maximum of 75 nm.

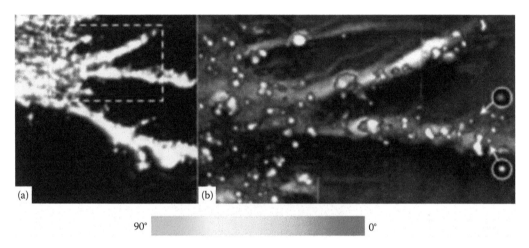

90°   0°

FIGURE 18.9 Organization of pathogen receptor DC-SIGN on the cell membrane of immature dendritic cells. (a) Confocal image (20 μm × 20 μm) of a dendritic cell stretched on fibronectin-coated glass expressing DC-SIGN on the membrane. (b) Combined topography (grey) and near-field fluorescence image (color) of the frame (12 μm × 7 μm) highlighted in (a). In both (a) and (b), the fluorescence signal is color coded according to the detected polarization (red for 0° and green for 90°). To illustrate the single-molecule detection sensitivity of the setup, the size as well as the intensity of two individual molecules are slightly enlarged and rescaled in the image [shown in circles in (b)]. Individual molecules are identified by their unique dipole emission (i.e., red and green color coding). The yellow color of most fluorescent spots results from additive emissions from multiple molecules with random in-plane orientation (combination of red and green) in one spot. (From de Bakker, B. I. et al., *Chemphyschem*, 8, 1473, 2007. Wiley-VCH Verlag GmbH & Co. KGaA. Reproduced with permission.)

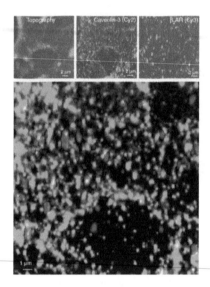

FIGURE 18.10 Dual color NSOM imaging for colocalization studies. (Top panel) NSOM topography of neonatal cardiac myocytes (right) with fluorescence-labeled caveolin-3 (middle) and β2AR (left). (Bottom panel) NSOM fluorescent overlay image showing colocalization (yellow) of β2AR (red) and caveolin-3 (green) on the nm scale. (Reprinted by permission of Macmillan Publishers Ltd: Ianoul, A., *Nat. Chem. Biol.*, 1, 196, 2005. Copyright 2005.)

Milton Keynes UK
Ingram Content Group UK Ltd.
UKHW050309111024
449327UK00049B/398

9 780367 384210